A Short Course in Calculus

A Short Course in Calculus

with applications to management, life and social sciences

Bodh R. Gulati
Southern Connecticut State College

The Dryden Press
Hinsdale, Illinois

Library of Congress Catalog Card Number:
ISBN: 0-03-047466-3
Printed in the United States of America
123-056-98765432

Acquiring Editor: Feeny Lipscomb
Project Editor: Bernice Gordon
Art Director: Stephen Rapley
Production Manager: Peter Coveney

Text and cover design by Harry Voight

To my parents

Preface

A short course in calculus is beginning to play a significant role in business, economics, management, life and social sciences. This is evidenced by the fact that more and more colleges and universities require at least one semester of calculus as a prerequisite for entry into these fields.

Students in a short calculus course have had a wide range of mathematical experiences. Some were afraid of mathematics and elected as few courses in high school as possible, some did not consider mathematics to be of any practical significance and therefore managed to forget most of what they'd learned, while others have a knowledge of intermediate algebra. It is for this wide variety of students that this text is intended. We have adopted a *down-to-earth* approach which presents the fundamentals in a lucid and straightforward manner to facilitate an intuitive understanding of the basic concepts. Although the mathematical background needed is slight, we have not compromised theoretical accuracy in the text.

Because of the varying background of the students, an applied calculus text has to be flexible. We have introduced several distinguishable features to reach this goal. First, more algebraic detail than is usual has been included. Second, Chapter 1 includes brief sections on sets, real numbers, and coordinate systems. More time is spent on inequalities, absolute value and algebra of straight lines-topics that often trouble a begining calculus student. Chapter 5 provides a detailed coverage of exponential and logarithmic functions before their derivatives are introduced. Additional review of algebra covering products, factoring, quadratic equations, rational expressions, integral and rational exponents, and rationalizing square roots is provided in Appendix A. The topics are arranged in a manner so that a student with two years of high school algebra can skim (or even skip) this material.

Each chapter opens with an introductory section explaining its scope. The important ideas in each section that need emphasis are presented in the form of propositions and then illustrated by a variety of solved examples. The applications are drawn from business, economics, management, life and social sciences and are carefully integrated throughout the text to maintain interest and stimulation.

The only way to learn calculus is by solving problems. We have provided an abundance of worked-out examples (approximately 300) to help the student gain a firm grasp of the principles presented in the text. The text contains more than 1500 carefully graded problems. In addition, there are review exercises at the end of each chapter. Answers to odd-numbered problems and review exercises are given at the end of the text. Solutions to the even-numbered problems are available in the Instructor's Manual.

There is sufficient text material for a one-semester course or a sequence of two-quarter courses. It has been the experience of many instructors (including myself) that material carefully selected from the first six chapters provides a good, one-semester, one-variable calculus course for students of management, life and social sciences. Chapter 7 on calculus of more than one variable may be covered, if time permits.

A word on hand-held calculators is in order. Most of the students who study mathematics own or have access to a scientific hand calculator. Not only useful in computational problems, a calculator can be employed as a learning device. For example, a student will develop a "feel" for limits more quickly if he can literally see the limit being approached. In Chapters 2 and 3, there are several tables which make use of the calculator to illustrate particular limits. In addition, the calculator has been used in several more places in the text for computational purposes. Some problems in Chapter 5 involve continuous compounding while others involve exponential functions. These problems might be difficult to attempt without the help of an electronic calculator. The same statement holds true for some problems on differential equations in Chapter 6. Although a calculator is not required for the vast majority of problems in the text, its use will facilitate the learning process. It will, therefore, be an advantage to own a hand-held scientific calculator that has common (log) and natural (ln) logarithm keys, exponential (y^x) and (e^x) keys, and a root key ($\sqrt[y]{x}$) which allows easy calculations of powers and roots.

New Haven, Connecticut
August, 1980.

Bodh R. Gulati

To the Student

A few words of advice and perspective are in order. The concepts and techniques of calculus you will encounter in this book are relatively easy. The difficulties, if any, will actually be problems of algebra. We expect that you have two years of high school algebra to draw upon as you progress through the text. With lesser preparation, you will have to work harder. In any event, be sure to read Chapter 1 on Preliminaries and Appendix A on Review of Algebra because they contain material that is used in the remainder of the book.

The mathematics in this book is not difficult, but some concepts are new and require patience and perseverence in order to be mastered. We hope that you find our style and exposition to be lucid. But do not expect to read this or any other mathematics book like a novel. Mathematics is not a spectator's sport; you will have to get involved.

Expect to read and reread the text thoughtfully. Make a habit of having a paper and pencil at hand to answer questions raised in the text, work through the exercises, and create examples of your own. This will reinforce the learning process. Test your own understanding by working through the review exercises placed at the end of each chapter. Do not go on to new material until you understand the previously covered material, or at least know what you do not understand. And do not hesitate to ask questions of your instructor. Remember that you are in the class to learn and if you have questions not yet answered in the class, be sure to ask the teacher for additional help. Many schools offer a mathematics lab or some form of tutorial service. If your school does, take full advantage of it.

We do not claim to encompass all of the calculus in this book, but we do believe that you can gain a significant appreciation of many applications of calculus from our book. The only requirement is that you give yourself a chance to learn what mathematics can do for you.

Good Luck.

B. R. G.

Acknowledgments

I wish to express my deep appreciation to my colleagues at Southern Connecticut State College whose suggestions, comments, and criticism have helped me in this pleasant undertaking. Special gratitude is expressed to Dr. J. Philip Smith for his helpful and penetrating insights while reviewing the manuscript. I would also like to convey my special thanks to the following reviewers for their assistance.

M. Khurshid Ali, *California State College*
Helen Baril, *Quinnnipiac College*
John Bishir, *North Carolina State University*
Louis F. Bush, *San Diego City College*
Thomas A. Carnevale, *Essex Community College*
Charles C. Cleaver, *South Dakota State University*
William E. Coppage, *Wright State University*
Walter Czarule, *Framingham State College*
Richard Grassel, *University of New Mexico*
George K. Goff, *Oklahoma State University*
G. S. Gill, *Brigham Young University*
Mal Golden, *University of Miami*
H. Eugene Hall, *DeKalb Community College*
Joseph A. Izzo, *University of Vermont*
Jack J. Motta, *Santa Fe Community College*
Donald E. Myers, *University of Arizona*
Harold M. Ness, *University of Wisconsin*
Richard Semmler, *North Virginia Community College*
David Shochat, *Santa Monica College*

In addition, I would like to single out as especially helpful the line-by-line reviews of the manuscript by Professors Garrett Etgen of the University of Houston and Salley Ann Folley of the University of New Hampshire.

I am grateful to George Telecki who proposed the project and to the staff of Dryden Press for their cooperation and valuable assistance. Notable among these are Bernice Gordon, Project Editor and Feeny Lipscomb, Acquiring Editor, whose patient prodding and amiable advice helped this project. Also I'd like to thank Alan Wendt, who carefully monitored the book's design program. The book could not have become a reality without the aid and assistance of each of these people, and to them I offer my sincere thanks.

I must not forget my indebtedness to my wife, Santosh, and to my children, Anita and Sunil, who during the writing of the book good-naturedly accepted a husband and a father who was absent even while he was present.

B. R. Gulati

Contents

A Short Course in Calculus

1 Preliminaries*

1.1 The Real Numbers

The numbers most familiar to the reader, namely

$$1, 2, 3, 4, \ldots$$

are called the *natural numbers*. If we add or multiply two natural numbers, we again obtain a natural number. Certain basic rules of elementary algebra are so familiar that the reader probably applies them without even thinking about them. We state some of them explicitly. If a, b, and c are natural numbers, then

1.1	$a + b = b + a$	Commutativity for Addition
1.2	$ab = ba$	Commutativity for Multiplication
1.3	$a + (b + c) = (a + b) + c$	Associativity for Addition
1.4	$(ab)c = a(bc)$	Associativity for Multiplication
1.5	$a(b + c) = ab + ac$	Distributive Property
1.6	$a \cdot 1 = 1 \cdot a = a$	Multiplicative Identity

Based on these properties, we can justify other statements about the natural numbers (also known as the positive integers). The operation of subtraction, for example, can be defined in terms of the familiar operation of addition.

* This chapter may be skimmed or skipped depending on the needs of students.

Definition 1.1

If a and b are any natural numbers and c is a natural number such that $a + c = b$, then we write $b - a = c$.

For example, if $a = 5$ and $b = 8$, then $8 - 5 = c$ if there exists a natural number c such that $5 + c = 8$. In this case, $c = 3$. But what about $5 - 8 = c$? Clearly, no natural number satisfies this expression. Similarly, there exists no natural number which satisfies the equation $5 - 5 = c$. This demonstrates that subtraction of one natural number from another does not *always* produce a natural number. Fortunately, this difficulty can be overcome by introducing the negative integers $-1, -2, -3, \ldots$ and the number zero (0). Taken together, the negative integers, zero, and the positive integers form the system of *all* integers; namely

$$\ldots, -3, -2, -1, 0, 1, 2, 3, \ldots$$

In this system every equation of the form $a + c = b$ always has a solution. Further, there exists a unique number 0 in this system with the property that for every number a

1.7 $$a + 0 = a$$

We refer to zero as an *identity element* under the operation of addition. Also, for each integer a there exists another integer $(-a)$ such that

1.8 $$a + (-a) = (-a) + a = 0$$

The integer $(-a)$ is called the *additive inverse* of a. We have extended the system of natural numbers to the system of integers in which subtraction is always possible. Thus, we have the following definition.

Definition 1.2

If a and b are any integers and c is an integer such that $a + c = b$, then we write $b - a = c$.

Thus, given any two integers a and b, we can subtract a from b; that is, we can always find an integer c such that $a + c = b$. But can we divide a by b if a and b are integers? In other words, is there an integer d such that $a = bd$? If $a = 12$ and $b = 3$, then certainly there exists an integer $d = 4$ such that $a = bd$. But what if $a = 12$ and $b = 5$? Obviously, there is no integer d satisfying the equation $12 = 5d$. This means that if we want a number system in which exact division is always possible, we must extend the system of integers so as to include fractions such as $\frac{2}{3}$, $\frac{3}{5}$, and $-\frac{4}{7}$. This system is called the system of rational numbers.

Definition 1.3

A *rational number* is a number that can be expressed in the form a/b where a and b are integers and $b \neq 0$.

Why do we make the restriction, $b \neq 0$? Suppose that a is divisible by zero. Then what is the number d such that $a/0 = d$? Note that $a/0 = d$ means that $a = 0 \cdot d$. If $a = 0$, then d could be any number—a situation we do not wish to permit. On the other hand, if $a \neq 0$, then there is no d for which $a = 0 \cdot d$. Hence, in both cases, there is no satisfactory number d such that $a/0 = d$. The expression $a/0$ cannot be defined and division by zero is not permitted. This leads to an important property in the system of rational numbers: for any number $a \neq 0$, there exists another number $(1/a)$ such that

1.9 $$a \cdot \left(\frac{1}{a}\right) = \left(\frac{1}{a}\right) \cdot a = 1.$$

The number $(1/a)$, also written as a^{-1}, is called the *multiplicative inverse* of the number a.

We now state some of the basic rules that govern the operations with rational numbers:

1.10 $$\frac{a}{1} = a$$

1.11 $$\frac{a}{b} = \frac{c}{d} \text{ if and only if } ad = bc, \qquad b \neq 0,\ d \neq 0$$

1.12 $$\frac{ac}{bc} = \frac{a}{b}, \qquad b \neq 0,\ c \neq 0$$

1.13 $$\frac{a}{b} + \frac{c}{d} = \frac{ad + bc}{bd}, \qquad b \neq 0,\ d \neq 0$$

1.14 $$\frac{a}{b} - \frac{c}{d} = \frac{ad - bc}{bd}, \qquad b \neq 0,\ d \neq 0$$

1.15 $$\frac{a}{b} \cdot \frac{c}{d} = \frac{ac}{bd}, \qquad b \neq 0,\ d \neq 0$$

1.16 $$\frac{a}{b} \div \frac{c}{d} = \frac{ad}{bc}, \qquad b \neq 0,\ c \neq 0,\ d \neq 0$$

The rational numbers, which include positive integers, negative integers, zero, and fractions, form a system in which the operations of addition, multiplication, and division (by non-zero numbers) are always possible. It appears that the system of rational numbers we have developed is all that we need in mathematics. However, this is not true since there are other numbers, for example, $\sqrt{3}$, $-\sqrt{5}$, π, e and infinitely many more, which cannot be represented in the form a/b where a and b are integers and $b \neq 0$. These numbers are called *irrational* numbers; together, the rational numbers and the irrational numbers form the real number system, R.

A geometric representation of the real number system may provide further insight into what the real numbers are. On a straight line taken to be horizontal for convenience, we choose an arbitrary point 0 as the origin and mark off an arbitrary unit length on the line segment (see Figure 1.1). The unit point is usually placed to the right of the origin. With this unit of length, we represent positive and negative integers by a set of points equally spaced on the line, placing positive integers to the right and negative integers to the left of the origin. By dividing each of the segments into two equal parts we can associate fractions such as $-\frac{3}{2}$, $-\frac{1}{2}$, $\frac{1}{2}$, $\frac{3}{2}$, with particular points. In general, we can represent the fraction m/n by dividing the unit length into n equal parts and counting off m of these parts. It may appear that the process of representing rationals by marking off the corresponding points would eventually cover the entire line. This, however, is the not the case because there are infinite numbers of points on the line which do not correspond to any rational number. There is no rational number, for instance, that can express the exact length of the hypotenuse of a right triangle that has legs of unit length.

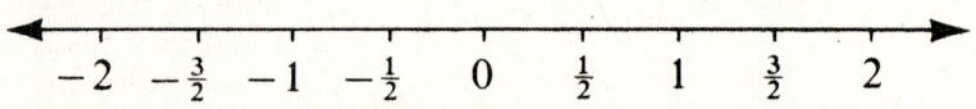

Figure 1.1

Consider a right triangle OAB with legs OA and AB of unit length (see Figure 1.2). Then, according to the famous theorem of Pythagoras, the length of the hypotenuse OB is given by

$$x^2 = 1^2 + 1^2$$

$$x^2 = 2$$

$$x = \sqrt{2}$$

It can be shown that the distance $OB = OC = \sqrt{2}$ *cannot* be represented in the form a/b where a and b are integers and $b \neq 0$. Consequently, the number $\sqrt{2}$ is not rational, yet it does correspond to a point on the number line.

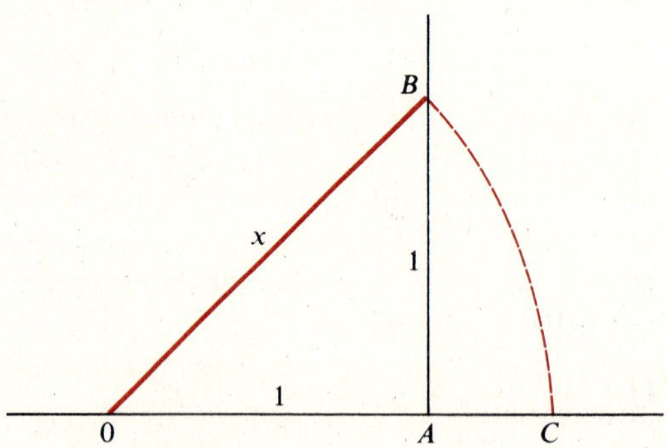

Figure 1.2

Definition 1.4

The collection of all rational numbers and irrational numbers is called the *real number system.*

Figure 1.3 will help you picture the hierarchy of the real number system we have developed so far. We wish to remark that there exists a one-to-one correspondence between the set of real numbers and the set of points on the number line. This means that to every point on the number line we can associate one and only one real number, and conversely, to each real number there corresponds one and only one point on the number line. When a point on the number line is associated with a real number, the number is called the *coordinate* of the point and the point is called the *graph* of the number.

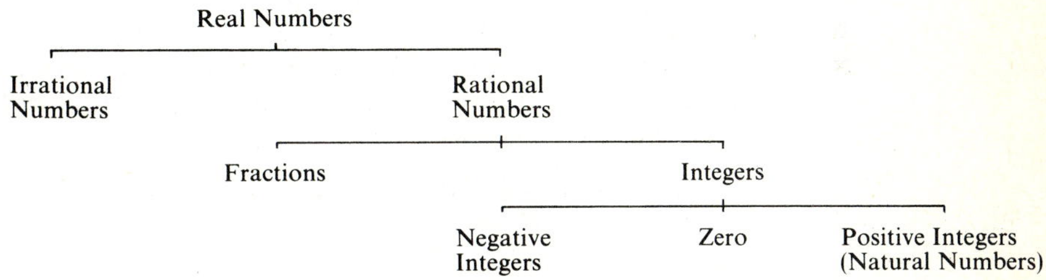

Figure 1.3

Exercises

Classify each of the following statements as true or false.

1. Every natural number is a rational number.

2. Every rational number is a natural number.

3. Every rational number is an integer.

4. Some integers are not rationals.

5. Some rationals are not integers.

6. The sum of two rationals is rational.

7. The product of two rationals is rational.

8. The division of an integer by zero is sometimes permitted.

9. Corresponding to every irrational number, there is a point on the real number line.

10. Corresponding to every point on the real number line, there is an irrational number.

11. There exists a one-to-one correspondence between the rational numbers and the points on the real line.

Classify each of the real numbers as rational or irrational.

12. $\frac{2}{3}$	**15.** $\sqrt{3}$	**18.** π	**21.** $\sqrt[4]{3}$
13. $\sqrt{2}$	**16.** $\dfrac{22}{15}$	**19.** $2e$	**22.** $\dfrac{4\pi}{3}$
14. -4	**17.** $\sqrt{9}$	**20.** $\sqrt[3]{4}$	**23.** $\dfrac{3e}{5}$

1.2 Inequalities

In elementary algebra and geometry, we study equalities almost exclusively. As we progress further in mathematics, and particularly in calculus, we shall see that the study of inequalities is both interesting and useful.

We begin our concern with inequalities among real numbers by recalling that positive numbers are associated with points to the right of zero on the real number line, while negative numbers are associated with those to the left of zero. A number a is greater than 0 if and only if a is positive. In symbols, we write

$$a > 0$$

where the symbol $>$ is read as "is greater than." Similarly, a number a is less than 0 if and only if a is negative. The statement is symbolized by

$$a < 0$$

where the symbol $<$ is interpreted as "is less than."

Definition 1.5

Given any two real numbers a and b, we say that a is greater than b ($a > b$), or equivalently, b is less than a ($b < a$), if $a - b$ is a positive number. The symbols $<$ and $>$ are called *inequality signs.*

This definition is related to our geometric idea that larger numbers lie to the right of smaller ones on the number line. Figure 1.1 (see page 4) shows, for instance, the point 2 to the right of -1 because $2-(-1)$ is the positive number 3. This means, according to Definition 1.5, that 2 is greater than -1 and -1 is less than 2. Let us now state briefly some basic properties of inequalities.

1.17 For any real number a, one and only one of the following relations is true:

$$a < 0, \quad \text{or } a = 0, \quad \text{or } a > 0$$

1.18 If $a > 0$ and $b > 0$, then

$$a + b > 0$$

and

$$ab > 0$$

1.19 If $a < 0$ and $b < 0$, then

$$a + b < 0$$

and

$$ab > 0$$

1.20 If $a < 0$ and $b > 0$, then

$$ab < 0$$

The following statements concerning inequalities can be justified by using statements (1.17) through (1.20) and Definition 1.5.

1.21 If $a < b$ and $b < c$, then $a < c$.

1.22 If $a < b$, then $a + c < b + c$ for any number c, positive, negative, or zero.

1.23 If $a < b$ and c is any *positive* number, then

$$ac < bc$$

1.24 If $a < b$ and c is any negative number, then

$$ac > bc$$

Thus, if $c = -1$ and $a < b$, then

$$-a > -b$$

To prove (1.21), note that $a < b$ and $b < c$ are equivalent to the statement that $b - a$ and $c - b$ are positive numbers. Their sum

$$(b - a) + (c - b) = c - a$$

is positive by statement (1.18). The statement $c - a > 0$ means that $a < c$ by Definition 1.5.

Property (1.23) follows from the fact that $a < b$ implies that $b - a$ is a positive number. Since $c > 0$ and the product of two positive numbers is positive, it follows that

$$(b - a)c > 0$$

or

$$bc - ac > 0$$

That is,

$$bc > ac$$

The verification of the properties (1.22) and (1.24) is left as an exercise for the reader.

The notation $a \geq b$ means that a is greater than or equal to b. Similarly, the notation $a \leq b$ implies that a is less than or equal to b. Note that properties of $<$ and $>$ may be extended to the relations $\leq$ and $\geq$, respectively.

We will now use some of the properties in the solution of an inequality involving a single unknown. By *a solution* of an inequality we mean a real number that satisfies the inequality, and by *solution set* we mean the set of all real numbers that satisfy the given inequality. Solving an inequality is thus a matter of describing the solution set of the inequality as explicitly as possible. Let us look at some examples.

Example 1.1

Solve the following inequality for x:

$$5x - 3 < 12$$

Solution

By property (1.22) we can add 3 to both sides of the inequality and obtain an equivalent statement

$$5x - 3 + 3 < 12 + 3$$

or

$$5x < 15$$

Now, multiplying both sides by $\frac{1}{5}$ (Property 1.23), we have

$$x < 3$$

The statement $x < 3$ is interpreted to mean that the solution set of the inequality $5x - 3 < 12$ consists of all real numbers less than 3. Thus, if we substitute any real number less than 3, say $x = 1$, into the inequality $5x - 3 < 12$, we obtain

$$5(1) - 3 < 12$$

That is,

$$2 < 12$$

which is a true statement.

It is convenient to give a graphical interpretation for the solution of inequalities. By the graph of a set of real numbers, we mean the set of all points on the real number line that correspond to the numbers. represent graphically the solution set of an inequality, we simply darken the appropriate portion of the number line. The graph corresponding to the solution set in the above example, for instance, consists of all points to the left of the point with coordinate 3. This is displayed in Figure 1.4. The open circle at 3 represents the fact that this number is not included in the solution set.

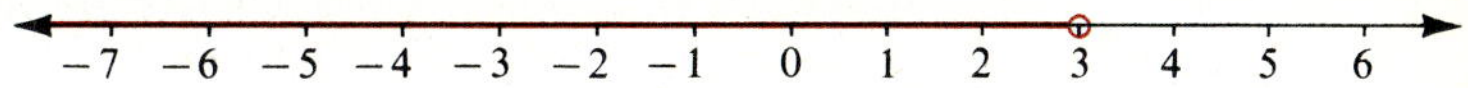

Figure 1.4

Example 1.2

Solve the following inequality for x:

$$5x + 16 \leq 3x + 25$$

and represent the solution set graphically.

Solution

Adding $(-3x)$ to both sides of the inequality, we obtain

$$(-3x) + 5x + 16 \leq (-3x) + 3x + 25$$
$$2x + 16 \leq 25$$

Next, we add (-16) to both sides and get the inequality

$$2x + 16 + (-16) \leq 25 + (-16)$$
$$2x \leq 9$$

Finally, we multiply both sides by $(\frac{1}{2})$, and we have

$$x \leq \frac{9}{2}$$

Thus, the solution set consists of all real numbers less than or equal to $\frac{9}{2}$. This is represented graphically in Figure 1.5. Note that the heavy dot indicates that the real number $\frac{9}{2}$ is included in the solution set.

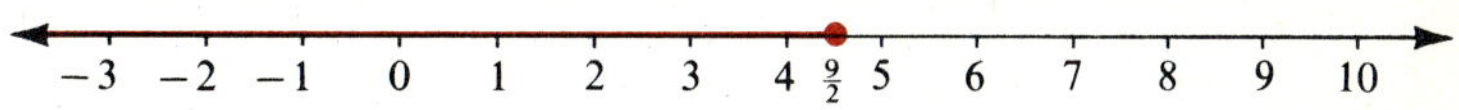

Figure 1.5

Example 1.3

Solve the following inequality for x:

$$4x - 32 \leq 7x - 17$$

and represent the solution set graphically.

Solution

Adding 32 to both sides, we obtain

$$4x \leq 7x + 15$$

Now we add $(-7x)$ to both sides and get

$$-3x \leq 15$$

Finally, we multiply both sides of this inequality by $(-\frac{1}{3})$. Recall from (1.24) that multiplication by a negative number reverses the direction of the inequality. Thus, we obtain

$$x \geq -5$$

which represents the fact that the solution set consists of all real numbers greater than or equal to -5. This is illustrated in Figure 1.6.

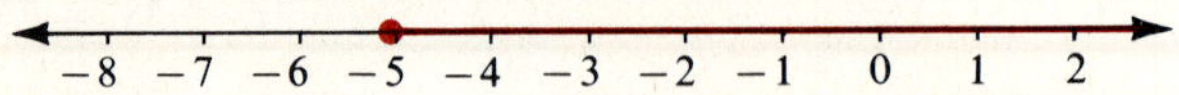

Figure 1.6

Note that in solving linear inequalities, we use exactly the same steps as in solving the linear equation $ax + b = c$, $a \neq 0$. We shall now consider inequalities of a slightly different type.

Example 1.4

Solve the following inequality for x:

$$x(x - 3) > 0$$

and represent the solution set graphically.

Solution

The sign of the factors x and $x - 3$ and the sign of the product $x(x - 3)$ are shown in Figure 1.7. The product of two real numbers is positive if and only if both numbers have the same sign. Hence, the inequality $x(x - 3) > 0$ holds only for those real numbers that satisfy the inequality $x < 0$ or the inequality $x > 3$, as illustrated in Figure 1.7. Note that the two points 0 and 3 are not included in the solution set.

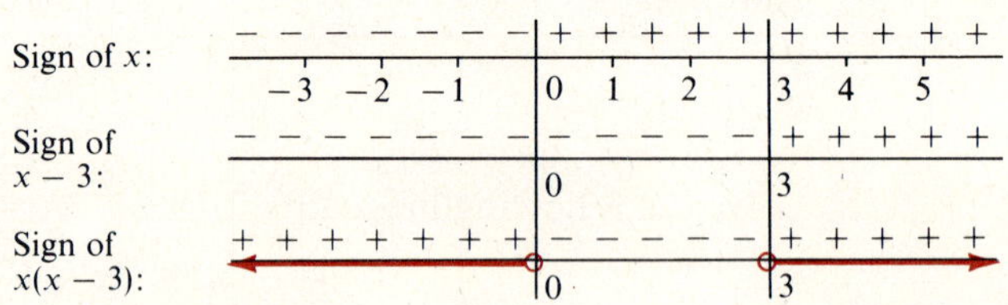

Figure 1.7

Example 1.5

Solve the following inequality for x

$$x^2 - 6x + 8 < 0$$

and represent the solution set graphically.

Solution

First, we factor* the left side of the inequality and obtain

$$(x - 2)(x - 4) < 0$$

Figure 1.8 shows the signs of each factor along with the signs of the expression itself. The factor $(x - 2)$ is positive for $x > 2$ and negative for $x < 2$. Similarly, $(x - 4)$ is positive for $x > 4$ and negative for $x < 4$. The product of two numbers is negative if and only if these numbers are of opposite signs. From the figure, it is clear that the solution set consists of real numbers between 2 and 4. Note that the endpoints are not included in the solution set.

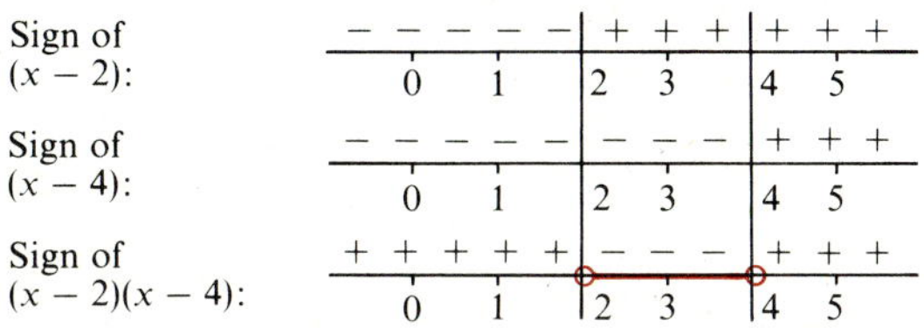

Figure 1.8

Example 1.6

Solve the following inequality for x:

$$x^2 + 4x - 5 \geq 0$$

and represent the solution set graphically.

Solution

Factoring the left side of the inequality, we obtain

$$(x + 5)(x - 1) \geq 0$$

Figure 1.9 shows the signs of each factor along with the given expression. The product of two numbers is positive if and only if both numbers are of the same sign. It is clear that $x^2 + 4x - 5 \geq 0$ if $x \leq -5$ or $x \geq 1$ as illustrated in the figure. The endpoints -5 and 1 are included in the solution set.

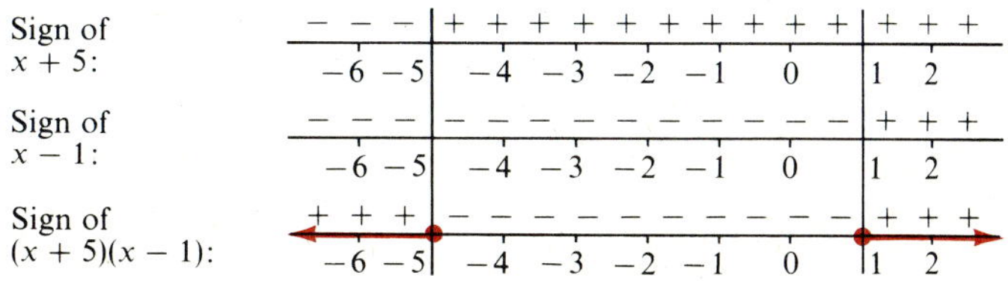

Figure 1.9

* A brief review on factorization is given in the Appendix A.

Example 1.7

Solve the following inequality for x:

$$x(x - 2)(x - 4) < 0$$

and represent the solution set graphically.

Solution

The signs of each factor along with the signs of the given expression are shown in Figure 1.10. The product of multiple factors is negative if and only if the number of negative factors is odd. Reading from the figure, we observe that $x(x - 2)(x - 4) < 0$ if and only if $x < 0$ (when three factors are negative) or $2 < x < 4$ (when one factor is negative).

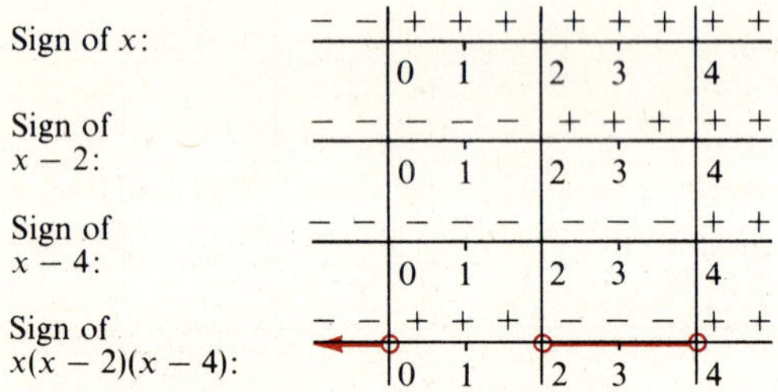

Figure 1.10

Exercises

Solve the following inequalities for x and represent the solution set graphically.

1. $x + 3 < 8$

2. $x - 2 < 4$

3. $x - 5 > -2$

4. $x + 1 > -4$

5. $2x - 1 < 9$

6. $3x + 2 > 4$

7. $4x - 5 \leq 7$

8. $5x + 4 \geq -6$

9. $\frac{1}{3}x + 2 \leq 5$

10. $\frac{1}{2}x + 1 \geq 4$

11. $3x - 8 < 5(2 - x)$

12. $13 - 5x < 1 + x$

13. $3(1 - 2x) \leq 2(x + 5)$

14. $3(5 - x) \leq 4(2 + x)$

15. $7(2 - 3x) \geq 3x + 8$

16. $\frac{1}{3}(x - 5) \geq \frac{3}{4}(2x + 1)$

Solve each of the following inequalities for x.

17. $x(x - 1) > 0$

18. $x(2 - x) < 0$

19. $x(4 - x) > 0$

20. $x(2x + 1) \geq 0$

21. $x^2 - 3x + 2 > 0$

22. $x^2 + 3x - 4 < 0$

23. $x^2 - 5x + 6 < 0$

24. $x^2 - 5x - 14 \geq 0$

25. $x^2 - x - 2 \geq 0$

26. $x^2 - 2x - 8 \leq 0$

27. $x(x - 1)(x - 4) > 0$

28. $(x + 1)(x - 2)(x - 3) < 0$

Prove each of the following statements.

29. If $a < b$ and $c > 0$, then $a/c < b/c$.

30. If $a + c < b + c$, then $a < b$.

31. If $ac < bc$ and $c < 0$, then $a > b$.

32. If a, b, c, and d are all positive numbers, and if $a < b$ and $c < d$, then $ac < bd$.

1.3 Absolute Value and Distance

We now introduce the geometric concept of distance in our real number system. Let x be the coordinate of a point on the number line. Then the number of units between this point and the origin O is called the *absolute value* of x and is denoted by the symbol $|x|$. Thus $|3| = 3$, $|-4| = 4$, $|4| = 4$, and so on. This leads us to suggest that the absolute value of a number gives only its distance from the origin and not its direction.

It is helpful to formulate a definition of absolute value that does not depend on a geometric notion. Of course, this definition gives the same result as the geometric concept.

Definition 1.6

For any real number x,

$$|x| = \begin{cases} x, & \text{if } x \geq 0 \\ -x, & \text{if } x < 0 \end{cases}$$

From the definition, we observe that $|5| = 5$ since $5 \geq 0$, while $|-5| = -(-5) = 5$ since $-5 < 0$. These examples illustrate that the absolute value of a number is never negative.

We next show how the concept of absolute value is used to measure the distance between any two points, say A and B, on the real line.

Definition 1.7

If x_1 and x_2 are the coordinates for two points A and B on the real line, then the distance between A and B is $|x_1 - x_2|$.

Thus, if $x_1 = -2$ and $x_2 = 6$ are coordinates of points A and B, respectively, then the distance between A and B, as illustrated in Figure 1.11, is given by

$$|x_1 - x_2| = |-2 - 6| = |-8| = 8$$

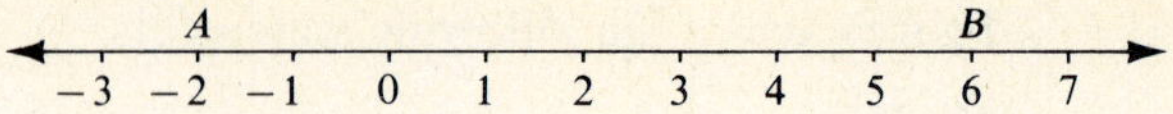

Figure 1.11

Observe that if we had interchanged x_1 and x_2, so that $x_1 = 6$ and $x_2 = -2$, then the result would have been the same because

$$|x_1 - x_2| = |6 - (-2)| = |8| = 8$$

Another important use of absolute value is in equations and inequalities.

Example 1.8

Solve for x:

$$|x - 5| = 4$$

Solution

Using Definition 1.6, we have

$$|x - 5| = \begin{cases} x - 5 & \text{if } x - 5 \geq 0 \\ -(x - 5) & \text{if } x - 5 < 0 \end{cases}$$

Consider first the case $x - 5 \geq 0$; here, $|x - 5| = x - 5$ and $x - 5 = 4$ implies that $x = 9$. Next, if $x - 5 < 0$, then $|x - 5| = -(x - 5)$ and

$$\begin{aligned} -(x - 5) &= 4 \\ -x + 5 &= 4 \\ -x &= -1 \\ x &= 1 \end{aligned}$$

Thus, the equation $|x - 5| = 4$ is satisfied only if $x = 1$ or $x = 9$. The result also makes sense using Definition 1.7. To solve $|x - 5| = 4$ means finding all points x whose distance from 5 is 4 units. There are clearly two such points: 1 and 9. This can be illustrated graphically as shown in Figure 1.12.

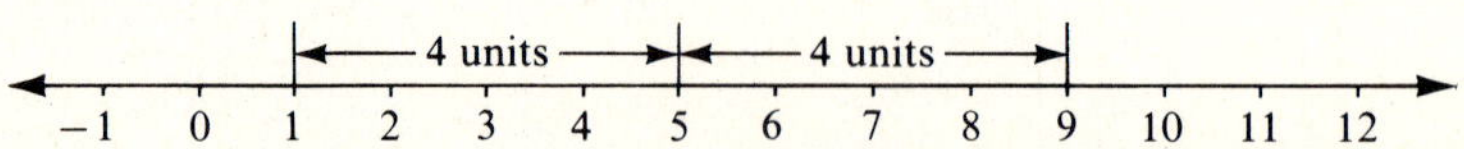

Figure 1.12

As a check, we observe that if $x_1 = 5$, and $x_2 = 9$, then

$$|x_1 - x_2| = |5 - 9| = |-4| = 4$$

and if $x_1 = 5$ and $x_2 = 1$, then

$$|x_1 - x_2| = |5 - 1| = |4| = 4$$

Example 1.9

Solve for x:

$$|2x - 3| = -2$$

Solution

Since the absolute value of a real number is never negative, we conclude that this equation has no solution.

Next, we use Definition 1.6 to determine the real numbers that satisfy an inequality such as $|x| < 3$. First, consider $x \geq 0$. Then $|x| = x$ and $x < 3$. Thus, the numbers $0 \leq x < 3$ satisfy the inequality. Second, consider $x < 0$, then $|x| = -x$ and $-x < 3$. Recalling that the expression $-x < 3$ is equivalent to $x > -3$, we have the numbers $-3 < x < 0$ that satisfy the equation. Combining these two results, we obtain

$$-3 < x < 3$$

Thus, all real numbers greater than -3 and less than 3 satisfy the inequality $|x| < 3$. This can be interpreted geometrically as shown in Figure 1.13. Note that the numbers -3 and 3 are *not* included in the solution set.

Figure 1.13

The inequality $|x| \leq 3$, on the other hand, consists of all real numbers between -3 and 3 including the numbers -3 and 3 as shown in Figure 1.14.

Figure 1.14

This discussion leads us to the following proposition which we state without proof.

Proposition 1.1

Let a be any positive real number. Then the inequality $|x| \leq a$ holds if and only if $-a \leq x \leq a$.

This proposition implies that x is located within a units of the origin on the number line. Let us now illustrate its use in the following examples.

Example 1.10

Find all real numbers x satisfying the inequality

$$|x - 4| \leq 3$$

Solution

The inequality $|x - 4| \leq 3$ is satisfied if and only if

$$-3 \leq x - 4 \leq 3$$

Adding 4 to each term, we have

$$1 \leq x \leq 7$$

Thus, all real numbers between 1 and 7, inclusive, satisfy the given inequality.

Example 1.11

Solve the following inequality for all real numbers x:

$$|4 - 5x| < 6$$

Solution

This inequality is equivalent to

$$-6 < 4 - 5x < 6$$

Subtracting 4 from each term, we have

$$-10 < -5x < 2$$

Note that multiplication of each term by $(-\frac{1}{5})$ reverses the direction of each of the inequalities, that is,

$$2 > x > -\tfrac{2}{5}$$

or

$$-\tfrac{2}{5} < x < 2$$

This implies that all real numbers between $-\frac{2}{5}$ and 2 satisfy the inequality $|4 - 5x| < 6$.

How do we interpret an inequality such as $|x| \geq 4$? Note that by Definition 1.6, if $x \geq 0$, then $|x| = x$ and $x \geq 4$. Thus the numbers for which $x \geq 4$ satisfy the inequality. If $x < 0$, then $|x| = -x$ and $-x \geq 4$. This means that $x \leq -4$ also satisfies the inequality. Combining these two results, we have

$$x \leq -4 \quad \text{or} \quad x \geq 4$$

This can be interpreted geometrically as shown in Figure 1.15.

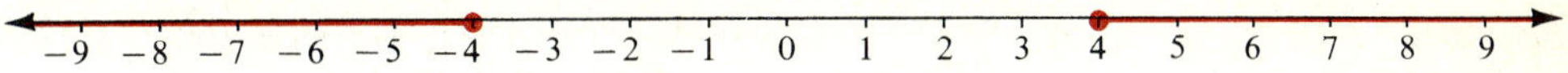

Figure 1.15

Clearly, x is *at least* 4 units away from the origin.

The inequality $|x| > 4$, on the other hand, consists of all real numbers less than -4 or greater than 4 as shown in Figure 1.16.

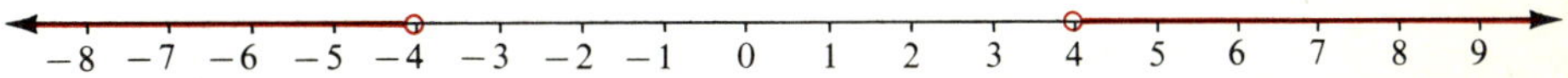

Figure 1.16

The endpoints -4 and 4 are not included in the solution set. Intuitively, we can say that x is located *more than* 4 units away from the origin on the number line. Thus, we have the following proposition.

Proposition 1.2 Let a be any positive real number. Then the inequality $|x| \geq a$ holds if and only if $x \leq -a$ or $x \geq a$.

Example 1.12 Solve the following inequality for all real numbers x:

$$|3x - 4| \geq 5$$

and represent the solution set graphically.

Solution

The inequality $|3x - 4| \geq 5$ holds if and only if

$$3x - 4 \geq 5$$

or

$$3x - 4 \leq -5$$

Solving each of these for x, we have $x \geq 3$ and $x \leq -\frac{1}{3}$, respectively. This may be represented geometrically as shown in Figure 1.17. All reals less than or equal to $-\frac{1}{3}$ or greater than or equal to 3 satisfy the given inequality.

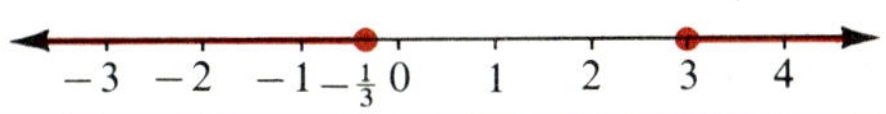

Figure 1.17

Exercises

Solve for x.

1. $|x - 2| = 3$

2. $|2x - 3| = 4$

3. $|3x - 2| = -1$

4. $|3x - 2| = 6$

5. $|8 - 3x| = 4$

6. $|3x - 8| = 4$

Solve each of the following inequalities for x and represent the solution set graphically.

7. $|x| \le 1$

8. $|2x| \le 1$

9. $|3x| > 1$

10. $|x - \frac{1}{2}| \ge 1$

11. $|x - 2| \le 3$

12. $|2x - 3| \le 5$

13. $|x - 3| > 2$

14. $|3x - 6| > 12$

15. $|2x - 4| > 6$

16. $|3x - 4| < 8$

17. $\left|\frac{x+2}{3}\right| \le 2$

18. $\left|\frac{x}{2} - 3\right| < 1$

19. $\left|\frac{x}{3} - 2\right| \le 2$

20. $\left|\frac{x}{4} + 1\right| \ge 3$

1.4 The Language of Sets

A set is an aggregate or collection of objects of some kind. These objects are called the *elements* or members of the set. There are many examples of sets with which the reader is familiar. For example, the reader is an element of the set of students in his or her class. A student not in the class is not a member of this set.

A set must be *well-defined* in the sense that we can determine whether or not an element belongs to that collection. Some examples of well-defined sets are:

The set of points on the real line

The set of rationals between 0 and 1

The set of natural numbers less than 6

The set of real numbers

Each of the above sets is so clearly defined that we can easily determine whether or not an element belongs to a given collection. Note that the set of *smart* students in a class, the set of *older* folks in the state of Vermont, the set of *attractive* people on the college campus, and the set of *rich* attorneys in New York are all examples of sets that are not well-defined. (Why?)

Sets are designated customarily by capital letters $A, B, C, \ldots$ while members of a set are denoted by small letters $a, b, c, \ldots$. If a set consists of the numbers 2, 4, 6, 8, 10, we write

$$A = \{2, 4, 6, 8, 10\}$$

In case a set has a large number of elements, we may abbreviate the listing of its members. For example, we may denote the set of letters of the alphabet by

$$\{a, b, c, \ldots, y, z\}$$

We interpret the dots to mean that the unlisted letters d through x, inclusive, are members of the set.

Essentially, a set may be specified in two different ways. One way is to list all the elements of a set; another is to enclose in braces a defining property so as to determine whether or not a given element belongs to that set. Consider, for example, the set of natural numbers less than or equal to 4. We may either write

$$A = \{1, 2, 3, 4\}$$

or denote this set by the following notation:

$$A = \{x \mid x \text{ is a natural number and } x \leq 4\}$$

and read this notation: "The set of all those numbers x such that x is a natural number and x is less than or equal to 4." The vertical line in this notation is read "such that." The letter x of the alphabet in the notation is called a *variable*. For the most part, we will use the second method of defining a set, namely,

$$A = \{x \mid x \text{ has property T}\}$$

and interpret A to be the set of all elements x which have property T.

To express the fact that a is a member of the set A, we write

$$a \in A$$

where the symbol $\in$ is read "is an element of" or "belongs to." Thus, if Q is the set of rational numbers, then $\frac{1}{3} \in Q$. To denote the fact that $\sqrt{2}$ is not a member of Q, we write

$$\sqrt{2} \notin Q$$

Definition 1.8 A set with no elements is called the *null set* or the *empty set*; it is denoted by $\{\ \}$ or $\varnothing$.

The set of all children who weigh more than 1500 pounds and the set of all natural numbers greater than 5 and less than 6 example are of empty sets.

Let us now consider the sets

$$P = \{x \mid x \text{ is a positive integer}\}$$

and

$$Q = \{x \mid x \text{ is a rational number}\}$$

How do these sets P and Q compare? Perhaps an obvious statement is that every element of the set P is also an element of the set Q.

Definition 1.9

A set A is a subset of a set B if and only if every element of A is also an element of B.

The notation $A \subseteq B$ or $B \supseteq A$ means that "A is a subset of B" or "B contains A." We have seen that $A \subseteq B$ means that if $x \in A$, then $x \in B$, but if there exists some element $x \in A$ which does not belong to B, then A is *not* a subset of the set B.

Example 1.13

In the set P of positive integers,

$$\{1,2,3\} \subseteq \{1,2,3,4\}$$
$$\{2,4,6,8\} \subseteq \{2,4,6,8,10\}$$
$$\{5,6,7\} \subseteq \{5,6,7\}$$

but

$$\{1,2,3\} \nsubseteq \{1,2,4\}$$

Note that the set A is a subset of itself, $A \subseteq A$, since every element of A is an element of A and the definition of a subset is satisfied. We also consider the null set as a subset of every set since there is no element of $\varnothing$ that does not belong to A; therefore, $\varnothing \subseteq A$. Further, it is possible to have $A \subseteq B$ and $B \subseteq A$; these inclusions are both true if and only if the sets A and B have exactly the same elements. In that case, we say that $A = B$.

Certain subsets of the real number system are so important to the study of calculus that we give them special consideration. These subsets, called *intervals*, are defined below.

Definition 1.10

If a and b are any real numbers and $a < b$, then the *open interval* from from a to b, written (a, b), is the collection of all numbers greater than a and less than b.

In other words,

$$(a,b) = \{x \mid a < x < b\}$$

Geometrically, the open interval (a, b) may be illustrated on the real line as shown in Figure 1.18. The end points a and b are not included in the interval.

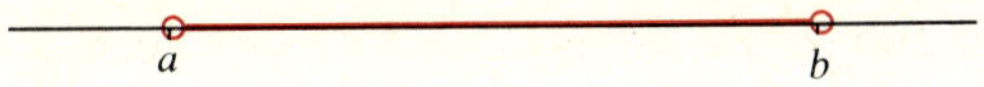

Figure 1.18

Definition 1.11 If a and b are any real numbers and $a \le b$, then the *closed interval* from a to b, written $[a, b]$, is the collection of all numbers greater than or equal to a and less than or equal to b.

Thus,

$$[a,b] = \{x \mid a \le x \le b\}$$

and is illustrated on the real line in Figure 1.19.

Figure 1.19

An interval which contains the endpoint a but not b is said to be *half-open on the right*. That is,

$$[a,b) = \{x \mid a \le x < b\}$$

Similarly, an interval which contains the endpoint b but not a is said to be *half-open on the left*. In other words,

$$(a,b] = \{x \mid a < x \le b\}$$

The concept of an interval can be extended to cover some unusual cases. Suppose that the interval has no upper limit, as is the case for the set of all reals greater than or equal to 4; $\{x \mid x \ge 4\}$. This will be represented by the symbol $[4, \infty)$. Note that the symbol ∞, which is read *infinity*, is not a number and is interpreted simply to mean "no upper limit." Similarly, the symbol $(-\infty, 3]$ represents the set $\{x \mid x \le 3\}$ and consists of all real numbers less than or equal to 3. The symbol $-\infty$, which is read "negative infinity" is interpreted to mean "no lower limit."

Operations of Sets

We have introduced the concept of a set and notation for describing sets. We have also discussed some of the important subsets on the real line. Now we ask how the sets under discussion interact with each other. To answer this question, we introduce the operations of union and intersection of sets.

Definition 1.12 Let A and B be subsets of the real number system R. The *union* of A and B, denoted by $A \cup B$, is the set of all elements that belong to A or to B or to both.

In symbols,

$$A \cup B = \{x | x \in A \text{ or } x \in B\}$$

Example 1.14

Let A and B be subsets of the real number system R. If

$$A = \{2, 3, 5, 7, 8\} \quad \text{and} \quad B = \{1, 2, 3, 4, 6, 8\}$$

then

$$A \cup B = \{1, 2, 3, 4, 5, 6, 7, 8\}$$

Example 1.15

Let A and B be subsets of the real number system R and

$$A = \{x | x < 1\} \quad \text{and} \quad B = \{x | -1 \leq x \leq 3\}$$

Describe the set $A \cup B$ as a collection of points of R.

Solution

The set A consists of all real numbers less than 1. Thus, a point belongs to A if and only if it lies to the left of the point 1 on the real line. Similarly, B is the set of all real numbers between -1 and 3, both inclusive. Then $A \cup B$ is the set of all real numbers less than or equal to 3 as illustrated in Figure 1.20.

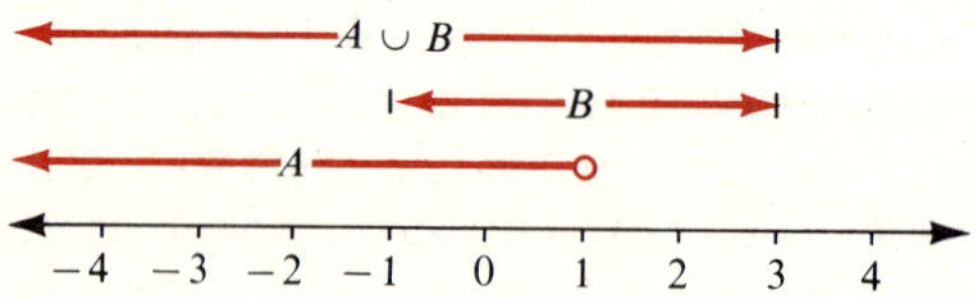

Figure 1.20

Definition 1.13

Let A and B be subsets of the real number system R. The *intersection* of A and B, denoted by $A \cap B$, is the set of all elements that are common to both A and B. In symbols,

$$A \cap B = \{x | x \in A \text{ and } x \in B\}$$

Example 1.16

If

$$A = \{1, 2, 3, 6, 7, 8\} \quad \text{and} \quad B = \{3, 7, 8, 9, 10\}$$

then

$$A \cap B = \{3, 7, 8\}$$

Example 1.17

Let A and B be subsets of R and let

$$A = \{x | x > -3\} \quad \text{and} \quad B = \{x | x < 3\}$$

Describe the set $A \cap B$ as a collection of points of R.

Solution

Since the set A consists of all real numbers greater than -3, a point belongs to A if and only if it lies to the right of the point -3 on the real line. Similarly, a point belongs to B if and only if it lies to the left of the point 3. Then $A \cap B$ is the set of all points that lie between -3 and 3. In other words,

$$A \cap B = \{x \mid -3 < x < 3\}$$

Figure 1.21 illustrates A, B, and $A \cap B$ geometrically.

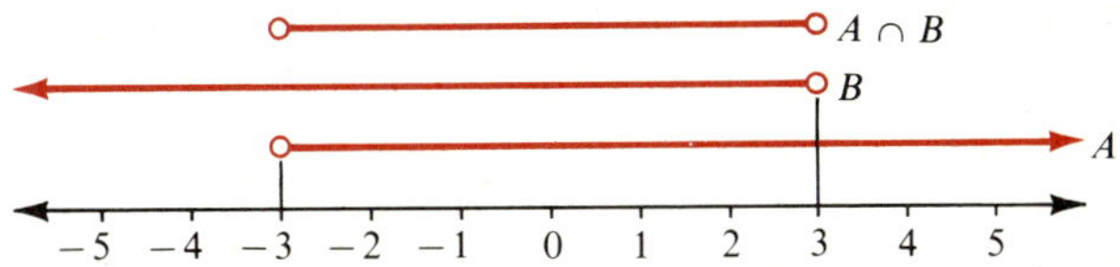

Figure 1.21

Cartesian Product of Two Sets

We recall that if $A = \{a, b, c, d\}$, then $\{a, b\}$ and $\{b, a\}$ are subsets of A; furthermore, they are the same subset since the order in which the elements in the set are listed is immaterial. At times, the order in which information is recorded plays an important role. Consider, for example, a table which records heights and weights of freshman students by listing a pair of members with the understanding that the first and second members of the pair represent, respectively, the height (in inches) and weight (in pounds). Thus the entry (62, 105) means that there is a student who is 62 inches tall and weighs 105 pounds. Note that this is completely different from the entry (105, 62) which means that there is a student who is 105 inches tall and weighs 62 pounds—an impossibility. As another illustration, consider an auditor describing the financial status of an airline by listing a pair of numbers so that the first and second members of the pair represent, respectively, the assets and the liabilities in millions of dollars. The pair (6, 0) means that the airline is in an excellent financial position while the pair (0, 6) implies that the airline is on the verge of bankruptcy. Thus, the pairs (6, 0) and (0, 6) convey very different information because of the order.

We consider it worth our while to point out that the two symbols { } and ( ) are different notations—one denotes a set and the other represents an ordered pair. While the notations for an open interval and for an ordered pair are the same, we should be able to distinguish them by the context.

Definition 1.14

Let A and B be two sets. Then an *ordered pair* (a, b), where $a \in A$ and $b \in B$, is a listing of two elements in a certain order. The elements a and b are called, respectively, the first and the second component in the ordered pair.

Definition 1.15

Two ordered pairs (a, b) and (c, d) are said to be equal if and only if $a = c$ and $b = d$.

Consider now two sets A and B and *all* the ordered pairs which can be formed by selecting the first component of the pair from the first set and the second component from the second set. The set of all ordered pairs thus obtained is called the cartesian product of the sets A and B and is written as $A \times B$.

Definition 1.16

If A and B are two sets, then the *cartesian product* of A and B, denoted by $A \times B$, is the set of all ordered pairs (a, b), where $a \in A$ and $b \in B$. In symbols,

$$A \times B = \{(a, b) \mid a \in A, b \in B\}$$

Example 1.18

Let $A = \{1, 2, 3, 4\}$ and $B = \{6, 7\}$. List the elements in each of the following cartesian products.

a. $A \times B$ **b.** $B \times A$

Solution

a. $A \times B = \{(1, 6), (2, 6), (3, 6), (4, 6), (1, 7), (2, 7), (3, 7), (4, 7)\}$, and

b. $B \times A = \{(6, 1), (6, 2), (6, 3), (6, 4), (7, 1), (7, 2), (7, 3), (7, 4)\}$

Notice that $A \times B$ and $B \times A$ have the same number of ordered pairs, but in general,

$$A \times B \neq B \times A$$

Exercises

1. Given that $A = \{1, 2, 3, 4, 5, 6, 7, 8\}$, state whether each of the following statements is true.
   a. $5 \in A$
   b. $12 \in A$
   c. $0 \notin A$
   d. $5 \subseteq A$
   e. $\varnothing \in A$
   f. $\varnothing \subseteq A$
   g. $A \subseteq A$
   h. $\{1, 2, 3\} \subseteq \{1, 2, 3, 4\}$
   i. $\{1, 2, 3\} \in \{1, 2, 3, 4\}$
   j. $\{1, 2, 3\} \subseteq A$
2. Let $A = \{1, 3, 5, 7, 9\}$, $B = \{2, 4, 5, 6, 8\}$, and $C = \{3, 6, 7, 9\}$. Determine the elements in the following sets.
   a. $A \cup B$
   b. $A \cap B$
   c. $A \cup C$
   d. $B \cup C$
   e. $B \cap C$
   f. $A \cup (B \cap C)$
   g. $A \cap (B \cup C)$
   h. $(A \cap B) \cup (A \cap C)$

3. Let $A = \{1, 5, 6, 10, 11, 12\}$, $B = \{2, 3, 7, 8, 9, 10\}$ and $C = \{2, 5, 8, 9, 12\}$. Show that
   a. $A \cap (B \cup C) = (A \cap B) \cup (A \cap C)$
   b. $A \cup (B \cap C) = (A \cup B) \cap (A \cup C)$

4. Describe each of the following sets as a collection of points on a number line R and represent each set graphically.
   a. $\{x|x > 2\} \cup \{x|x < 3\}$
   b. $\{x|x > 4\} \cap \{x|x < 6\}$
   c. $\{x|x < -5\} \cap \{x|x > 1\}$

5. Describe each of the following sets as a collection of points on a number line R and represent each set graphically.
   a. $\{x|1 \leq x < 3\} \cup \{x|2 < x \leq 5\}$
   b. $\{x|-1 < x < 4\} \cap \{x|2 \leq x \leq 6\}$
   c. $\{x|3 < x < 5\} \cap \{x|2 < x < 4\}$
   d. $\{x|-2 < x \leq 3\} \cup \{x|1 \leq x < 7\}$
   e. $\{x|-2 < x \leq 3\} \cap \{x|1 \leq x < 7\}$

6. Let $A = \{2, 3, 4\}$ and $B = \{5, 6\}$. List the elements in each cartesian product.
   a. $A \times B$ b. $B \times A$ c. $A \times A$ d. $B \times B$

7. Let $A = \{2, 3, 4, 5\}$ and $B = \{6, 7, 8\}$. List the elements in each cartesian product.
   a. $A \times B$ b. $B \times A$ c. $A \times A$ d. $B \times B$

8. For the set

$$A = \{1, 2, 3, 4, 5, 6\}$$

   find a subset of $A \times A$ such that
   a. the sum of the two numbers in each ordered pair is 7
   b. the two elements in each ordered pair are equal.

1.5 Coordinate System

In Section 1.1, we stated that a one-to-one correspondence exists between the real number system R and the points on the real number line. We shall now examine a one-to-one correspondence between the points in the plane and the set of ordered pairs of real numbers. Note that the set of all ordered pairs of real numbers is the cartesian product $R \times R$.

We choose an arbitrary point 0 in the plane and call it the origin. Through 0, we draw a pair of perpendicular lines, one horizontal and the other vertical. The horizontal line is called the *x-axis* and the vertical line is called the *y-axis*. The axes together are called the *coordinate axes* and the plane in which the axes lie is called the coordinate plane. An arbitrary unit of length is selected and a number scale is marked on the x-axis, positive to the right of the origin and negative to the left. Similarly, a scale is marked along the vertical axis with positive numbers marked upward and negative numbers marked downward. Usually the units measured along the x-axis have the same length as the units along the y-axis as shown in Figure 1.22. However, this is *not* necessary in every instance and for convenience, different scales may be used on the axes. The coordinate axes divide the plane into four parts, called *quadrants*, numbered counter-clockwise as indicated in Figure 1.22.

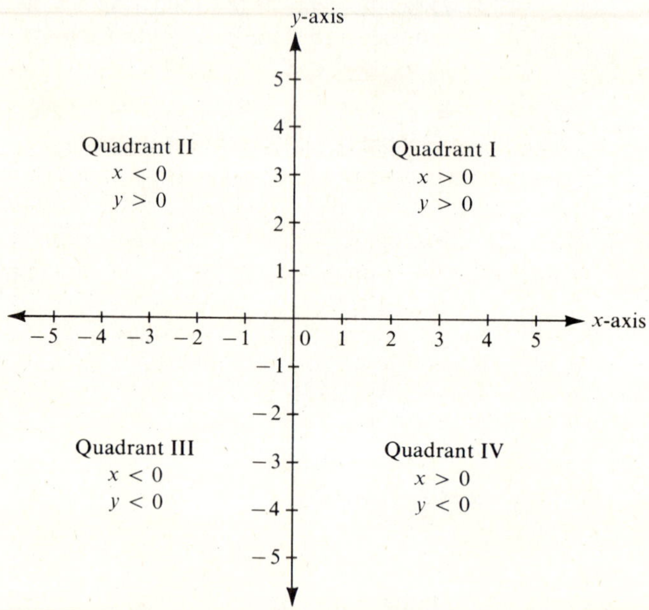

Figure 1.22

We are now ready to set up a one-to-one correspondence between the set of ordered pairs of real numbers and the points of the plane. Let P be a point in the plane (Figure 1.23). We draw two lines through P, one perpendicular to the x-axis and the other perpendicular to the y-axis. The intersection with the horizontal axis is at the point A while the intersection with the vertical axis is at the point B. If a is the real number associated with the point A on the horizontal axis, then $a > 0$ if A is to the right of the origin and $a < 0$ if A is to the left of the origin. Similarly, if b is the real number associated with the point B on the vertical axis, then B is above the x-axis if $b > 0$ and below the x-axis if $b < 0$. The ordered pair associated with the point P in the plane is (a, b). The number a is called the *x coordinate* or *abscissa* of P, and the number b is called the *y coordinate* or *ordinate* of P. Thus P is a point with coordinates (a, b).

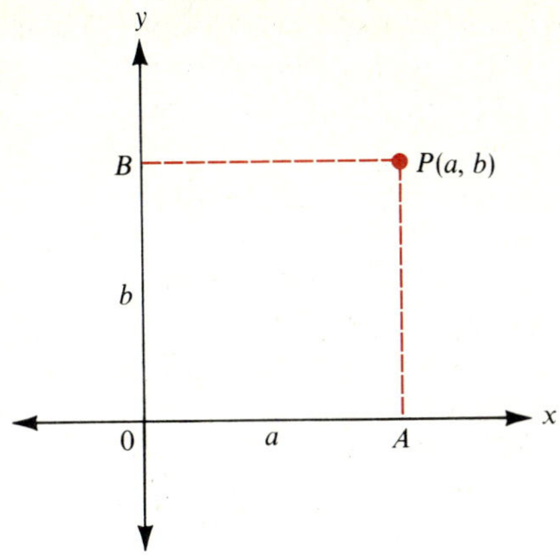

Figure 1.23

The symbol $P(a, b)$ means the point P with coordinates (a, b). To *graph* a point $P(a, b)$ means to locate, in the plane, the point P with coordinates (a, b). Figure 1.24 shows, for instance, the location of points P, Q, R, and S. Since the coordinate axes divide the plane into four quadrants, we can determine the quadrant of a point from the signs of its coordinates. A point with two positive coordinates lies in Quadrant I; a point (c, d) with negative c and positive d lies in Quadrant II; a point with both negative coordinates is located in Quadrant III;

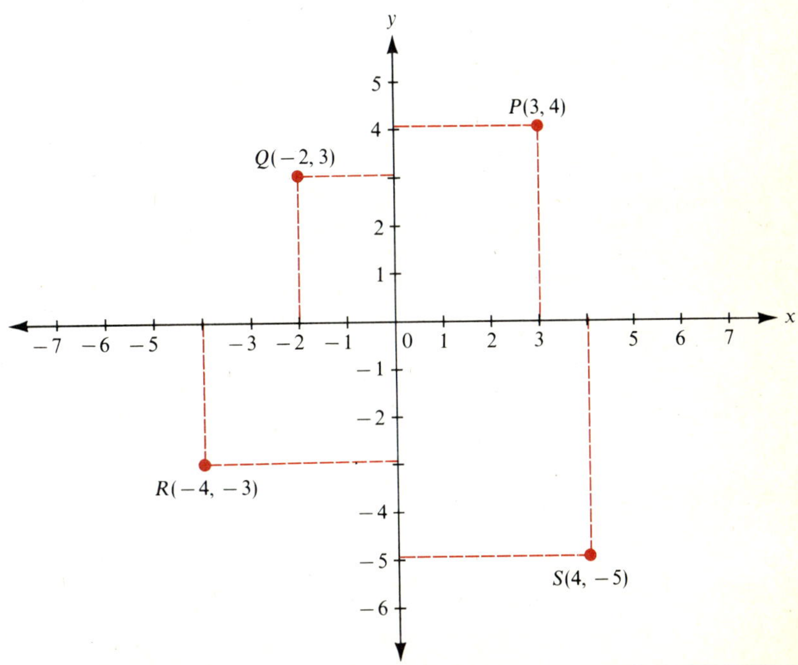

Figure 1.24

and a point with positive c and negative d is located in Quadrant IV. Thus if (c, d) is any ordered pair of real numbers, we start at the origin and go along the x-axis until we reach $|c|$ units from the origin. Then we go $|d|$ units along the y-axis, upward if $d > 0$ and downward if $d < 0$. The point so located has coordinates (c, d). We have noted earlier that to each point in the plane there corresponds a unique ordered pair of real numbers, and conversely, corresponding to each ordered pair of real numbers there exists exactly one point in the plane.

Let us now use the coordinate system to determine the distance between two points. Let x_1 and x_2 be the x-coordinates of the points A and B that have the same y-coordinate. The horizontal distance between them, by Definition 1.6, is $|x_1 - x_2|$. Consider the pairs of points in Figure 1.25. The distance between $A(2, 3)$ and $B(7, 3)$ is $|2 - 7| = |-5| = 5$ units and the distance between $C(-5, 5)$ and $D(6, 5)$ is $|-5 - 6| = |-11| = 11$ units. Similarly, if y_1 and y_2 are the y-coordinates of two points having the same x-coordinate, then the distance between them is $|y_1 - y_2|$. This means that the vertical distance between $P(-2, 2)$ and $Q(-2, 6)$ is simply $|2 - 6| = |-4| = 4$ units while the vertical distance between $R(4, 4)$ and $S(4, -5)$ is $|4 - (-5)| = 9$ units. Recall that distance is always non-negative.

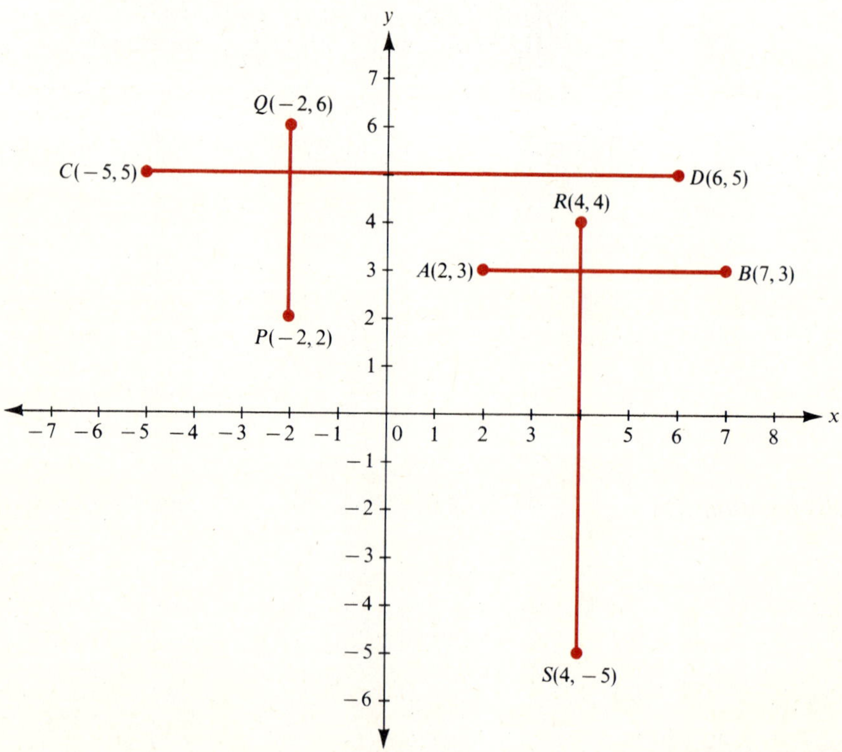

Figure 1.25

Let us now extend the concept of distance between points to two points $A(x_1, y_1)$ and $B(x_2, y_2)$ which have neither the same x-coordinate nor the same y-coordinate. (See Figure 1.26.)

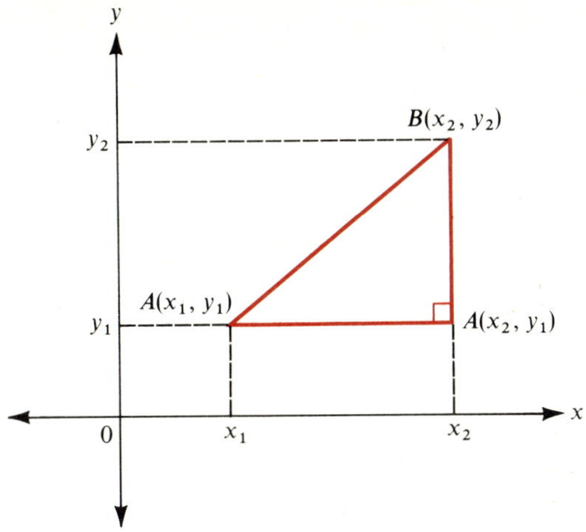

Figure 1.26

Draw a line through A parallel to the x-axis and another line through B parallel to the y-axis. If N is the point of intersection, then the coordinates of N are (x_2, y_1). The triangle ANB so formed is a right triangle with a right angle at N and the sides AN and NB are of length $|x_2 - x_1|$ and $|y_2 - y_1|$, respectively. If d_{AB} denotes the distance from A to B, then by the Pythagorean Theorem, we have

$$d_{AB}^2 = d_{AN}^2 + d_{NB}^2$$

or

$$\begin{aligned} d_{AB} &= \sqrt{d_{AN}^2 + d_{NB}^2} \\ &= \sqrt{|x_2 - x_1|^2 + |y_2 - y_1|^2} \\ &= \sqrt{(x_2 - x_1)^2 + (y_2 - y_1)^2} \end{aligned}$$

Hence, we have established the following proposition.

Proposition 1.3

The distance from $A(x_1, y_1)$ to $B(x_2, y_2)$ is equal to

$$\sqrt{(x_2 - x_1)^2 + (y_2 - y_1)^2}$$

It can be shown that this proposition holds for $x_1 = x_2$ or $y_1 = y_2$. We will leave this as an exercise for the reader.

Example 1.19

Find the distance between $A(8, 10)$ and $B(20, 15)$.

Solution

Letting $x_1 = 8$, $y_1 = 10$, $x_2 = 20$, and $y_2 = 15$, by Proposition 1.3, we have

$$\begin{aligned} d_{AB} &= \sqrt{(20-8)^2 + (15-10)^2} \\ &= \sqrt{(12)^2 + (5)^2} \\ &= \sqrt{144 + 25} \\ &= \sqrt{169} \\ &= 13 \end{aligned}$$

If we interchange (x_1, y_1) and (x_2, y_2), d_{AB} is still the same.

Let $A(x_1, y_1)$ and $B(x_2, y_2)$ be two points in the plane. Then the following properties hold.

1. $d_{AB} = 0$ if $B = A$
2. $d_{AB} > 0$ if $B \neq A$
3. $d_{AB} = \sqrt{(x_2 - x_1)^2 + (y_2 - y_1)^2} = \sqrt{(x_1 - x_2)^2 + (y_1 - y_2)^2} = d_{BA}$

Exercises

1. Graph the following points.
   a. (2, 4) b. (−3, 8) c. (−4, −5) d. (0, 1) e. (0, 3) f. (12, 5) g. (5, 12) h. (−5, −12)
2. Determine the coordinates of the following points in Figure 1.27.
   a. Point A b. Point B c. Point C d. Point D e. Point P f. Point Q
3. Determine the distance between each of the following pairs of points without using Proposition 1.3.
   a. (2, 4) and (8, 4)
   b. (3, 5) and (3, −4)
   c. (8, 6) and (8, 12)
   d. (8, 7) and (8, −7)
   e. (12, 5) and (−8, 5)
   f. (4, −5) and (−10, −5)
4. Determine the distance between the following pairs of points.
   a. (0, 0) and (3, 4)
   b. (6, 2) and (10, 5)
   c. (3, 4) and (9, 12)
   d. (8, 5) and (8, −10)
   e. (−2, 1) and (10, 6)
   f. (−8, −9) and (16, −2)
   g. (5, 6) and (5, 12)
   h. (4, −3) and (20, 9)
5. Some household furniture is to be shipped from town A to town B. It can be shipped directly by train at a cost of \$1.25 per mile or by moving van via town C at a cost of \$1.00 per mile. Assuming that towns A, B, and C are located as shown below in Figure 1.28, which way is more economical?

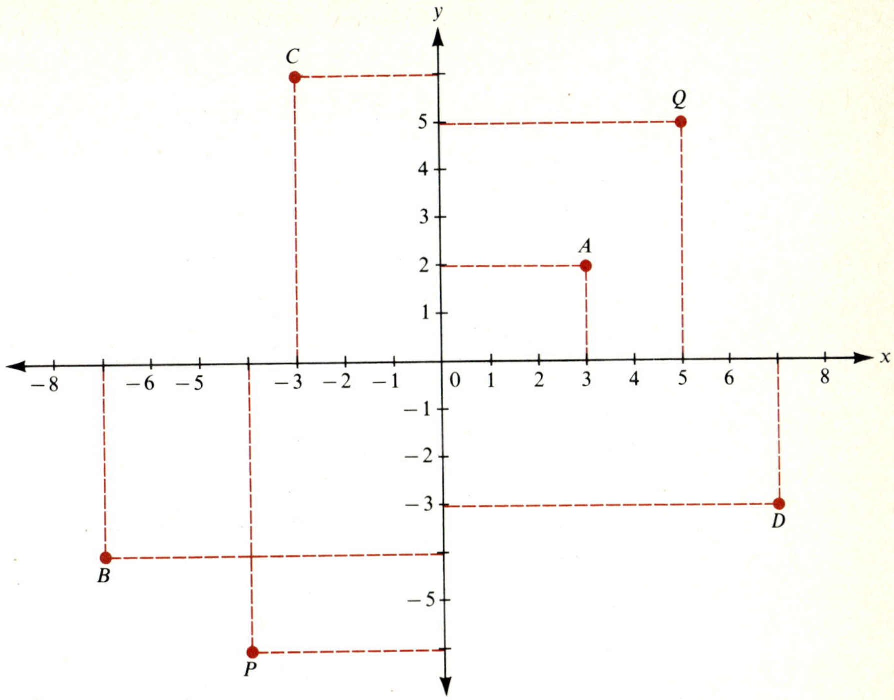

Figure 1.27

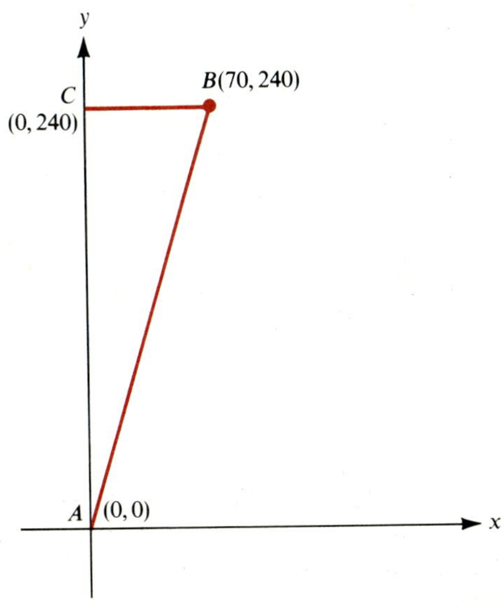

Figure 1.28

6. A new automobile is to be shipped to town F from a plant located in town D. It can be shipped directly by a freight train at a cost of \$1.30 per mile or by truck via town E at a cost of \$1.25 per mile. Assuming that the coordinates of D, E, and F are (20, 11), (200, 11) and (200, 30), respectively, determine which mode of transportation is more economical.

1.6 Straight Lines

The equation $y = mx + b$, where m and b are fixed real numbers, represents a straight line. Conversely, the equation of any non-vertical line can be expressed in the form $y = mx + b$.

The coefficient of x determines what we call the *slope* of the line. This concept is important in calculus because it measures the rate at which changes are taking place—the rate at which the population is increasing in a certain town, the rate at which the number of bacteria is multiplying in a certain culture, the rate at which enrollment is dropping in the liberal arts colleges, the rate at which unemployment is increasing in the private sector of the economy, the rate at which the economy is recovering from a recession, and so on. Consider the sales for three companies A, B, and C and compare the rate at which their volumes of sales have increased or decreased from 1976 to 1979. The following table shows the sales volumes of these companies in 1976 and 1979. Assuming that the volumes of sales for each of these companies follow a straight line trend, we plot two points corresponding

	Sales in	
Company	1976	1979
A	\$145,000	\$163,000
B	\$135,000	\$141,000
C	\$140,000	\$122,000

to each company and then connect them by means of a straight line as displayed in Figure 1.29. Observe that two points are sufficient to graph a straight line.

The sales for Company A have increased from \$145,000 in 1976 to \$163,000 in 1979. Thus, the average annual increase in sales is \$18,000/3 = \$6,000. The sales for Company B have increased by \$6,000 over this three-year period; thus, the average annual increase in sales is \$2,000. This accounts for the fact that the slope for Company B's line is not as steep as for Company A. Nevertheless, the upward trend of Companies A and B reflect their growth. The sales for Company C, on the other hand, have decreased from \$140,000 in 1976 to \$122,000 in 1979. This is reflected by the line segment which sinks downward thus showing a drop in sales during the three-year period. The slope of the line can also be described as a measure of its steepness.

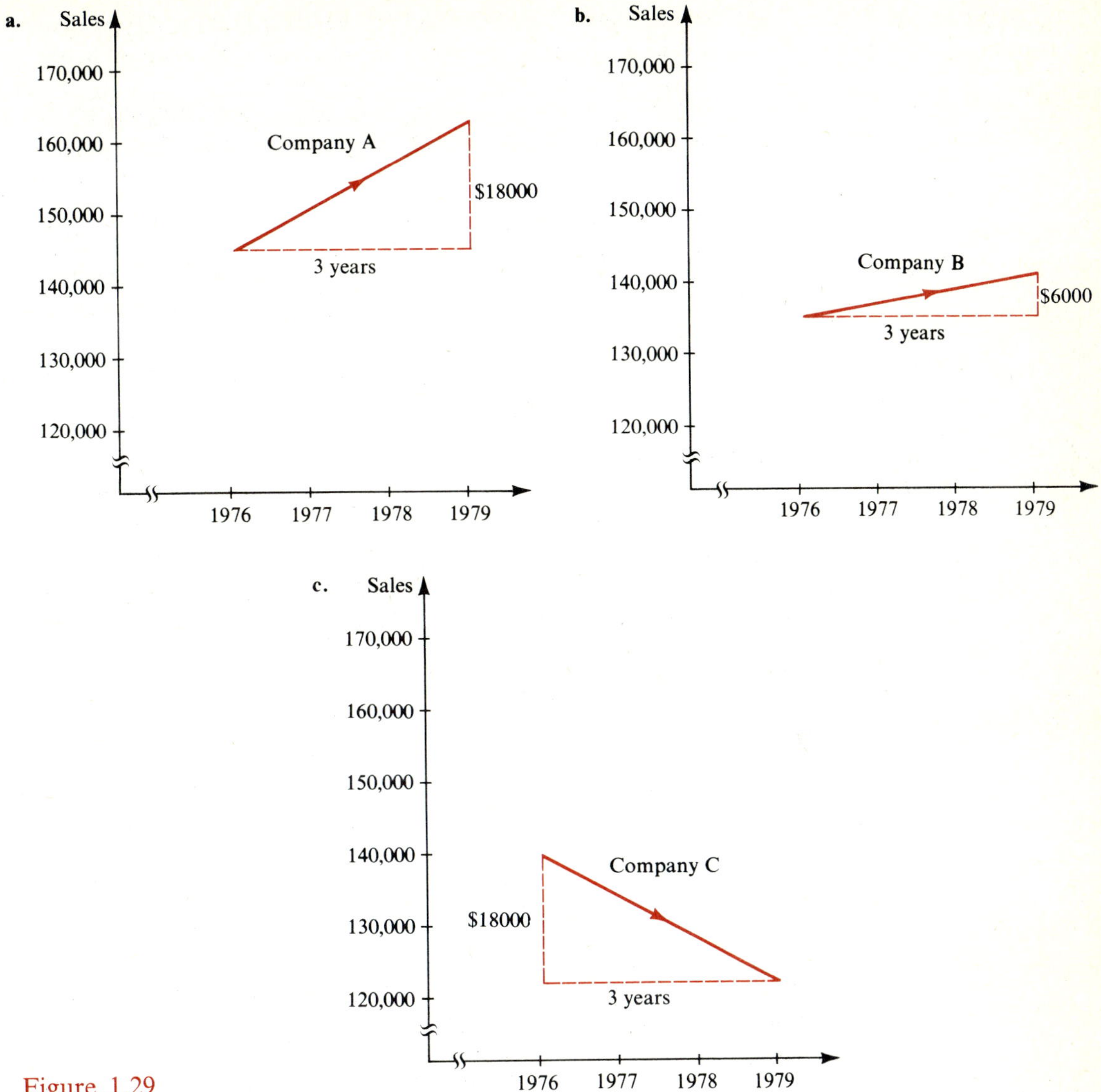

Figure 1.29

Formally, we shall define the slope of a line as follows.

Definition 1.17

Let $\mathscr{L}$ be a non-vertical line joining two distinct points A and B with coordinates (x_1, y_1) and (x_2, y_2), respectively. Then the *slope* m of the line $\mathscr{L}$ is defined by

1.25 $$m = \frac{y_2 - y_1}{x_2 - x_1}, \qquad x_1 \neq x_2$$

An immediate consequence of this definition is that if a line rises sharply to the right, as illustrated in Figure 1.29(a), the difference in the y coordinates of any two points is large when compared to the difference in x coordinates. This situation yields a large value of m. In case a line slants downward from left to right (see Figure 1.29(c)), the numerator and denominator of the slope have opposite signs; consequently, the slope of such a line is negative. If the line is parallel to the x-axis, all points on this horizontal line have the same y-coordinate, that is, $y_1 = y_2$ (see Figure 1.30). Thus the difference between any two y coordinates on this line is zero; consequently, the slope m is zero. A vertical line, on the other hand, does not have any slope. This is clear from the fact that the x coordinates of any two points on a vertical line are the same (see Figure 1.31); that is, $x_1 = x_2$. The denominator of the fraction $y_2 - y_1/x_2 - x_1$ would be zero. Thus the slope m of a vertical line is not defined. We wish to emphasize that a horizontal line has zero slope while a vertical line is said to have an undefined slope.

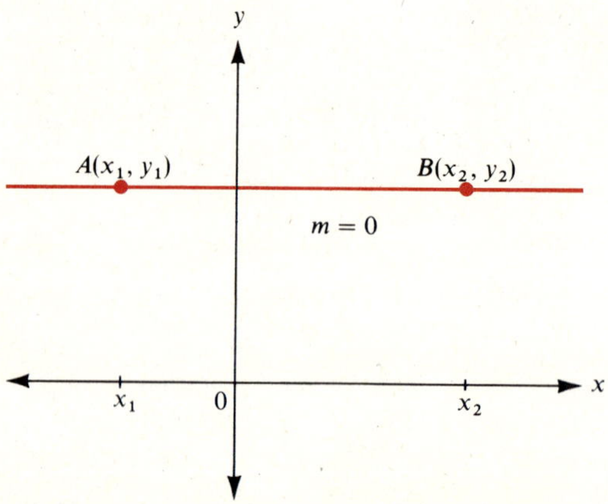

Figure 1.30

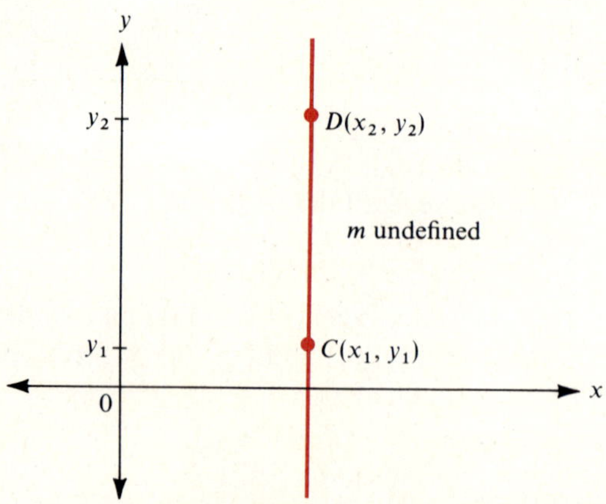

Figure 1.31

Example 1.20

Find the slope of the line through A and B for each pair of points below:

a. $A(3, 5)$, $B(5, 11)$

b. $A(8, -10)$, $B(3, -2)$

c. $A(6, 8)$, $B(-15, 8)$

Solution

a. Let

$$x_1 = 3, \quad y_1 = 5; \qquad x_2 = 5, \quad y_2 = 11$$

then,

$$m = \frac{y_2 - y_1}{x_2 - x_1} = \frac{11 - 5}{5 - 3} = \frac{6}{2} = 3$$

Note that had we used $B(5, 11)$ as (x_1, y_1) and $A(3, 5)$ as (x_2, y_2), the resulting slope m would be the same.

b. Let

$$x_1 = 8, \quad y_1 = -10; \qquad x_2 = 3, \quad y_2 = -2$$

then

$$m = \frac{y_2 - y_1}{x_2 - x_1} = \frac{-2 - (-10)}{3 - 8} = -\frac{8}{5}$$

c. The y-coordinates of the two points are equal, hence the slope of this line is zero.

The slope of a line does not depend on the pair of points used to calculate it. Let A, B, and C be any three points on a non-vertical straight line (see Figure 1.32). For convenience, we will assume the slope to be positive, an analogous argument for negative slope also holds true. Observe that the right triangles APB and BQC are similar. (Why?) Thus, the lengths of the corresponding sides of similar triangles are in proportion. Hence,

$$\frac{d_{PB}}{d_{AP}} = \frac{d_{QC}}{d_{BQ}}.$$

Now,

$$d_{PB} = y_2 - y_1, \, d_{AP} = x_2 - x_1$$
$$d_{QC} = y_3 - y_2, \, d_{BQ} = x_3 - x_2$$

Thus,

$$\frac{y_2 - y_1}{x_2 - x_1} = \frac{y_3 - y_2}{x_3 - x_2}$$

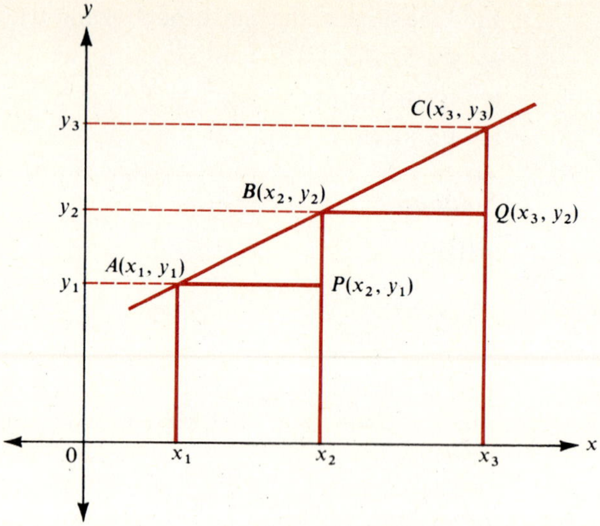

Figure 1.32

But the left and right sides of the equation are the slopes of AB and BC, respectively. Therefore, the slopes of AB and BC are equal. Further, the slope of a line is independent of the pair of points chosen.

Equation of a Straight Line

Finding the slope of a line is just one phase of determining the equation of the line. We will now see how to determine that equation. We begin with lines that are parallel to an axis.

Proposition 1.4

The equation of a straight line parallel to the x-axis and $|b|$ units from it is $y = b$. If b is positive, the line is above the x-axis and if b is negative, the line is below the x-axis.

This means that the equation $y = 2$ represents a line parallel to the x-axis and two units above it. Similarly, the equation $y = -2$ is a line also parallel to the x-axis but two units below it (see Figure 1.33). Lines

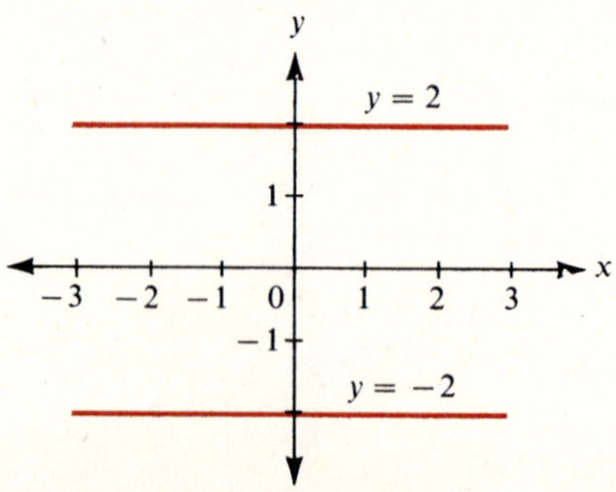

Figure 1.33

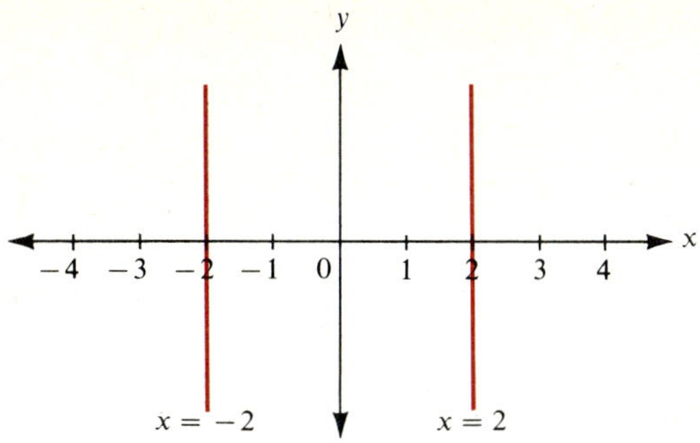

Figure 1.34

parallel to the y-axis are treated in a similar manner. The equation $x = 2$ represents a line parallel to the y-axis and two units to the right of it. Similarly, the equation $x = -2$ is a line parallel to the y-axis but two units to the left of it (see Figure 1.34). In general, we have the following proposition.

Proposition 1.5

The equation of a straight line parallel to the y-axis and $|a|$ units from it is $x = a$. If a is positive, the line is to the right of the y-axis and if a is negative, the line is to the left of the y-axis.

Next, we determine the equation of a non-vertical line. Where this line meets the y-axis at some point, the x-coordinate is zero. The y-coordinate of this point is called the *y-intercept* of the line and is generally denoted by b. Thus the coordinates of the point C are $(0, b)$. If $D(x, y)$ is any other point on this line, then the slope m is obtained by applying Definition 1.7 to the points $C(0, b)$ and $D(x, y)$ (see Figure 1.35).

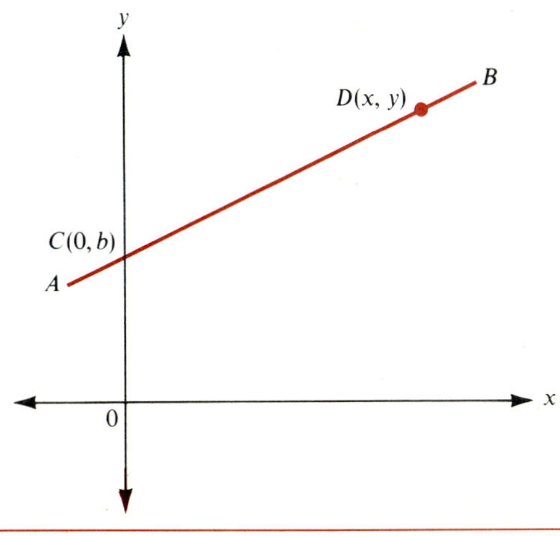

Figure 1.35

Thus,

$$m = \frac{y - b}{x - 0}$$

that is,

$$y - b = mx$$

or

$$y = mx + b$$

Conversely, suppose that we are given any point (x_1, y_1) satisfying the equation $y = mx + b$. Observe that $(0, b)$ also satisfies the equation and that the slope of the line connecting $(0, b)$ and (x_1, y_1) is

$$\frac{y_1 - b}{x_1 - 0} = \frac{(mx_1 + b) - b}{x_1 - 0} = \frac{mx_1}{x_1} = m, \qquad x_1 \neq 0$$

Because any other point (x_2, y_2) satisfying $y = mx + b$ will also lie on the line of slope m through $(0, b)$, we conclude that *all* points satisfying $y = mx + b$ lie on the straight line with slope m and y intercept b. Thus we have established the following proposition.

Proposition 1.6

Let $\mathscr{L}$ be the line passing through $(0, b)$ and having slope m. Then a point (x, y) lies on $\mathscr{L}$ if and only if it satisfies the equation

1.26 $$y = mx + b$$

Note that any non-vertical line has a slope m and a y-intercept b. Thus the equation of the line passing through $(0, 3)$ with slope 4 (that is, $m = 4$ and $b = 3$) is given by

$$y = 4x + 3$$

It is because of the prominence of m and b that the equation (1.26) is called the *slope-intercept* form of the equation of a straight line. Observe that if the non-vertical line passes through the origin, then $b = 0$ and the equation (1.26) becomes

$$y = mx$$

Example 1.21

Determine the slope and y-intercept of the line

$$2y = 3x + 6$$

Solution

Rewriting this equation in the slope-intercept form (1.26), we have

$$y = \tfrac{3}{2}x + 3$$

Comparing this equation with (1.26), we see that $m = \frac{3}{2}$ and $b = 3$. Thus the slope is $\frac{3}{2}$ and this line intersects the y-axis at the point (0, 3).

The slope of the line also helps us determine whether or not two distinct lines are parallel. We state (without proof) the following proposition.

Proposition 1.7 Two distinct (non-vertical) lines are parallel if and only if their slopes are equal.

Example 1.22 Determine whether or not the following lines are parallel.

a. $4x + 5y - 8 = 0$ and $3x - 2y + 10 = 0$

b. $2x - 3y + 4 = 0$ and $6x - 9y + 25 = 0$

Solution

a. Rewriting $4x + 5y - 8 = 0$ and $3x - 2y + 10 = 0$ in the slope-intercept form (1.26), we have

$$y = -\tfrac{4}{5}x + \tfrac{8}{5} \quad \text{and} \quad y = \tfrac{3}{2}x + 5$$

respectively. These lines have slopes $-\frac{4}{5}$ and $\frac{3}{2}$, respectively. Because these slopes are not equal, it follows that the lines are *not* parallel.

b. Expressing $2x - 3y + 4 = 0$ in the form (1.26), we get

$$y = \tfrac{2}{3}x + \tfrac{4}{3}$$

which implies that the slope of this line is $m_1 = \frac{2}{3}$. Similarly, solving $6x - 9y + 25 = 0$ for y, we have

$$y = \tfrac{6}{9}x + \tfrac{25}{9}$$

We note that the slope of this line $m_2 = \frac{6}{9} = \frac{2}{3}$. Because the slopes of these lines are equal, we conclude that the lines are parallel.

We now state (without proof) the following proposition.

Proposition 1.8 Two distinct (non-vertical) lines are perpendicular to each other if and only if the product of their slopes equals -1.

Example 1.23 Determine whether or not the following lines are perpendicular:

a. $4x + 3y - 6 = 0$ and $3x - 4y + 12 = 0$

b. $3x + 2y + 8 = 0$ and $5x - 3y + 10 = 0$

Solution

a. Solving $4x + 3y - 6 = 0$ for y, we get

$$y = -\tfrac{4}{3}x + 2$$

Thus, the line has slope $m_1 = -\tfrac{4}{3}$. Similarly, the line $3x - 4y + 12 = 0$ can be written as

$$y = \tfrac{3}{4}x + 3$$

so that this line has slope $m_2 = \tfrac{3}{4}$. Because $m_1m_2 = (-\tfrac{4}{3})(\tfrac{3}{4}) = -1$, it follows that the lines are perpendicular to each other.

b. The lines $3x + 2y + 8 = 0$ and $5x - 3y + 10 = 0$ expressed in the slope-intercept form (1.26) are

$$y = -\tfrac{3}{2}x - 4 \quad \text{and} \quad y = \tfrac{5}{3}x + \tfrac{10}{3}$$

The slope of the lines are $m_1 = -\tfrac{3}{2}$ and $m_2 = \tfrac{5}{3}$, respectively. Since $m_1m_2 \neq -1$, we conclude that lines are *not* perpendicular to each other.

Suppose that the slope m of a straight line and a point (x_1, y_1) on the line are known. If (x, y) is any other point on the line, then using (1.25), we get

$$\frac{y - y_1}{x - x_1} = m, \qquad x \neq x_1$$

and

$$y - y_1 = m(x - x_1)$$

Thus we have established the following proposition.

Proposition 1.9

The equation of a straight line passing through (x_1, y_1) with slope m is

1.27
$$y - y_1 = m(x - x_1).$$

Since equation (1.27) uses a given point (x_1, y_1) and slope m of the line, it is called the *point-slope* form of the equation of a line.

Example 1.24

Find an equation of the line which passes through the point (3, 5) and has slope $m = 2$.

Solution

Note that

$$x_1 = 3, \quad y_1 = 3, \quad \text{and} \quad m = 2.$$

Using (1.27), the required equation of the line is

$$y - 5 = 2(x - 3)$$

which simplifies to

$$y = 2x - 1$$

Example 1.25

Find the equation of the line passing through $A(5,6)$ and $B(3,2)$.

Solution

The slope of the line joining two points $A(5,6)$ and $B(3,2)$ is

$$m = \frac{2-6}{3-5} = \frac{-4}{-2} = 2$$

Choosing one of the given points as (x_1, y_1), say (5, 6), we use (1.27) and obtain

$$y - 6 = 2(x - 5)$$

which simplifies to

$$y = 2x - 4$$

Note that had we used (3, 2) as (x_1, y_1), the resulting equation would be the same.

We have considered the equations of straight lines in all possible positions. A line is either parallel to the x-axis, parallel to the y-axis, or meets the axes in fixed points. In the first case, its equation is of the form $y = b$ (Proposition 1.4), in the second case, the equation is of the form $x = a$ (Proposition 1.5), and in the third case, it is represented by $y = mx + b$, where m and b are fixed real numbers (Proposition 1.6). All of these equations can be rewritten into the general form $Ax + By + C = 0$ where A, B, and C are real numbers and A and B cannot both be zero. Note that the first equation is in y alone, the second is in x alone, and third contains both x and y.

Conversely, it can be shown that every equation of the form

$$Ax + By + C = 0, \quad A \neq 0 \quad \text{or} \quad B \neq 0$$

where A, B, and C are fixed real numbers represents a straight line.

Exercises

1. Find the slope of the lines passing through the points A and B.

a. $A(3,4)$, $B(4,3)$

b. $A(-2,2)$, $B(5,5)$

c. $A(-5,2)$, $B(5,6)$

d. $A(6,20)$, $B(6,30)$

e. $A(4,5)$, $B(8,13)$

f. $A(3,2)$, $B(6,8)$

g. $A(8,4)$, $B(10,4)$

h. $A(9,1)$, $B(6,5)$

2. Find the slope of the sides of the triangle whose vertices are $P(-4, 0)$, $Q(1, 1)$, and $R(3, -1)$.

3. Find the equation of each of the following lines.
a. Parallel to the x-axis and 3 units above it.
b. Parallel to the x-axis and 2 units below it.
c. Parallel to the y-axis and 5 units to the right of it.
d. Parallel to the y-axis and 4 units to the left of it.

4. Find the slope of each of the following lines.
a. $x = 4$
b. $y = x + 3$
c. $2x - 3y = 5$
d. $y = 5$
e. $y = 2x + 3$
f. $4x + 5y - 7 = 0$

5. Find the slope and the y-intercept of each of the following lines.
a. $3x - 2y + 12 = 0$
b. $x - y = 0$
c. $x + 5y - 10 = 0$
d. $2x - 5y + 10 = 0$
e. $3x + 2y - 15 = 0$
f. $-3x + 4y + 7 = 0$

6. Determine whether or not each of the following pairs of lines are parallel.
a. $2x + 3y - 8 = 0$ and $3x + y - 2 = 0$
b. $3x - 2y + 1 = 0$ and $4x - 5y + 8 = 0$
c. $x + 2y + 3 = 0$ and $3x + 6y + 2 = 0$
d. $3x + y + 6 = 0$ and $6x - 2y + 3 = 0$

7. Determine whether or not each of the following pairs of lines are perpendicular.
a. $2x + 3y - 6 = 0$ and $3x - 2y + 4 = 0$
b. $4x - 5y + 20 = 0$ and $5x + 4y - 30 = 0$
c. $3x - 4y + 12 = 0$ and $x - 2y + 15 = 0$
d. $6x + 7y - 42 = 0$ and $7x - 6y + 25 = 0$

8. Find the slope-intercept form of the equation of each line.
a. through $P(0, 3)$ with slope $m = 2$
b. through $P(0, -3)$ with slope $m = -\frac{2}{3}$
c. through $P(3, 4)$ with slope $m = 0$
d. through $P(1, 2)$ with slope $m = \frac{1}{4}$

9. Find the slope-intercept form of the equation of each line.
a. through $P(-2, -3)$ with slope $m = -2$
b. through $P(3, -4)$ with slope $m = \frac{3}{4}$
c. through $P(4, 5)$ with slope $m = \frac{2}{3}$
d. through $P(4, 5)$ and parallel to the line $2x - 3y - 3 = 0$
e. through $P(1, 2)$ and parallel to the line $3x + 5y - 12 = 0$
f. through $P(2, 3)$ and perpendicular to the line $x - 2y + 4 = 0$
g. through $P(-2, 5)$ and perpendicular to the line $3x + 4y - 6 = 0$

10. Find the equation of the line which passes through the following pairs of points.

a. $P(5, 2)$ and $Q(4, -1)$
b. $P(5, 6)$ and $Q(3, 2)$
c. $P(2, 3)$ and $Q(5, 6)$
d. $P(-1, 2)$ and $Q(3, 8)$
e. $P(0, 0)$ and $Q(1, 1)$
f. $P(-2, -3)$ and $Q(-7, -9)$
g. $P(7, 6)$ and $Q(9, 8)$
h. $P(-4, -7)$ and $Q(-10, -13)$

11. Show that the lines

$$a_1x + b_1y + c_1 = 0 \quad \text{and} \quad a_2x + b_2y + c_2 = 0$$

are parallel if and only if $a_1b_2 - a_2b_1 = 0$.

12. Show that the lines

$$a_1x + b_1y + c_1 = 0 \quad \text{and} \quad a_2x + b_2y + c_2 = 0$$

are perpendicular to each other if and only if $a_1a_2 + b_1b_2 = 0$.

1.7 Applications

Apartment buildings, office equipment, furniture, machines and other business assets depreciate in value over a number of years. This loss in value is computed generally by using the Linear Depreciation Method. As the name suggests, the method assumes that equipment with an estimated n years of use loses $1/n$ of its original cost every year. Thus, if C is the original cost of a piece of equipment depreciated linearly over n years, then the total depreciation after x years is $(C/n)x$. The value y of this equipment at the end of x years is

1.28
$$\begin{aligned} y &= C - \frac{C}{n} \cdot x \\ &= C\left(1 - \frac{x}{n}\right) \qquad 0 \le x \le n \end{aligned}$$

Example 1.26

A commercial building worth \$80,000 in 1978 is assumed to depreciate linearly over 20 years of its life. Determine the value in

a. 1984

b. 1989

Solution

Here,

$$C = \$80{,}000, \quad n = 20 \text{ years}$$

The value of the building in 1984 is obtained by letting $x = 6$ in equation (1.28). Thus,

$$y = 80{,}000(1 - \tfrac{6}{20}) = \$56{,}000$$

Similarly, the value of this asset in 1989 is obtained by substituting $x = 11$ in equation (1.28). This yields

$$y = 80{,}000(1 - \tfrac{11}{20}) = \$36{,}000$$

Example 1.27

The sales in a certain department store approximate a straight line trend. The sales were $45,000 in 1972 and $75,000 in 1978. Determine the equation of the straight line representing this increase in sales. Then, assuming that this linear trend continues, estimate the sales in 1981.

Solution

Let x represent time in years, with $x = 0$ representing 1972. Then $x = 6$ represents the year 1978. Since sales are linearly related to time, we need to determine an equation of the straight line passing through (0, 45000) and (6, 75000). The slope of this line is

$$m = \frac{75000 - 45000}{6 - 0} = \frac{30000}{6} = 5000$$

Thus the equation of the straight line is

$$y = 5000x + 45000$$

where y denotes the sales in year x.

To find the estimated sales in 1981 we substitute $x = 9$ in the above equation. Then

$$\begin{aligned} y &= 5000(9) + 45000 \\ &= \$90{,}000 \end{aligned}$$

Exercises

1. A community college is offering a course in Principles of Accounting. The operating cost for this course is $750 plus $15 for each student registered. What is the operating cost if 30 students are enrolled in this course?
2. A local newspaper is planning to conduct a survey to study the voting habits of the residents in a certain community. The operating cost is $1,000 plus $2 for each family polled.
   a. What is the total cost if 2,450 families are to be polled?
   b. How many families can be polled if there is $5,000 available in the budget?

3. A truck rental agency in a town leases trucks and charges $20 a day plus 50¢ a mile. What is the total charge if the truck is used for 38 miles during one day?

4. A new machine was installed in the plant in 1978 at an expense of $100,000. If this machine is depreciated linearly over 25 years, determine its value in
   a. 1993
   b. 2000

5. An apartment building worth $300,000 in 1975 is being depreciated linearly over 50 years. Determine the value of this building in 1994.

6. An electric typewriter purchased on December 1, 1978, for $750 is being depreciated linearly over a period of 5 years. Determine the value of this typewriter on June 30, 1980.

7. A new machine purchased for $50,000 in 1970 has a scrap value of $3,000 at the end of 1980.
   a. Determine the equation of the straight line which expresses the value of the machine at the end of n years.
   b. What was the value of the machine at the end of 1973?

8. The Hill Publishing Company must set a price for a new book. The past sales experience of the company with similar books reveals the following information:

Number of Copies Sold	Price of the Book
40,000	$8
60,000	$6

   Assuming that the relationship is linear, determine the equation of the line which expresses the price of the book in terms of the number of copies sold. What price should be set for the book if the management expects to sell 75,000 copies?

1.8 Circles*

Definition 1.18

A *circle* is the collection of all points in the plane that are at a fixed distance r from a fixed point C. The fixed point is called the *center* and the fixed distance is called the *radius* of the circle.

* This section may be deferred until Chapter 7.

To find the equation of a circle, we suppose that $C(h, k)$ is the center of the circle and r is its radius. If $P(x, y)$ is an arbitrary point in the plane as shown in Figure 1.36, then P lies on the circle if and only if $d_{CP} = x$.

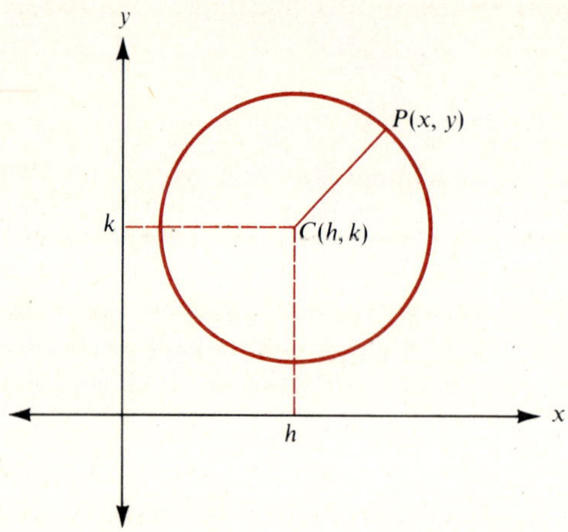

Figure 1.36

Using the distance formula (Proposition 1.3), we get

$$\sqrt{(x-h)^2 + (y-k)^2} = r$$

or

1.29 $$(x-h)^2 + (y-k)^2 = r^2, \qquad r > 0$$

which is called the *standard equation of the circle* with its center at the point (h, k) and radius r.

In particular, if $h = 0$ and $k = 0$, then (1.29) reduces to

1.30 $$x^2 + y^2 = r^2$$

which is an equation of the circle of radius r with center at the origin.

Example 1.28

Find an equation of the circle which has center C at the origin and radius $r = 5$.

Solution

Substituting $r = 5$ in the equation (1.30), we have

$$x^2 + y^2 = 25$$

The graph of this circle is shown in Figure 1.37.

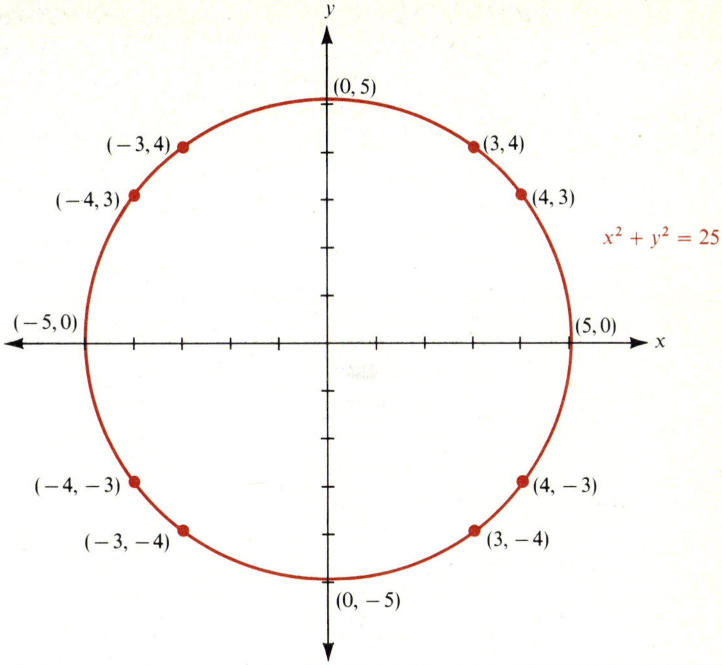

Figure 1.37

Example 1.29

Find an equation of the circle with center $C(3, 2)$ and radius $r = 4$.

Solution

Substituting $h = 3$, $k = 2$, and $r = 4$ in the equation (1.29), we obtain

$$(x - 3)^2 + (y - 2)^2 = 16$$

which on simplification yields

$$x^2 + y^2 - 6x - 4y - 3 = 0$$

Example 1.30

Find the center and radius of the circle with the equation

$$x^2 + y^2 - 10x + 4y + 4 = 0$$

Solution

First, we rearrange the given equation in the form

$$(x^2 - 10x \qquad) + (y^2 + 4y \qquad) = -4$$

and then complete the squares* within each pair of parentheses. To complete the square of an expression $x^2 + ax$ in the first pair of parentheses we add $(a/2)^2$, the square of one-half the coefficient of x, to both sides of the equation. Similarly for the expression $y^2 + by$ in the second pair of parentheses, we add $(b/2)^2$ to both sides of the equation. Thus we obtain

$$(x^2 - 10x + 25) + (y^2 + 4y + 4) = -4 + 25 + 4$$

or

$$(x - 5)^2 + (y + 2)^2 = 25$$

which is of the form (1.29). The center of the circle is $(5, -2)$ and the radius $r = 5$.

Exercises

In problems 1 through 6, find the equation of each circle with center C and radius r.

1. $C(0, 0)$, $r = 6$
2. $C(0, 0)$, $r = 8$
3. $C(4, 2)$, $r = 5$
4. $C(3, 7)$, $r = 9$
5. $C(-4, 1)$, $r = 3$
6. $C(5, -3)$, $r = 10$

In problems 7–12, find the center and the radius of each circle.

7. $x^2 + y^2 - 8x - 14y - 16 = 0$
8. $x^2 + y^2 + 6x - 10y + 9 = 0$
9. $x^2 + y^2 + 6x - 12y + 20 = 0$
10. $x^2 + y^2 + 4x - 6y - 1 = 0$
11. $x^2 + y^2 + 14x + 8y + 29 = 0$
12. $x^2 + y^2 + 8x - 12y + 52 = 0$

Exercises for Review

1. Classify each of the following statements as true or false.
   a. Every natural number is an integer.
   b. Every integer is a natural number.
   c. $\sqrt{2}$ is an irrational number.
   d. The product of two irrationals is irrational.
   e. There exists a one-to-one correspondence between the rational numbers and the points on the real line.
   f. There exists a one-to-one correspondence between the real numbers and the points on the real line.

* See Appendix A.

2. Solve each of the following inequalities for x.
   a. $3x + 7 > 5$
   b. $4x + 21 \leq 2x + 26$
   c. $5x - 14 \geq 9x - 7$
   d. $\frac{1}{2}(3 - 4x) \leq \frac{1}{3}(2 + 3x)$

3. Solve each of the following inequalities for x.
   a. $x^2 > 2x$
   b. $x^2 \leq 2x + 8$
   c. $x^2 + 2x \geq 3$
   d. $x(x - 2)(x - 4) > 0$

4. Solve for x.
   a. $|x - 1| = 3$
   b. $|x - 2| + 4 = 7$
   c. $|2x + 1| = 7$
   d. $|x - 4| + 7 = 2$

5. Solve for x.
   a. $|3x| < 6$
   b. $|x + 1| > 2$
   c. $|2x - 3| \leq 5$
   d. $\left|\frac{x}{2} - 3\right| < 1$

6. Describe each of the following sets as a collection of points.
   a. $\{x|0 < x < 4\} \cap \{x|1 < x < 5\}$
   b. $\{x|0 \leq x < 7\} \cup \{x|2 < x \leq 8\}$
   c. $\{x|-1 < x < 7\} \cap \{x|2 < x < 9\} \cap \{x|0 < x < 3\}$

7. For the set

$$A = \{1, 2, 3, 4, 5\}$$

   find a subset of $A \times A$ such that
   a. the sum of two numbers in each ordered pair is 5;
   b. the two elements in each ordered pair are equal.

8. Find the distance between each of the following pairs of points.
   a. $(-3, 7)$ and $(-3, -3)$
   b. $(2, 3)$ and $(10, 9)$
   c. $(-5, -2)$ and $(7, 3)$
   d. $(20, 11)$ and $(44, 18)$

9. Find the slope of the line passing through each pair of points.
   a. $(2, 7)$ and $(5, 13)$
   b. $(-1, 9)$ and $(-2, 11)$
   c. $(4, 8)$ and $(4, 16)$
   d. $(8, 4)$ and $(16, 4)$

10. A line with slope 7 passes through the points $(3, 2)$ and $(4, y)$. Determine the value of y.

11. Determine the slope and y-intercept of each of the following lines.
    a. $2x - 3y + 6 = 0$
    b. $3x + 2y - 16 = 0$
    c. $x - 3y + 4 = 0$

12. Find the equation of each line.

a. through the point $P(3, 4)$ with slope $m = 2$

b. through the points $P(7, 6)$ and $Q(9, 8)$

c. through the point $P(3, 5)$ and parallel to the line $2x - 3y - 3 = 0$

d. through the point $P(-2, 5)$ and perpendicular to the line $3x + 4y - 12 = 0$

13. A rental agency in a town leases cars for \$25.95 a day plus 25 cents a mile. What is the total charge if the car is used for 50 miles during one day?

14. An electric typewriter purchased on September 1, 1979, for \$840 is being depreciated linearly over a period of 5 years. Determine the value of this typewriter on April 30, 1981.

15. Find the center and the radius of the circle with the equation

$$x^2 + y^2 - 10x - 12y - 110 = 0$$

16. Prove each of the following statements.

a. If $ac < bc$ and $c > 0$, then $a < b$.

b. If a, b, c, and d are all positive numbers and if $a < b$ and $c < d$, then $ac < bd$.

2 Functions, Limit and Continuity

2.1 Introduction

Scientists strive to establish relationships between quantities that can be measured or observed. In the physical sciences, the relationship between the pressure and volume of a gas at a certain temperature is expressed by Boyle's Law. Civil engineers develop formulas which predict the number of inches a beam will bend when subject to certain weights or loads; a psychologist describes the relationship between the average reaction time of an adult and the quantity of a given stimulus; physicians are interested in the relationship between blood pressure and heart attack, between smoking and lung cancer, and the like; a sociologist may look for a relationship between a person's economic status and integration into society or between an increase in crime and an increase in urban population. In education, researchers study how an IQ measure relates to achievement in the elementary grades, high school, or college. Economists strive to determine the relationships between the demand and the price of a product, between consumption of a commodity and its supply in the market, and so on.

This chapter presents the essential features of quantitative relationships that are central in the study of calculus. The chapter includes two especially important topics, the limit and continuity of functions.

2.2 Functions and Equations

The equation $A = \pi r^2$ expresses the relationship between the area A of a circle and its radius r. This formula, or rule, which associates a unique value of A with each positive real number r is a function.

As another illustration, consider the equation $s = 16t^2$ which describes the position of a falling body. If an object is released from rest, then one second after its release it will fall 16 feet from its starting point, two seconds after its release, 64 feet, three seconds after its release, 144 feet, and so on. Thus if we know the time t (in seconds) after the release of an object, the formula $s = 16t^2$ gives the number of feet the object will fall from its starting point. Observe that for *each* value of t there is one and only one value of s that corresponds to that value of t.

Let us now consider an equation $y = 4x + 3$. For each real number x there is one and only one real number y that corresponds to that value of x. The rule, or correspondence, is: multiply the value of x by 4 and then add 3. For example, if $x = \frac{1}{2}$, thcn $y = 5$; if $x - 1$, then $y = 7$; if $x = \frac{5}{4}$, then $y = 8$; if $x = 2$, then $y = 11$; if $x = 5$, then $y = 23$; if $x = -3$, then $y = -9$; if $x = -7$, then $y = -25$, and so on. This is an intuitive description of a function. A more rigorous definition is as follows:

Definition 2.1

Let A and B be two non-empty sets. Then a function from A to B is a rule which assigns to *each* element $x \in A$ one and only one element $y \in B$. This element y is called the *image* of x.

Figure 2.1 illustrates the concept of a function. The set A is called the *domain* of the function while the subset of B which consists of the images of *all* the elements of A is called the *range* of the function. We may use any kind of objects to make up the domain and the range of the function. However, most of the functions we encounter in calculus are real-valued functions of a real variable; that is, the domain and the range of the functions are sets of real numbers. The letter x which represents an arbitrary number from the domain is called an *independent variable* and the corresponding letter representing an element y in the range is called the *dependent variable,* since its value depends on the number assigned to x. When two variables x and y are related in this manner, we say that y is a function of x. Thus the area of a circle is a function of its radius, the distance traveled by an object when released from rest is a function of time, and so on.

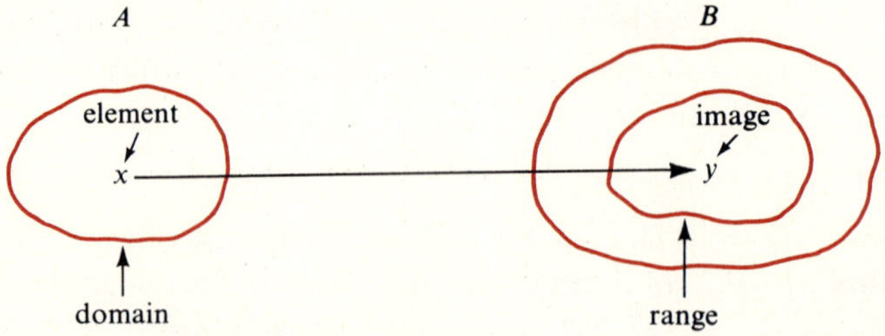

Figure 2.1

We shall denote functions by small letters: $f, g, h, \ldots$. If f is a function, then the number which f assigns to an element x in the domain is denoted by $f(x)$ or simply by y. The symbol $f(x)$ is read "f of x" or "f evaluated at x" or "the value of f at the point x." To point out the usefulness of this mathematical symbolism, suppose that we want to describe the demand for a commodity in a symbolic form. If p denotes the price of a commodity and q represents the quantity of that commodity demanded by consumers, then we can express this quantitative relationship by $q = f(p)$. This simply means that the quantity in demand depends, in some way, on the price of the commodity. Likewise, we may write $A = f(r)$ where A denotes the area of a circle and r designates its radius. The symbolic expression $s = g(t)$ is interpreted to mean that the distance s traveled is a function of time t.

To specify a function, we must give its domain D_f and the rule f which determines the range R_f by assigning to each element x from the domain one and only one value $f(x)$. The rule of correspondence is generally expressed in the form of a mathematical equation, as illustrated in the next few examples.

Example 2.1

Let f be a function defined by

$$f(x) = 3x + 2$$

with the set $\{1, 2, 3, 4\}$ as its domain. Then

$$f(1) = 3(1) + 2 = 5$$
$$f(2) = 3(2) + 2 = 8$$
$$f(3) = 3(3) + 2 = 11$$
$$f(4) = 3(4) + 2 = 14$$

The range R_f is the set $\{5, 8, 11, 14\}$.

Example 2.2

Let f be a function defined by

$$f(x) = x^2 + 3$$

with the set $\{-2, -1, 0, 1, 2, 3\}$ as its domain D_f. Then

$$f(-2) = (-2)^2 + 3 = 7 \qquad f(1) = 1^2 + 3 = 4$$
$$f(-1) = (-1)^2 + 3 = 4 \qquad f(2) = 2^2 + 3 = 7$$
$$f(0) = 0^2 + 3 = 3 \qquad f(3) = 3^2 + 3 = 12$$

The range R_f is the set $\{3, 4, 7, 12\}$.

Example 2.3

Let f be a function defined by

$$f(x) = x^3 + 2$$

with the set $\{-3, -2, -1, 0, 1, 2, 3, 4\}$ as its domain. Then

$$\begin{aligned} f(-3) &= (-3)^3 + 2 = -25 & f(1) &= 1^3 + 2 = 3 \\ f(-2) &= (-2)^3 + 2 = -6 & f(2) &= 2^3 + 2 = 10 \\ f(-1) &= (-1)^3 + 2 = 1 & f(3) &= 3^3 + 2 = 29 \\ f(0) &= 0^3 + 2 = 2 & f(4) &= 4^3 + 2 = 66 \end{aligned}$$

The range R_f is the set $\{-25, -6, 1, 2, 3, 10, 29, 66\}$.

Example 2.4

Let f be a function defined by

$$f(x) = 3x + 1$$

whose domain D_f is the set of real numbers. Then for some arbitrary choices from the domain, we have

$$\begin{aligned} f(-4) &= 3(-4) + 1 = -11 & f(1) &= 3(1) + 1 = 4 \\ f(-3) &= 3(-3) + 1 = -8 & f(\sqrt{2}) &= 3\sqrt{2} + 1 \\ f(-1) &= 3(-1) + 1 = -2 & f(2) &= 3(2) + 1 = 7 \\ f(0) &= 3(0) + 1 = 1 & f(\tfrac{8}{3}) &= 3(\tfrac{8}{3}) + 1 = 9 \\ f(\tfrac{1}{3}) &= 3(\tfrac{1}{3}) + 1 = 2 & f(\pi) &= 3\pi + 1 \\ f(\tfrac{2}{3}) &= 3(\tfrac{2}{3}) + 1 = 3 & f(6) &= 3(6) + 1 = 19 \end{aligned}$$

and so on. Intuitively, the range is the set of all real numbers.

Although the domain is an essential part of the function, it is frequently not mentioned, especially when the rule of correspondence is given by a mathematical equation. *The general convention is that unless the domain is stated explicitly, it is understood to consist of all real numbers for which the expression in x is defined.* The rule f that associates each element x in the domain with one and only value $f(x)$ determines the range.

Example 2.5

Let f be a function defined by

$$f(x) = \sqrt{9 - x^2}$$

Observe that this rule of correspondence does not yield a real number for $x > 3$ or $x < -3$. For example, if $x = 4$, then

$$\sqrt{9 - x^2} = \sqrt{9 - 16} = \sqrt{-7}$$

which is not a real number. Therefore, the domain D_f is the set of real numbers between -3 and 3, both inclusive. The rule f assigns to each $x \in D_f$ one and only one real number $\sqrt{9 - x^2}$ in the range. Since $x^2 \geq 0$ for each real number x, it follows that $9 - x^2 \leq 9$ and consequently, $\sqrt{9 - x^2} \leq 3$. One can observe intuitively that the range R_f consists of non-negative real numbers between 0 and 3, both inclusive.

It may be instructive to think of a function as a machine that produces a unique output for a given input. If f is a function, then we can interpret the domain of the function as the set of all those elements that can be entered into the machine. If we enter an element x in the machine, it produces $f(x)$ as its output. The range of the machine is, therefore, the set of all possible outputs. As an example, consider the function

$$f(x) = 3x^2 + 2$$

with the set of real numbers as its domain. If a real number x is entered, the machine squares this input, multiplies the result by 3 and then adds 2. Thus, for some arbitrary choices from the domain, we have

$$\begin{aligned} f(-4) &= 3(-4)^2 + 2 = 50 & f(1) &= 3(1)^2 + 2 = 5 \\ f(-3) &= 3(-3)^2 + 2 = 29 & f(\sqrt{2}) &= 3(\sqrt{2})^2 + 2 = 8 \\ f(-2) &= 3(-2)^2 + 2 = 14 & f(\sqrt{3}) &= 3(\sqrt{3})^2 + 2 = 11 \\ f(-1) &= 3(-1)^2 + 2 = 5 & f(2) &= 3(2)^2 + 2 = 14 \\ f(0) &= 3(0)^2 + 2 = 2 & f(\pi) &= 3\pi^2 + 2 \end{aligned}$$

Since x^2 is non-negative for each real number x, it follows that $3x^2 \geq 0$ and consequently $3x^2 + 2 \geq 2$. Intuitively, the range of the function is the set of all real numbers greater than or equal to 2.

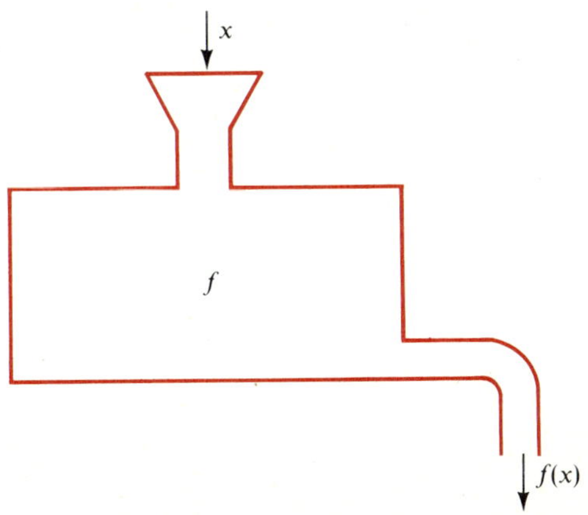

Figure 2.2

Notice that in the notation $f(x)$, the independent variable x is a "dummy"; for example, $f(x) = 3x^2 + 2$, $f(a) = 3a^2 + 2$, $f(b) = 3b^2 + 2$, $f(c) = 3c^2 + 2$, $f(2d) = 3(2d)^2 + 2 = 12d^2 + 2$, $f(h + 1) = 3(h + 1)^2 + 2$, all represent the same function. Thus, no matter what is put in the machine, that quantity is squared, the result is multiplied by 3 and 2 is added. It is sometimes helpful to think of this function as:

$$f(\quad) = 3(\quad)^2 + 2$$

where any symbol may be placed in both the parentheses to represent an element of the domain.

Example 2.6

Let f be a function defined by

$$f(x) = \sqrt{x} + 4$$

Determine

a. $f(\pi)$ **b.** $f(9)$ **c.** $f(a + b)$ **d.** $f(a) + f(b)$

Solution

a. $f(\pi) = \sqrt{\pi} + 4$
b. $f(9) = \sqrt{9} + 4 = 3 + 4 = 7$
c. $f(a + b) = \sqrt{a + b} + 4$
d. $f(a) + f(b) = \sqrt{a} + 4 + \sqrt{b} + 4 = \sqrt{a} + \sqrt{b} + 8$

Observe carefully that in general

$$f(a + b) \neq f(a) + f(b)$$

Example 2.7

Let f be a function defined by the equation

$$f(x) = x^2 + 3x + 4$$

If a and h are real numbers and $h \neq 0$, find

$$\frac{f(a + h) - f(a)}{h}$$

Solution

$$\begin{aligned} f(a + h) &= (a + h)^2 + 3(a + h) + 4 \\ &= (a^2 + 2ah + h^2) + (3a + 3h) + 4 \\ f(a) &= a^2 + 3a + 4 \\ f(a + h) - f(a) &= (a^2 + 2ah + h^2) + (3a + 3h) + 4 - (a^2 + 3a + 4) \\ &= 2ah + 3h + h^2 \end{aligned}$$

Thus,

$$\frac{f(a+h)-f(a)}{h}=\frac{2ah+3h+h^2}{h}=2a+3+h, \qquad h\neq 0$$

Functions whose domains are restricted occur frequently in business and economics. Consider the following example.

Example 2.8

The market demand function of a particular product is given by the equation

$$f(p)=200-10p$$

where $f(p)$ denotes the quantity demanded per year in thousands of units and p denotes the price per unit. Some corresponding values of p and $f(p)$ are

$$\begin{array}{rr} f(0)=200 & f(12)=80 \\ f(1)=190 & f(15)=50 \\ f(2)=180 & f(18)=20 \\ f(3)=170 & f(18.5)=15 \\ \vdots & \\ f(9.5)=105 & f(19.5)=5 \\ f(10)=100 & f(20)=0 \end{array}$$

Since demand is never negative, it follows that $f(p)$ must be nonnegative. The price per unit is charged in dollars and cents, so the price p may assume the values in the set $\{0, 0.01, 0.02, \ldots, 19.98, 19.99, 20.00\}$. If $p > 20$ then $200 - 10p$ would be a negative number. Thus the domain D_f of the function f consists of two-place decimal numbers between 0 and 20, inclusive. It can be shown that the range R_f is the set $\{0, 0.1, 0.2, \ldots, 199.8, 199.9, 200\}$.

Observe that for each value p in the set D_f in Example 2.8 there is one and only one value $f(p)$ in the set R_f, and conversely, for each value $f(p)$ in the set R_f there is one and only one element in the domain D_f. The relationship between the elements of the two sets thus constitutes what is commonly known as a *one-to-one function.* Other examples of one-to-one functions are the relationships between husbands and wives in a monogamous society, employees of a company and each one's social security number, the 50 states of the United States and the governors, the set of all real numbers and the points on the number line, and so on. Before we get carried away, let us pause and provide a formal definition of a one-to-one function.

Definition 2.2

Let A and B be two non-empty sets. A function from A to B is called a *one-to-one function* if for each $f(x)$ in the range R_f there is one and only one value of x in the domain D_f.

In other words, if $f(a) = f(b)$ for elements a and b in the domain of f, then $a = b$.

Example 2.9

If $f(x) = 4x + 3$, where x is a real number, show that f is one-to-one.

Solution

Suppose that a and b are two real numbers in the domain of the function. Then $4a + 3$ and $4b + 3$ are the corresponding images in the range. If $f(a) = f(b)$, then

$$4a + 3 = 4b + 3$$
$$4a = 4b$$
$$a = b$$

Therefore, f is one-to-one.

Example 2.10

If $f(x) = 3x^2 + 2$, where x is a real number, show that f is not one-to-one.

Solution

Let a and $(-a)$ be two non-zero real numbers in the domain of f. Then

$$f(a) = 3a^2 + 2$$

and

$$f(-a) = 3(-a)^2 + 2 = 3a^2 + 2$$

Since two different numbers a and $(-a)$ in D_f have the same image, namely $3a^2 + 2$, we conclude that f is not a one-to-one function.

To give a concrete example of a function which is not one-to-one, consider a set of 30 students who have taken a final examination in a chemistry course and whose tests have been graded A, B, C, D, or F. Note that for each test there corresponds one and only one letter grade;

thus the definition for a function is satisfied. The domain is the set of 30 tests, and the range consists of only those letter grades that have actually been assigned. The function is not one-to-one since *more than one test* corresponds to some letter grade in the range.

As an illustration of a real-valued function which is not one-to-one, consider a travel agency in Boston which advertises a group-rate 5-day cruise to Nova Scotia. The agency makes reservations for 50 couples on the boat. The charge is \$600 per couple with an additional charge of \$20 per couple for each subsequent cancellation. If x is the number of cancellations received prior to the departure of the boat, then $(50 - x)$ couples plan to be on the boat and each pays \$600 plus \$20x. Thus the total receipts of the travel agency are given by the function

$$f(x) = \$(50 - x)(600 + 20x)$$

which simplifies to

$$f(x) = \$20(1500 + 20x - x^2)$$

The domain D_f is the set

$$\{0, 1, 2, \ldots, 49, 50\}$$

Let us now compute $f(x)$ for some values of $x \in D_f$. For example,

$$\begin{aligned} f(0) &= \$30{,}000 & f(15) &= \$31{,}500 \\ f(5) &= \$31{,}500 & f(20) &= \$30{,}000 \\ f(10) &= \$32{,}000 & f(25) &= \$27{,}500 \end{aligned}$$

and so on. Again, we can verify that for each value x in the domain there is a unique value $f(x)$ so that the definition of the function is satisfied. However, two values of x in the domain, for example, $x = 0$ and $x = 20$ correspond to the same value $f(x) = \$30{,}000$. Similarly, $x = 5$ and $x = 15$ in the domain correspond to the same value $f(x) = \$31{,}500$ in the range. Thus we have some elements in the range of the function which correspond to more than one element in the domain. Thus the function is not one-to-one.

So far, we have considered a function as a rule of correspondence that pairs with each element of its domain one and only one element of its range. This shows that a function is actually a set of ordered pairs so that no two ordered pairs have the same first element. The following is, therefore, an *alternate* definition.

Definition 2.3 Let A and B be two non-empty sets. Then a function from A to B is a set of ordered pairs (x, y), where $x \in A$ and $y \in B$ such that every element of A appears as a first element of exactly one ordered pair.

Example 2.11

The set of ordered pairs

$$\{(1,5),(2,7),(3,9),(4,11),(5,13),(6,15)\}$$

is a function with the set $A = \{1,2,3,4,5,6\}$ as its domain and the set $B = \{5,7,9,11,13,15\}$ as its range. Figure 2.3 describes this function.

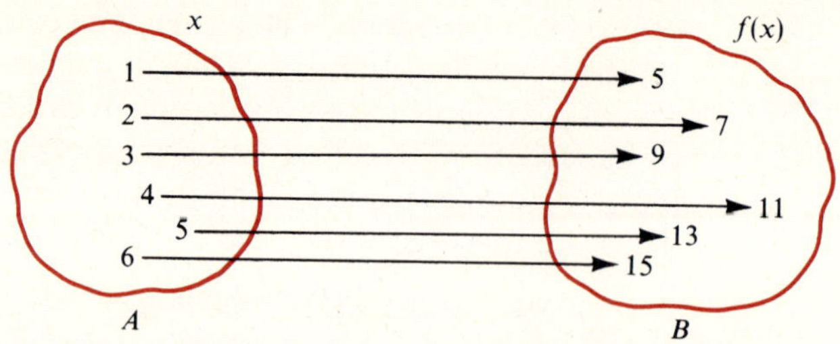

Figure 2.3

Another way to describe this function is

$$\{(x,f(x))\,|\,f(x) = 2x + 3, \qquad x \in A\}$$

where $A = \{1,2,3,4,5,6\}$. Note that this function is one-to-one.

Example 2.12

The set of ordered pairs

$$\{(-3,19),(-1,3),(1,3),(3,19)\}$$

is a function with the set $A = \{-3, -1, 1, 3\}$ as its domain and the set $B = \{3, 19\}$ as its range. The rule of correspondence that pairs elements of the domain with the elements of the range is described in Figure 2.4.

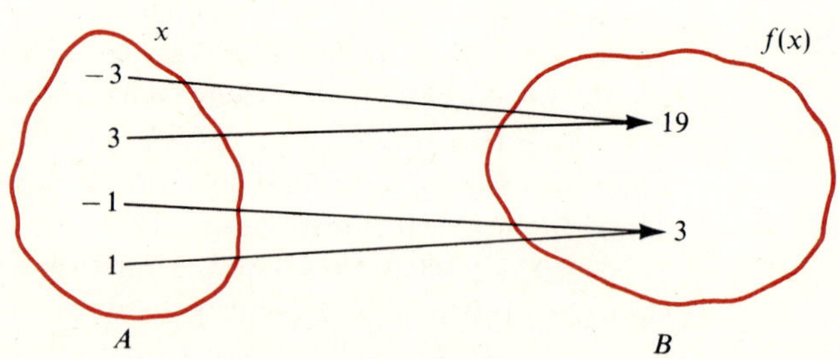

Figure 2.4

We may express this function as a mathematical equation:

$$\{(x,f(x))\,|\,f(x) = 2x^2 + 1, \qquad x \in A\}$$

Note that $f(-3) = f(3) = 19$, but $-3 \neq 3$: hence this function is not one-to-one.

Example 2.13

The set of ordered pairs

$$\{(2,4),(3,5),(4,6),(3,7)\}$$

is *not* a function because an element 3 in the domain is paired with two elements 5 and 7 in the range as illustrated in Figure 2.5. This violates Definition 2.3.

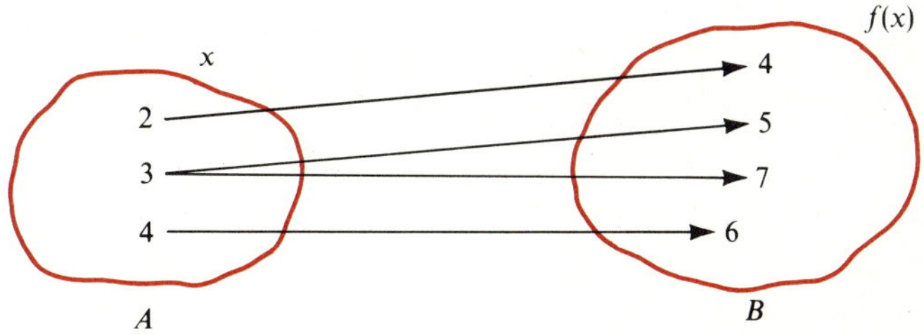

Figure 2.5

Notice that in order for a set of ordered pairs to be a function from set A to set B, each element of A must appear as a first element in one and only one of the ordered pairs.

Exercises

1. Tell whether the situations described below constitute functions.
a. the correspondence between the State Colleges in Connecticut and their presidents;
b. the correspondence between each book in a library and the number of pages in the book;
c. the correspondence between each state and the number of representatives in the House of Representatives;
d. the correspondence between a driver and his or her driving license number;
e. the correspondence between each product sold in a department store and its price;
f. the correspondence between you and your weight;
g. the correspondence between the set of ordered pairs of real numbers and the points in the plane;
h. the correspondence between each resident of Boston and that person's telephone number.

2. State which of the functions of Exercise 1 are one-to-one.

3. From Figure 2.6, determine whether or not the rules of correspondence from set A to set B are functions.

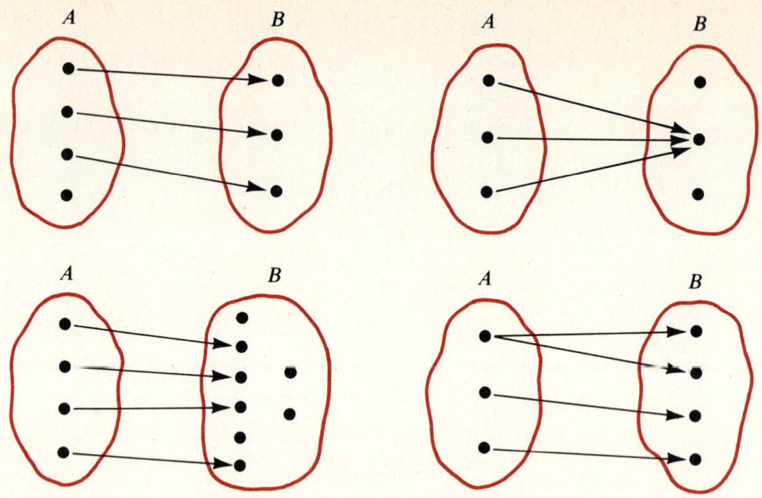

Figure 2.6

4. Determine whether or not the following set of ordered pairs are functions:
   a. $\{(1,1),(2,2),(3,3),(4,4),(5,5),(6,6)\}$
   b. $\{(-1,1),(1,1),(-2,4),(2,4),(3,9)\}$
   c. $\{(1,-1),(1,1),(4,-2),(4,2),(9,3)\}$
   d. $\{(1,1),(2,1),(3,1),(4,1),(5,1),(6,1)\}$
   e. $\{(1,1),(2,8),(3,27),(4,64)\}$
   f. $\{(1,1),(8,2),(27,3),(64,4)\}$
   g. $\{(2,3),(3,4),(4,5),(5,6)\}$
   h. $\{(4,5),(5,6),(5,7),(5,8),(5,9)\}$
5. State which of the functions in Exercise 4 are one-to-one.
6. Given that

$$f(x) = 2x^2 + 3x + 1,$$

determine each of the following.

   a. $f(0)$
   b. $f(-1)$
   c. $f(1)$
   d. $f(-2)$
   e. $f(3)$
   f. $f(\frac{1}{2})$
7. Given that

$$g(x) = x^3 + x^2 + x + 6,$$

determine each of the following.

   a. $g(-1)$
   b. $g(-2)$
   c. $g(0)$
   d. $g(1)$
   e. $g(2)$
   f. $g(\pi)$

8. Given that

$$h(x) = \frac{15 + 20x}{x},$$

determine each of the following.

a. $h(1)$
b. $h(2)$
c. $h(3)$
d. $h(4)$
e. $h(5)$
f. $h(6)$

9. Given that

$$f(x) = 2x + 3,$$

determine each of the following.

a. $f(a)$
b. $f(1/a)$
c. $f(a + b)$
d. $f(a) + f(b)$
e. $f(a + 3)$
f. $f\left(\frac{1}{a + b}\right)$

10. Given that

$$f(x) = 2x^2 + 3x + 6,$$

determine each of the following.

a. $f(a)$
b. $f(-a)$
c. $f(\sqrt{a})$
d. $f(a + h)$
e. $f(a + h) - f(a)$
f. $\frac{f(a + h) - f(a)}{h}, \quad h \neq 0$

11. Assuming that $g(x) = -2x + 40$ gives the number of students in a large class who require x hours of study weekly, determine the number of students who require the following levels of study.

a. $x = 0$
b. $x = 1$
c. $x = 2$
d. $x = 4$
e. $x = 5$
f. $x = 10$

12. Suppose that the demand d for a certain product is given by

$$d(p) = 100 - 4p^2$$

where p is the price in dollars per unit. Using an increment of \$1, determine the demand $d(p)$ for $p = 0$ to $p = 5$, inclusive.

Assuming that the domain is the set $\{-3, -2, -1, 0, 1, 2, 3\}$, determine the range for each of the functions in Exercises 13–18.

13. $f(x) = x + 1$
14. $f(x) = x^2 + 4$
15. $f(x) = -x$
16. $f(x) = 1 - x^2$
17. $f(x) = x^3$
18. $f(x) = \sqrt{9 - x^2}$

In each of the Exercises 19–26, determine the largest subset of real numbers that can serve as the domain of the function.

19. $f(x) = x + 2$

20. $f(x) = x^2 + 1$

21. $f(x) = \sqrt{x - 1}, \quad x \geq 1$

22. $f(x) = 3$

23. $f(x) = x^2 - 4$

24. $f(x) = 4 - x^2$

25. $f(x) = \dfrac{1}{x - 2}, \quad x \neq 2$

26. $f(x) = \dfrac{x - 1}{x + 1}, \quad x \neq -1$

27. The market demand function of a particular product is given by

$$f(p) = 100 - 4p$$

where $f(p)$ represents the quantity demanded (in units) and p denotes the price per unit (in dollars). Determine the domain and the range of f.

28. A high school has arranged a dance for a senior class. The charge is \$5.00 per couple if 40 couples attend and then decreases by \$0.05 for each additional couple above the minimum of 40. Express the total revenue, $R(x)$, as a function of x, the number of additional couples above the minimum.

29. A travel agency has chartered a flight to Rome for high school students. The charge is \$200 with an additional charge of \$2.50 for each subsequent cancellation. The plane has a seating capacity of 100 passengers. Express the total revenue, $R(x)$, as a function of x, the number of cancellations received prior to the departure date.

30. A real estate office handles an apartment complex with 150 units. Past experience shows that all units are occupied if the rent is \$200 per month. For each \$10 increase in rent, five units become vacant. Express the revenue function, $R(x)$, as a function of x, the number of units unoccupied.

2.3 Some Useful Functions

When a function is given by a mathematical equation, it is possible to study the simultaneous behavior of x and y by plotting the set of points (x, y) where x represents an arbitrary value selected from the domain of the function and y is the corresponding value determined by the rule of correspondence. *We shall follow the convention that if the domain consists of only a finite number of points, then the points plotted in the plane represent the entire graph; but if the domain consists of real numbers in an interval, then the graph consists of a solid curve drawn through the plotted points. If the domain of the function consists of rational numbers within a specified interval, then the graph will be drawn as a dotted curve.* In this section, we shall consider certain elementary functions and their graphs.

Definition 2.4

Any function of the form $f(x) = mx + b$, where m and b are fixed real numbers, is called a *linear function.*

The word "linear" implies that the graph of this function is a straight line with slope m. (See Section 1.6)

We will look at some linear functions and their graphs in the next few examples.

Example 2.14

Consider a function

$$f(x) = 2x + 3$$

with domain as the set of integers between -2 and 2, inclusive. This function gives us the set of ordered pairs in the table below. The first

x	-2	-1	0	1	2
$f(x)$	-1	1	3	5	7

component is an element of the domain while the second component is the corresponding element in the range. Plotting the ordered pairs $(-2, -1)$, $(-1, 1)$, $(0, 3)$, $(1, 5)$, and $(2, 7)$, we have the graph in Figure 2.7. Had the domain instead consisted of all real numbers, then the

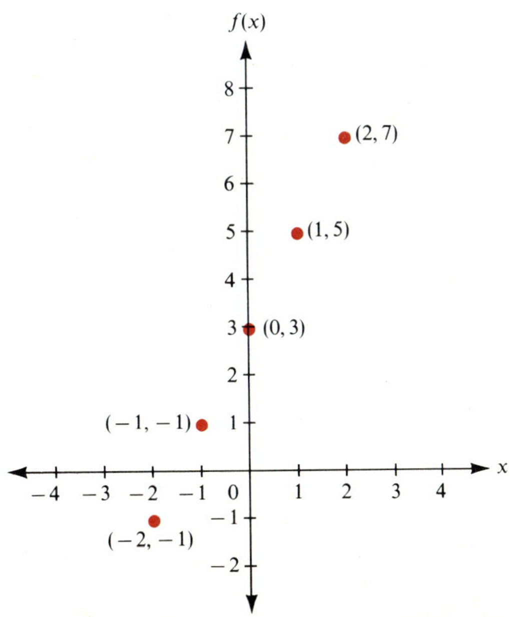

Figure 2.7

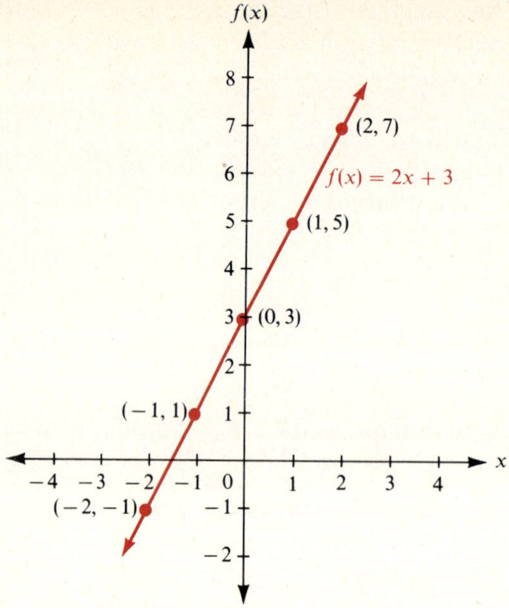

Figure 2.8

graph would be the solid line drawn through the plotted points as shown in Figure 2.8.

Example 2.15

The demand for a certain product is given by

$$f(x) = 100 - 20x$$

where $f(x)$ represents the demand (in units) of the product at a unit price of x (in dollars and cents). Graph $f(x)$.

Solution

Since the price per unit is charged in dollars and cents, the variable x may assume the values $0, 0.01, 0.02, 0.03, \ldots, 4.98, 4.99, 5.00$. If $x > 5$, then $100 - 20x$ would be a negative number. Thus the domain of the function f consists of two-place decimal numbers between 0 and 5, inclusive. The values of $f(x)$ for some values of x are given in the following table.

x	$f(x) = 100 - 20x$
0	100
1	80
2	60
3	40
4	20
5	0

The related set of ordered pairs (0, 100), (1, 80), (2, 60), (3, 40), (4, 20), and (5, 0) determine a set of points that are plotted in Figure 2.9. The dotted line represents the graph of $f(x)$.

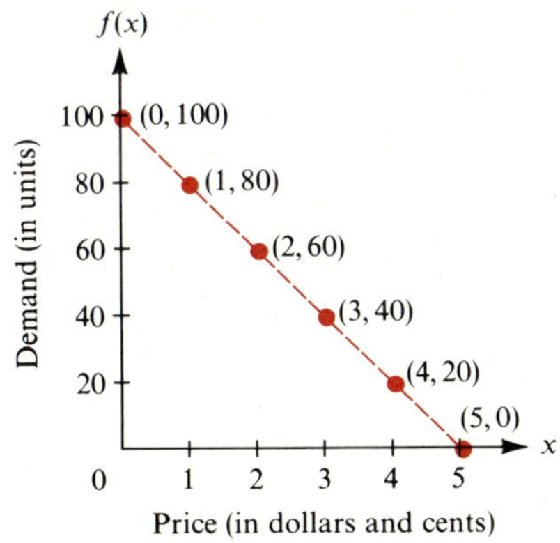

Figure 2.9

Example 2.16

Plot the graph of

$$f(x) = \begin{cases} 5 - x & \text{for} \quad x < 2 \\ 2x - 1 & \text{for} \quad x \geq 2 \end{cases}$$

where the domain of f is the set of all real numbers.

Solution

Note that the function has its domain separated into disjoint sets, for each of which it is defined differently. The values of $f(x)$ for some values of x are given for each set in the following tables.

$f(x) = 5 - x, x < 2$

x	$f(x)$
-3	8
-2	7
-1	6
0	5
1	4

$f(x) = 2x - 1, x \geq 2$

x	$f(x)$
2	3
3	5
4	7
5	9
6	11

The graph consists of two straight lines—one passing through the points $(-3, 8)$, $(-2, 7)$, . . . , $(1, 4)$ and the other passing through the points $(2, 3)$, $(3, 5)$, . . . , $(6, 11)$ as displayed in Figure 2.10.

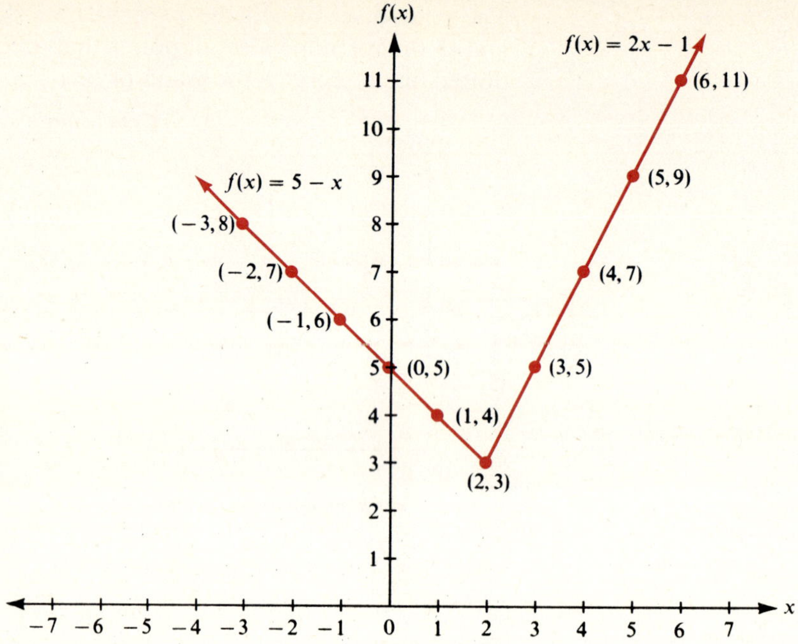

Figure 2.10

Example 2.17

Plot the graph of

$$f(x) = |x|$$

Solution

The domain D_f of the function f is the set of all real numbers while R_f is the set of non-negative reals. (Why?) The values of $f(x)$ for some arbitrary values of x are given in the table below:

x	$f(x) = \|x\|$
-3	3
-2	2
-1	1
0	0
1	1
2	2
3	3

Plotting the ordered pairs $(-3, 3)$, $(-2, 2)$, . . . , $(3, 3)$, we have the graph in Figure 2.11.

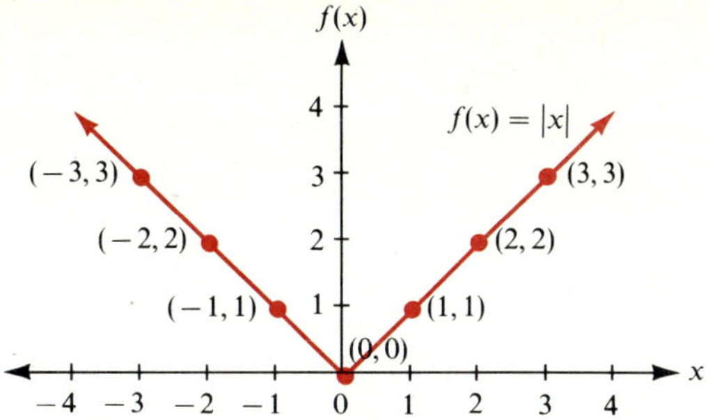

Figure 2.11

Example 2.18

Plot the graph of

$$f(x) = \sqrt{x}$$

Solution

The domain D_f of the function f is the set of non-negative real numbers, since $f(x)$ does not yield a real number for $x < 0$. The values of $f(x)$ for some values of x are given in the table below:

x	0	1	2	3	4	5
$f(x)$	0	1	$\sqrt{2}$	$\sqrt{3}$	2	$\sqrt{5}$

Plotting the ordered pairs $(0, 0)$, $(1, 1)$, . . . , $(5, \sqrt{5})$ leads us to the graph show in Figure 2.12. Since the graph lies above the x-axis, we conclude

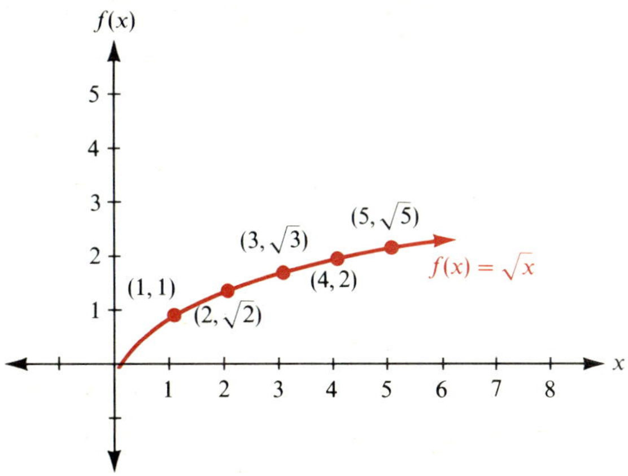

Figure 2.12

that the range of the function f is the set of non-negative reals.

Example 2.19

Plot the graph of

$$f(x) = \sqrt{9 - x^2}$$

Solution

Recall from Example 2.5 that the domain of the function f is the set of real numbers between -3 and 3, inclusive, while the range R_f is the set of real numbers between 0 and 3, inclusive. Thus the graph of f must be in the first two quadrants. The following table lists some points $(x, f(x))$ on the graph shown in Figure 2.13. Observe that the graph of the function f is an upper semi-circle.

x	-3	-2	-1	0	1	2	3
$f(x)$	0	$\sqrt{5}$	$\sqrt{8}$	3	$\sqrt{8}$	$\sqrt{5}$	0

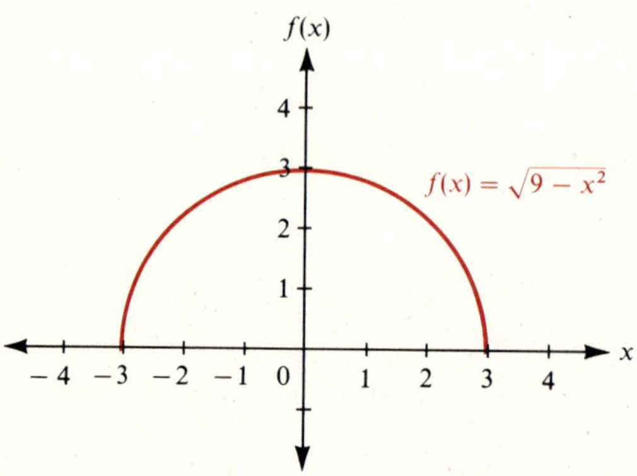

Figure 2.13

Definition 2.5

Any function of the form

$$f(x) = ax^2 + bx + d, \qquad a \neq 0$$

where a, b, and d are real numbers is called a *quadratic function.* If $a = 0$ and $b \neq 0$, then we have $f(x) = bx + d$, which is a linear function.

The simplest types of quadratic functions are $f(x) = x^2$ for $a = 1$, $b = d = 0$ and $g(x) = -x^2$ for $a = -1$, $b = d = 0$.

Example 2.20

Plot the graph of

a. $f(x) = x^2$

b. $g(x) = -x^2$

Solution

Since we cannot list all possible ordered pairs of real numbers satisfying the above equations, we will use a partial list of points to determine the shapes of these curves. The values of $f(x)$ and $g(x)$ for some values of x, are given in the table below:

x	-4	-3	-2	-1	0	1	2	3	4
$f(x)$	16	9	4	1	0	1	4	9	16
$g(x)$	-16	-9	-4	-1	0	-1	-4	-9	-16

Plotting the related ordered pairs and joining them by a smooth curve, we have the graph of $f(x)$ and $g(x)$ in Figures 2.14 and 2.15, respectively.

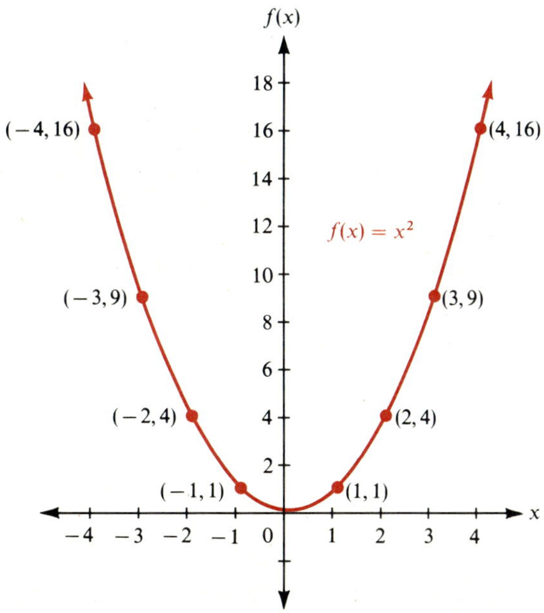

Figure 2.14

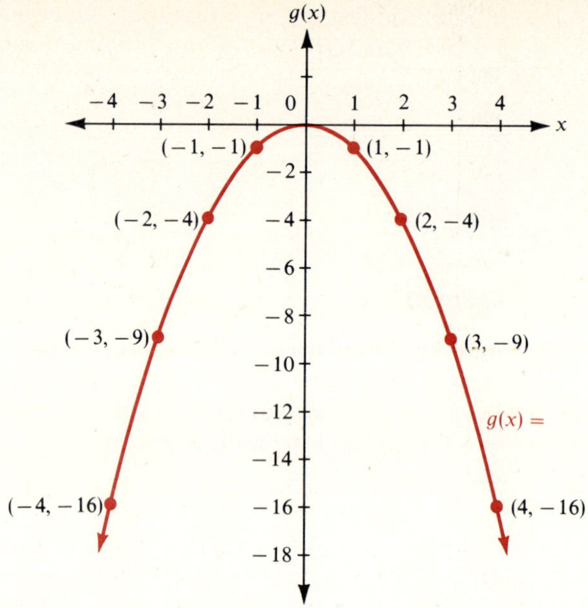

Figure 2.15

The graphs shown in Figures 2.14 and 2.15 are called *parabolas*. The point (0, 0) in each figure is called the *vertex* of the parabola. It can be shown that the graph of every quadratic function is a parabola that opens upward if $a > 0$ and opens downward if $a < 0$.

Let us now look at some illustrative examples.

Example 2.21

Due to raw material shortages and higher labor costs, it has become increasingly expensive to drill for oil. In fact, the profit (in millions of dollars) is determined by the equation

$$P(x) = 80x - 10x^2 \qquad 0 \leq x \leq 8$$

where x is the amount of oil produced (in millions of barrels) and $P(x)$ is the corresponding profit (in millions of dollars). Graph the function $P(x)$ and determine the level of production that would yield maximum profit.

Solution

The following table shows the value of $P(x)$ for some values of x in the closed interval $[0, 8]$.

x	0	1	2	3	4	5	6	7	8
$P(x)$	0	70	120	150	160	150	120	70	0

Plotting the ordered pairs (0, 0), (1, 70), (2, 120), . . . , (8, 0) and joining them by a smooth curve, we have the graph of $P(x)$ as shown in Figure 2.16.

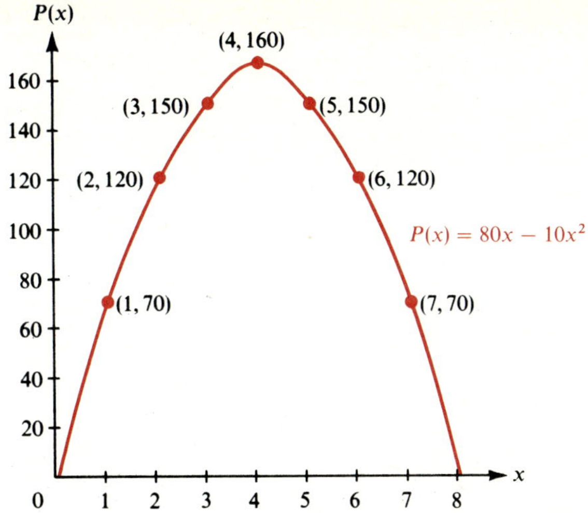

Figure 2.16

It appears as if the maximum profit of 160 million dollars is obtained by producing 4 million barrels of nearly inaccessible oil. Later on, we shall be able to apply some calculus concepts to verify that this is indeed true.

Let us now return to the example in the preceding section of a travel agency that advertises a five-day cruise to Nova Scotia. The total revenue received by the travel agency is given by

$$f(x) = \$20(1500 + 20x - x^2) \qquad x = 0, 1, 2, \ldots, 50$$

where x is the number of cancellations received prior to the departure of the boat.

The domain of f is the set of non-negative integers between 0 and 50, both inclusive. The values of $f(x)$ for several values of x in the domain are given in the following table.

Number of Cancellations	Receipts $f(x)$
0	\$30,000
5	31,500
10	32,000
15	31,500
20	30,000
25	27,500
30	24,000
35	19,500
40	14,000
45	7,500
50	0

Plotting the ordered pairs (0, 30000), (5, 31500), . . . , (45, 7500) and (50, 0), we have the graph of $f(x)$ in Figure 2.17. Notice that the maximum value of $f(x)$ seems to occur at $x = 10$. Thus the maximum receipts of the travel agency are \$32,000 when 10 cancellations are received.

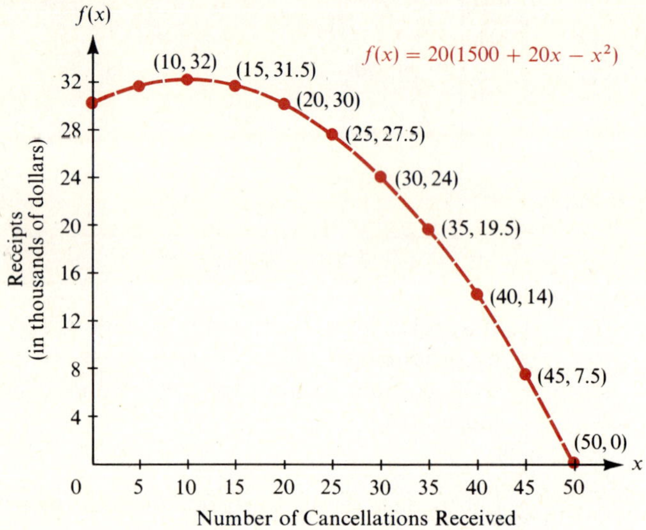

Figure 2.17

Definition 2.6 Any function of the form $f(x) = x^n$, where n is a positive integer, is called a *power function.*

For $n = 1$, the graph of $f(x)$ is a straight line. For $n = 2$, the graph of $f(x) = x^2$ is a parabola that opens upward (see Figure 2.14). We suggest that the readers draw the graphs of $f(x) = x^4$ and $f(x) = x^6$ and convince themselves that the shapes of their graphs are similar to that of $f(x) = x^2$.

Example 2.22 Plot the graph of

$$f(x) = x^3$$

Solution

The function $f(x)$ has as its domain the set of all real numbers. Unable to list all ordered pairs satisfying the equation $f(x) = x^3$, we will use only a partial list of points to determine the shape of the curve. The values of $f(x)$ for some values of x are given in the table below:

x	-2	$-\frac{3}{2}$	-1	$-\frac{1}{2}$	0	$\frac{1}{2}$	1	$\frac{3}{2}$	2
$f(x)$	-8	$-\frac{27}{8}$	-1	$-\frac{1}{8}$	0	$\frac{1}{8}$	1	$\frac{27}{8}$	8

Plotting the ordered pairs $(-2, -8)$, $(-\frac{3}{2}, -\frac{27}{8})$, $(-1, -1)$, $(-\frac{1}{2}, -\frac{1}{8})$, $(0, 0)$, $(\frac{1}{2}, \frac{1}{8})$, $(1, 1)$, $(\frac{3}{2}, \frac{27}{8})$ and $(2, 8)$ and joining them by a smooth curve, we have the graph in Figure 2.18.

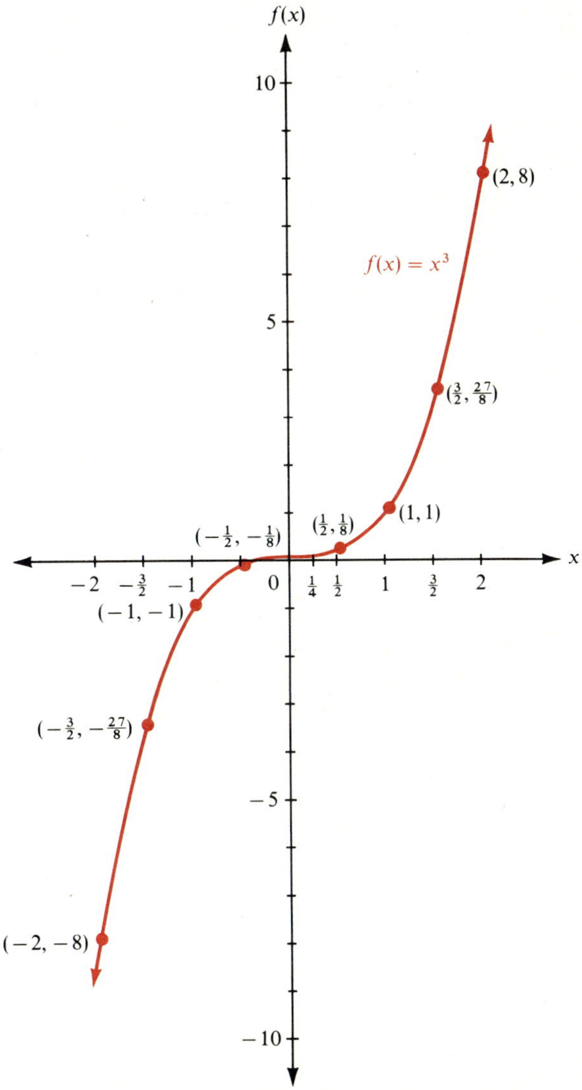

Figure 2.18

The graph of $f(x) = x^n$ for n odd and greater than 3 is similar to Figure 2.18.

Definition 2.7

Any function of the form

$$f(x) = a_n x^n + a_{n-1} x^{n-1} + \dots + a_2 x^2 + a_1 x + a_0, \qquad a_n \neq 0$$

where n is a positive integer or zero and $a_0, a_1, \dots, a_n$ are real numbers is called a polynomial of degree n.

If $n = 0$, we have a trivial polynomial of a constant function. If $n = 1$, the polynomial is of first degree. Such polynomials are the linear functions that were studied in Section 1.6. If $n = 2$, the polynomial is of second degree and is called a *quadratic function*; these were discussed in Examples 2.20 and 2.21. In the case where all values of a except a_n are zero and $a_n = 1$, we obtain the power function, $f(x) = x^n$.

Example 2.23

Plot the graph of

$$f(x) = x^3 - 3x^2 - 9x + 20$$

Solution

This polynomial function of degree 3 has as its domain the set of all real numbers. Clearly, we cannot list all possible ordered pairs satisfying this polynomial. However, it is necessary that we find a sufficient number of points to determine the shape of the curve.

The following table shows the values of $f(x)$ for some values of x.

x	-3	-2	-1	0	1	2	3	4	5
$f(x)$	-7	18	25	20	9	-2	-7	0	25

Plotting the ordered pairs $(-3, -7)$, $(-2, 18)$, ..., $(4, 0)$ and $(5, 25)$ and then joining them by a smooth curve, we have the graph of $f(x)$ as displayed in Figure 2.19.

In general it is very difficult to decide whether or not the graph drawn through the relatively few plotted points actually represents the polynomial of degree 3 or more. It is conceivable that the actual graph may be somewhat different from the one we have shown in Figure 2.19.

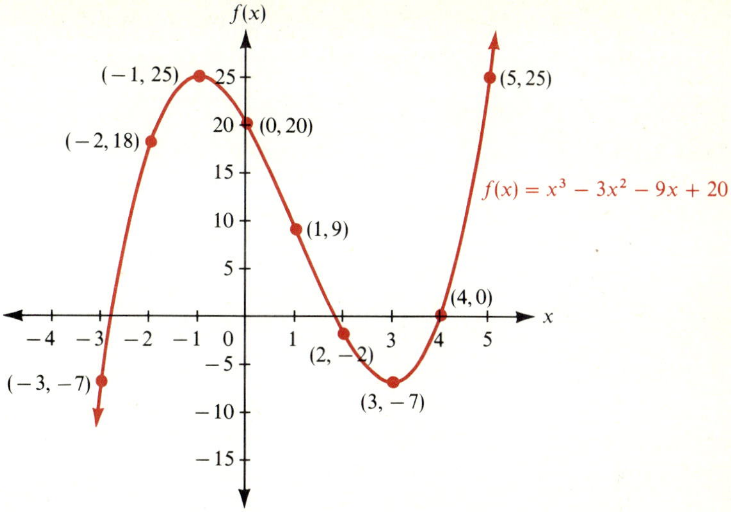

Figure 2.19

For the time being, we will restrict ourselves to polynomial functions whose graphs are smooth curves drawn on the basis of a relatively few but sufficient number of points. Later, in Chapter 4, we will discuss some techniques of calculus which will reduce the number of points necessary to draw an accurate graph.

Definition 2.8

Let $f(x)$ and $g(x)$, where $g(x) \neq 0$, be two polynomials. A function of the form

$$r(x) = \frac{f(x)}{g(x)}$$

is called a *rational function.*

Some examples of rational functions are:

$$f(x) = \frac{x+1}{x-1}, \qquad x \neq 1$$

$$f(x) = \frac{3x^2 - x + 5}{x+1}, \qquad x \neq -1$$

$$f(x) = \frac{1}{x}, \qquad x \neq 0$$

Remember that we must exclude from the domain of the function all those real numbers for which the denominator is zero.

The following examples introduce the concepts of vertical and horizontal asymptotes. Asymptotes are often helpful for graphing rational functions.

Example 2.24

Plot the graph of

$$f(x) = \frac{1}{x+1}$$

Solution

The function $f(x)$ has as its domain the set of real numbers except $x = -1$. At $x = -1$, the denominator $(x + 1)$ is zero and the function $f(x)$ is not defined. However, we can examine the behavior of $f(x)$ for values of x sufficiently close to -1. For $x > -1$, $(x + 1) > 0$ and therefore $f(x) > 0$. Further,

$$\begin{aligned} f(-0.999) &= 1000 & f(1) &= 0.50 \\ f(-0.99) &= 100 & f(3) &= 0.25 \\ f(-0.90) &= 10 & f(9) &= 0.10 \\ f(-0.50) &= 2 & f(49) &= 0.02 \\ f(0) &= 1 & f(99) &= 0.01 \end{aligned}$$

This means that as x gets closer and closer to -1 from the right, $(x + 1)$ gets closer and closer to zero and $f(x)$ gets larger and larger.

For $x < -1$, $(x + 1)$ is negative and hence $f(x) < 0$. Also

$$\begin{aligned} f(-101) &= -0.01 & f(-2) &= -1 \\ f(-51) &= -0.02 & f(-1.5) &= -2 \\ f(-26) &= -0.04 & f(-1.1) &= -10 \\ f(-5) &= -0.25 & f(-1.01) &= -100 \\ f(-3) &= -0.50 & f(-1.001) &= -1000 \end{aligned}$$

As x approaches -1 from the left, $(x + 1)$ approaches zero and $f(x)$ becomes a large negative number. Because $|f(x)|$ becomes large as x approaches -1, we call the vertical line $x = -1$ a *vertical asymptote* for the graph.

It may also be noticed that as x increases and moves to the right away from -1, $f(x)$ decreases and approaches zero. Similarly, as x moves farther away from -1 in the negative direction, $f(x)$ again approaches 0. Since $f(x)$ gets closer and closer to zero as $|x|$ increases, we call the x-axis a *horizontal asymptote* for the graph. This is displayed in Figure 2.20.

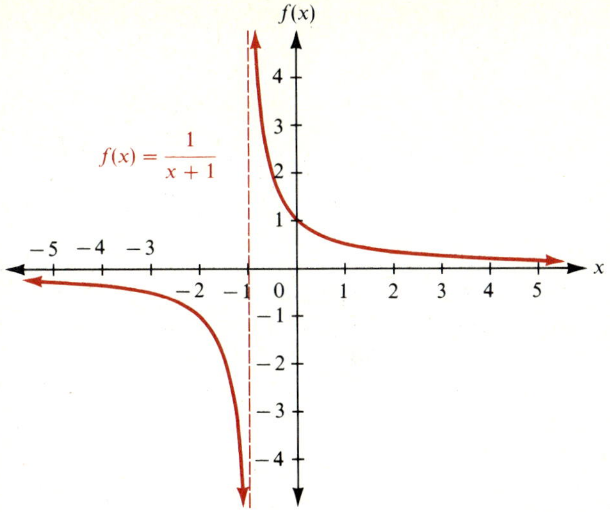

Figure 2.20

Example 2.25

The manager of a chain of bookstores has determined that the supply s of certain paperbacks is approximated by

$$s(p) = 30 - \frac{60}{p-1}$$

where $s(p)$ denotes the supply at the price p (in dollars) per carton of books.

a. Find the supply for $p = 3, 4, 5, 6, 7, 11, 16, 21, 31$.

b. Graph $s(p)$ by plotting the points obtained in part (a) and joining them by a dotted curve.

Solution

a. The value of $s(p)$ for each of the given values of p is given below:

p	3	4	5	6	7	11	16	21	31
$s(p)$	0	10	15	18	20	24	26	27	28

b. Plotting the ordered pairs (3, 0), (4, 10), (5, 15), (6, 18), (7, 20), (11, 24), (16, 26), (21, 27) and (31, 28) and joining them by a dotted curve, we have the graph in Figure 2.21.

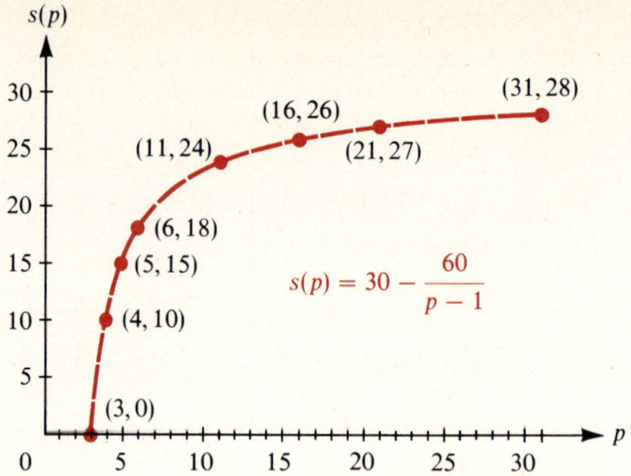

Figure 2.21

Thus far we have considered functions which are relatively easy to graph and have somewhat familiar forms. Consider next the greatest integer function

$$f(x) = [\![x]\!]$$

where $[\![x]\!]$ denotes the greatest integer not exceeding x. This means that

$$[\![\tfrac{1}{3}]\!] = 0,\ [\![\tfrac{4}{3}]\!] = 1,\ [\![2.5]\!] = 2,\ [\![3]\!] = 3$$
$$[\![-\tfrac{1}{3}]\!] = -1,\ [\![-\tfrac{5}{2}]\!] = -3,\ [\![-3.4]\!] = -4,\ [\![-4.5]\!] = -5,$$

and so on. The graph of this function for $-4 \leq x \leq 4$ is shown in Figure 2.22. Note that the open circle placed at the right end of each line segment means that the right endpoint of that segment is not a part of the graph.

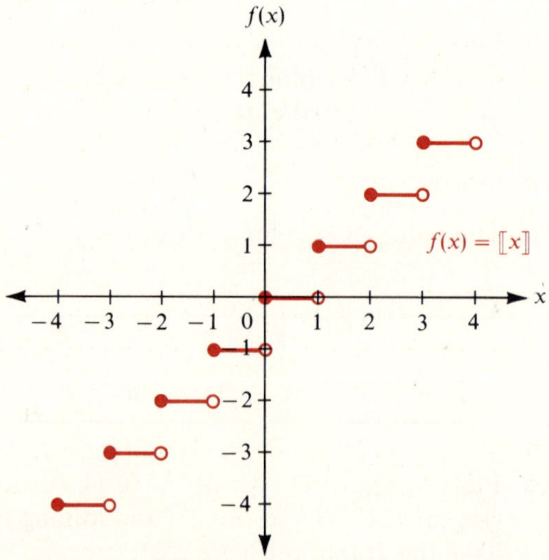

Figure 2.22

The graph consists of short intervals of horizontal lines. Note that the function has a constant value in each interval and that each interval is closed on the left and open on the right.

As a final illustration, consider a graph where $f(x)$ remains constant for some values of x.

Example 2.26

The management of a parking lot charges its customers daily according to the formula

$$f(x) = \begin{cases} \$1.50 + \$0.50(x - 2), & x = 2, 3, 4, \dots, 8 \\ \$4.50, & 8 < x \leq 24 \end{cases}$$

where x is the number of hours (or part thereof) that the car is parked and $f(x)$ is the amount assessed for parking the car. Finally, there is a minimum charge of \$1.50 and a maximum charge of \$4.50.

a. Determine the cost for parking up to 2, 3, 4, . . . , 24 hours.

b. Graph $f(x)$.

c. Find from the graph the cost for parking for $3\frac{1}{2}$ hours and for $4\frac{3}{4}$ hours.

Solution

a. Apparently the management charges \$1.50 for the first two hour period and 50¢ for each additional hour or part thereof. The parking schedule is given in the table below.

Time	Parking Cost $f(x)$
first two hours	\$1.50
third hour	2.00
fourth hour	2.50
fifth hour	3.00
sixth hour	3.50
seventh hour	4.00
eighth through twenty-fourth hour	4.50

b. The graph of $f(x)$ is given in Figure 2.23.

c. The parking for $3\frac{1}{2}$ hours costs \$2.50 while the cost for $4\frac{3}{4}$ hours is \$3.00.

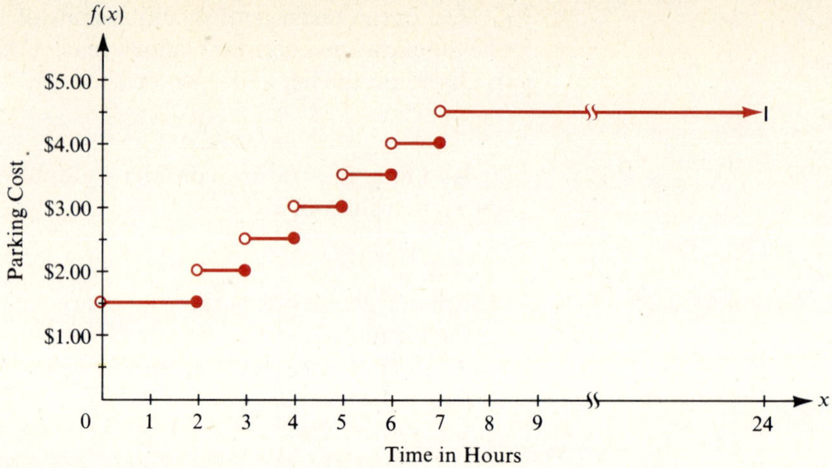

Figure 2.23

Exercises

Graph each of the following functions.

1. $f(x) = \sqrt{x - 1}, \quad x \geq 1$
2. $f(x) = \dfrac{|x|}{x}, \quad x \neq 0$
3. $f(x) = 4x + 5$
4. $f(x) = x^2 + 2$
5. $f(x) = (x + 2)^2$
6. $f(x) = x^2 + 2x$
7. $f(x) = x^3 - 9x^2 + 24x$
8. $f(x) = -x^3$
9. $f(x) = x^3 - 6x^2 + 9x$
10. $f(x) = x^4 - 8x^2$
11. $f(x) = \dfrac{1}{x + 2}$
12. $f(x) = \dfrac{1}{(x + 2)^2}$
13. $f(x) = \begin{cases} 7 - 2x & x < 2 \\ 4x - 5 & x \geq 2 \end{cases}$
14. $f(x) = \begin{cases} x & \text{for} \quad x < 0 \\ x^2 & \text{for} \quad x \geq 0 \end{cases}$
15. The demand for a certain product is given by

$$g(p) = 100 - \tfrac{5}{4}p$$

where $g(p)$ is the demand in units at price p. Graph $g(p)$ using the values $p = 0, 20, 40, 60$, and 80.

16. A travel agency has arranged a trip to Niagara Falls by bus for a senior class. It has rented a bus with a seating capacity of 80 students with an understanding that at least 30 students sign up for the tour. The fare is \$60 per student if the minimum requirement of 30 students is met and the fare will decrease by \$1 per student for each additional student who signs up.

a. Determine the revenue function, $R(x)$, where x is the number of additional students who sign up. Give the domain of $R(x)$.
b. Calculate $R(x)$ for $x = 0, 5, 10, 15, 20, 25, 30, 35$, and 40.
c. Graph $R(x)$.
d. Determine the value of x for which $R(x)$ is maximum.

17. A travel agency advertises an all-expense paid tour to Washington, D.C., for college students. The agency charters one car on the train with a seating capacity of 50 passengers. The fare is \$100 per student plus an additional \$5 for each empty seat.

a. Determine the revenue function, $R(x)$, where x is the number of empty seats on the train. Give the domain of $R(x)$.
b. Calculate $R(x)$ for $x = 0, 5, 10, 15, 20, 25, 30, 35$, and 40.
c. Graph $R(x)$.
d. Determine the value of x for which $R(x)$ is maximum.

18. A cab charges \$0.50 for the first quarter mile of transportation and \$0.10 for each additional quarter mile or part thereof. What will a cab cost a passenger who needs to travel:

a. one-half mile?
b. one mile?
c. one and a half miles?
d. two miles?
e. two and a half miles?

19 An attendant in a parking lot charges \$2.00 for the first hour and \$0.75 for each additional half-hour or part thereof. What will a driver pay for parking

a. two hours?
b. three hours?
c. three and a half hours?
d. four hours?

2.4 Composite and Inverse* Functions

When f and g are two real valued functions, we may define a new function h, called the composite of g on f.

Definition 2.9

The composite of g on f, denoted by $g \circ f$, is defined by

$$(g \circ f)x = g[f(x)]$$

This implies that we first determine $f(x)$ under the rule f and then apply the rule g to the element thus obtained. Observe in Figure 2.24 that $f(x) \in B$ and that B is the domain of the function g. Thus the domain of the composite function $g \circ f$ consists of only those elements x for which $f(x)$ is in the domain of g.

* The study of inverse functions may be postponed until Chapter 5.

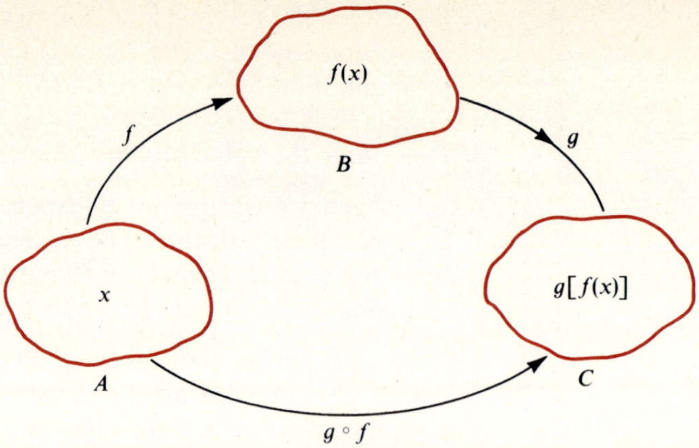

Figure 2.24

Consider Figure 2.25. If an element x is entered into the machine f, the output from this machine becomes the input for the machine g. In turn, machine g yields $g[f(x)]$ as the final output.

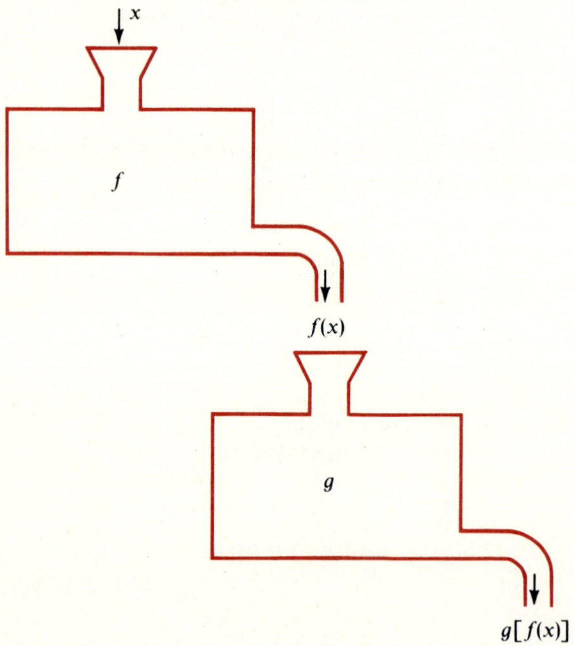

Figure 2.25

Example 2.27

Let f be defined by a function

$$f(x) = x + 2$$

with the set $A = \{1, 2, 3, 4, 5\}$ as its domain. Determine $g[f(x)]$ for each $x \in A$ if $g(x) = 2x + 5$ and the domain of g is the range of f.

Solution

First, we compute $f(x)$ for each $x \in A$. Since

$$f(1) = 3, f(2) = 4, f(3) = 5, f(4) = 6 \text{ and } f(5) = 7$$

then the range of the function f is the set $B = \{3, 4, 5, 6, 7\}$ as illustrated in Figure 2.26. The function g is a correspondence from set B to set C. We compute $g[f(x)]$ for each element in B. Since $g(x) = 2x + 5$, we have

$$g(3) = 11, g(4) = 13, g(5) = 15, g(6) = 17 \text{ and } g(7) = 19.$$

Thus, the range of the function $g \circ f$ is the set $C = \{11, 13, 15, 17, 19\}$.

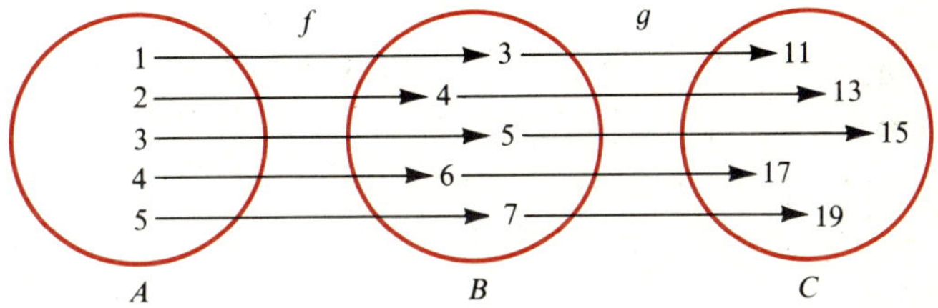

Figure 2.26

Alternatively, we can determine a formula for the composite function $g \circ f$ and then compute $g[f(x)]$ for each $x \in A$. Recall that since $g(x) = 2x + 5$, we have

$$\begin{aligned}(g \circ f)x = g[f(x)] &= 2[f(x)] + 5\\ &= 2(x + 2) + 5\\ &= 2x + 9\end{aligned}$$

Now

$$\begin{aligned}g[f(1)] &= 2.1 + 9 = 11\\ g[f(2)] &= 2.2 + 9 = 13\\ g[f(3)] &= 2.3 + 9 = 15\\ g[f(4)] &= 2.4 + 9 = 17\\ g[f(5)] &= 2.5 + 9 = 19\end{aligned}$$

Thus, the range of the composite function $h = g[f(x)]$ is precisely the same we established in the Figure 2.26 above.

As a practical application of the composite function, assume that the demand of a product is given by $d(p)$ where p is the price at which the product is sold. However, this selling price p depends, in turn, on the cost of producing that product. Thus demand is a function of price and price is a function of cost. In other words

$$(d \circ p)c = d[p(c)]$$

Example 2.28

The demand for a product is given by

$$d(p) = 100 - p^2, \qquad 0 < p \leq 10$$

where p is the selling price of the product. If p is given by the linear function

$$p(c) = 3c + 5$$

then the demand in terms of cost c is given by

$$\begin{aligned}(d \circ p)c &= d[p(c)] \\ &= 100 - (3c + 5)^2 \\ &= 100 - (9c^2 + 30c + 25) \\ &= -3(3c^2 + 10c - 25)\end{aligned}$$

Example 2.29

Let $f(x) = x^2 + 1$ and $g(x) = 3x + 4$. Determine each of the following:

a. $g[f(x)]$

b. $f[g(x)]$

Solution

a. $$\begin{aligned}g[f(x)] &= 3[f(x)] + 4 \\ &= 3(x^2 + 1) + 4 \\ &= 3x^2 + 7\end{aligned}$$

b. $$\begin{aligned}f[g(x)] &= [g(x)]^2 + 1 \\ &= (3x + 4)^2 + 1 \\ &= 9x^2 + 24x + 17\end{aligned}$$

Example 2.29 demonstrates the fact that in general

$$g[f(x)] \neq f[g(x)]$$

However, if the function f is one-to-one, we can define a function g so that $g[f(x)]$ and $f[g(x)]$ are equal. For each $a \in A$, if f carries $a \in A$ to $f(a) \in B$, then g is defined to be the function which brings $f(a) \in B$ back to $a \in A$ as shown in Figure 2.27. This implies that $g[f(a)] = a$ for each $a \in A$.

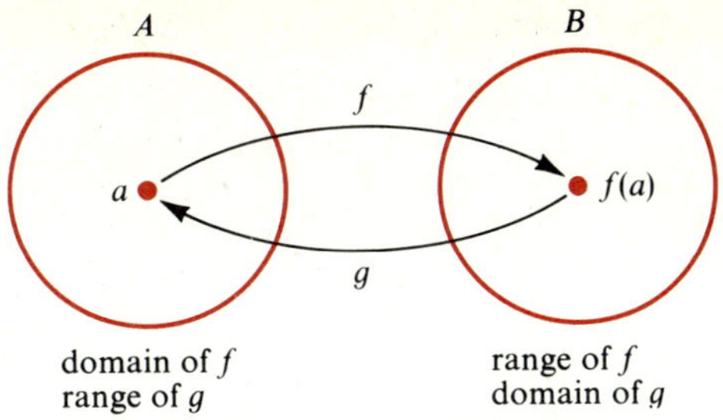

Figure 2.27

Definition 2.10

Let f be a one-to-one function with the domain A and range B. The function g with domain B and range A is called the *inverse* of f if

$$g[f(x)] = x \quad \text{for} \quad x \in A.$$

The *inverse function* g of the function f is denoted by f^{-1}. Thus,

$$f^{-1}[f(x)] = x \quad \text{for} \quad x \in A.$$

One can also show that

$$f[f^{-1}(x)] = x \quad \text{for} \quad x \in B.$$

Example 2.30

The functions $f(x) = 3x - 2$ and $g(x) = (x + 2)/3$ are inverses of each other, since

$$g[f(x)] = \frac{f(x) + 2}{3} = \frac{(3x - 2) + 2}{3} = x$$

and

$$f[g(x)] = 3g(x) - 2 = 3\left(\frac{x + 2}{3}\right) - 2 = x.$$

Thus, the inverse function of f is

$$f^{-1}(x) = \frac{x + 2}{3}.$$

Also,

$$g^{-1}(x) = 3x - 2.$$

Example 2.31

Show that

$$f(x) = \sqrt{x - 2}, x \geq 2 \quad \text{and} \quad g(x) = x^2 + 2, x \geq 0$$

are inverses of each other.

Solution

We need to check whether or not

$$g[f(x)] = f[g(x)] = x$$

Observe that

$$\begin{aligned} g[f(x)] &= [f(x)]^2 + 2, \qquad f(x) \geq 0 \quad \text{for} \quad x \geq 2 \\ &= (\sqrt{x-2})^2 + 2 \\ &= (x-2) + 2 \\ &= x \end{aligned}$$

and

$$\begin{aligned} f[g(x)] &= \sqrt{g(x) - 2}, \qquad g(x) \geq 2 \quad \text{for} \quad x \geq 0 \\ &= \sqrt{(x^2+2) - 2} = \sqrt{x^2} \\ &= x \quad \text{if } x \geq 0 \end{aligned}$$

This proves that f and g are inverses of each other.

Example 2.32

Let $f(x) = 2x + 3$. Does $f^{-1}(x)$ exist? If so, find f^{-1}.

Solution

To show that f^{-1} exists, it is necessary to demonstrate that f is one-to-one. Suppose that a and b are two elements in the domain of f and $f(a) = f(b)$. Then

$$2a + 3 = 2b + 3$$

which imples that $a = b$. Thus f is a one-to-one function.

If f^{-1} is to be the inverse function of f, then

$$f[f^{-1}(x)] = x$$

which implies that

$$2[f^{-1}(x)] + 3 = x$$

Solving for $f^{-1}(x)$, we have

$$f^{-1}(x) = \frac{x-3}{2}$$

Let us now compare the graphs of $f(x) = 2x + 3$ and $f^{-1}(x) = (x-3)/2$ as shown in Figure 2.28.

The ordered pairs $(-3, -3)$, $(0, 3)$, $(2, 7)$, and $(3, 9)$ are elements of the function f while its inverse f^{-1} contains the ordered pairs $(-3, -3)$, $(3, 0)$, $(7, 2)$, $(9, 3)$. Note that each ordered pair (a, b) of the original function f has been replaced by an ordered pair (b, a) of the inverse function. Geometrically the graphs of $f(x)$ and $f^{-1}(x)$ are symmetric about the line $y = x$. *Thus if f is a one-to-one function, its inverse f^{-1}*

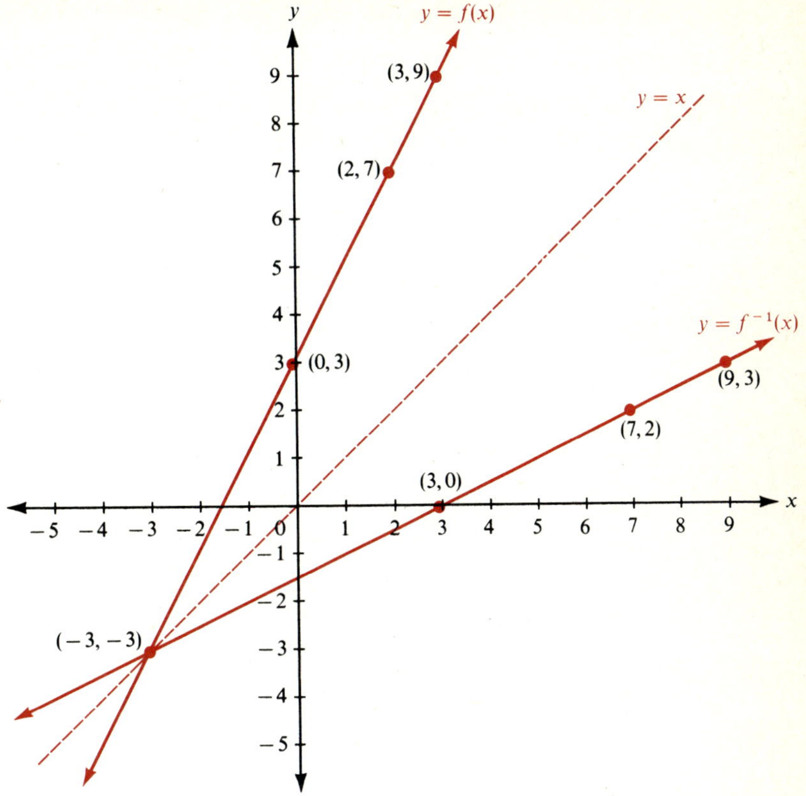

Figure 2.28

is obtained simply by interchanging the roles of x and y in the original expression, and then solving for y. Consider, for example, a function

$$f = \{(x, y) \mid y = 3x - 4\}$$

Then

$$f^{-1} = \{(x, y) \mid x = 3y - 4\}$$
$$= \left\{(x, y) \mid y = \frac{x + 4}{3}\right\}$$

Exercises

1. Given that $f(x) = 3x + 2$ and $g(x) = 2x + 3$, determine the values of $g[f(x)]$ for $x = 1, 2, 3, 4, 5$, and 6.
2. Given that $f(x) = 4x + 5$ and $g(x) = 5x + 2$, determine the values of $f[g(x)]$ for $x = 1, 2, 3, 4, 5$, and 6.
3. Given that $f(x) = 3x + 4$ and $g(x) + 4x + 3$, determine the values of $g[f(x)]$ and $f[g(x)]$ for $x = 1, 2, 3, 4$, and 5.

4. Determine $d[p(c)]$ in each of the following composite functions.

a. $d(p) = 100 - 4p^2$, $p(c) = 5c$

b. $d(p) = 1000 - p^3$, $p(c) = c + 4$

c. $d(p) = 100 - \frac{p^2}{10}$, $p(c) = 2c + 1$

d. $d(p) = \frac{500}{p}$, $p(c) = c + 1$

Determine $g[f(x)]$ and $f[g(x)]$ in problems 5 through 16.

5. $f(x) = 3x + 5$, $g(x) = x - 5$

6. $f(x) = 2x + 1$, $g(x) = 2x - 3$

7. $f(x) = 3x + 1$, $g(x) = \frac{x + 4}{3}$

8. $f(x) = x$, $g(x) = 1/x$

9. $f(x) = x^2 + 1$, $g(x) = 3x$

10. $f(x) = x^2$, $g(x) = 1 - x$

11. $f(x) = 2x^2 + 5$, $g(x) = 3x - 1$

12. $f(x) = x + 1/x$, $g(x) = 6x - 5$

13. $f(x) = x^2 + 2x + 3$, $g(x) = 1 - x^3$

14. $f(x) = x^2 - 3$, $g(x) = \sqrt{x + 3}$, $\quad x \geq -3$

15. $f(x) = x^2 + 2$, $g(x) = \sqrt{x - 1}$, $\quad x \geq 1$

16. $f(x) = x^2$, $g(x) = \frac{x}{x + 1}$, $\quad x \neq -1$

In each of the following problems 17–20, show that functions f and g are inverses of each other.

17. $f(x) = 2x + 3$, $g(x) = \frac{x - 3}{2}$

18. $f(x) = x^2 + 4$, $g(x) = \sqrt{x - 4}$, $\quad x \geq 4$

19. $f(x) = x^3 - 1$, $g(x) = (x + 1)^{1/3}$

20. $f(x) = \frac{2x}{x + 1}$, $\quad x \neq -1$; $\quad g(x) = \frac{x}{2 - x}$, $\quad x \neq 2$

In each of the following problems 21–24, decide whether the function f is one-to-one. If it is, determine f^{-1}.

21. $f(x) = 4x - 1$

22. $f(x) = 3x$

23. $f(x) = x + 3$

24. $f(x) = x^3$

In each of the following problems 25–32, determine the inverse function f^{-1}.

25. $f(x) = 3 - 2x$

26. $f(x) = 2x$

27. $f(x) = 3x - 2$

28. $f(x) = 2x^2, \qquad x \geq 0$

29. $f(x) = \sqrt{x - 3}, \qquad x \geq 3$

30. $f(x) = \sqrt{x^2 - 6}$

31. $f(x) = \sqrt{9 - x^2}, \qquad 0 \leq x \leq 3$

32. $f(x) = \dfrac{1}{x + 1}, \qquad x \neq -1, x \neq 0$

2.5 Limit of a Function

The limit of a function is perhaps the single most important concept of calculus. Limits are used to examine the behavior of a function as a variable *approaches* but never equals a given number. We introduce the concept of a limit in an informal manner.

Consider a function $y = f(x)$ and a line $x = a$ as shown in Figure 2.29. What is the value of $f(x)$ as x approaches a through values of x *less than* a? Does it, in some sense, get closer and closer to some particular number, say L_1? If so, we say that the *left-hand limit* exists and we write

$$\lim_{x \to a^-} f(x) = L_1$$

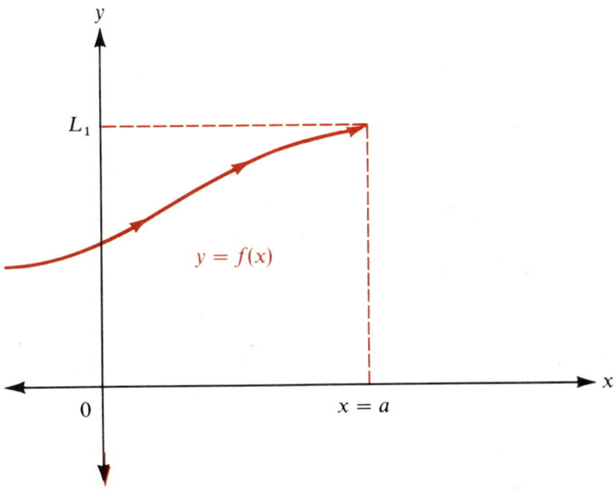

Figure 2.29

The symbol $x \to a^-$ means that x approaches a through values of x less than a. Now consider an analogous situation and ask what is the value of $f(x)$ as x approaches a through values of x *greater than* a as shown in Figure 2.30. If $f(x)$ approaches some particular number, say L_2, then we say that the *right-hand limit* exists and we write

$$\lim_{x \to a^+} f(x) = L_2$$

The symbol $x \to a^+$ implies that x approaches a through values of x greater than a.

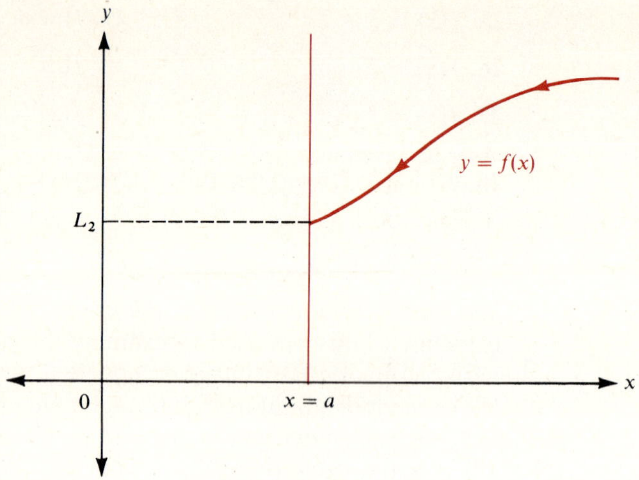

Figure 2.30

If $L_1 = L_2 = L$, then the limit of $f(x)$, as x approaches a, exists and we write

$$\lim_{x \to a} f(x) = L$$

In essence, as x approaches a from either side, the limit of the function $f(x)$ equals L.

In the following examples we find the limits by computing the values of $f(x)$ as x approaches a through values of $x < a$ and through values of $x > a$.

Example 2.33

Evaluate $\lim_{x \to 3} f(x)$ where

$$f(x) = \begin{cases} x + 1 & \text{for} \quad x < 3 \\ 7 - x & \text{for} \quad x > 3 \end{cases}$$

Graph $f(x)$.

Solution

First we determine the values of $f(x)$ as x approaches 3 through values of $x < 3$. Note that $f(x) = x + 1$ for $x > 3$ and we have

$f(2.8) = 3.8$ $\qquad f(2.99) = 3.99$

$f(2.9) = 3.9$ $\qquad f(2.999) = 3.999$

$f(2.95) = 3.95$ $\qquad f(2.9999) = 3.9999$

This means that as x approaches 3 from the left, $f(x)$ gets closer and closer to 4. Thus the left-hand limit of $f(x)$ equals 4. Now observe the values of $f(x)$ as x approaches 3 through values of $x > 3$. Clearly, if $f(x) = 7 - x$ for $x > 3$, then

$$\begin{array}{ll} f(3.2) = 3.8 & f(3.05) = 3.95 \\ f(3.15) = 3.85 & f(3.005) = 3.995 \\ f(3.1) = 3.9 & f(3.0005) = 3.9995 \end{array}$$

As x gets closer and closer to 3 from the right, $f(x)$ gets closer and closer to the real number 4. This means that the right-hand limit of $f(x)$ equals 4. Since both the left-hand and the right-hand limits of $f(x)$ exist and are equal, it follows that

$$\lim_{x \to 3} f(x) = 4$$

The graph of $f(x)$ is shown in Figure 2.31. Note that the function has a limit as x approaches 3 but the function is not defined at $x = 3$.

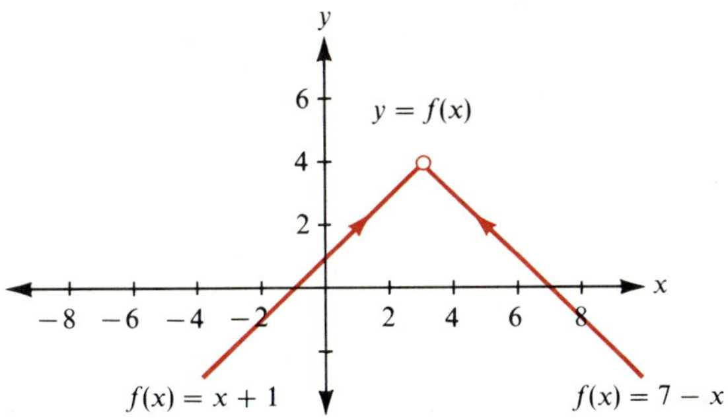

Figure 2.31

Example 2.34

Let $f(x)$ be defined as follows:

$$f(x) = \begin{cases} 1 - x, & x < 1 \\ 2, & x = 1. \\ x - 1, & x > 1 \end{cases}$$

Graph $f(x)$ and determine whether or not $\lim_{x \to 1} f(x)$ exists.

Solution

The graph of $f(x)$ is shown in Figure 2.32. The open circle at the point $(1, 0)$ is not a part of the graph. As x approaches 1 through values of $x < 1$, $f(x)$ approaches 0. Thus the left-hand limit exists. As x gets closer and closer to 1 through values of $x > 1$, $f(x)$ again approaches the number 0. Thus,

$$\lim_{x \to 1^-} f(x) = 0 \quad \text{and} \quad \lim_{f(x) \to 1^+} f(x) = 0$$

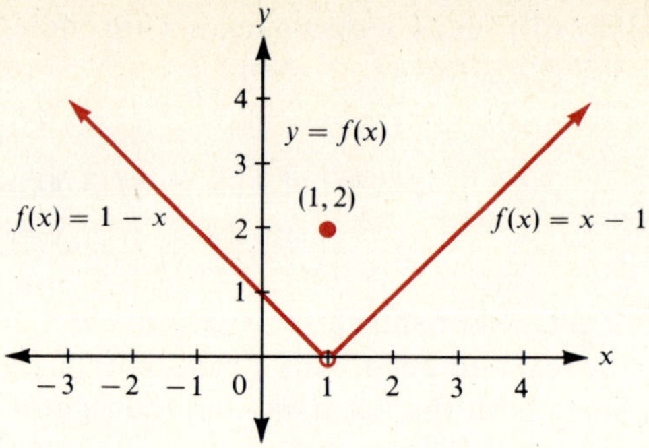

Figure 2.32

Since both the left-hand and the right-hand limits exist and are equal, we conclude that

$$\lim_{x \to 1} f(x)$$

exists. We must emphasize that we can *define* the value of the function at $x = 1$ to be any value we please. That defined value may have nothing to do with the limit of the function.

Example 2.35

Evaluate $\lim_{x \to 1} f(x)$ where

$$f(x) = (x + 2)^2, \qquad -2 \leq x \leq 2$$

Solution

As before, we first investigate the behavior of the function $f(x)$ for values of x less than 1. Note that

$$\begin{array}{ll} f(0.6) = 6.76 & f(0.95) = 8.7025 \\ f(0.7) = 7.29 & f(0.99) = 8.9401 \\ f(0.8) = 7.84 & f(0.995) = 8.970025 \\ f(0.9) = 8.41 & f(0.999) = 8.994001 \end{array}$$

Thus as x approaches 1 through values of $x < 1$, $f(x)$ gets closer and closer to 9. Thus, the left-hand limit of $f(x)$ equals 9; that is

$$\lim_{x \to 1^-} f(x) = 9$$

Now we consider the values of $f(x)$ as x approaches 1 through values of x greater than 1. Observe that

$$\begin{array}{ll} f(1.5) = 12.25 & f(1.1) = 9.61 \\ f(1.4) = 11.56 & f(1.05) = 9.3025 \\ f(1.2) = 10.24 & f(1.01) = 9.0601 \\ f(1.15) = 9.9225 & f(1.001) = 9.006001 \end{array}$$

and so on. It is clear that as x gets closer and closer to 1 through values of x greater than 1, $f(x)$ gets closer and closer to 9. This suggests that the right-hand limit of $f(x)$ equals 9; that is

$$\lim_{x \to 1^+} f(x) = 9$$

Since both the right- and the left-hand limits exist and are equal, we conclude that

$$\lim_{x \to 1} (x + 2)^2 = 9$$

Example 2.36

Evaluate $\lim_{x \to 1} f(x)$ where

$$f(x) = \frac{x^2 - 1}{x - 1}, \qquad x \neq 1$$

Solution

Again, we need to study the behavior of the function $f(x)$ for values of $x < 1$. We see that

$$\begin{array}{ll} f(0.8) = 1.8 & f(0.98) = 1.98 \\ f(0.85) = 1.85 & f(0.99) = 1.99 \\ f(0.9) = 1.9 & f(0.995) = 1.995 \\ f(0.95) = 1.95 & f(0.999) = 1.999 \end{array}$$

and so on. The values of $f(x)$ get closer and closer to 2 as x moves through values of $x < 1$. Thus the left-hand limit of $f(x)$ exists and equals 2. In other words,

$$\lim_{x \to 1^-} f(x) = 2$$

Similarly, we observe that

$$\begin{array}{ll} f(1.2) = 2.2 & f(1.05) = 2.05 \\ f(1.15) = 2.15 & f(1.01) = 2.01 \\ f(1.1) = 2.1 & f(1.001) = 2.001 \end{array}$$

Note that as x approaches 1 through values of $x > 1$, $f(x)$ gets closer and closer to 2. This means that

$$\lim_{x \to 1^+} f(x) = 2$$

Since

$$\lim_{x \to 1^-} f(x) = \lim_{x \to 1^+} f(x)$$

it follows that

$$\lim_{x \to 1} f(x) = 2$$

The same result could have been derived more quickly by using elementary algebra. Note that for $x \neq 1$,

$$f(x) = \left(\frac{x^2 - 1}{x - 1}\right) = \frac{(x + 1)(x - 1)}{(x - 1)} = x + 1$$

This provides a simpler device to calculate the values of $f(x)$ as x approaches 1. The graph of $f(x)$ is shown in Figure 2.33. The circle at the point $(1, 2)$ indicates that this point is not a part of the graph.

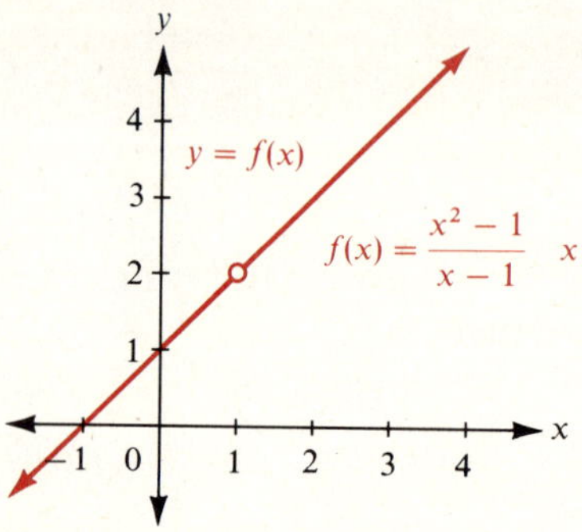

Figure 2.33

We wish to reinforce the fact that the value of the function $f(x)$ at $x = a$ has no effect on the existence or non-existence of the limit. In order for the limit of $f(x)$ to exist (as x approaches a) and be equal to L, it is necessary that both the left-hand and the right-hand limits at a exist and be equal to L.

We now consider situations in which the limit of a function does not exist.

Example 2.37

Consider the graph of the function

$$f(x) = \begin{cases} 1, & x \leq 2 \\ 2, & x > 2 \end{cases}$$

in Figure 2.34. Clearly, the left-hand limit exists and equals 1. However, we see that as x approaches 2 through values of $x > 2$, the right-hand limit exists and equals 2. Thus,

$$\lim_{x \to 2^-} f(x) = 1 \quad \text{and} \quad \lim_{x \to 2^+} f(x) = 2$$

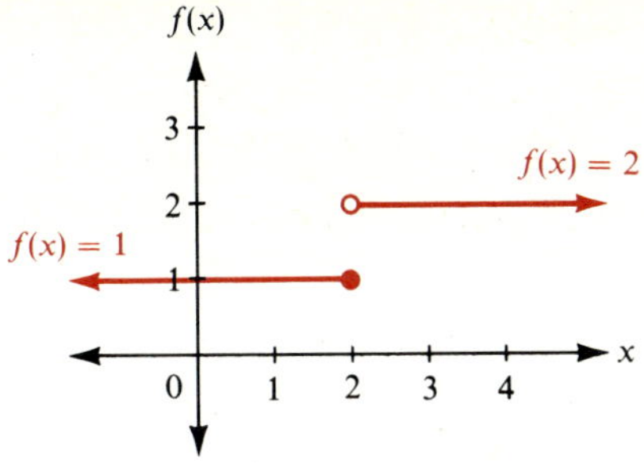

Figure 2.34

Because $f(x)$ approaches two different numbers depending on whether $x \to 2$ from the left or from the right, we conclude that

$$\lim_{x \to 2} f(x)$$

does not exist.

Example 2.38

Evaluate $\lim_{x \to 0} f(x)$ where

$$f(x) = \frac{|x|}{x}, \qquad x \neq 0$$

Solution

Recall from the definition of the absolute value in Chapter 1 that for $x > 0$, $|x| = x$ and thus

$$f(x) = \frac{x}{x} = 1, \qquad x > 0$$

For $x < 0$, $|x| = -x$, and we have

$$f(x) = \frac{-x}{x} = -1, \qquad x < 0$$

This means that as x approaches 0 through the values of $x < 0$, $f(x)$ approaches -1, while as x approaches 0 through values of $x > 0$, $f(x)$ approaches 1. Thus

$$\lim_{x \to 0^-} f(x) = -1 \quad \text{and} \quad \lim_{x \to 0^+} f(x) = 1$$

Since

$$\lim_{x \to 0^-} f(x) \neq \lim_{x \to 0^+} f(x)$$

we conclude that $\lim_{x \to 0} f(x)$ does not exist.

Example 2.39

Evaluate $\lim_{x \to 2} f(x)$ where

$$f(x) = \frac{1}{(x-2)^2}, \qquad x \neq 2$$

Solution

Observe that the function is defined for all real numbers x except $x = 2$ and that $x = 2$ is the vertical asymptote for the graph of $f(x)$. To examine the behavior of $f(x)$ for values of $x < 2$, note that

$$\begin{aligned} f(-0.5) &= 0.16 & f(1.8) &= 25 \\ f(0) &= 0.25 & f(1.9) &= 100 \\ f(1) &= 1 & f(1.99) &= 10{,}000 \\ f(1.5) &= 4 & f(1.999) &= 1{,}000{,}000 \end{aligned}$$

Now we observe the values of $f(x)$ for $x > 2$. Clearly,

$$\begin{aligned} f(4) &= 0.25 & f(2.1) &= 100 \\ f(3) &= 1 & f(2.01) &= 10^4 \\ f(2.5) &= 4 & f(2.001) &= 10^6 \\ f(2.2) &= 25 & f(2.0001) &= 10^8 \end{aligned}$$

The graph of the function $f(x)$ is shown in Figure 2.35.

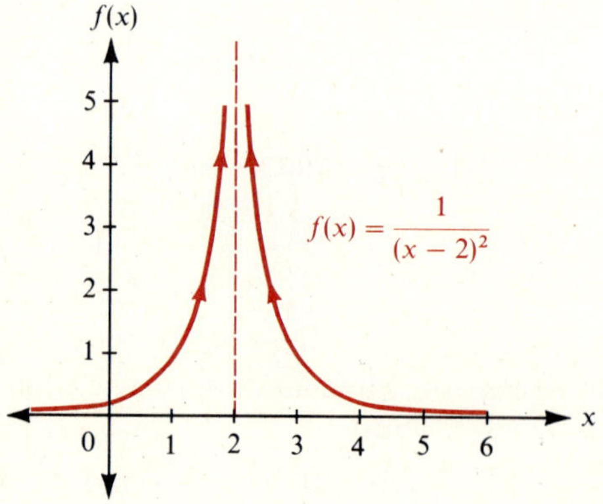

Figure 2.35

Note that as $x \to 2$ through values of $x < 2$, $f(x)$ becomes larger and larger but does not approach any particular number. Thus, the left-hand limit fails to exist. Similarly, as x approaches 2 through values of $x > 2$, again $f(x)$ gets larger and larger. This means that the right-hand limit does not exist. Hence, it follows that

$$\lim_{x \to 2} f(x)$$

does not exist.

Exercises

Determine the limit, if it exists, in each of the following problems. (Do not graph the functions.)

1. $\lim_{x \to 2} (2x + 3)$
2. $\lim_{x \to 3} (3x + 1)$
3. $\lim_{x \to 3} (4x^2 + 3)$
4. $\lim_{x \to 1} (x + 1)^2$
5. $\lim_{x \to 2} \left(\frac{x^2 - 4}{x - 2}\right)$
6. $\lim_{x \to 1} \frac{1}{(x - 1)^2}$
7. $\lim_{x \to 2} \left(\frac{1}{x - 2}\right)$
8. $\lim_{x \to 1} \left(\frac{x^3 - 1}{x - 1}\right)$
9. $\lim_{x \to -1} \left(\frac{x^2 - 1}{x + 1}\right)$
10. $\lim_{x \to 2} \left(\frac{x^2 - x - 2}{x - 2}\right)$
11. $\lim_{x \to 3} \frac{x^2 - 5x + 6}{x - 3}$
12. $\lim_{x \to 4} \frac{x^2 - 7x + 12}{x - 4}$
13. $\lim_{x \to 1} \left(\frac{|x - 1|}{x - 1}\right)$
14. $\lim_{x \to 3} \frac{|x - 3|}{3(x - 3)}$
15. $\lim_{x \to 2} \frac{x - 2}{3|x - 2|}$
16. $\lim_{x \to 2} \left(\frac{1}{x^2}\right)$
17. $\lim_{x \to 2} f(x)$ where

$$f(x) = \begin{cases} 2x - 1, & x > 2 \\ 9 - 3x, & x < 2 \end{cases}$$

18. $\lim_{x \to 1} f(x)$ where

$$f(x) = \begin{cases} 2x, & x > 1 \\ x^2 + 1, & x < 1 \end{cases}$$

19. $\lim_{x \to 2} f(x)$ where

$$f(x) = \begin{cases} 4x + 1, & x > 2 \\ 13 - 2x, & x < 2 \end{cases}$$

20. $\lim_{x \to 1} f(x)$ where

$$f(x) = \begin{cases} x + 1, & x < 1 \\ 3 - x, & x > 1 \end{cases}$$

Graph the functions specified in each of the following and use the graphs to determine the limit of $f(x)$, if it exists.

21. $\lim_{x\to -1} f(x)$ where

$$f(x) = \begin{cases} 2, & x > -1 \\ 1 + x, & x < -1 \end{cases}$$

22. $\lim_{x\to 3} f(x)$ where

$$f(x) = \begin{cases} 3, & x > 3 \\ 1 + x, & x < 3 \end{cases}$$

23. $\lim_{x\to -1} f(x)$ where

$$f(x) = \begin{cases} 3 + x^2, & x < -1 \\ 3x, & x > -1 \end{cases}$$

24. $\lim_{x\to 1} f(x)$ where

$$f(x) = \begin{cases} x + 2, & x > 1 \\ 2x, & x < 1 \end{cases}$$

25. $\lim_{x\to -1} f(x)$ where

$$f(x) = \begin{cases} (x+1)^2, & x < -1 \\ x, & x > -1 \end{cases}$$

26. $\lim_{x\to 2} f(x)$ where

$$f(x) = \begin{cases} |x|, & -2 \le x < 2 \\ 2, & x > 2 \end{cases}$$

27. $\lim_{x\to 3} f(x)$ where

$$f(x) = \begin{cases} x - 1 & x < 3 \\ 9 - 2x & x > 3 \end{cases}$$

28. $\lim_{x\to 2} f(x)$ where

$$f(x) = \begin{cases} 2x, & x < 2 \\ 3x + 1, & x > 2 \end{cases}$$

2.6 Properties of Limits

The methods for evaluating the limit of $f(x)$ for values of x near a but not at a were time consuming, cumbersome, and awkward. We can simplify the evaluation of a limit if we use some general properties of limits. The following properties are presented in the form of propositions; their proofs, however, are omitted.

Proposition 2.1

If $f(x) = mx + b$, where m and b are constants, then

$$\lim_{x\to a} f(x) = ma + b$$

As an immediate consequence of this proposition, we have the following special cases.

a. If $f(x) = b$, where b is a constant, then

$$\lim_{x\to a} f(x) = b$$

That is, the limit of a constant is that constant.

b. If $f(x) = x$, then

$$\lim_{x\to a} f(x) = a$$

Proposition 2.2

If $\lim_{x\to a} f(x) = L$, then for any constant c,

$$\lim_{x\to a} [cf(x)] = c[\lim_{x\to a} f(x)] = cL$$

Proposition 2.3

If $\lim_{x\to a} f(x) = L$ and $\lim_{x\to a} g(x) = M$, then

$$\lim_{x\to a} [f(x) \pm g(x)] = \lim_{x\to a} f(x) \pm \lim_{x\to a} g(x) = L \pm M$$

This proposition states that the limit of the sum (or difference) of two functions is the sum (or difference) of their limits, provided that the limits exist.

Proposition 2.4

If $\lim_{x\to a} f(x) = L$ and $\lim_{x\to a} g(x) = M$, then

$$\lim_{x\to a} [f(x)g(x)] = [\lim_{x\to a} f(x)][\lim_{x\to a} g(x)] = \mathrm{LM}$$

In words, the limit of the product of two functions is equal to the product of their limits, provided that the limits exist.

Corollary

If $\lim_{x\to a} f(x) = L$, then for any positive integer n, $\lim_{x\to a} [f(x)]^n = L^n$. This is easily seen since

$$\begin{aligned}\lim_{x\to a} [f(x)]^n &= \lim_{x\to a} \Big\{ \underbrace{f(x)\cdot f(x)\cdots f(x)}_{n \text{ factors}} \Big\} \\ &= \underbrace{[\lim_{x\to a} f(x)][\lim_{x\to a} f(x)]\cdots[\lim_{x\to a} f(x)]}_{n \text{ factors}} \\ &= \underbrace{\mathrm{L}\cdot\mathrm{L}\cdots\mathrm{L}}_{n \text{ factors}} \\ &= L^n\end{aligned}$$

Proposition 2.5

If $\lim_{x\to a} f(x) = L$ and $\lim_{x\to a} g(x) = M \neq 0$, then

$$\lim_{x\to a} \left[\frac{f(x)}{g(x)}\right] = \frac{\lim_{x\to a} f(x)}{\lim_{x\to a} g(x)} = \frac{L}{M}$$

To illustrate these propositions, let us consider the following examples.

Example 2.40

Evaluate $\lim_{x\to 2} (3x + 4)$.

Solution

Using Proposition 2.1, we have

$$\begin{aligned}\lim_{x\to 2} (3x + 4) &= 3(2) + 4 \\ &= 10\end{aligned}$$

Example 2.41

Evaluate $\lim_{x\to 3}(2x^2+3x+7)$.

Solution

$$\lim_{x\to 3}(2x^2+3x+7)=\lim_{x\to 3}(2x^2)+\lim_{x\to 3}(3x)+\lim_{x\to 3}7 \qquad \textbf{(Proposition 2.3)}$$

$$\begin{aligned}\lim_{x\to 3}(2x^2) &= 2\lim_{x\to 3}x^2 && \textbf{(Proposition 2.2)}\\ &= 2\lim_{x\to 3}(x\cdot x)\\ &= 2(\lim_{x\to 3}x)(\lim_{x\to 3}x) && \textbf{(Proposition 2.4 Corollary)}\\ &= 2\cdot 3\cdot 3\\ &= 18\end{aligned}$$

$$\begin{aligned}\lim_{x\to 3}(3x) &= 3\lim_{x\to 3}x && \textbf{(Proposition 2.2)}\\ &= 3\cdot 3 = 9 && \textbf{(Proposition 2.1a)}\end{aligned}$$

and

$$\lim_{x\to 3}7 = 7 \qquad \textbf{(Proposition 2.1b)}$$

Thus,

$$\lim_{x\to 3}(2x^2+3x+7) = 18+9+7 = 34$$

Example 2.42

Evaluate $\lim_{x\to 4}[(x^2+3)(x+1)]$.

Solution

$$\lim_{x\to 4}[(x^2+3)(x+1)] = \lim_{x\to 4}(x^2+3)\cdot\lim_{x\to 4}(x+1) \qquad \textbf{(Proposition 2.4)}$$

Now,

$$\begin{aligned}\lim_{x\to 4}(x^2+3) &= \lim_{x\to 4}x^2+\lim_{x\to 3}3 && \textbf{(Proposition 2.3)}\\ &= 16+3 = 19 && \textbf{(Proposition 2.1)}\end{aligned}$$

and

$$\begin{aligned}\lim_{x\to 4}(x+1) &= \lim_{x\to 4}x+\lim_{x\to 4}1 && \textbf{(Proposition 2.3)}\\ &= 4+1 = 5 && \textbf{(Proposition 2.1)}\end{aligned}$$

Thus,

$$\lim_{x\to 4} [(x^2 + 3)(x + 1)] = (19)(5) = 95$$

Example 2.43

Evaluate $\lim_{x\to 2} \left[\dfrac{(4x^2 + 7x + 6)}{(2x^2 + x - 1)}\right]$.

Solution

$$\lim_{x\to 2} \left(\frac{4x^2 + 7x + 6}{2x^2 + x - 1}\right)$$

$$= \frac{\lim_{x\to 2} (4x^2 + 7x + 6)}{\lim_{x\to 2} (2x^2 + x - 1)} \qquad \textbf{(Proposition 2.5)}$$

$$= \frac{\lim_{x\to 2} (4x^2) + \lim_{x\to 2} (7x) + \lim_{x\to 2} 6}{\lim_{x\to 2} (2x^2) + \lim_{x\to 2} (x) - \lim_{x\to 2} 1} \qquad \textbf{(Proposition 2.3)}$$

$$= \frac{4 \lim_{x\to 2} x^2 + 7 \lim_{x\to 2} x + \lim_{x\to 2} 6}{2 \lim_{x\to 2} x^2 + \lim_{x\to 2} x - \lim_{x\to 2} 1} \qquad \textbf{(Proposition 2.2)}$$

$$= \frac{4(4) + 7(2) + 6}{2(4) + 2 - 1} \qquad \textbf{(Proposition 2.1)}$$

$$= \frac{16 + 14 + 6}{8 + 2 - 1}$$

$$= 4$$

The next examples introduce

$$\lim_{h\to 0} \left[\frac{f(x + h) - f(x)}{h}\right].$$

The interpretation of this limit for calculus is reserved for the next chapter.

Example 2.44

Let f be a function defined by

$$f(x) = 2x^2 + 5x + 3$$

If h is a real number and $h \neq 0$, evaluate

$$\lim_{h\to 0} \left[\frac{f(x + h) - f(x)}{h}\right]$$

Solution

Observe that

$$\begin{aligned} f(x+h) &= 2(x+h)^2 + 5(x+h) + 3 \\ &= 2(x^2 + 2xh + h^2) + 5(x+h) + 3 \\ &= 2x^2 + 4xh + 2h^2 + 5x + 5h + 3 \\ f(x) &= 2x^2 + 5x + 3 \\ f(x+h) - f(x) &= (2x^2 + 4xh + 2h^2 + 5x + 5h + 3) - (2x^2 + 5x + 3) \\ &= 4xh + 2h^2 + 5h \end{aligned}$$

Thus,

$$\begin{aligned} \frac{f(x+h) - f(x)}{h} &= \frac{4xh + 2h^2 + 5h}{h}, \qquad h \neq 0 \\ &= \frac{h(4x + 2h + 5)}{h}, \qquad h \neq 0 \\ &= 4x + 2h + 5 \end{aligned}$$

and

$$\begin{aligned} \lim_{h \to 0} \left[\frac{f(x+h) - f(x)}{h} \right] &= \lim_{h \to 0} (4x + 2h + 5) \\ &= \lim_{h \to 0} (4x + 5) + \lim_{h \to 0} (2h) \qquad \textbf{(Proposition 2.3)} \end{aligned}$$

Now

$$\lim_{h \to 0} (4x + 5) = 4x + 5 \qquad \textbf{(Proposition 2.1)}$$

$$\begin{aligned} \lim_{h \to 0} (2h) &= 2 \lim_{h \to 0} (h) \qquad \textbf{(Proposition 2.2)} \\ &= 2(0) = 0 \end{aligned}$$

We conclude that

$$\lim_{h \to 0} \left[\frac{f(x+h) - f(x)}{h} \right] = 4x + 5$$

Exercises

Using the limit propositions, evaluate each of the following problems.

1. $\lim_{x \to 1} (4x + 3)$
2. $\lim_{x \to -2} (3x + 11)$
3. $\lim_{x \to 3} (2x - 6)$
4. $\lim_{x \to 1} (2x + 5)$
5. $\lim_{x \to 2} (3x^2 + 2x + 1)$
6. $\lim_{x \to 4} (x^2 - 2x + 3)$

7. $\lim_{x \to -1} (x^3 + x^2 + x + 1)$

8. $\lim_{x \to -2} (2x^3 - 8x^2 + 3x + 4)$

9. $\lim_{x \to 2} (2x + 3)(3x + 4)$

10. $\lim_{x \to 3} (x^2 + x + 1)(x + 2)$

11. $\lim_{x \to 1} (x^2 + 2x + 1)(2x + 3)$

12. $\lim_{x \to 2} (x^3 + x - 2)(4x - 1)$

13. $\lim_{x \to 1} \left(\frac{4x + 5}{2x + 1}\right)$

14. $\lim_{x \to -2} \left(\frac{x}{3x + 8}\right)$

15. $\lim_{x \to 0} \left(\frac{2x^2 - 3x}{x}\right)$

16. $\lim_{x \to 0} \left(\frac{x^3 + x^2 + x}{2x}\right)$

17. $\lim_{x \to 2} \left(\frac{x^3 - 8}{x - 2}\right)$

18. $\lim_{x \to 3} \left(\frac{3x^3 - 81}{x - 3}\right)$

19. $\lim_{x \to 3} \left(\frac{x^2 - 7x + 12}{x - 3}\right)$

20. $\lim_{x \to -2} \left(\frac{x^2 - 4}{x + 2}\right)$

Evaluate the limit, if it exists, in each of the following problems.

21. $\lim_{x \to 2} \left(\frac{x^2 - 4}{x - 2}\right)$

22. $\lim_{x \to 1} \left(\frac{x^2 - 3x + 2}{x - 1}\right)$

23. $\lim_{x \to 3} \left(\frac{x^2 - 4x + 3}{x - 3}\right)$

24. $\lim_{x \to 4} \left(\frac{x^2 - 6x + 8}{x - 4}\right)$

25. $\lim_{x \to 2} \left(\frac{x^2 - x - 2}{x - 2}\right)$

26. $\lim_{x \to 1} \left(\frac{x^3 - 1}{x - 1}\right)$

27. $\lim_{x \to 2} \frac{|x - 2|}{x - 2}$

28. $\lim_{x \to 1} \frac{x - 2}{|x - 2|}$

29. $\lim_{x \to 2} \left(\frac{x^2 - 6x + 8}{x - 2}\right)$

30. $\lim_{x \to 2} \left(\frac{x + 3}{x - 2}\right)$

Evaluate

$$\lim_{h \to 0} \left[\frac{f(x + h) - f(x)}{h}\right]$$

in each of the following problems.

31. $f(x) = 5x + 4$

32. $f(x) = 3 - 4x$

33. $f(x) = 3x + 2$

34. $f(x) = 7x + 1000$

35. $f(x) = x^2 + 4x$

36. $f(x) = 3x^2 + 7x$

37. $f(x) = x^3$

38. $f(x) = 2x^3 + x^2 + x + 1$

2.7 Limits as x Approaches Infinity

We have so far evaluated $\lim_{x \to a} f(x)$ where a is a fixed real number. What is the limit of

$$f(x) = \frac{2x + 3}{x + 4}$$

as x increases without bound and approaches positive infinity, written $x \to +\infty$? Clearly, the domain of f is the set of all real numbers except $x = -4$. We can show intuitively that by choosing x large enough $f(x)$ can be made as close to 2 as we please. Consider, for example, the values of $f(x)$ for sufficiently large values of x. Note that

$$\begin{array}{rr} f(10) = 1.6428571 & f(1000) = 1.9950199 \\ f(50) = 1.9074074 & f(5000) = 1.9990007 \\ f(100) = 1.9519230 & f(10000) = 1.9995001 \\ f(500) = 1.9900793 & f(100000) = 1.9999500 \end{array}$$

Thus as x gets larger and larger, $f(x)$ approaches the number 2. Hence, we conclude that

$$\lim_{x \to +\infty} f(x) = 2$$

We may also be interested in finding the limit of $f(x)$ as x decreases without bound and approaches negative infinity, written $x \to -\infty$. What is the limit, for instance, of

$$f(x) = \frac{2x + 3}{x + 4}$$

when $x \to -\infty$? Observe that

$$\begin{array}{rr} f(-10) = 2.8333333 & f(-1000) = 2.0050200 \\ f(-50) = 2.1086956 & f(-5000) = 2.0010008 \\ f(-100) = 2.0520833 & f(-10000) = 2.0005002 \\ f(-500) = 2.0100806 & f(-100000) = 2.0000500 \end{array}$$

This means that $f(x)$ can be made as close to 2 as we like by choosing smaller and smaller negative values of x. In other words,

$$\lim_{x \to -\infty} \left(\frac{2x + 3}{x + 4} \right) = 2.$$

We now state a proposition that is useful in evaluating $\lim f(x)$ when x approaches positive or negative infinity.

Proposition 2.6

Let c be a constant and n any positive real number. Then

a. $\lim_{x \to +\infty} \left(\frac{c}{x^n} \right) = 0$

b. $\lim_{x \to -\infty} \left(\frac{c}{x^n} \right) = 0$

To apply this proposition, consider again the function

$$f(x) = \frac{2x+3}{x+4}$$

Since the numerator and denominator do not have limits as x approaches positive or negative infinity, we cannot use the limit proposition 2.5 for quotients. However, we simplify the problem by dividing the numerator and denominator by the highest power of x in the fraction, in this case x^1 or x, and obtain

$$\begin{aligned}\lim_{x\to+\infty}\left(\frac{2x+3}{x+4}\right) &= \lim_{x\to+\infty}\left(\frac{2+3/x}{1+4/x}\right)\\ &= \frac{\lim_{x\to+\infty}(2+3/x)}{\lim_{x\to+\infty}(1+4/x)} \qquad \textbf{(Proposition 2.5)}\\ &= \frac{2+0}{1+0} \qquad \textbf{(Proposition 2.6)}\\ &= 2\end{aligned}$$

Example 2.45

Evaluate $\lim_{x\to+\infty} f(x)$ where

$$f(x) = \sqrt{x^2+x} - x, \qquad x > 0$$

Solution

For $x > 0$, we can rationalize* $f(x)$ and obtain

$$\begin{aligned}\sqrt{x^2+x} - x &= \frac{\sqrt{x^2+x}-x}{1}\,\frac{\sqrt{x^2+x}+x}{\sqrt{x^2+x}+x}\\ &= \frac{(x^2+x)-x^2}{\sqrt{x^2+x}+x}\\ &= \frac{x}{\sqrt{x^2+x}+x}\\ &= \frac{x}{\sqrt{x^2(1+1/x)}+x}\\ &= \frac{x}{x\sqrt{1+1/x}+x}\\ &= \frac{x}{x[\sqrt{1+1/x}+1]}\\ &= \frac{1}{\sqrt{1+1/x}+1}\end{aligned}$$

* A brief review of rationalizing expressions is given in Appendix A.

Now

$$\lim_{x\to+\infty} (\sqrt{x^2+x}-x) = \lim_{x\to+\infty} \left(\frac{1}{\sqrt{1+1/x}+1}\right)$$
$$= \frac{1}{\sqrt{1+0}+1}$$
$$= \tfrac{1}{2}$$

Example 2.46

Evaluate $\lim_{x\to+\infty} f(x)$ where

$$f(x) = \frac{ax^2+bx+c}{dx^2+ex+f}$$

where $a, b, c, d, e,$ and f are real numbers and $d \neq 0$, $e \neq 0$ and $f \neq 0$.

Solution

Dividing both the numerator and denominator by x^2, the *highest power of x in the fraction*, we obtain

$$\lim_{x\to+\infty} f(x) = \lim_{x\to+\infty} \left(\frac{a+b/x+c/x^2}{d+e/x+f/x^2}\right)$$
$$= \frac{\lim_{x\to+\infty}(a+b/x+c/x^2)}{\lim_{x\to+\infty}(d+e/x+f/x^2)}$$
$$= \frac{a+0+0}{d+0+0}$$
$$= \frac{a}{d}, \qquad d \neq 0$$

Example 2.47

Evaluate $\lim_{x\to+\infty} f(x)$ where

$$f(x) = \left(\frac{ax^2+b}{cx+d}\right)$$

where a, b, c, and d are real numbers and $c \neq 0$ and $d \neq 0$.

Solution

Dividing both the numerator and the denominator by x^2, the highest power of x in the fraction, we obtain

$$\lim_{x\to+\infty} f(x) = \left(\frac{a+b/x^2}{c/x+d/x^2}\right)$$

As $x \to +\infty$, the denominator approaches 0. Since a zero value would violate Proposition 2.5, we conclude that $\lim_{x \to +\infty} f(x)$ does not exist.

The limit of a function, as x increases or decreases without bound, can be used to determine *horizontal asymptotes* which in turn can be used to sketch the graph of a rational function. To graph the rational function

$$r(x) = \frac{4x + 1}{2x - 5}$$

for instance, we divide both the numerator and the denominator by the highest power of x in the fraction, in this case x^1 or x, and write

$$r(x) = \frac{4 + 1/x}{2 - 5/x}$$

Since

$$\lim_{x \to +\infty} r(x) = 2 \quad \text{and} \quad \lim_{x \to -\infty} r(x) = 2$$

we conclude that $y = 2$ is a horizontal asymptote. Remember that if

$$r(x) = \frac{f(x)}{g(x)}, \qquad g(x) \neq 0$$

and $\lim_{x \to +\infty} r(x) = k$ or $\lim_{x \to -\infty} r(x) = k$, where k is any fixed real number, then $y = k$ is a horizontal asymptote of the graph of $r(x)$.

Exercises

Determine the limit, if it exists, in each of the following problems.

1. $\lim_{x \to +\infty} \left(\frac{2x + 3}{x} \right)$

2. $\lim_{x \to +\infty} \left(\frac{x + 1}{x - 1} \right)$

3. $\lim_{x \to +\infty} \left(\frac{x - 5}{2x + 3} \right)$

4. $\lim_{x \to +\infty} \left(\frac{2x - 5}{x^2} \right)$

5. $\lim_{x \to +\infty} \left(\frac{x + 3}{x - 2} \right)$

6. $\lim_{x \to +\infty} \left(\frac{2x}{3x + 1} \right)$

7. $\lim_{x \to +\infty} \left(\frac{3x + 2}{2x + 3} \right)$

8. $\lim_{x \to -\infty} \left(\frac{3x + 2}{2x^2 + 3} \right)$

9. $\lim_{x \to -\infty} \left(\frac{2x^2 + 3}{3x + 2} \right)$

10. $\lim_{x \to -\infty} \left(\frac{x^2 + 6x + 1}{3x^2 + 4x + 8} \right)$

11. $\lim_{x \to +\infty} \left(\frac{2x + 1}{2x^2 + 3x + 1} \right)$

12. $\lim_{x \to -\infty} \left(\frac{4x^2}{x^3 + 1} \right)$

13. $\lim_{x \to +\infty} \left(\frac{x^2 + 4}{x + 4}\right)$

14. $\lim_{x \to +\infty} \left(\frac{3x + 4}{x^2 + 2x + 1}\right)$

15. $\lim_{x \to -\infty} \left(\frac{x + 1}{x + 2}\right)^2$

16. $\lim_{x \to -\infty} \left(\frac{2x - 1}{3x + 2}\right)^2$

17. $\lim_{x \to +\infty} \left(\frac{4x + 3}{2x + 1}\right)^3$

18. $\lim_{x \to +\infty} \left(\frac{x^2 - x + 3}{2x^2 + 5x + 1}\right)^2$

19. $\lim_{x \to +\infty} \left(\frac{4x^3 + 3x^2 + 2x + 5}{8x^3 + 16x^2 - x + 21}\right)^2$

20. $\lim_{x \to +\infty} \left(\frac{2x^2 - x + 4}{x^4 + x^2 + 1}\right)^3$

21. $\lim_{x \to +\infty} (\sqrt{x^2 - 1} - x)$

22. $\lim_{x \to +\infty} (\sqrt{x^2 + 1} - x)$

23. $\lim_{x \to +\infty} (\sqrt{x^2 + x} - x)$

24. $\lim_{x \to +\infty} (\sqrt{x^2 - x} - x)$

25. Evaluate $\lim_{x \to +\infty} f(x)$ where

$$f(x) = \frac{ax^2 + bx + c}{dx^2 + ex + f}$$

and a, b, c, d, e, and f are real numbers, and $e \neq 0$, $f \neq 0$ but $d = 0$.

26. The market price for x gallons of crude oil is given by

$$p(x) = \$\left(\frac{x^2 + 1}{2x^2 + 3}\right)$$

Determine the price $p(x)$ when x gets larger and larger.

27. The demand function of a certain product is given by

$$f(x) = \frac{400}{x^2}$$

where $f(x)$ represents the demand (in thousands of units) per week and x denotes the price (in dollars) per unit. Determine the demand $f(x)$ when x gets larger and larger.

28. An increase in the number of bacteria in a certain culture follows approximately the function

$$f(t) = \frac{10000t^2}{t^2 + 10}, \qquad t \geq 0$$

Determine $\lim_{t \to +\infty} f(t)$.

29. The potency of a certain drug follows approximately the function

$$f(t) = \frac{t + 50}{t^2 + 1}$$

where $f(t)$ represents the number of milligrams per cubic centimeter and t is the elapsed storage time in months. Determine the potency when t gets larger and larger. Is the result surprising?

30. A known quantity of a new drug is injected into the bloodstream of a patient and the concentration $f(t)$ of the drug is recorded continuously for t hours after injection. If the function f is given by

$$f(t) = \frac{10t}{2t^2 + 5t + 8} \qquad t \geq 0$$

determine the percentage of concentration of the drug in the bloodstream after (a) $t = 1$ hour, (b) $t = 2$ hours, and (c) $t = 4$ hours. Determine also $\lim_{t \to +\infty} f(t)$.

Determine the horizontal asymptotes for each of the following problems.

31. $f(x) = \dfrac{2x}{x + 1}$

32. $f(x) = \dfrac{3x + 5}{x - 1}$

33. $f(x) = \dfrac{x^3 + 1}{3x^2 + 4}$

34. $f(x) = \dfrac{x^2 + x + 1}{2x^2 + 3x + 8}$

2.8 Continuity

In Sections 2.5 and 2.6 we analyzed the meaning of

$$\lim_{x \to a} f(x) = L$$

and in so doing we paid no attention to the value of the function at $x = a$. In fact the function was not even defined at $x = a$ for several expressions. In this section, the limit of $f(x)$ as x approaches a and the value of the function at $x = a$ both play a significant role.

Consider an airline that charges full fare for passengers 12 years of age or over, one-half fare for those under 12, and one-tenth fare for infants under 2. In symbols,

$$f(x) = \begin{cases} \frac{1}{10} & 0 < x < 2 \\ \frac{1}{2} & 2 \leq x < 12 \\ 1 & x \geq 12 \end{cases}$$

The graph of $f(x)$ is shown in Figure 2.36. Note that

$$\lim_{x \to 2^-} f(x) = \tfrac{1}{10} \quad \text{and} \quad \lim_{x \to 2^+} f(x) = \tfrac{1}{2}$$

Hence, we conclude that $\lim_{x \to 2} f(x)$ does not exist. Similarly, $\lim_{x \to 12} f(x)$ does not exist. Figure 2.36 shows a jump or a break in the graph

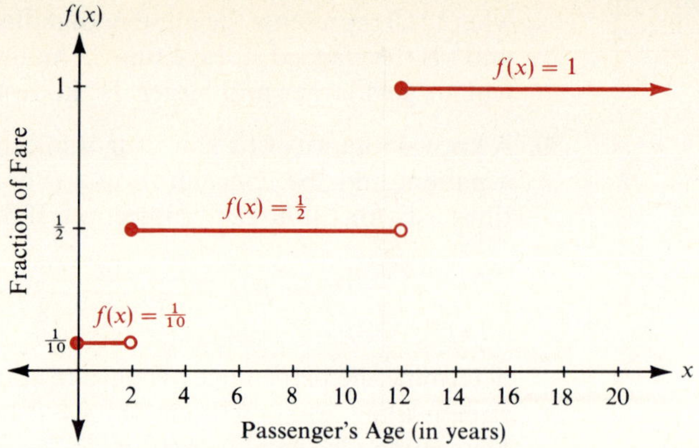

Figure 2.36

both at $x = 2$ and $x = 12$. Formally, we say that the function is *discontinuous* at these points. Intuitively, we say that a function is continuous if we can draw its graph without taking the pencil off the paper.

Another example of a discontinuous function is given by the following:

$$f(x) = \frac{x^2 - 4}{x - 2}, \qquad x \neq 2$$

Note that the function is not defined at 2. Using the methods explained in earlier sections, we find that

$$\begin{aligned}\lim_{x \to 2} \left(\frac{x^2 - 4}{x - 2}\right) &= \lim_{x \to 2} \frac{(x + 2)(x - 2)}{(x - 2)} \\ &= \lim_{x \to 2} (x + 2) \\ &= 4\end{aligned}$$

Thus the limit of $f(x)$ exists at $x = 2$. However, the function is discontinuous at $x = 2$, because its graph (see Figure 2.37) has a break at this point and the function is not defined at $x = 2$. But if we define $f(2) = 4$, then the function is continuous at $x = 2$, and if we define $f(2)$ to have any value other than 4, the function will be discontinuous at $x = 2$. This discussion leads us to the following definition.

Definition 2.11

A function f is *continuous* at $x = a$ if

$$\lim_{x \to a} f(x) = f(a)$$

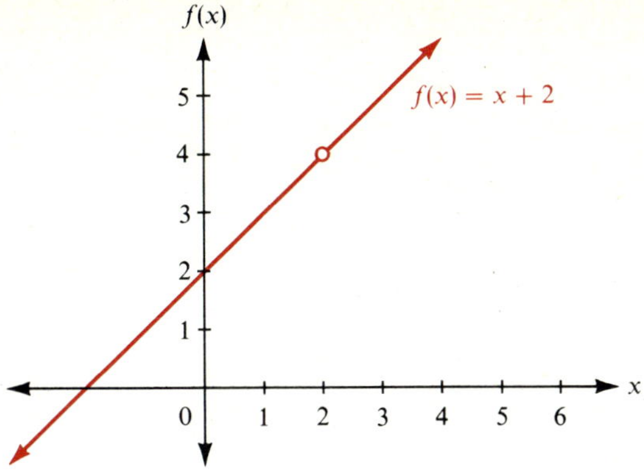

Figure 2.37

This means that in order for a function $f(x)$ to be continuous at $x = a$, the following three conditions must be satisfied.

1. $f(a)$ is defined
2. $\lim_{x \to a} f(x)$ exists
3. $\lim_{x \to a} f(x) = f(a)$

If $f(x)$ is not continuous at $x = a$, we say that $f(x)$ is discontinuous at $x = a$, or that there is a discontinuity at point a.

Definition 2.12

A function $f(x)$ is said to be *continuous over an open interval* if it is continuous at *every* point in that interval.

Let us consider a few more examples to reinforce the concept of continuity.

Example 2.48

Let

$$f(x) = \begin{cases} 3x + 2, & x \neq 2 \\ 8, & x = 2 \end{cases}$$

Determine whether or not $f(x)$ is continuous at $x = 2$.

Solution

Note that

$$\lim_{x \to 2} (3x + 2) = 3(2) + 2 = 8 \qquad \textbf{(Proposition 2.1)}$$

Because $f(2)$ is defined and also equals 8, we conclude that $f(x)$ is continuous at $x = 2$.

Example 2.49

Let $f(x)$ be defined as follows:

$$f(x) = \begin{cases} 6x, & 0 < x < 5 \\ 25 + 2x, & x \geq 5 \end{cases}$$

Is $f(x)$ continuous at $x = 5$?

Solution

If x approaches 5 through values of $x < 5$, $f(x)$ gets closer and closer to 30. This means that

$$\lim_{x \to 5^-} f(x) = 30$$

If x approaches 5 through values of $x > 5$, $f(x)$ gets closer and closer to 35. That is,

$$\lim_{x \to 5^+} f(x) = 35$$

Because

$$\lim_{x \to 5^-} f(x) \neq \lim_{x \to 5^+} f(x)$$

we conclude that $\lim_{x \to 5} f(x)$ does not exist. The second condition in Definition 2.11 is violated and hence $f(x)$ is discontinuous at $x = 5$.

Example 2.50

Let $f(x)$ be defined as follows.

$$f(x) = \begin{cases} x + 1, & x < 2 \\ 1, & x = 2 \\ 5 - x, & x > 2 \end{cases}$$

Graph $f(x)$ and determine whether or not $f(x)$ is continuous at $x = 2$.

Solution

The graph of $f(x)$ is shown in Figure 2.38. Note that

$$\lim_{x \to 2^-} (x + 1) = 3 \quad \text{and} \quad \lim_{x \to 2^+} (5 - x) = 3$$

Thus, $\lim_{x \to 2} f(x) = 3$. Because $f(2) \neq 3$, the function is discontinuous at $x = 2$. However, if the definition of f was originally stated so that $f(2) = 3$, then

$$\lim_{x \to 2} f(x) = f(2)$$

and the function would be continuous at $x = 2$.

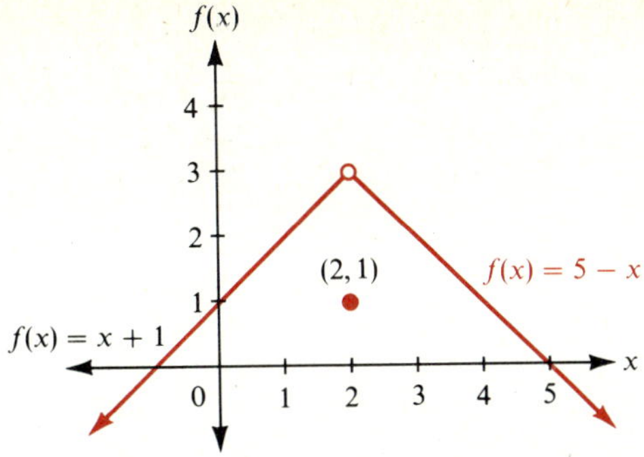

Figure 2.38

Continuous functions have certain properties which we state below. Their proofs, however, are omitted.

Proposition 2.7 If $f(x) = c$, where c is a constant, then $f(x)$ is continuous for all x.

Proposition 2.8 If $f(x) = x^n$ where n is any positive integer, then $f(x)$ is continuous for all x.

Proposition 2.9 If $f(x)$ is continuous at any point x_0 in its domain and c is a constant, then $cf(x)$ is also continuous at x_0.

Proposition 2.10 If $f(x)$ and $g(x)$ are continuous at a point x_0, then the functions $F(x)$, $G(x)$, and $H(x)$ defined by

$$F(x) = f(x) + g(x)$$
$$G(x) = f(x) - g(x)$$
$$H(x) = f(x) \cdot g(x)$$

are also continuous at x_0.

This means that the sum, difference, and product of two continuous functions is continuous.

Proposition 2.11 Any polynomial function

$$p(x) = a_n x^n + a_{n-1} x^{n-1} + \cdots + a_2 x^2 + a_1 x + a_0$$

where n is a positive integer or zero and $a_0, a_1, a_2, \ldots, a_n$ are real numbers is continuous for all x.

Proposition 2.12 If $p(x)$ and $q(x)$ are polynomial functions in x, then the rational function

$$R(x) = \frac{p(x)}{q(x)}$$

is continuous at x_0 if and only if $q(x_0) \neq 0$.

As a simple illustration, recall that $f(x) = 1/x$ is not defined at $x = 0$ and hence is discontinuous at that point.

Exercises

In problems 1 through 6, determine whether or not the function is continuous at the point indicated.

1. $f(x) = \begin{cases} 2x + 3, & x \leq 2 \\ 7, & x > 2 \end{cases}; \qquad x = 2$

2. $f(x) = \begin{cases} 3x + 1, & x \leq 3 \\ 4x - 2, & x > 3 \end{cases}; \qquad x = 3$

3. $f(x) = \begin{cases} 3x + 2, & x \leq 1 \\ 5x, & x > 1 \end{cases}; \qquad x = 1$

4. $f(x) = \begin{cases} x^2, & x \leq 0 \\ x, & x > 0 \end{cases}; \qquad x = 0$

5. $f(x) = |x - 1|; \qquad x = 1$

6. $f(x) = |x - 3|; \qquad x = 3$

In problems 7 through 12, determine whether or not the function is continuous at all points in the domain.

7. $f(x) = \begin{cases} 2x + 3, & x \neq 3 \\ 9, & x = 3 \end{cases}$

8. $f(x) = \begin{cases} x^2 + 2, & x \neq 1 \\ 3, & x = 1 \end{cases}$

9. $f(x) = \begin{cases} x + 2, & x \leq 3 \\ x - 2, & x > 3 \end{cases}$

10. $f(x) = |2x - 1|$

11. $f(x) = \begin{cases} \left(\dfrac{x^2 - 1}{x - 1}\right), & x \neq 1 \\ 3, & x = 1 \end{cases}$

12. $f(x) = \begin{cases} \left(\dfrac{x^2 - 3x + 2}{x - 2}\right), & x \neq 2 \\ 2, & x = 2 \end{cases}$

In problems 13 through 16, determine all values of x for which the function is discontinuous.

13. $f(x) = \dfrac{1}{x - 1}$

14. $f(x) = \dfrac{1}{(x - 2)(x - 3)}$

15. $f(x) = \dfrac{x^2 - 1}{x^2 - 7x + 12}$

16. $f(x) = \dfrac{(x - 3)}{x^2 + 5x + 6}$

17. A cost function is defined by

$$C(x) = \begin{cases} \$5x, & 0 < x < 5 \\ \$20 + x, & 5 \leq x < 10 \\ \$15 + \frac{3}{2}x, & 10 < x \end{cases}$$

where x is the number of pounds of a certain material sold. Determine whether or not $C(x)$ is continuous at $x = 5$ and $x = 10$.

18. A moving van line offers a discount in freight rates on a large shipment. The cost function is

$$C(x) = \begin{cases} \$0.08x, & 0 < x \leq 250 \\ \$10 + 0.03x, & 250 < x \leq 500 \\ \$12 + 0.02x, & 500 < x \end{cases}$$

where x is the number of pounds of goods shipped and $C(x)$ is the cost of shipping. Determine the points where $C(x)$ is discontinuous.

19. An attendant in a parking lot charges \$1.00 for the first hour and \$0.50 for each additional hour or part thereof subject to a maximum of \$4.00. Sketch a graph showing the parking rates and determine the points at which the graph is discontinuous.

20. The current postage rate is \$0.15 up to and including one ounce and \$0.13 for each additional ounce or fraction thereof up to 10 ounces. Sketch a graph showing the postage for a letter weighing up to 7 ounces and determine the points at which the graph is discontinuous.

Exercises for Review

1. Define each of the following concepts:
   a. function
   b. one-to-one function
   c. inverse function
   d. composite function
   e. limit of a function
   f. continuous function

2. Given that
$$f(x) = x^2 - 3x + 4$$
determine each of the following:
   a. $f(0)$
   b. $f(-2)$
   c. $f(2)$
   d. $f(3)$
   e. $f(5)$
   f. $\dfrac{f(a+h) - f(a)}{h}$

3. Given that
$$f(x) = \frac{x^2 + 1}{x + 2}$$
determine each of the following:
   a. $f(0)$
   b. $f(1)$
   c. $f(2)$
   d. $f(-1)$
   e. Does $f(-2)$ exist?

4. Determine the largest subset of real numbers that can serve as the domain of the following functions:
   a. $f(x) = 3x + 4$
   b. $f(x) = x^2 + 3$
   c. $f(x) = \sqrt{x - 2}$
   d. $f(x) = \sqrt{4 - x^2}$

5. For each of the following functions, determine the largest subset of real numbers that can serve as the domain and then find the range:
   a. $f(x) = 2x + 3$
   b. $f(x) = |x - 1|$
   c. $f(x) = \sqrt{16 - x^2}$
   d. $f(x) = \dfrac{1}{x - 2}$

6. Graph each of the following functions:
   a. $f(x) = x^2 + 1$
   b. $f(x) = (x + 1)^2$
   c. $f(x) = 1 - x^2$
   d. $f(x) = |x - 2|$
   e. $f(x) = \sqrt{x - 3}$
   f. $f(x) = \sqrt{16 - x^2}$
   g. $f(x) = \dfrac{1}{x - 2}$
   h. $f(x) = x^3$

7. A charter flight charges a fare of \$200 per person plus \$5 for each unsold seat on the plane. The plane has a seating capacity of 100 passengers.
   a. Determine the total revenue, $R(x)$, as a function of x, the number of empty seats on the plane, and give the domain of the function.
   b. Calculate $R(x)$ for $x = 10, 20, 30, 40, 50, 60, 70,$ and 80.

c. Plot $R(x)$.
d. Determine the value for x for which $R(x)$ is maximum.

8. Determine $g[f(x)]$ and $f[g(x)]$ for each of the following functions.
a. $f(x) = x^2 + x + 1, \quad g(x) = x - 3$
b. $f(x) = x + 1, \quad g(x) = x^2$
c. $f(x) = x^2 - 9, \quad g(x) = \sqrt{x + 9}$
d. $f(x) = x^2, \quad g(x) = \dfrac{x}{x + 1}$

9. Determine the inverse function f^{-1} for each of the following functions.
a. $f(x) = \dfrac{x + 3}{2}$
b. $f(x) = 2x - 3$
c. $f(x) = \sqrt{2x - 3}, \quad x \geq \frac{3}{2}$
d. $f(x) = x^3 + 2$

10. Find the limit, if it exists, for each of the following functions.
a. $\lim\limits_{x \to 3} \dfrac{x^2 - 9}{x - 3}$
b. $\lim\limits_{x \to -1} \dfrac{x^3 + x^2 + x + 1}{x + 1}$
c. $\lim\limits_{x \to 1} \dfrac{|x - 1|}{x - 1}$
d. $\lim\limits_{x \to +\infty} \dfrac{3x + 4}{4x - 1}$
e. $\lim\limits_{x \to +\infty} \dfrac{x^2 + x + 1}{x^3 + x^2 + x + 1}$
f. $\lim\limits_{x \to +\infty} (\sqrt{x^2 + x} - x)$

11. Evaluate

$$\lim_{h \to 0} \left[\frac{f(x + h) - f(x)}{h}\right]$$

for each of the following functions.
a. $f(x) = 5x$
b. $f(x) = 4 - 3x$
c. $f(x) = x^2 + 4$
d. $f(x) = x^3 + 2$

12. A cost function is defined by

$$C(x) = \begin{cases} \$2400, & 0 < x \leq 40 \\ \$80x - \frac{1}{2}x^2, & 40 < x \leq 80 \\ \$40x, & 80 < x \end{cases}$$

where x is the number of tons of certain material purchased. Discuss whether or not $C(x)$ is continuous at $x = 40$ and $x = 80$.

13. Determine all values of x for which each of the following functions is discontinuous.
a. $f(x) = \dfrac{3}{x - 4}$
b. $f(x) = \dfrac{1}{x^2 - 1}$
c. $f(x) = \dfrac{x - 2}{x^2 + 3x + 2}$
d. $f(x) = \dfrac{x^2 - 1}{x - 2}$

14. A cost function is defined by

$$C(x) = \begin{cases} \$0.60x & 0 < x \le 100 \\ \$0.50x & 100 < x \le 500 \\ \$0.40x & 500 < x \end{cases}$$

where x is the number of pounds of goods shipped and $C(x)$ is the cost of shipping. Graph $C(x)$ and determine the points for which $C(x)$ is discontinuous.

15. Give an example of a function which is continuous for all points in its domain.

16. Give an example of a function which is discontinuous for some points in its domain.

3 The Derivative

3.1 Introduction

In this chapter we shall apply the concept of limit to introduce the idea of derivative. The derivative is a useful tool in constructing mathematical models that have meaningful applications. For example, if the criterion is to choose the highest return on our financial investment, we can develop a model which will tell us what the return is likely to be if a certain action is taken. The same rationale applies to problems involving optimum selling prices or minimum production costs.

3.2 Instantaneous Rate of Change

The derivative measures the *rate* at which change is taking place—the rate at which enrollment is dropping in a liberal arts college, the rate at which the number of bacteria is increasing in a certain culture, the rate at which crime is increasing in an urban population, the rate at which a drug injected into the bloodstream of a patient is losing its effectiveness, the rate at which a coal-burning electrical generating plant is polluting the surrounding air, the rate at which a student is learning a new language, the rate at which the purchasing power of the dollar is dwindling, the rate at which the production of compact cars is increasing, and so on.

Consider, for example, a manufacturer whose revenue (in millions of dollars) is given by the function

3.1 $$f(x) = 20x - x^2, \qquad 0 \le x \le 20$$

where x is the number of articles (in thousands) that the firm produces. The graph of $f(x)$ is shown in Figure 3.1.

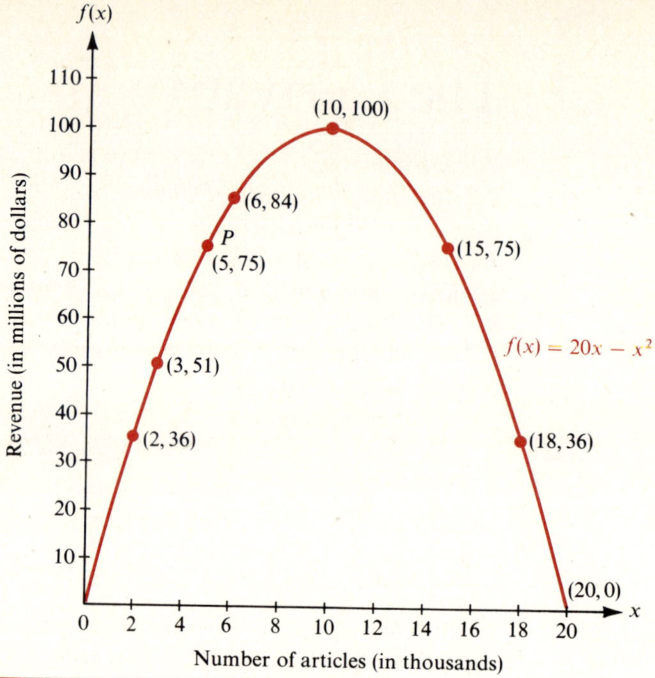

Figure 3.1

That x cannot exceed 20 is an indication of the production capacity of the plant. We wish to study the behavior of this function in the vicinity of some particular value of x; specifically, whether or not the revenue is increasing or decreasing at that particular level of production. This kind of information would be essential for an industry planning to expand its operations, because it would not be economic to expand beyond this level. In general,

$$\frac{f(b) - f(a)}{b - a}$$

which is known as the *difference quotient*, may be used to find the *average* rate of change of a function.

Consider the value of the function at $x = 5$. Note that

$$f(5) = 20(5) - 5^2 = 75$$

If the production is increased to $x = 6$, then the corresponding revenue is

$$f(6) = 20(6) - 6^2 = 84$$

Thus the revenue increases by 9 when the production level is increased from $x = 5$ to $x = 6$. We say that the average rate of increase in revenue is

$$\frac{f(6) - f(5)}{6 - 5} = \frac{84 - 75}{1} = 9 \text{ units}$$

Since we are primarily interested in the behavior of $f(x)$ in the neighborhood of the point $P(5, 75)$, let us see what happens when the production level increases by a smaller amount, say, from $x = 5$ to $x = 5.5$. Note that

$$f(5.5) = 20(5.5) - (5.5)^2 = 79.75$$

Thus when production increases from $x = 5$ to $x = 5.5$, the average rate of increase in revenue is

$$\frac{f(5.5) - f(5)}{5.5 - 5.0} = \frac{79.75 - 75}{5.5 - 5.0} = 9.50 \text{ units}$$

What is the average rate of increase in revenue when production increases from $x = 5$ to $x = 5.1$? Since $f(5) = 75$, we need to compute $f(5.1)$. Substituting $x = 5.1$ in (3.1), we obtain

$$f(5.1) = 20(5.1) - (5.1)^2 = 75.99$$

Thus the average rate of increase for this interval is

$$\frac{f(5.1) - f(5.0)}{5.1 - 5.0} = \frac{75.99 - 75}{0.1} = 9.90 \text{ units}$$

We have observed that an increase in production from $x = 5$ to $x = 6$ is associated with the average rate of increase in revenue of 9 units; an increase from $x = 5$ to $x = 5.5$ results in an average rate of increase in revenue of 9.5 units; and an increase from $x = 5$ to $x = 5.1$ corresponds to an average rate of increase in revenue of 9.9 units. To get a still better approximation, we let production increase from $x = 5$ to $x = 5 + \Delta x$, where $\Delta x > 0$, and examine the corresponding average rate of increase in revenue as $\Delta x \to 0^+$. This is shown in Table 3.1.

Table 3.1
$f(x) = 20x - x^2$

Δx	$5 + \Delta x$	$f(5 + \Delta x)$	$\Delta y = f(5 + \Delta x) - f(5)$	$\frac{\Delta y}{\Delta x}$
1	6	84	9	9
0.5	5.5	79.75	4.75	9.50
0.1	5.1	75.99	0.99	9.90
0.01	5.01	75.0999	0.0999	9.99
0.001	5.001	75.009999	0.009999	9.999
0.0001	5.0001	75.00099999	0.00099999	9.9999

The notation Δx (Δ is pronounced "delta") indicates a change in the variable x; it does not mean Δ times x. The ratio $\Delta y/\Delta x$ in the last column reflects the average rate of increase in the production level from $x = 5$ to $x = 5 + \Delta x$. As the increase Δx in the production level gets smaller and smaller, the corresponding increase in revenue, Δy, also gets smaller. The difference quotient

$$\frac{\Delta y}{\Delta x} = \frac{f(5 + \Delta x) - f(5)}{\Delta x}$$

gets closer and closer to 10 as $\Delta x \to 0^+$. This is equivalent to the statement that the right-hand limit of $\Delta y/\Delta x$ equals 10; that is,

$$\lim_{\Delta x \to 0^+} \left(\frac{\Delta y}{\Delta x}\right) = \lim_{\Delta x \to 0^+} \left[\frac{f(5 + \Delta x) - f(5)}{\Delta x}\right] = 10$$

Now we let the production level change from $x = 5$ to $x = 5 + \Delta x$ where $\Delta x < 0$ and investigate the behavior of $\Delta y/\Delta x$ as Δx approaches 0 from the left. This is shown in Table 3.2.

Table 3.2
$f(x) = 20x - x^2$

Δx	$5 + \Delta x$	$f(5 + \Delta x)$	$\Delta y = f(5 + \Delta x) - f(5)$	$\left(\frac{\Delta y}{\Delta x}\right)$
−1	4	64	−11	11
−0.5	4.5	69.75	−5.25	10.50
−0.1	4.9	73.99	−1.01	10.10
−0.01	4.99	74.8999	−0.1001	10.01
−0.001	4.999	74.989999	−0.010001	10.001
−0.0001	4.9999	74.99899999	−0.00100001	10.0001

As Δx gets closer and closer to 0 from the left, the difference quotient $\Delta y/\Delta x$ gets closer and closer to 10. This means that the left-hand limit of $\Delta y/\Delta x$ is 10; that is,

$$\lim_{\Delta x \to 0^-} \left(\frac{\Delta y}{\Delta x}\right) = \lim_{\Delta x \to 0^-} \left[\frac{f(5 + \Delta x) - f(5)}{\Delta x}\right] = 10$$

Since both the right- and left-hand limits exist and are equal, we conclude that

$$\lim_{\Delta x \to 0} \left(\frac{\Delta y}{\Delta x}\right) = \lim_{\Delta x \to 0} \left[\frac{f(5 + \Delta x) - f(5)}{\Delta x}\right] = 10$$

and this represents the *instantaneous rate of change* in the revenue function $f(x)$ at $x = 5$. To determine the instantaneous rate of change in $f(x)$ at some other level of production, say $x = 8$, we need to repeat the above steps using $x = 8$ instead of $x = 5$. However, we will develop a general definition of the instantaneous rate of change at any arbitrary point $(x, f(x))$.

When production level changes from a fixed level x to the level $x + \Delta x$, the corresponding change in revenue is

$$\Delta y = f(x + \Delta x) - f(x)$$

and the average rate of change in revenue is

$$\frac{\Delta y}{\Delta x} = \frac{f(x + \Delta x) - f(x)}{\Delta x}.$$

To get the instantaneous rate of change in revenue, we examine the limit of $\Delta y/\Delta x$ as Δx approaches zero.

Definition 3.1

If $y = f(x)$, then the derivative of the function f at the point x is defined by

3.2 $$\lim_{\Delta x \to 0} \left[\frac{f(x + \Delta x) - f(x)}{\Delta x}\right]$$

The limit is generally denoted by $f'(x)$ (read "f prime of x,") or by dy/dx.

The derivative dy/dx or $f'(x)$ is a another function defined for every value of x for which $f(x)$ is defined and for which the limit in (3.2) exists. The process of finding the derivative of a function is called *differentiation.*

Note that dy/dx, which is a symbol for $\lim_{\Delta x \to 0} (\Delta y/\Delta x)$, represents the fact that we are differentiating a variable y with respect to a variable x. Other notations in common use are

$$D_x(y), \quad D_x[f(x)], \quad \frac{d}{dx} f(x) \quad \text{or} \quad \frac{df}{dx}$$

To apply Definition 3.1 to the revenue function $f(x) = 20x - x^2$, we proceed as follows:

$$\begin{aligned} f(x) &= 20x - x^2 \\ f(x + \Delta x) &= 20(x + \Delta x) - (x + \Delta x)^2 \\ &= 20x + 20\,\Delta x - x^2 - 2x(\Delta x) - (\Delta x)^2 \end{aligned}$$

Then,

$$\begin{aligned} \Delta y &= f(x + \Delta x) - f(x) \\ &= 20x + 20(\Delta x) - x^2 - 2x(\Delta x) - (\Delta x)^2 - 20x + x^2 \\ &= 20(\Delta x) - 2x(\Delta x) - (\Delta x)^2 \end{aligned}$$

Therefore,

$$\frac{\Delta y}{\Delta x} = \frac{20(\Delta x) - 2x(\Delta x) - (\Delta x)^2}{\Delta x}$$

$$= 20 - 2x - (\Delta x)$$

Taking the limit of $\Delta y/\Delta x$ as $\Delta x \to 0$, we have

$$\frac{dy}{dx} = \lim_{\Delta x \to 0} \left(\frac{\Delta y}{\Delta x}\right)$$

$$= \lim_{\Delta x \to 0} (20 - 2x - \Delta x)$$

$$= 20 - 2x$$

Thus for our revenue function $f(x) = 20x - x^2$, the instantaneous rate of change in the vicinity of x is

$$\frac{dy}{dx} = 20 - 2x$$

The instantaneous rate of change in the vicinity of $x = 8$ can now be computed simply:

$$\left(\frac{dy}{dx}\right)_{x=8} = 20 - 2(8) = 4$$

At $x = 10$, the instantaneous rate of change is

$$\left(\frac{dy}{dx}\right)_{x=10} = 20 - 2(10) = 0$$

while at $x = 12$, the instantaneous rate of change is

$$\left(\frac{dy}{dx}\right)_{x=12} = 20 - 2(12) = -4$$

This suggests that the revenue function $f(x)$ can be increased to its maximum value when the level of production is at $x = 10$ since the instantaneous rate of change at that point is zero. Any increase in production beyond $x = 10$ results in reduced revenue and is no longer feasible. Later, we will be able to verify that this is indeed the case.

Definition 3.2

Let $s(t)$ denote the distance traveled by a moving object in t seconds. Then the instantaneous velocity after t seconds is defined by

3.3
$$s'(t) = \frac{ds}{dt} = \lim_{\Delta t \to 0} \left[\frac{s(t + \Delta t) - s(t)}{\Delta t}\right]$$

Example 3.1

A ball is thrown vertically in the air. Its height s measured after t seconds is given by

$$s(t) = 96t - 16t^2$$

where s is the height of the ball measured in feet above the ground at time t seconds. Find the instantaneous velocity at $t = 2.5$ seconds. How long will it take the ball to reach its highest point?

Solution

$$\begin{aligned} s(t + \Delta t) &= 96(t + \Delta t) - 16(t + \Delta t)^2 \\ &= 96(t + \Delta t) - 16(t^2 + 2t \cdot \Delta t + (\Delta t)^2) \\ &= 96t + 96\,\Delta t - 16t^2 - 32t \cdot \Delta t - 16(\Delta t)^2 \\ s(t) &= 96t - 16t^2 \end{aligned}$$

Thus,

$$\begin{aligned} s(t + \Delta t) - s(t) &= [96t + 96\,\Delta t - 16t^2 - 32t \cdot \Delta t - 16(\Delta t)^2] \\ &\quad - [96t - 16t^2] \\ &= 96(\Delta t) - 32t \cdot \Delta t - 16(\Delta t)^2 \\ &= \Delta t(96 - 32t - 16\Delta t) \end{aligned}$$

and

$$\begin{aligned} \frac{\Delta s}{\Delta t} &= \frac{\Delta t(96 - 32t - 16\Delta t)}{\Delta t} \\ &= 96 - 32t - 16(\Delta t) \end{aligned}$$

Taking the limit when $\Delta t \to 0$, we obtain

$$\begin{aligned} s'(t) = \frac{ds}{dt} &= \lim_{\Delta t \to 0} \left(\frac{\Delta s}{\Delta t}\right) \\ &= \lim_{\Delta t \to 0} (96 - 32t - 16\Delta t) \\ &= 96 - 32t \end{aligned}$$

Thus the instantaneous velocity at $t = 2.5$ seconds is given by

$$s'(2.5) = \left(\frac{ds}{dt}\right)_{t=2.5} = 96 - 32(2.5) = 16 \text{ feet per second}$$

The ball will reach its highest point when the instantaneous velocity $s'(t) = 0$. Solving the equation

$$96 - 32t = 0$$

for t, we conclude that $t = 3$. Thus the ball will reach its highest point three seconds after it is thrown vertically into the air.

Observe that the instantaneous velocity ds/dt given by (3.3) is the derivative of the distance s with respect to the time t. We must emphasize that although Definition 3.1 involved the variables x and y, there is no change in the concept when x and y are replaced by t and s, respectively. In other words, we can think of the derivative as a rate of change of y with respect to x, rate of change of s with respect to t, and so on.

Example 3.2

A metal plate used by a pollution inspector has its heat content given by a function

$$f(t) = 300 - 10t - t^2, \qquad 0 \le t \le 5$$

where t is the time measured in seconds. How fast is the plate cooling when $t = 2$ seconds?

Solution

We must calculate the instantaneous rate of change $f'(t)$. We have

$$\begin{aligned}
f'(t) &= \lim_{\Delta t \to 0} \left[\frac{f(t+\Delta t) - f(t)}{\Delta t}\right] \\
&= \lim_{\Delta t \to 0} \frac{[300 - 10(t+\Delta t) - (t+\Delta t)^2] - [300 - 10t - t^2]}{\Delta t} \\
&= \lim_{\Delta t \to 0} \frac{[300 - 10t - 10\Delta t - t^2 - 2t \cdot \Delta t - (\Delta t)^2] - [300 - 10t - t^2]}{\Delta t} \\
&= \lim_{\Delta t \to 0} \frac{[-10(\Delta t) - 2t \cdot \Delta t - (\Delta t)^2]}{\Delta t} \\
&= \lim_{\Delta t \to 0} \frac{\Delta t(-10 - 2t - \Delta t)}{\Delta t} \\
&= \lim_{\Delta t \to 0} (-10 - 2t - \Delta t) \\
&= -10 - 2t
\end{aligned}$$

Thus,

$$f'(t) = -10 - 2t$$

At $t = 2$ seconds,

$$\begin{aligned}
f'(2) &= -10 - 2(2) \\
&= -14
\end{aligned}$$

Since $f'(2)$ is negative, we conclude that the metal plate is losing its heat. The instantaneous heat loss at time $t = 2$ seconds is 14 degrees.

Example 3.3

Consider a commodity such as the amount of crude oil which may be produced in the Gulf Coast in a given year. The total cost of production, $c(x)$, is associated with the output level x. The *marginal cost* of the commodity at the production level x is defined to be the derivative $c'(x)$.

For example, suppose

3.4 $$c(x) = ax^2 + bx + d$$

where a, b, and d are constants. Then the marginal cost at level x is

$$\begin{aligned} c'(x) &= \lim_{\Delta x \to 0} \left[\frac{c(x + \Delta x) - c(x)}{\Delta x}\right] \\ &= \lim_{\Delta x \to 0} \left[\frac{a(x + \Delta x)^2 + b(x + \Delta x) + d - ax^2 - bx - d}{\Delta x}\right] \\ &= \lim_{\Delta x \to 0} \frac{a[x^2 + 2x\Delta x + (\Delta x)^2 - x^2] + b[x + (\Delta x) - x]}{\Delta x} \\ &= \lim_{\Delta x \to 0} \frac{a[2x(\Delta x) + (\Delta x)^2] + b(\Delta x)}{\Delta x} \\ &= \lim_{\Delta x \to 0} (2ax + a(\Delta x) + b) \end{aligned}$$

Thus,

3.5 $$c'(x) = 2ax + b$$

We have assumed that the amount of crude oil is a *continuous* variable and that the change in this amount from x to $(x + \Delta x)$ may be infinitesimal. But in the production of tables, chairs, refrigerators, radios, etc., a change of one unit is the smallest change possible. Suppose, for example, that a firm produces x color television sets in one week. This level of production depends upon several factors, such as the capital invested, the number of workers available, the frequency of equipment breakdown, and so on. Assume that the cost of production for producing x color television sets is given by the cost function

3.6 $$c(x) = 10{,}000 + 200x - \frac{x^2}{2}$$

If 100 color television sets come off the assembly line in one week, the cost is

$$c(100) = \$\left(10{,}000 + 200(100) - \frac{(100)^2}{2}\right) = \$25{,}000$$

but if 101 color television sets are produced, the cost is

$$c(101) = \$\left(10{,}000 + 200(101) - \frac{(101)^2}{2}\right) = \$25{,}099.50$$

Thus the cost of an additional set is

$$c(101) - c(100) = \$99.50$$

Let us now calculate the cost of an additional set by evaluating the marginal cost $c'(x)$ at $x = 100$. Applying (3.6) to (3.4) we observe that

$$a = -\tfrac{1}{2}, \quad b = 200, \quad d = 10{,}000$$

From (3.5) we have

$$\begin{aligned} c'(x) &= 2ax + b \\ &= 2(-\tfrac{1}{2})x + 200 \\ &= 200 - x \end{aligned}$$

and

$$c'(100) = 200 - 100 = 100$$

which is *approximately* the same as $c(101) - c(100) = \$99.50$. In general,

$$c'(x) = \lim_{\Delta x \to 0} \left[\frac{c(x + \Delta x) - c(x)}{\Delta x}\right]$$

except in the case of a product manufactured in *discrete* units (positive integers only). In such cases, $\Delta x = 1$ is the smallest change possible and

$$c'(x) \approx \frac{c(x + 1) - c(x)}{(x + 1) - x} = \frac{c(x + 1) - c(x)}{1} = c(x + 1) - c(x)$$

Thus for a commodity produced in discrete units, the marginal cost $c'(x)$ is an approximation for the cost of the $(x + 1)$-st item produced.

Another concept in economics is that of *marginal revenue*, which is defined as the increase in total revenue resulting from the sale of an additional unit. It can be shown that the *marginal revenue* $R'(x)$ is approximately the revenue received for the $(x + 1)$-st item, assuming that the first x items have already been sold. In other words,

$$R'(x) \approx \left[\frac{R(x + 1) - R(x)}{(x + 1) - x}\right] = \frac{R(x + 1) - R(x)}{1}$$

Similarly, the *marginal profit* $P'(x)$ is approximately the profit realized from the sale of an additional unit. That is,

$$P'(x) \approx \left[\frac{P(x + 1) - P(x)}{(x + 1) - x}\right] = \frac{P(x + 1) - P(x)}{1}$$

Example 3.4 The profit function of a department store is given by

3.7 $$P(x) = 200x + \frac{x^2}{5}$$

where x is the number of articles sold in one day. Find the marginal profit at $x = 50$ and at $x = 95$.

Solution

To find the marginal profit, we use the derivative $P'(x)$. Applying (3.4) to (3.7) we note that

$$a = \tfrac{1}{5}, \quad b = 200, \quad d = 0$$

Thus,

$$\begin{aligned} P'(x) &= 2ax + b \\ &= \tfrac{2}{5}x + 200 \end{aligned}$$

The marginal profit at $x = 50$ is

$$P'(50) = \tfrac{2}{5}(50) + 200 = \$220$$

which is approximately the same as $P(51) - P(50)$, since

$$P(51) - P(50) = 10{,}720.20 - 10{,}500 = \$220.20$$

Similarly, the marginal profit at $x = 95$ is

$$P'(95) = \tfrac{2}{5}(95) + 200 = \$238$$

and this is approximately the same as $P(96) - P(95)$, since

$$P(96) - P(95) = 21{,}043.20 - 20{,}805.00 = \$238.20$$

In the following examples we give a systematic procedure for obtaining derivatives—a method we call the *five-step rule.*

Example 3.5

Given $f(x) = x^3$, find $f'(x)$.

Solution

Step 1. Write the function f at x.

$$f(x) = x^3$$

Step 2. Find $f(x + \Delta x)$.

$$\begin{aligned} f(x + \Delta x) &= (x + \Delta x)^3 \\ &= x^3 + 3x^2(\Delta x) + 3x(\Delta x)^2 + (\Delta x)^3 \end{aligned}$$

Step 3. Subtract $f(x)$ from $f(x + \Delta x)$.

$$f(x + \Delta x) - f(x) = 3x^2(\Delta x) + 3x(\Delta x)^2 + (\Delta x)^3$$

Step 4. Form the difference-quotient, $[f(x+\Delta x)-f(x)]/\Delta x$, and simplify.

$$\frac{f(x+\Delta x)-f(x)}{\Delta x}=3x^2+3x(\Delta x)+(\Delta x)^2$$

Step 5. Find $\lim_{\Delta x\to 0}[f(x+\Delta x)-f(x)]/\Delta x$.

$$\lim_{\Delta x\to 0}\frac{f(x+\Delta x)-f(x)}{\Delta x}=\lim_{\Delta x\to 0}[3x^2+3x(\Delta x)+(\Delta x)^2]$$
$$=3x^2$$

We must reinforce the fact that in the limiting process, x is fixed and Δx is the variable.

Example 3.6

Given $f(x)=1/x$, find $f'(x)$ by the five-step rule.

Solution

1. $f(x)=\dfrac{1}{x}, \qquad x\neq 0$

2. $f(x+\Delta x)=\dfrac{1}{x+\Delta x}, \qquad x+\Delta x\neq 0$

3. $f(x+\Delta x)-f(x)=\dfrac{1}{x+\Delta x}-\dfrac{1}{x}=\dfrac{x-x-\Delta x}{x(x+\Delta x)}=-\dfrac{\Delta x}{x(x+\Delta x)}$

4. $\dfrac{f(x+\Delta x)-f(x)}{\Delta x}=-\dfrac{1}{x(x+\Delta x)}$

5. $f'(x)=\lim_{\Delta x\to 0}\left[\dfrac{f(x+\Delta x)-f(x)}{\Delta x}\right]=\lim_{\Delta x\to 0}\left[-\dfrac{1}{x(x+\Delta x)}\right]$

$$=-\frac{1}{x^2}, \qquad x\neq 0$$

Observe that the first three steps are purely mechanical and can be carried out in a routine manner. The last two steps frequently require some skill in algebraic manipulation.

Example 3.7

Given $f(x)=\sqrt{x}$, $x>0$, find $f'(x)$.

Solution

1. $f(x)=\sqrt{x}, \qquad x>0$
2. $f(x+\Delta x)=\sqrt{x+\Delta x}, \qquad x+\Delta x>0$
3. $f(x+\Delta x)-f(x)=\sqrt{x+\Delta x}-\sqrt{x}=\dfrac{\sqrt{x+\Delta x}-\sqrt{x}}{1}$

We rationalize the numerator as follows:

$$\frac{\sqrt{x+\Delta x}-\sqrt{x}}{1}\,\frac{\sqrt{x+\Delta x}+\sqrt{x}}{\sqrt{x+\Delta x}+\sqrt{x}}=\frac{(x+\Delta x)-x}{\sqrt{x+\Delta x}+\sqrt{x}};\qquad x>0,\quad x+\Delta x>0$$

$$=\frac{\Delta x}{\sqrt{x+\Delta x}+\sqrt{x}}$$

4. $\dfrac{f(x+\Delta x)-f(x)}{\Delta x}=\dfrac{1}{\sqrt{x+\Delta x}+\sqrt{x}}$

5. $f'(x)=\lim\limits_{\Delta x\to 0}\left[\dfrac{f(x+\Delta x)-f(x)}{\Delta x}\right]=\lim\limits_{\Delta x\to 0}\left[\dfrac{1}{\sqrt{x+\Delta x}+\sqrt{x}}\right]$

$$=\frac{1}{\sqrt{x}+\sqrt{x}}=\frac{1}{2\sqrt{x}},\qquad x>0$$

Example 3.8

Given $f(x)=1/\sqrt{x}$, $x>0$, find $f'(x)$.

Solution

1. $f(x)=\dfrac{1}{\sqrt{x}},\qquad x>0$

2. $f(x+\Delta x)=\dfrac{1}{\sqrt{x+\Delta x}},\qquad x+\Delta x>0$

3. $f(x+\Delta x)-f(x)=\dfrac{1}{\sqrt{x+\Delta x}}-\dfrac{1}{\sqrt{x}}$

$$=\frac{\sqrt{x}-\sqrt{x+\Delta x}}{\sqrt{x}\,\sqrt{x+\Delta x}}$$

4. $\dfrac{f(x+\Delta x)-f(x)}{\Delta x}=\dfrac{\sqrt{x}-\sqrt{x+\Delta x}}{(\Delta x)\sqrt{x}\,\sqrt{x+\Delta x}}$

Now we need to rationalize the numerator. Observe that

$$\frac{\sqrt{x}-\sqrt{x+\Delta x}}{(\Delta x)\sqrt{x}\,\sqrt{x+\Delta x}}\cdot\frac{\sqrt{x}+\sqrt{x+\Delta x}}{\sqrt{x}+\sqrt{x+\Delta x}}=\frac{1}{(\Delta x)\sqrt{x}\,\sqrt{x+\Delta x}}\cdot\frac{x-(x+\Delta x)}{\sqrt{x}+\sqrt{x+\Delta x}}$$

$$=\frac{1}{(\Delta x)\sqrt{x}\,\sqrt{x+\Delta x}}\cdot\frac{-\Delta x}{\sqrt{x}+\sqrt{x+\Delta x}}$$

$$=\frac{-1}{\sqrt{x}\,\sqrt{x+\Delta x}\,[\sqrt{x}+\sqrt{x+\Delta x}]}$$

5. $f'(x) = \lim_{\Delta x \to 0} \left[\frac{f(x + \Delta x) - f(x)}{\Delta x}\right]$

$= \lim_{\Delta x \to 0} \left[\frac{-1}{\sqrt{x}\sqrt{x + \Delta x}[\sqrt{x} + \sqrt{x + \Delta x}]}\right]$

$= \frac{-1}{\sqrt{x}\sqrt{x}(\sqrt{x} + \sqrt{x})} = -\frac{1}{x^{1/2} \cdot x^{1/2}(2x^{1/2})}$

$= -\frac{1}{2x^{3/2}}, \qquad x > 0$

Thus,

$$f'(x) = -\frac{1}{2x^{3/2}}, \qquad x > 0$$

Exercises

Find the derivative of each of the following functions. Use the five-step rule.

1. $f(x) = 3x$
2. $f(x) = 3x^2$
3. $f(x) = 10 - 5x$
4. $f(x) = x^2 + 5x + 6$
5. $f(x) = x^3 + 8$
6. $f(x) = 8$
7. $f(x) = \frac{1}{x^2}$
8. $f(x) = \frac{1}{x + 4}$
9. $f(x) = \frac{1}{x - 4}$
10. $f(x) = \frac{1}{4 - x}$
11. $f(x) = \frac{1}{x^3}$
12. $f(x) = \frac{1}{x^2 + 1}$
13. $f(x) = \frac{x}{x + 1}$
14. $f(x) = \frac{x}{x - 1}$
15. $f(x) = \sqrt{x}$
16. $f(x) = \sqrt{x + 2}$
17. $f(x) = \frac{1}{\sqrt{x + 2}}$
18. $f(x) = \frac{2}{\sqrt{x + 3}}$
19. $f(x) = \frac{1}{\sqrt{x + 1}}$
20. $f(x) = \frac{3}{\sqrt{x - 1}}$
21. A particle moves along a line according to the formula

$$s = 200t - 5t^2, \qquad 0 \le t \le 20$$

where s (in feet) is the distance traveled in t seconds. Find the instantaneous velocity of the particle at

a. $t = 2$ **b.** $t = 4$ **c.** $t = 10$ **d.** $t = 15$

22. A ball is thrown vertically into the air. Its height s (in feet) at time t (in seconds) is given by the formula

$$s = 256t - 16t^2$$

Find the instantaneous velocity of the ball at

a. $t = 3$ **b.** $t = 5$ **c.** $t = 6$

d. How long will it take the ball to reach its highest point?

23. The production function for a firm producing a certain commodity is given by:

$$f(x) = 100 + 80x - 5x^2$$

where $f(x)$ denotes the total cost (in thousands of dollars) and x represents the amount of commodity (in hundreds of units) produced per week. Determine the instantaneous rate of change at

a. $x = 5$ **b.** $x = 7$ **c.** $x = 8$ **d.** $x = 10$

24. Assume that the market demand function for unskilled labor in the United States is given by:

$$f(x) = \frac{400}{x}$$

where $f(x)$ represents labor hours (in thousands) per month and x denotes the wage rate (in dollars per hour). Determine the marginal demand at:

a. $x = 5$ **b.** $x = 10$ **c.** $x = 15$ **d.** $x = 20$

25. The manager of a chain of bookstores has determined that the supply of certain paperbacks is given approximately by

$$S(p) = 320 - \frac{64}{p - 1}$$

where $S(p)$ denotes the supply at the price p (in dollars) per carton of books. Determine the marginal supply at:

a. $p = 2$ **b.** $p = 3$ **c.** $p = 5$ **d.** $p = 9$

26. The market demand function for a certain commodity is given by

$$f(p) = 400 - 20\sqrt{p}$$

where $f(p)$ is the quantity demanded at price p. Determine the marginal demand at

a. $p = 9$ **b.** $p = 16$ **c.** $p = 25$ **d.** $p = 36$

27. The number of bacteria in a certain culture after t hours is given by

$$f(t) = t^3 + 20t + 100$$

Determine the instantaneous rate of change in the number of bacteria at

a. $t = 2$ hours **b.** $t = 5$ hours

28. The number of new items a person learns in t hours is given by the formula

$$f(t) = 24t + 3t^2 + 8\sqrt{t}$$

Determine the rate of change after

a. $t = 1$ hour **b.** $t = 4$ hours

29. A psychological study of welders at an industrial plant calculates job satisfaction on a scale of 0 to 10 according to the formula

$$f(t) = -0.1t^2 + t + 6.5$$

where t is recorded in weeks. How fast is the worker satisfaction increasing or decreasing after 4 weeks on the job?

30. The cortex of a sea urchin egg was exposed to a certain chemical. Its mean thickness was described by the function

$$f(t) = 4.1 - 0.2t + 0.02t^2 - 0.001t^3$$

where t was the number of days the egg remained exposed. How fast was the thickness of the egg changing after

a. $t = 3$ days **b.** $t = 6$ days **c.** $t = 9$ days

3.3 Geometric Interpretation of the Derivative

In this section we develop equations for describing the tangent to a curve $y = f(x)$ at a point P. For a circle, we may intuitively know that any line having exactly one point in common with it is a tangent line, while any line having two points in common with it is a secant line. Consider, for example, the tangent line to a circle at a point P as shown in Figure 3.2. The tangent is perpendicular to the radius OP of the circle.

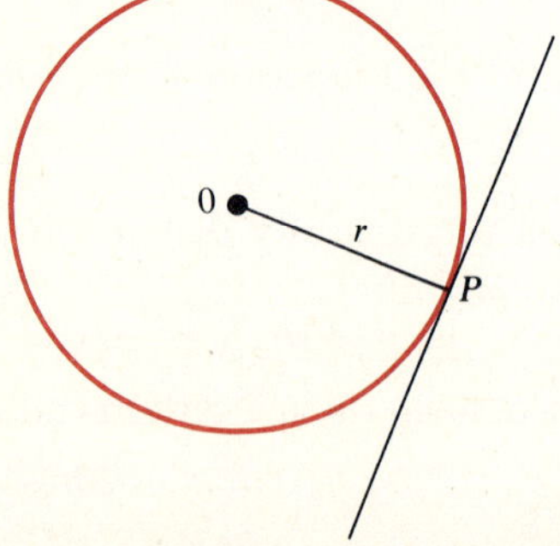

Figure 3.2

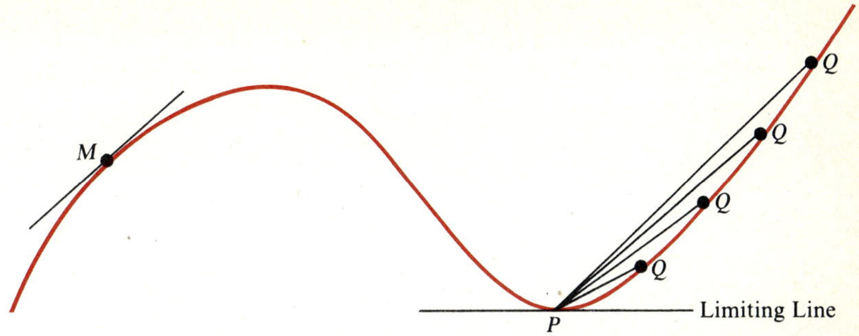

Figure 3.3

For a general curve as in Figure 3.3, let us now ask whether the line is tangent to the curve at point M. Our intuition may conclude that this is the tangent line, but is it really? If it is, how do we describe the tangent line to this curve at point P? The description requires the general notion of a limit for a curve. Perhaps we can get some idea of a tangent line at P by considering several secant lines joining P to a point Q of the curve "near" P. Observe that as we move Q closer and closer to P, the secant line joining P to Q is getting closer and closer to a limiting line. Thus it seems reasonable to consider that the tangent line to the curve at a point P is the limiting position of a secant line joining P to some other point Q of the curve, as Q moves closer and closer to P.

Suppose that $P(x_0, f(x_0))$ is a point on the curve $y = f(x)$ and we wish to find the tangent at P (see Figure 3.4). Let $Q(x_0 + \Delta x, f(x_0 + \Delta x))$ be another point "near" P on the curve. Note that while Figure 3.4 shows Δx to be positive, it need not be so—it may be negative; however, Δx cannot be zero since point Q is different from P. The slope of the secant line PQ is:

$$\frac{\Delta y}{\Delta x} = \frac{f(x_0 + \Delta x) - f(x_0)}{(x_0 + \Delta x) - x_0} = \frac{f(x_0 + \Delta x) - f(x_0)}{\Delta x}$$

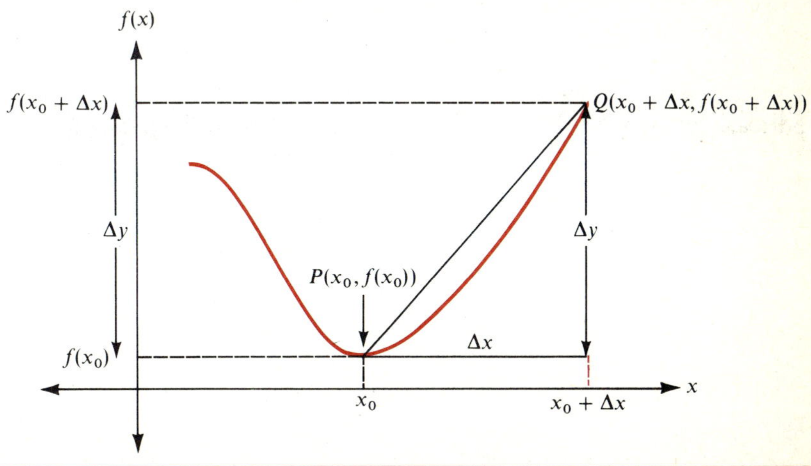

Figure 3.4

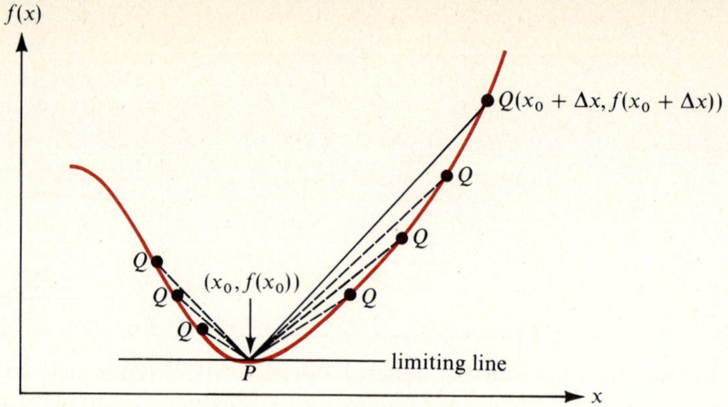

Figure 3.5

As Q moves closer and closer to P, Δx approaches zero from the right and secant PQ rotates clockwise about point P. Intuitively, PQ approaches a limiting line as $\Delta x \to 0^+$. A similar statement holds if Q is located on the left of P and Δx approaches 0 from the left. This limiting line is the tangent to the curve $y = f(x)$ at the point P. Figure 3.5 shows several positions for secant PQ and its limiting position as $\Delta x \to 0$ from the right as well as from the left. Thus if a function possesses a derivative at $x = x_0$, then

$$\lim_{\Delta x \to 0} \left[\frac{f(x_0 + \Delta x) - f(x_0)}{\Delta x}\right] = \left(\frac{dy}{dx}\right)_{x = x_0} = f'(x_0)$$

suggests that $f'(x_0)$ or $(dy/dx)_{x=x_0}$ is the *slope of the tangent* to the curve $y = f(x)$ at the point $P(x_0, f(x_0))$.

Definition 3.3 A function $f(x)$ is said to be differentiable at $x = x_0$ if and only if it has a derivative at that point.

Definition 3.4 A function $f(x)$ is said to be differentiable if and only if it has a derivative at every point in the domain of f.

Definition 3.5 Let $y = f(x)$ be a function that is differentiable at the point $x = x_0$. The *tangent to the curve* $y = f(x)$ at $P(x_0, f(x_0))$ is the line through P with slope $f'(x_0)$.

Thus, the equation of the tangent to the curve $y = f(x)$ at the point $P(x_0, f(x_0))$ is:

$$y - y_0 = m(x - x_0)$$

where $y_0 = f(x_0)$ and $m = f'(x_0)$ is the slope of the tangent line at the point P.

Example 3.9

Find the equation of the line which is tangent to the curve $y = x^2 + 1$ at the point $(3, f(3))$.

Solution

In the problem $f(x) = x^2 + 1, x_0 = 3$. Thus, $f(x_0) = f(3) = 10$. We need to determine whether or not $f(x)$ is differentiable at $x = 3$. In other words, does $\lim_{\Delta x \to 0} [f(3 + \Delta x) - f(3)]/\Delta x$ exist? Note that

$$\begin{aligned}\frac{f(3 + \Delta x) - f(3)}{\Delta x} &= \frac{[(3 + \Delta x)^2 + 1] - 10}{\Delta x} \\ &= \frac{9 + 6(\Delta x) + (\Delta x)^2 + 1 - 10}{\Delta x} \\ &= \frac{6(\Delta x) + (\Delta x)^2}{\Delta x} \\ &= \frac{\Delta x(6 + \Delta x)}{\Delta x} \\ &= 6 + \Delta x\end{aligned}$$

Hence,

$$\lim_{\Delta x \to 0} \left[\frac{f(3 + \Delta x) - f(3)}{\Delta x}\right] = \lim_{\Delta x \to 0} (6 + \Delta x) = 6$$

and $f(x)$ is differentiable at $x_0 = 3$. Thus the slope of the tangent line is $m = 6$ and the equation of the tangent line is

$$y - 10 = 6(x - 3)$$

That is

$$y = 6x - 8$$

Exercises

Find the equation of the tangent to each of the following curves at the point $(x_0, f(x_0))$.

1. $f(x) = x^2 - x; \quad x_0 = 2$
2. $f(x) = x^2 + 2x + 1; \quad x_0 = -1$
3. $f(x) = x^2 - x + 1; \quad x_0 = 0$
4. $f(x) = x^2 - 2x + 3; \quad x_0 = 2$
5. $f(x) = x^3; \quad x_0 = 1$

6. $f(x) = x^3 + x^2 + x - 1; \qquad x_0 = -1$

7. $f(x) = \sqrt{x}, \quad x \geq 0; \qquad x_0 = 4$

8. $f(x) = \sqrt{x + 2}, \quad x \geq -2; \qquad x_0 = 7$ [Hint: Use the five-step rule to find $f'(x)$]

9. $f(x) = \sqrt{2x}, \quad x \geq 0; \qquad x_0 = 2$ [Hint: $\sqrt{2x} = \sqrt{2}\sqrt{x}$]

10. $f(x) = x + \sqrt{x}, \quad x \geq 0; \qquad x_0 = 4$

3.4 Differentiability and Continuity

It is convenient at this point to relate the concepts of continuity and differentiability. To say that a function is *continuous* at $x = a$ means that

$$\lim_{x \to a} f(x) = f(a)$$

To say that a function is *differentiable* at $x = a$ implies that

$$f'(a) = \lim_{\Delta x \to 0} \left[\frac{f(a + \Delta x) - f(a)}{\Delta x} \right]$$

exists. Note that these two concepts are different.

Proposition 3.1

If a function $f(x)$ is differentiable at $x = a$, then $f(x)$ is continuous at that point.

In other words, if a function $f(x)$ is not continuous at $x = a$, then $f(x)$ is not differentiable at that point.

The proposition assures us that differentiability implies continuity. The converse, however, is not true; this is shown in the following example.

Example 3.10

The function

$$f(x) = |x|$$

is continuous at $x = 0$. (Why?) The graph of $f(x)$ is shown in Figure 3.6. However, $f(x)$ is not differentiable at $x = 0$ because

$$\frac{f(0 + \Delta x) - f(0)}{\Delta x} = \frac{f(\Delta x)}{\Delta x} = \frac{|\Delta x|}{\Delta x} = \begin{cases} 1 & \text{for } \Delta x > 0 \\ -1 & \text{for } \Delta x < 0 \end{cases}$$

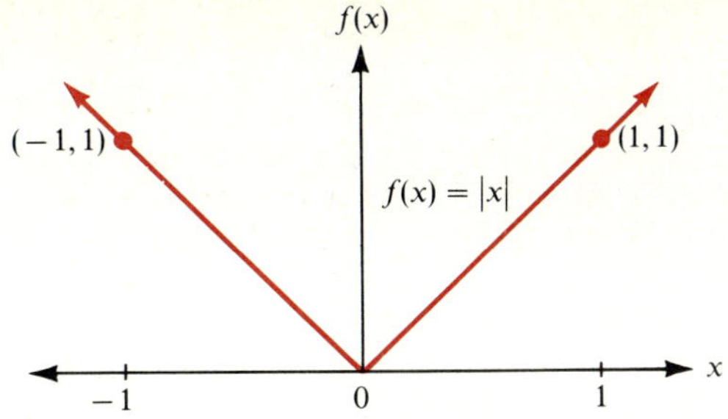

Figure 3.6

Thus,

$$\lim_{\Delta x \to 0^+} \frac{f(0 + \Delta x) - f(0)}{\Delta x} = 1, \qquad \lim_{\Delta x \to 0^-} \frac{f(0 + \Delta x) - f(0)}{\Delta x} = -1$$

Since the right- and left-hand limits are not equal, we conclude that

$$\lim_{\Delta x \to 0} \frac{f(0 + \Delta x) - f(0)}{\Delta x}$$

does not exist.

Example 3.11

Let

$$f(x) = \begin{cases} x + 2, & \text{if } x \neq 2 \\ 5, & \text{if } x = 2 \end{cases}$$

a. Is $f(x)$ continuous at $x = 2$?

b. Is $f(x)$ differentiable at $x = 2$?

Solution

a. Note that

$$\lim_{x \to 2} f(x) = \lim_{x \to 2} (x + 2) = 2 + 2 = 4 \qquad \textbf{(Proposition 2.5)}$$

Since $f(2) \neq 4$, we conclude that $f(x)$ is discontinuous at $x = 2$.

b. The function is *not* differentiable at $x = 2$ since $f(x)$ is not continuous at that point (Proposition 3.1). We could also see this by observing that

$$\lim_{\Delta x \to 0} \frac{f(2 + \Delta x) - f(2)}{\Delta x}$$

does not exist. Note that

$$f(2 + \Delta x) = (2 + \Delta x) + 2 = 4 + \Delta x \text{ and } f(2) = 5$$

Thus,

$$\lim_{\Delta x \to 0} \frac{f(2 + \Delta x) - f(2)}{\Delta x} = \lim_{\Delta x \to 0} \left(\frac{4 + \Delta x - 5}{\Delta x}\right) = \lim_{\Delta x \to 0} \left(1 - \frac{1}{\Delta x}\right)$$

This limit does not exist since $(1 - (1/\Delta x))$ gets large and negative as Δx approaches 0 from the right and large and positive as Δx approaches 0 from the left.

Exercises

In problems 1–6, determine whether or not the function is (a) continuous and (b) differentiable at the specified point.

1. $f(x) = \begin{cases} x + 5, & x \neq 4 \\ 10, & x = 4 \end{cases}$

$x = 4$

2. $f(x) = \begin{cases} 2x + 1, & x \neq 3 \\ 9, & x = 3 \end{cases}$

$x = 3$

3. $f(x) = \begin{cases} 3x + 4, & x \neq 2 \\ 10, & x = 2 \end{cases}$

$x = 2$

4. $f(x) = \begin{cases} 2x + 7, & x \neq 1 \\ 9, & x = 1 \end{cases}$

$x = 1$

5. $f(x) = |x - 1|, \quad x = 1$

6. $f(x) = |2x - 1|, \quad x = \frac{1}{2}$

7. Let $f(x)$ be defined as follows:

$$f(x) = \begin{cases} x + 1, & 0 \leq x < 3 \\ 1, & x = 3 \\ 7 - x, & 3 < x \leq 5 \end{cases}$$

Is $f(x)$ differentiable at $x = 3$?

8. Let $f(x)$ be defined as follows:

$$f(x) = \begin{cases} \dfrac{x^2 - 4}{x - 2}, & 0 \leq x < 2 \\ 4, & x = 2 \\ 6 - x, & 2 < x \leq 4 \end{cases}$$

a. Is $f(x)$ continuous at $x = 2$?

b. Is $f(x)$ differentiable at $x = 2$?

9. Let $f(x)$ be defined as follows:

$$f(x) = \begin{cases} \dfrac{x^2 - 9}{x - 3}, & 0 \leq x < 3 \\ 6, & x = 3 \\ 7 - 2x, & 3 < x \leq 6 \end{cases}$$

a. Is $f(x)$ continuous at $x = 3$?

b. Is $f(x)$ differentiable at $x = 3$?

10. Let $f(x)$ be defined as follows:

$$f(x) = \begin{cases} 3x + 2, & 0 \leq x < 4 \\ 15 - 3x, & 4 \leq x < 6 \end{cases}$$

Is $f(x)$ differentiable at $x = 4$? Explain.

3.5 Some Fundamental Rules of Differentiation

The use of the five-step rule for finding the derivative of a function can be a lengthy and complicated process. This section develops some simple rules of differentiation so that the derivatives of certain simple functions can be written down at sight.

Proposition 3.2 If $f(x) = c$, where c is a constant, then $f'(x) = 0$ for all x.

The graph of the function $f(x) = c$ is a line parallel to the y-axis (see Figure 3.7) and all points on a horizontal line have the same y coordinate. This means that the difference in y coordinates of all pairs of points on the line is zero; hence, the slope of the line is zero.

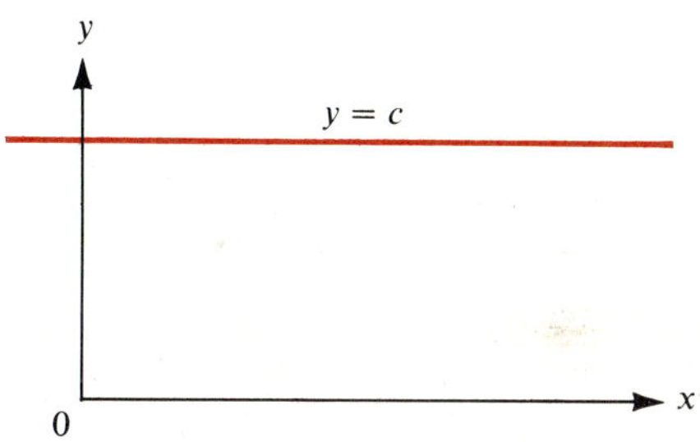

Figure 3.7

Alternatively, we may use Definition 3.1. Since $f(x) = c$ and $f(x + \Delta x) = c$, we obtain

$$\begin{aligned} f'(x) &= \lim_{\Delta x \to 0} \left[\frac{f(x + \Delta x) - f(x)}{\Delta x} \right] \\ &= \lim_{\Delta x \to 0} \left[\frac{c - c}{\Delta x} \right] \\ &= 0 \end{aligned}$$

Next, we use the definition of the derivative to differentiate some other special cases. Let $f(x) = x$. Then

$$\begin{aligned} f'(x) &= \lim_{\Delta x \to 0} \frac{f(x + \Delta x) - f(x)}{\Delta x} \\ &= \lim_{\Delta x \to 0} \frac{(x + \Delta x) - x}{\Delta x} \\ &= \lim_{\Delta x \to 0} (1) \\ &= 1 \end{aligned}$$

Next, if $f(x) = x^2$, then

$$\begin{aligned} f'(x) &= \lim_{\Delta x \to 0} \frac{f(x + \Delta x) - f(x)}{\Delta x} \\ &= \lim_{\Delta x \to 0} \frac{(x + \Delta x)^2 - x^2}{\Delta x} \\ &= \lim_{\Delta x \to 0} \frac{2x(\Delta x) + (\Delta x)^2}{\Delta x} = \lim_{\Delta x \to 0} \frac{\Delta x(2x + \Delta x)}{\Delta x} \\ &= \lim_{\Delta x \to 0} [2x + (\Delta x)] \\ &= 2x \end{aligned}$$

The definition of derivative was used in Example 3.5 to show that if $f(x) = x^3$, then

$$f'(x) = 3x^2$$

This pattern suggests the following proposition, commonly known as the *power rule*, which we state without proof.

Proposition 3.3 If $f(x) = x^n$ where n is a positive integer, then $f'(x) = nx^{n-1}$.

Thus if $f(x) = x^5$, we can immediately write down that $f'(x) = 5x^4$ and if $f(x) = x^{20}$, then $f'(x) = 20x^{19}$. What is even more interesting is that Proposition 3.3 holds for *any* real number n. For example, if

$$f(x) = \frac{1}{x}, \qquad x \neq 0$$

then we first write

$$f(x) = \frac{1}{x} = x^{-1}$$

and apply Proposition 3.3. Thus we obtain

$$\begin{aligned} f'(x) &= (-1)x^{-1-1} \\ &= (-1)x^{2} \\ &= -\frac{1}{x^2}, \qquad x \neq 0 \qquad \textbf{(Example 3.6)} \end{aligned}$$

Similarly, for $x > 0$, if

$$f(x) = \sqrt{x} = x^{1/2}$$

then

$$\begin{aligned} f'(x) &= \tfrac{1}{2}x^{1/2-1} \\ &= \tfrac{1}{2}x^{-1/2} \\ &= \frac{1}{2\sqrt{x}} \qquad \textbf{(Example 3.7)} \end{aligned}$$

Again, if

$$f(x) = \frac{1}{\sqrt{x}}$$

we must first convert the radical to a fractional exponent and write

$$f(x) = x^{-1/2}$$

Now we can apply Proposition 3.3 and obtain

$$\begin{aligned} f'(x) &= (-\tfrac{1}{2})x^{(-1/2)-1} \\ &= -\tfrac{1}{2}x^{-3/2} \\ &= -\frac{1}{2x^{3/2}}, \qquad x \neq 0 \qquad \textbf{(Example 3.8)} \end{aligned}$$

Proposition 3.4

Assume that $f(x)$ is differentiable. If $g(x) = cf(x)$ where c is a constant, then

$$g'(x) = c \cdot f'(x)$$

Proof

If

$$g(x) = cf(x)$$

then

$$g(x + \Delta x) = cf(x + \Delta x)$$

and

$$\frac{g(x+\Delta x)-g(x)}{\Delta x}=\frac{cf(x+\Delta x)-cf(x)}{\Delta x}$$
$$=c\cdot\left[\frac{f(x+\Delta x)-f(x)}{\Delta x}\right]$$

and finally,

$$g'(x)=\lim_{\Delta x\to 0}\frac{g(x+\Delta x)-g(x)}{\Delta x}$$
$$=\lim_{\Delta x\to 0} c\left[\frac{f(x+\Delta x)-f(x)}{\Delta x}\right]$$
$$=c\cdot\lim_{\Delta x\to 0}\left[\frac{f(x+\Delta x)-f(x)}{\Delta x}\right]$$
$$=c\cdot f'(x)$$

Thus, if $f(x)=8x^5$, then $f'(x)=8(5x^4)=40x^4$.

Next, we consider the derivative of the sum (or difference) of any finite number of functions. We have seen many propositions about the limit of the sum, difference, product and quotient of two functions. From our experience in Chapter 2 one might suggest that since the limit of the sum (or difference) of two functions is the sum (or difference) of their limits, then the derivative of the sum (or difference) of two functions is equal to the sum (or difference) of their derivatives.

Proposition 3.5

Assuming that $f(x)$ and $g(x)$ have derivatives, if $h(x)=f(x)+g(x)$, then
$$h'(x)=f'(x)+g'(x)$$

Proof

To prove this proposition, we use Definition 3.1. Then

$$\lim_{\Delta x\to 0}\frac{h(x+\Delta x)-h(x)}{\Delta x}$$
$$=\lim_{\Delta x\to 0}\frac{[f(x+\Delta x)+g(x+\Delta x)]-[f(x)+g(x)]}{\Delta x}$$
$$=\lim_{\Delta x\to 0}\frac{[f(x+\Delta x)-f(x)]+[g(x+\Delta x)-g(x)]}{\Delta x}$$
$$=\lim_{\Delta x\to 0}\frac{f(x+\Delta x)-f(x)}{\Delta x}+\lim_{\Delta x\to 0}\frac{g(x+\Delta x)-g(x)}{\Delta x}$$
$$=f'(x)+g'(x)$$

Corollary

Let $f(x)$ and $g(x)$ be two differentiable functions. If $k(x) = f(x) - g(x)$, then

$$k'(x) = f'(x) - g'(x)$$

Example 3.12

Find the derivative of

$$f(x) = 3x^4 + 4x^3 - 2x^2 + 7x$$

Solution

The quantity x^4 has the derivative $4x^3$, as we know from Proposition 3.3. Using this fact and Proposition 3.4, we find that

$$\frac{d}{dx}(3x^4) = 12x^3,$$

$$\frac{d}{dx}(4x^3) = 12x^2,$$

$$\frac{d}{dx}(2x^2) = 4x,$$

$$\frac{d}{dx}(7x) = 7$$

Using Proposition 3.5 and its corollary, we have

$$f'(x) = 12x^3 + 12x^2 - 4x + 7$$

Exercises

Find the derivative of each of the following functions *without* using the five-step rule.

1. $f(x) = x^4$

2. $f(x) = 3x^5$

3. $f(x) = 4x^8$

4. $f(x) = 5\sqrt{x}$

5. $f(x) = 2x^{3/2}, \quad x \geq 0$

6. $f(x) = 8x^{2/3}, \quad x \geq 0$

7. $f(x) = \frac{1}{x}, \quad x \neq 0$

8. $f(x) = \frac{1}{x^2}, \quad x \neq 0$

9. $f(x) = x^3 + \frac{1}{x^3}, \quad x \neq 0$

10. $f(x) = x^2 + \frac{1}{x^2} + 2, \quad x \neq 0$

11. $f(x) = x^3 + 6x^2 + 7$

12. $f(x) = x^5 + 3x^3 + 2x + 1$

13. $f(x) = \sqrt{x} + \frac{1}{\sqrt{x}}, \; x > 0$

14. $f(x) = x^3 + 8x^2 + \sqrt{x} + \dfrac{1}{2x}, \qquad x > 0$

15. $f(x) = 5x^3 + 4x^2 + \dfrac{1}{x}, \qquad x \neq 0$

16. $f(x) = 6x^4 + 7x^3 + 8x^2 + \dfrac{1}{\sqrt{x}}, \qquad x > 0$

17. $f(x) = 3x^5 + 4x^4 + 3x^3 + \dfrac{3}{\sqrt{x}}, \qquad x > 0$

18. $f(x) = \sqrt{2x} + \dfrac{1}{\sqrt{2x}}, \qquad x > 0$

19. $f(x) = x^{5/4} + x^{3/2} + x^{1/2}, \qquad x \geq 0$

20. $f(x) = x^{2/3} + x^{1/3} + x^{-1/2}, \qquad x > 0$

For each of the following functions, find

a. $f'(x)$ **b.** $f'(2)$ **c.** $f\ (3)$ **d.** $f'(4)$

21. $f(x) = x^2 + 5x$

22. $f(x) = x^3$

23. $f(x) = \sqrt{x}, \qquad x > 0$

24. $f(x) = \sqrt{x} + \dfrac{1}{\sqrt{x}} + 1, \qquad x > 0$

25. $f(x) = \dfrac{1}{2\sqrt{x}}, \qquad x > 0$

26. $f(x) = \sqrt{x} - \dfrac{1}{\sqrt{x}}, \qquad x > 0$

27. A particle moves along a line according to the formula

$$s = t^3 + 3t^2 + 5t$$

where s is the distance (in feet) traveled in t seconds. Determine the instantaneous velocity at

a. $t = 3$ seconds **b.** $t = 5$ seconds

28. A rocket fired from the surface of the earth reaches a height of

$$h = t^4 + 3t^3 + 12t^2$$

feet above the level of the ground in time t seconds. Determine the instantaneous velocity of the rocket after

a. $t = 10$ seconds **b.** $t = 20$ seconds

29. The total cost of production in an industry is given as

$$c(x) = 0.03x^2 + 10x + 1200$$

where $c(x)$ denotes the total cost associated with production level x. Find $c'(x)$, the marginal cost of production.

30. The total cost of production at a plant is given by

$$c(x) = 0.005x^3 - 0.02x^2 + 30x + 3000$$

where $c(x)$ is the total cost associated with the total output x. Determine $c'(x)$, the marginal cost of production.

31. The investment function for the building industry as a whole is given by

$$I(r) = 200 - 10{,}000r^3$$

where $I(r)$ denotes the total investment in millions of dollars and r denotes the rate of interest. Determine the instantaneous rate of change when
a. $r = 0.12$ **b.** $r = 0.15$ **c.** $r = 0.18$

32. A colony of bacteria after t hours in a certain culture is given by the function

$$f(t) = 1000 + 4t^3 + 20\sqrt{t}$$

Determine the instantaneous rate of change in the number of bacteria when
a. $t = 1$ hour **b.** $t = 4$ hours

33. The number of fish in a lake t days after it was contaminated with untreated sewage was estimated to be

$$f(t) = 100{,}000 - 4t^2$$

How fast was the fish population declining after 30 days?

34. A fixed quantity of drug is injected into the bloodstream of a patient and the concentration $f(t)$ of the drug in the bloodstream is recorded continuously for t hours. The function f is defined by

$$f(t) = 2\sqrt{t}, \qquad t \geq 0$$

Determine the instantaneous rate of change in the concentration of the drug in the bloodstream after
a. $t = 1$ hour **b.** $t = 2$ hours **c.** $t = 3$ hours

35. A legionnaire's disease hits a certain town. The number of persons sick with the disease after t days is approximated by the function

$$f(t) = 1 + 72\sqrt{t}, \qquad t \geq 0$$

where t is measured in days from the outbreak of the disease. Determine the rate at which the disease is spreading after

a. $t = 1$ day
b. $t = 4$ days
c. $t = 9$ days
d. $t = 16$ days
e. $t = 36$ days

36. Two species are assumed to have t_1 and t_2 members, respectively, in a certain habitat. Both species feed on certain plants; in addition, species 1 feeds on species 2. Collected data indicates that the number of species 1 present is best approximated by the function

$$t_1 = f(t_2) = 2(t_2)^2 - 60t_2 + 40$$

a. Is the population of species 1 increasing or decreasing when $t_2 = 10$?
b. What is the value of t_2 when the instantaneous rate of change in t_1 equals zero?

3.6 The Product and Quotient Rules

Thus far we have limited ourselves to the derivatives of sums, differences, and constant multiples of functions. We have shown that the derivative of the sum (or difference) of two functions is the sum (or difference) of their derivatives. Is the derivative of the product of two differentiable functions equal to the product of their derivatives? In other words, does

3.8
$$\frac{d}{dx}[f(x)g(x)] \stackrel{?}{=} \frac{d}{dx}[f(x)] \cdot \frac{d}{dx}[g(x)]$$

We can easily show that this is *not* the case. Suppose, for instance, that $f(x) = 2x + 3$ and $g(x) = x + 2$. Then the product of the functions is

$$\begin{aligned} f(x) \cdot g(x) &= (2x + 3)(x + 2) \\ &= 2x^2 + 7x + 6 \end{aligned}$$

and the derivative of that product is

$$\begin{aligned} \frac{d}{dx}[f(x) \cdot g(x)] &= \frac{d}{dx}(2x^2 + 7x + 6) \\ &= \frac{d}{dx}(2x^2) + \frac{d}{dx}(7x) + \frac{d}{dx}(6) \\ &= 4x + 7 \end{aligned}$$

The derivatives of the functions are

$$\frac{d}{dx}[f(x)] = \frac{d}{dx}(2x + 3) = \frac{d}{dx}(2x) + \frac{d}{dx}(3) = 2,$$

and

$$\frac{d}{dx}[g(x)] = \frac{d}{dx}(x + 2) = \frac{d}{dx}(x) + \frac{d}{dx}(2) = 1$$

so that the product of the derivatives is

$$\frac{d}{dx}[f(x)] \cdot \frac{d}{dx}[g(x)] = 2$$

Thus it follows that (3.8) does *not* hold. What is the relationship between the derivative of two functions and their product? The answer lies in the following proposition.

Proposition 3.6

Let $f(x)$ and $g(x)$ be two differentiable functions and let $h(x) = f(x) \cdot g(x)$. Then

3.9 $$h'(x) = f'(x) \cdot g(x) + g'(x)f(x)$$

Proof

We use the five-step rule.

Step 1. $h(x) = f(x) \cdot g(x)$

Step 2. $h(x + \Delta x) = f(x + \Delta x)g(x + \Delta x)$

Step 3. $h(x + \Delta x) - h(x) = f(x + \Delta x)g(x + \Delta x) - f(x) \cdot g(x)$

At this point, we use a standard trick for this type of problem. We add and subtract the term $f(x)g(x + \Delta x)$. This enables us to write *Step 3* as follows:

$$\begin{aligned} h(x + \Delta x) - h(x) &= f(x + \Delta x)g(x + \Delta x) - f(x)g(x + \Delta x) \\ &\quad + f(x)g(x + \Delta x) - f(x) \cdot g(x) \\ &= [f(x + \Delta x) - f(x)]g(x + \Delta x) \\ &\quad + [g(x + \Delta x) - g(x)]f(x) \end{aligned}$$

Step 4. $$\frac{h(x + \Delta x) - h(x)}{\Delta x} = \left[\frac{f(x + \Delta x) - f(x)}{\Delta x}\right]g(x + \Delta x) + \left[\frac{g(x + \Delta x) - g(x)}{\Delta x}\right]f(x)$$

Step 5. $$\begin{aligned} h'(x) &= \lim_{\Delta x \to 0}\left[\frac{h(x + \Delta x) - h(x)}{\Delta x}\right] \\ &= \lim_{\Delta x \to 0}\left[\frac{f(x + \Delta x) - f(x)}{\Delta x}\right] \cdot \lim_{\Delta x \to 0} g(x + \Delta x) \\ &\quad + \lim_{\Delta x \to 0}\left[\frac{g(x + \Delta x) - g(x)}{\Delta x}\right] \cdot f(x) \end{aligned}$$

Since $g(x)$ is differentiable and hence continuous (Proposition 3.1), we have

$$\lim_{\Delta x \to 0} g(x + \Delta x) = g(x)$$

Using this fact, we conclude that

$$h'(x) = f'(x)g(x) + g'(x)f(x)$$

Example 3.13

If $h(x) = (x^2 + 5)(x^3 + 7x^2 + 8)$, find $h'(x)$.

Solution

Let $f(x) = x^2 + 5$ and $g(x) = x^3 + 7x^2 + 8$. Then

$$f'(x) = 2x \quad \text{and} \quad g'(x) = 3x^2 + 14x$$

Using the product rule, we have

$$\begin{aligned} h'(x) &= f'(x)g(x) + g'(x)f(x) \\ &= 2x(x^3 + 7x^2 + 8) + (3x^2 + 14x)(x^2 + 5) \\ &= (2x^4 + 14x^3 + 16x) + (3x^4 + 14x^3 + 15x^2 + 70x) \\ &= 5x^4 + 28x^3 + 15x^2 + 86x \end{aligned}$$

Example 3.14

If $h(x) = x^2(\sqrt{x} + 2)$, $x \geq 0$, find $h'(x)$.

Solution

Let $f(x) = x^2$ and $g(x) = x^{1/2} + 2$. Then

$$f'(x) = \frac{d}{dx}(x^2) = 2x \quad \text{and} \quad \begin{aligned} g'(x) &= \frac{d}{dx}(x^{1/2} + 2) \\ &= \frac{d}{dx}(x^{1/2}) + \frac{d}{dx}(2) \\ &= \tfrac{1}{2}x^{-1/2}, \qquad x > 0 \end{aligned}$$

Using the product rule, we obtain

$$\begin{aligned} h'(x) &= f'(x)g(x) + g'(x)f(x) \\ &= 2x(x^{1/2} + 2) + (\tfrac{1}{2}x^{-1/2})x^2 \\ &= 2x^{3/2} + 4x + \tfrac{1}{2}x^{3/2} = \tfrac{5}{2}x^{3/2} + 4x, \qquad x > 0 \end{aligned}$$

Proposition 3.7

If $h(x) = 1/g(x)$ where $g(x) \neq 0$ and $g(x)$ is differentiable, then $h(x)$ is differentiable and

3.10
$$h'(x) = -\frac{g'(x)}{[g(x)]^2}$$

Proof

We use the five-step rule.

Step 1. $h(x) = \dfrac{1}{g(x)}, \qquad g(x) \neq 0$

Step 2. $h(x + \Delta x) = \dfrac{1}{g(x + \Delta x)}, \qquad g(x + \Delta x) \neq 0$

Step 3.
$$h(x + \Delta x) - h(x) = \frac{1}{g(x + \Delta x)} - \frac{1}{g(x)}, \qquad g(x) \neq 0,\ g(x + \Delta x) \neq 0$$
$$= \frac{g(x) - g(x + \Delta x)}{g(x + \Delta x) \cdot g(x)}$$
$$= -\left(\frac{g(x + \Delta x) - g(x)}{g(x + \Delta x) \cdot g(x)}\right)$$

Step 4. $\dfrac{\Delta h}{\Delta x} = -\left(\dfrac{g(x + \Delta x) - g(x)}{\Delta x}\right) \cdot \left(\dfrac{1}{g(x + \Delta x) \cdot g(x)}\right)$

Step 5.
$$h'(x) = \lim_{\Delta x \to 0} \left(\frac{\Delta h}{\Delta x}\right) = \lim_{\Delta x \to 0} \left[-\left(\frac{g(x + \Delta x) - g(x)}{\Delta x}\right) \cdot \left(\frac{1}{g(x + \Delta x) \cdot g(x)}\right)\right]$$
$$= -\frac{1}{g(x)}\left[\lim_{\Delta x \to 0} \left(\frac{g(x + \Delta x) - g(x)}{\Delta x}\right)\right] \cdot \left[\lim_{\Delta x \to 0} \left(\frac{1}{g(x + \Delta x)}\right)\right]$$

Since $g(x)$ is differentiable and hence continuous, we have

$$\lim_{\Delta x \to 0} g(x + \Delta x) = g(x)$$

Using this fact, we conclude that

$$h'(x) = -\frac{1}{g(x)} \cdot g'(x) \cdot \frac{1}{g(x)}$$
$$= -\frac{g'(x)}{[g(x)]^2}$$

Example 3.15

Given that $h(x) = 1/(2x + 3)$, $x \neq -\frac{3}{2}$, find $h'(x)$.

Solution

Let $g(x) = 2x + 3$. Then $g'(x) = 2$. Using Proposition 3.7, we have

$$h'(x) = \frac{-2}{(2x + 3)^2}, \qquad x \neq -\tfrac{3}{2}$$

Next, we use the *product rule* (Proposition 3.6) to generalize Proposition 3.6 and develop what is commonly known as the *quotient rule.*

Proposition 3.8 Let $f(x)$ and $g(x)$ be two differentiable functions such that $g(x) \neq 0$. Then the function

$$h(x) = \frac{f(x)}{g(x)}$$

is differentiable and

3.11
$$h'(x) = \frac{g(x)f'(x) - f(x)g'(x)}{[g(x)]^2}$$

Proof

Observe that

$$h(x) = \frac{f(x)}{g(x)} = f(x) \cdot \frac{1}{g(x)}$$

Using Proposition 3.6, we get

$$h'(x) = f'(x)\frac{1}{g(x)} + \frac{d}{dx}\left(\frac{1}{g(x)}\right) \cdot f(x)$$

Note that

$$\frac{d}{dx}\left(\frac{1}{g(x)}\right) = -\frac{g'(x)}{[g(x)]^2} \qquad \textbf{(Proposition 3.7)}$$

Thus,

$$\begin{aligned} h'(x) &= \frac{f'(x)}{g(x)} + \left(-\frac{g'(x)}{[g(x)]^2}\right) \cdot f(x) \\ &= \frac{g(x)f'(x) - f(x)g'(x)}{[g(x)]^2} \end{aligned}$$

In simple notation, if

$$h(x) = \frac{N}{D}$$

where N and D stand for numerator and denominator, respectively, then

$$h'(x) = \frac{DN' - ND'}{D^2}$$

In words, the derivative of the quotient of two functions is the denominator times the derivative of the numerator, minus the numerator, times the derivative of the denominator, divided by the denominator squared.

Example 3.16

Given that $h(x) = (x + 1)/(x + 2)$, $x \neq -2$, find $h'(x)$.

Solution

Let $f(x) = x + 1$ and $g(x) = x + 2$. Then

$$f'(x) = 1 \quad \text{and} \quad g'(x) = 1$$

Using Proposition 3.8, we have

$$\begin{aligned} h'(x) &= \frac{(x + 2) \cdot 1 - (x + 1) \cdot 1}{(x + 2)^2} \\ &= \frac{1}{(x + 2)^2}, \qquad x \neq -2 \end{aligned}$$

Example 3.17

Given that $h(x) = (x^2 - 3x + 4)/(2x^2 + 7x + 1)$, find $h'(x)$.

Solution

Let $f(x) = x^2 - 3x + 4$ and $g(x) = 2x^2 + 7x + 1$. Then

$$f'(x) = 2x - 3 \quad \text{and} \quad g'(x) = 4x + 7$$

Applying the quotient rule, we obtain

$$\begin{aligned} h'(x) &= \frac{(2x^2 + 7x + 1)(2x - 3) - (x^2 - 3x + 4)(4x + 7)}{(2x^2 + 7x + 1)^2} \\ &= \frac{\begin{array}{l}(4x^3 - 6x^2 + 14x^2 - 21x + 2x - 3) \\ \quad - (4x^3 + 7x^2 - 12x^2 - 21x + 16x + 28)\end{array}}{(2x^2 + 7x + 1)^2} \\ &= \frac{(4x^3 + 8x^2 - 19x - 3) - (4x^3 - 5x^2 - 5x + 28)}{(2x^2 + 7x + 1)^2} \\ &= \frac{13x^2 - 14x - 31}{(2x^2 + 7x + 1)^2}, \qquad 2x^2 + 7x + 1 \neq 0 \end{aligned}$$

Example 3.18

Given that $h(x) = (\sqrt{x} + 1)/(x + 3)$, $x \neq -3$, find $h'(x)$.

Solution

Let $f(x) = x^{1/2} + 1$ and $g(x) = x + 3$. Then

$$f'(x) = \tfrac{1}{2}x^{-1/2} \quad \text{and} \quad g(x) = 1$$

Using Proposition 3.8, we obtain

$$\begin{aligned} h'(x) &= \frac{(x+3)\frac{1}{2}x^{-1/2} - (x^{1/2}+1)\cdot 1}{(x+3)^2} \\ &= \frac{(x+3) - 2x^{1/2}(x^{1/2}+1)}{2x^{1/2}(x+3)^2} \\ &= \frac{(x+3) - 2x - 2x^{1/2}}{2x^{1/2}(x+3)^2} \\ &= \frac{3 - x - 2x^{1/2}}{2x^{1/2}(x+3)^2}, \qquad x > 0 \text{ and } x \neq -3 \end{aligned}$$

Exercises

In problems 1–8, multiply the factors and then differentiate the result.

1. $f(x) = (x+2)(x+3)$

2. $f(x) = (x+1)(x^2+2)$

3. $f(x) = (x+3)(x+4)$

4. $f(x) = (2x+1)(2x-1)$

5. $f(x) = (2x+5)(3x+7)$

6. $f(x) = (5x+7)(2x-3)$

7. $f(x) = (3x+2)(2x-3)$

8. $f(x) = (x+1)\sqrt{x}$

In problems 9–24, differentiate each of the following functions by using the product rule.

9. $f(x) = (x^2+4)(3x^2+7)$

10. $f(x) = (x^2+x-1)(x^2+x+1)$

11. $f(x) = (x^2+2x+1)(x^2-2x+1)$

12. $f(x) = (x^3+3x^2+1)(x^3-x+4)$

13. $f(x) = (x^3+5x+1)(x^2-7x)$

14. $f(x) = (x^2+5x+7)(x^3+x^2+1)$

15. $f(x) = (x^4+x^3+x^2+1)(x^4-1)$

16. $f(x) = (x^4+x^2+1)(x^4-x^2+1)$

17. $f(x) = (x^2-2x+1)(x^3+x^2+x+1)$

18. $f(x) = (x-1)(x^4+x^3+x^2+x+1)$

19. $f(x) = (x+1)(2\sqrt{x}+3)$

20. $f(x) = (\sqrt{x} + 1)(\sqrt{x} + 3)$

21. $f(x) = \left(x^2 + \sqrt{x} + \frac{1}{x}\right)\left(x - \frac{1}{x}\right)$

22. $f(x) = \left(\sqrt{x} + \frac{1}{\sqrt{x}}\right)\left(\sqrt{x} - \frac{1}{\sqrt{x}} + 1\right)$

23. $f(x) = \sqrt{x}\left(x + \sqrt{x} + \frac{1}{\sqrt{x}}\right)$

24. $f(x) = (x^{3/2} + x^{1/2} + 1)(x^{3/2} - x^{1/2} - 4)$

Using the quotient rule, find dy/dx for each of the following functions.

25. $y = \frac{1}{x}, \quad x \neq 0$

26. $y = \frac{1}{x^2}, \quad x \neq 0$

27. $y = \frac{1}{x+2}, \quad x \neq -2$

28. $y = \frac{1}{2x+3}, \quad x \neq -\frac{3}{2}$

29. $y = \frac{x}{3x+2}, \quad x \neq -\frac{2}{3}$

30. $y = \frac{x}{2x+1}, \quad x \neq -\frac{1}{2}$

31. $y = \frac{2x-1}{2x+1}, \quad x \neq -\frac{1}{2}$

32. $y = \frac{1+x}{1-x}, \quad x \neq 1,$

33. $y = \frac{1-x^2}{1+x^2}$

34. $y = \frac{1+\sqrt{x}}{1-\sqrt{x}}, \quad x \neq 1, x \geq 0$

35. $y = \frac{1-x+x^2}{1+x+x^2}$

36. $y = \frac{2x}{1+x^2}$

37. $y = \dfrac{\sqrt{x}}{\sqrt{x}+1}$

38. $y = \dfrac{\sqrt{x}}{x^{3/2}+4}$

39. $y = \dfrac{x^3 + x^2 + x + 1}{x^4 - x^2 + 6}$

40. $y = \dfrac{x + \sqrt{x} + 1}{x - \sqrt{x} - 1}$

For each of the following functions, find
a. $f'(0)$ **b.** $f'(2)$ **c.** $f'(4)$ **d.** $f'(8)$

41. $f(x) = \dfrac{1}{x+1}, \quad x \neq -1$

42. $f(x) = \dfrac{x^2 + 2x + 3}{x+1}, \quad x \neq -1$

43. $f(x) = \dfrac{x-1}{x+2}, \quad x \neq -2$

44. $f(x) = \dfrac{3x+1}{2x-1}, \quad x \neq \dfrac{1}{2}$

45. $f(x) = \dfrac{x}{2x+1}, \quad x \neq -\dfrac{1}{2}$

46. $f(x) = \dfrac{2x-1}{2x+1}, \quad x \neq -\dfrac{1}{2}$

3.7 The Chain Rule

The preceding sections of this chapter have expanded the variety of of functions that we are able to differentiate. However, we still are unable to find derivatives of such functions as

$$\sqrt{x^2+1}, \quad (3x^2 + 4x + 7)^{98}, \quad \text{or} \quad \left(\frac{2x-5}{4x+3}\right)^8$$

In this section we develop what is commonly known as the *chain rule.* This rule is one of the most useful tools in the differentiation of composite functions. Recall from Chapter 2 that when $f(x)$ and $g(x)$ are two functions, we can obtain the composite function $h(x)$

$$h(x) = (g \circ f)x = g[f(x)]$$

Thus if g is a function of u, and u is a function of x, then the composite of the function $g[f(x)]$ is also a function of x. For example, if

$$g(u) = u^{1/2} \quad \text{and} \quad u = f(x) = x^2 + 1$$

then by substitution, we have

$$g[f(x)] = (x^2 + 1)^{1/2}$$

Similarly, if

$$g(u) = u^2 + 1 \quad \text{and} \quad u = f(x) = \frac{2x - 1}{3x + 2}, \qquad x \neq -\frac{2}{3}$$

then

$$g[f(x)] = \left(\frac{2x - 1}{3x + 2}\right)^2 + 1, \qquad x \neq -\frac{2}{3}$$

To find the derivative of the composite function, $g[f(x)]$, we need the following proposition.

Proposition 3.9

If g and f are differentiable, then $h(x) = g[f(x)]$ is differentiable and

3.12
$$h'(x) = g'[f(x)] \cdot f'(x)$$

In d/dx notation, we may express (3.12) as

3.13
$$\frac{dy}{dx} = \frac{dy}{du} \cdot \frac{du}{dx}$$

where $y = g(u)$ and $u = f(x)$.

Corollary

If $y = [f(x)]^n$ where n is any real number, then

3.14
$$\frac{dy}{dx} = n[f(x)]^{n-1} \cdot f'(x)$$

Example 3.19

Find dy/dx if $y = (3x^2 + 4x + 8)^7$

Solution

Note that we can write y as

$$y = [f(x)]^7 \quad \text{where} \quad f(x) = 3x^2 + 4x + 8$$

Then

$$f'(x) = 6x + 4$$

Using (3.14), with $n = 7$, we have

$$\frac{dy}{dx} = 7(3x^2 + 4x + 8)^6(6x + 4)$$

Example 3.20

Find dy/dx if $y = \sqrt{x^2 + 1}$

Solution

Here $y = u^{1/2}$ where $u = x^2 + 1$. Then

$$\frac{dy}{du} = \frac{1}{2}u^{-1/2} \quad \text{and} \quad \frac{du}{dx} = 2x$$

Applying (3.13), we have

$$\frac{dy}{dx} = \left(\frac{1}{2}u^{-1/2}\right) \cdot 2x = \frac{x}{\sqrt{x^2 + 1}}$$

Example 3.21

Find dy/dx if $y = [(2x - 5)/(4x + 3)]^8, \qquad x \neq -\frac{3}{4}$

Solution

We write y as

$$y = [f(x)]^8 \quad \text{where} \quad f(x) = \frac{2x - 5}{4x + 3}$$

Then using (3.14), with $n = 8$, we obtain

$$\frac{dy}{dx} = 8[f(x)]^7 \frac{d}{dx}\left(\frac{2x - 5}{4x + 3}\right)$$

To find the derivative of $(2x - 5)/(4x + 3)$, we use the quotient rule. Thus,

$$\frac{d}{dx}\left(\frac{2x - 5}{4x + 3}\right) = \frac{(4x + 3)\dfrac{d}{dx}(2x - 5) - (2x - 5)\dfrac{d}{dx}(4x + 3)}{(4x + 3)^2}$$

$$= \frac{(4x + 3) \cdot 2 - (2x - 5) \cdot 4}{(4x + 3)^2}$$

$$= \frac{8x + 6 - 8x + 20}{(4x + 3)^2}$$

$$= \frac{26}{(4x + 3)^2}, \qquad x \neq -\frac{3}{4}$$

Hence,

$$\frac{dy}{dx} = 8\left(\frac{2x-5}{4x+3}\right)^7 \frac{26}{(4x+3)^2} = \frac{8(2x-5)^7}{(4x+3)^7}\frac{26}{(4x+3)^2}$$

$$= \frac{208(2x-5)^7}{(4x+3)^9}, \qquad x \neq -\frac{3}{4}$$

Example 3.22

Find the derivative of

$$y = (3x+4)^6(2x+7)^8$$

Solution

Note that if $u(x) = (3x+4)^6$ and $v(x) = (2x+7)^8$, then by product formula (3.9), we have

3.15 $$\frac{dy}{dx} = u'(x)v(x) + v'(x)u(x)$$

To find $u'(x)$ and $v'(x)$, we use (3.14):

$$u'(x) = 6(3x+4)^5 \cdot 3 = 18(3x+4)^5$$
$$v'(x) = 8(2x+7)^7 \cdot 2 = 16(2x+7)^7$$

Substituting in the formula (3.15) for dy/dx, we obtain

$$\frac{dy}{dx} = 18(3x+4)^5(2x+7)^8 + 16(2x+7)^7(3x+4)^6$$
$$= 2(3x+4)^5(2x+7)^7[9(2x+7) + 8(3x+4)]$$
$$= 2(3x+4)^5(2x+7)^7(18x+63+24x+32)$$
$$= 2(3x+4)^5(2x+7)^7(42x+95)$$

Example 3.23

Find the equation of the tangent line to the curve

$$y = (3x^2+x+1)^2$$

at the point (1,25).

Solution

First, we must find the slope of the line tangent to the curve at $x = 1$. Remember that we can write y as

$$y = [f(x)]^2 \quad \text{where} \quad f(x) = 3x^2 + x + 1$$

Then

$$f'(x) = 6x + 1$$

Applying (3.14) with $n = 2$, we obtain

$$\frac{dy}{dx} = 2(3x^2 + x + 1)(6x + 1)$$

The slope of the tangent line at (1, 25) is

$$m = \left(\frac{dy}{dx}\right)_{x=1} = 70$$

Thus, the equation of the tangent line through (1, 25) is given by

$$\begin{aligned} y - 25 &= m(x - 1) \\ &= 70(x - 1) \\ &= 70x - 70 \end{aligned}$$

That is,

$$y = 70x - 45$$

Example 3.24

The demand for a product is given by the function

$$D(p) = \frac{10}{\sqrt{4p + 1}}, \qquad p > -\frac{1}{4}$$

where $D(p)$ represents the demand at the price p. Find the marginal demand at $p = 6$.

Solution

$$\begin{aligned} D(p) &= 10(4p + 1)^{-1/2} \\ D'(p) &= \frac{d}{dp}[10(4p + 1)^{-1/2}] \\ &= 10\frac{d}{dp}[(4p + 1)^{-1/2}] \qquad \textbf{(Proportion 3.4)} \\ &= 10(-\tfrac{1}{2})(4p + 1)^{-3/2}(4) \qquad \textbf{(Chain rule)} \\ &= -20(4p + 1)^{-3/2}, \qquad p > -\tfrac{1}{4} \end{aligned}$$

At $p = 6$,

$$D'(6) = -\tfrac{20}{125} = -0.16$$

The fact that $D'(p)$ is negative reflects the fact that the demand for a product is decreasing. In the next chapter, we will look more closely at these kinds of interpretations.

Exercises

Find dy/dx in each of the following problems 1–26.

1. $y = (2x + 3)^4$

2. $y = (4x + 7)^5$

3. $y = (6 - 3x^2)^3$

4. $y = (x^2 + 2x + 5)^4$

5. $y = (2x^2 + 3x + 1)^{-2}$

6. $y = (x^2 - 2x - 1)^{-3}$

7. $y = \sqrt{3x + 8}, \quad x \geq -\frac{8}{3}$

8. $y = \sqrt{5x - 3}, \quad x \geq \frac{3}{5}$

9. $y = \sqrt{x^2 + 1}$

10. $y = \sqrt{x^2 - 2x - 1}$

11. $y = x\sqrt{2x + 1}, \quad x \geq -\frac{1}{2}$

12. $y = x^2\sqrt{3x + 2}, \quad x \geq -\frac{2}{3}$

13. $y = (x^2 + 2)\sqrt{x}, \quad x \geq 0$

14. $y = (x^2 + 1)\sqrt{x^2 + 2}$

15. $y = x^2(3x + 7)^{10}$

16. $y = x^3(2x + 3)^{12}$

17. $y = (x^2 + 1)(2x - 1)^8$

18. $y = (x^3 + 2)(4x^2 + 7)^4$

19. $y = (2x + 3)^4(3x + 2)^5$

20. $y = (3x - 1)^4(3x + 1)^6$

21. $y = (x^2 + x + 1)^3(x^2 - x + 1)^4$

22. $y = (1 - x - x^2)^8(1 - x + x^2)^5$

23. $y = \left(\frac{2x - 1}{3x + 2}\right)^3, \quad x \neq -\frac{2}{3}$

24. $y = \left(\frac{2x + 1}{2x - 1}\right)^4, \quad x \neq \frac{1}{2}$

25. $y = \left(\frac{x + 1}{x - 1}\right)^{1/2}, \quad x \neq 1$

26. $y = \left(\frac{4x - 5}{3x + 1}\right)^{1/3}, \quad x \neq -\frac{1}{3}$

Find the equation of the line tangent to each of the following curves at the point specified.

27. $y = \frac{16}{2x^2 - 3x + 3}; \quad (1, 8)$

28. $y = \sqrt{x + 8}; \quad (1, 3)$

29. $y = \frac{x^2}{3x + 1}; \quad (1, \frac{1}{4})$

30. $y = \frac{x}{\sqrt{x^2 + 8}}; \quad (1, \frac{1}{3})$

31. $y = (x^2 - 3)^5(2x - 1)^3; \quad (2, 1)$

32. $y = \left(\frac{x + 1}{x - 1}\right)^2; \quad (3, 4)$

33. A particle moves along a line according to the equation

$$s = \left(\frac{t + 1}{t - 1}\right)^3, \qquad t > 1$$

where s is the distance traveled in feet in t seconds. Find the instantaneous rate of change of s with respect to t. Is ds/dt increasing or decreasing at $t = 3$?

34. A particle moves along a line according to the formula

$$s = \frac{27t}{\sqrt{t+1}}$$

where s (in feet) is the distance traveled in t seconds. Find the instantaneous velocity at $t = 8$ seconds.

35. The demand for a certain product is given by

$$D(p) = \sqrt{25 - 5p + 2p^2}$$

where p is the price in cents per unit. Find the marginal demand at $p = 5$ cents.

36. The demand $D(p)$ for a new product is given by

$$D(p) = \frac{500}{\sqrt{4p+5}}$$

where p is the price in dollars per unit. Find the marginal demand at $p = 5$.

37. The revenue from the sale of x items is given by

$$R(x) = 25x + \frac{x^2}{50}$$

Find the marginal revenue at
a. $x = 25$ **b.** $x = 75$

38. The revenue from the sale of x items is given by

$$R(x) = 5x + \frac{10}{(2x+1)}$$

Find the marginal revenue at
a. $x = 4$ **b.** $x = 12$

39. A fixed quantity of a drug is injected into the bloodstream of a patient and the concentration of the drug in the bloodstream is recorded continuously for t hours. The concentration is approximated by the function

$$f(t) = \frac{10t}{2t^2 + 5t + 8}$$

Determine the instantaneous rate of change of the drug concentration in the bloodstream. How fast is the drug losing its effectiveness at $t = 2$ hours?

40. The temperature of an air-conditioned apartment on a hot day is approximated by the function

$$f(t) = 90 - 10t + \frac{8}{t+1}, \qquad 0 \le t \le 3$$

Determine the instantaneous rate of change in the apartment temperature. How fast is the temperature falling at $t = 2$ hours?

3.8 Higher Derivatives

The function $f'(x)$ defined by

$$\lim_{\Delta x \to 0} \left[\frac{f(x + \Delta x) - f(x)}{\Delta x} \right]$$

is the derivative of the function $f(x)$. More specifically, it is the *first derivative* of the function $y = f(x)$ with respect to x. If $f'(x)$ has a derivative at a point x, then this new derivative is called the *second derivative* of $f(x)$ at that point and is denoted by $f''(x)$ or d^2y/dx^2. In other words,

3.16
$$f''(x) = \lim_{\Delta x \to 0} \left[\frac{f'(x + \Delta x) - f'(x)}{\Delta x} \right]$$

defines a new function on all points for which the limit (3.16) exists. If the second derivative also has a derivative at a point x, then the newest derivative is called the *third derivative* of $f(x)$ at that point. Various notations are used to denote the higher derivatives of a function $y = f(x)$. The first, second, third, . . . , n-th derivative of $y = f(x)$ can be represented by

$$y', y'', y''', \ldots, y^{(n)}$$

or by

$$f'(x), f''(x), f'''(x), \ldots, f^{(n)}(x)$$

or by

$$\frac{dy}{dx}, \frac{d^2y}{dx^2}, \frac{d^3y}{dx^3}, \ldots, \frac{d^ny}{dx^n}$$

or by

$$Dy, D^2y, D^3y, \ldots, D^ny$$

Example 3.25

Let $y = x^4$, then

$$\frac{dy}{dx} = 4x^3$$

$$\frac{d^2y}{dx^2} = \frac{d}{dx}\left(\frac{dy}{dx}\right) = \frac{d}{dx}(4x^3) = 12x^2$$

$$\frac{d^3y}{dx^3} = \frac{d}{dx}\left(\frac{d^2y}{dx^2}\right) = \frac{d}{dx}(12x^2) = 24x$$

and

$$\frac{d^4y}{dx^4} = \frac{d}{dx}\left(\frac{d^3y}{dx^3}\right) = \frac{d}{dx}(24x) = 24$$

Note that when $y = x^4$, each derivative higher than the fourth derivative is equal to zero.

Example 3.26

Find the fourth derivative of y if $y = (x + 3)/(x + 2)$, $x \neq -2$

Solution

To facilitate differentiation, we will use an algebraic trick.

$$y = \frac{x+3}{x+2} = \frac{(x+2)+1}{x+2} = 1 + \frac{1}{x+2} = 1 + (x+2)^{-1}$$

then

$$\frac{dy}{dx} = (-1)(x+2)^{-2}$$

$$\frac{d^2y}{dx^2} = (-1)(-2)(x+2)^{-3}$$

$$\frac{d^3y}{dx^3} = (-1)(-2)(-3)(x+2)^{-4}$$

and

$$\frac{d^4y}{dx^4} = (-1)(-2)(-3)(-4)(x+2)^{-5} = 24(x+2)^{-5}$$

$$= \frac{24}{(x+2)^5}, \qquad x \neq -2$$

We will encounter several interpretations of the second derivative in the next chapter. However, we may state in passing that the second derivative is the *instantaneous rate of change of the first derivative.* For example, if the first derivative represents the marginal revenue, then the second derivative gives the rate at which the marginal revenue is changing. Similarly, if the first derivative is the velocity of a particle, then the second derivative gives the rate of change of the velocity, or the *acceleration*, of the particle.

Example 3.27

A particle moves along a line according to the equation

$$s = 3t^2 + 5t + 10$$

where s is the distance traveled (in feet) by the particle in t seconds. Find the velocity and the acceleration at $t = 3$ seconds.

Solution

The velocity of the particle at time t is given by ds/dt while the acceleration is the derivative of ds/dt. Thus,

$$\frac{ds}{dt} = \frac{d}{dt}(3t^2 + 5t + 10) = 6t + 5$$

and

$$\frac{d^2s}{dt^2} = \frac{d}{dt}\left(\frac{ds}{dt}\right) = \frac{d}{dt}(6t + 5) = 6$$

Thus the velocity of the particle at time $t = 3$ is 23 feet per second and the acceleration is 6 feet per second.

Example 3.28

The demand for a product is given by the function

$$D(p) = \frac{10}{\sqrt{4p+1}}, \qquad p > -\frac{1}{4}$$

where $D(p)$ represents the demand at the price p. Find the rate at which the marginal demand is changing at $p = 6$.

Solution

The first derivative $D'(p)$ is the marginal demand at the price p while $D''(p)$ is the rate at which the marginal demand is changing. We need to determine $D''(p)$ at $p = 6$.

Recall from Example 3.24 that

$$D'(p) = -20(4p + 1)^{-3/2}, \qquad p > -\tfrac{1}{4}$$

Differentiating $D'(p)$, we get

$$\begin{aligned} D''(p) &= (-20)(-\tfrac{3}{2})(4p + 1)^{-5/2}(4) \qquad \textbf{(Chain rule)} \\ &= 120\,(4p + 1)^{-5/2}, \qquad p > -\tfrac{1}{4} \end{aligned}$$

At $p = 6$,

$$D''(6) = \frac{120}{3125} = 0.0384$$

Exercises

Find the first *two* derivatives for each of the following functions.

1. $y = x^7$
2. $y = \sqrt{x}, \quad x \geq 0$
3. $y = x^4 + 4x^3 + 6x^2$
4. $y = 4x^3 - 7x^2 + 6x + 12$
5. $y = x^5 + 4x^4 + 3x^3 + x^2$
6. $y = (x + 2)^3(x - 3)$
7. $y = \frac{1}{x}, \quad x \neq 0$
8. $y = \frac{1}{x\sqrt{x}}, \quad x > 0$
9. $y = \frac{1}{x + 1}, \quad x \neq -1$
10. $y = \frac{2}{x + 3}$
11. $y = \frac{x + 2}{x + 1}, \quad x \neq -1$
12. $y = \frac{2x + 3}{2x + 1}$

Find the first *four* derivatives for each of the following functions.

13. $y = (3x + 4)^5$
14. $y = (4x + 7)^6$
15. $y = (ax + b)^n$
16. $y = \frac{1}{x + 2}, \quad x \neq -2$
17. $y = \frac{x + 4}{x + 3}, \quad x \neq -3$
18. $y = \frac{x}{x + 1}, \quad x \neq -1$
19. $y = \frac{x}{x - 1}, \quad x \neq 1$
20. $y = \frac{1}{ax + b}, \quad x \neq -\frac{b}{a}$

21. A particle moves along a line according to the equation

$$s = 4t^3 + 12t^2 + 5t$$

where s is the distance traveled (in feet) by the particle in t seconds. Find the particle's velocity and acceleration at $t = 2$ seconds.

22. A ball is thrown vertically up into the air. Its height s (in feet) at time t (in seconds) is given by

$$s = 128t - 16t^2$$

Find the ball's velocity and acceleration at $t = 2$ and $t = 4$ seconds.

23. A ball is thrown vertically up into the air. Its height s (in feet) at time t (in seconds) is given by the equation

$$s = 192t - 16t^2$$

Find the velocity and acceleration of the ball at $t = 3$ and $t = 5$ seconds.

24. The demand for a certain product is given by

$$D(p) = \sqrt{25 - 5p + 2p^2}$$

where p is the price (in cents per unit). Find the rate at which the marginal demand is changing at $p = 5$.

25. The demand for a new product is related to the price p by the equation

$$D(p) = \frac{500}{\sqrt{4p + 5}}$$

Find the rate at which the marginal demand is changing at $p = 5$.

3.9 Implicit Functions

Thus far we have considered functions which can be expressed in terms of the independent variable. Such functions are called *explicit functions.* It sometimes happens that x and y are both involved in an equation and it is not possible to determine y *uniquely* in terms of x. Consider the following equations:

3.17a $$b^2x^2 + a^2y^2 = a^2b^2$$

3.17b $$x^2 + 2xy + 3y^2 + 5y = 8$$

3.17c $$y^3 + 3y^2x + 6yx^2 + x^3 = 57$$

None of the expressions can be solved for y *uniquely.* Further, each of the expressions can be transformed into the general form

$$f(x, y) = 0$$

For instance, equation (3.17b)

$$x^2 + 2xy + 3y^2 + 5y = 8$$

can easily be transformed to

$$x^2 + 2xy + 3y^2 + 5y - 8 = 0$$

Note that in all three equations, y was defined *implicitly* in terms of x.

Return to equation (3.17a)

$$b^2x^2 + a^2y^2 = a^2b^2$$

We can solve for y and obtain

$$y^2 = \frac{b^2}{a^2}(a^2 - x^2)$$

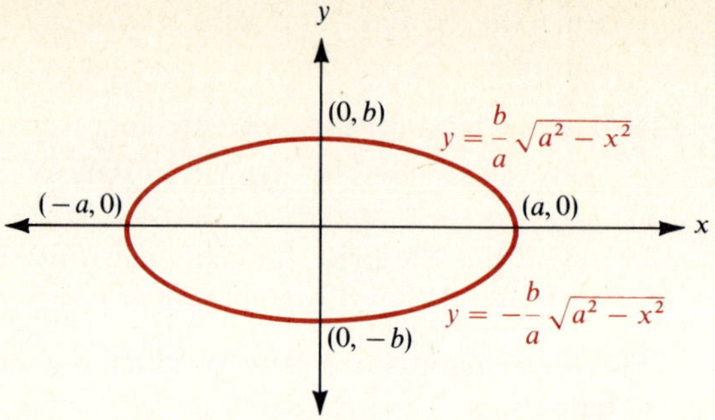

Figure 3.8

which means that

$$y = \frac{b}{a}\sqrt{a^2 - x^2} \quad \text{and} \quad y = -\frac{b}{a}\sqrt{a^2 - x^2}$$

Thus there are two functions of x, each restricted to the range $-a \leq x \leq a$. Geometrically, the first function corresponds to the upper half and the second to the lower half of an ellipse (see Figure 3.8). We would need additional information to determine which half of the ellipse to use in a particular problem.

Example 3.29 If

$$16x^2 + 25y^2 = 400$$

then

$$y = \tfrac{4}{5}\sqrt{25 - x^2} \quad \text{or} \quad y = -\tfrac{4}{5}\sqrt{25 - x^2}$$

Using the chain rule for

$$y = \tfrac{4}{5}\sqrt{25 - x^2} = \tfrac{4}{5}(25 - x^2)^{1/2}$$

we find that

$$\frac{dy}{dx} = \frac{4}{5}\left[\frac{1}{2}(25 - x^2)^{-1/2}(-2x)\right] = -\frac{4}{5}\frac{x}{\sqrt{25 - x^2}}$$

and for

$$y = -\tfrac{4}{5}\sqrt{25 - x^2} = -\tfrac{4}{5}(25 - x^2)^{1/2}$$

a similar computation yields

$$\frac{dy}{dx} = \frac{4}{5}\frac{x}{\sqrt{25 - x^2}}$$

If we want the derivative at a given point on the ellipse, we need to determine which of the two halves of the ellipse contains the point and use the derivative for that half. For example, if we need to find the derivative at the point $(3, \frac{16}{5})$, we note that this point lies on the upper semi-ellipse

$$y = \tfrac{4}{5}\sqrt{25 - x^2}$$

and

$$\frac{dy}{dx} = -\frac{4}{5}\frac{x}{\sqrt{25 - x^2}}$$

At $x = 3$,

$$\left(\frac{dy}{dx}\right)_{x=3} = -\frac{4}{5}\frac{3}{\sqrt{25 - 9}} = -\frac{3}{5}$$

This represents the slope of the tangent line to the ellipse at the point $(3, \frac{16}{5})$.

Sometimes, we may be able to differentiate without first solving for y in terms of x. This is illustrated in the following example.

Example 3.30

Find dy/dx for the equation $16x^2 + 25y^2 = 400$ without solving for y.

Solution

We think of y as a function of x and differentiate each term in the equation

$$16x^2 + 25y^2 = 400$$

with respect to x. Observe that

$$\frac{d}{dx}(16x^2) = 32x$$

$$\frac{d}{dx}(25y^2) = \frac{d}{dy}(25y^2) \cdot \frac{dy}{dx} \qquad \textbf{(Chain rule)}$$

$$= 50y\frac{dy}{dx}$$

and

$$\frac{d}{dx}(400) = 0$$

Thus, we have

$$32x + 50y\frac{dy}{dx} = 0$$

Solving for dy/dx, we get

$$50y\frac{dy}{dx} = -32x$$

$$\frac{dy}{dx} = -\frac{32x}{50y} = -\frac{16x}{25y}, \qquad y \neq 0$$

If $y = \frac{4}{5}\sqrt{25 - x^2}$ (Example 3.29), then

$$\frac{dy}{dx} = -\frac{4}{5}\frac{x}{\sqrt{5 - x^2}}$$

If $y = -\frac{4}{5}\sqrt{25 - x^2}$ (Example 3.29), then

$$\frac{dy}{dx} = \frac{4}{5}\frac{x}{\sqrt{25 - x^2}}$$

We underscore the fact that, in an implicit function, we continue to regard y as a function of x even though y may not be defined explicitly in terms of x.

Example 3.31

Find dy/dx if

$$x^{2/3} + y^{2/3} - a^{2/3} = 0$$

where a is a real number.

Solution

Differentiating each term with respect to x, we have

$$\frac{2}{3}x^{-1/3} + \frac{2}{3}y^{-1/3}\frac{dy}{dx} = 0, \qquad x, y \neq 0$$

Simplifying, we have

$$\frac{2}{3}y^{-1/3}\frac{dy}{dx} = -\frac{2}{3}x^{-1/3}$$

$$\frac{dy}{dx} = \left(\frac{3}{2}y^{1/3}\right)\left(-\frac{2}{3}x^{-1/3}\right)$$

$$= -\frac{y^{1/3}}{x^{1/3}}$$

$$= -\left(\frac{y}{x}\right)^{1/3}$$

Example 3.32

Find d^2y/dx^2 if

$$x^2 + 2xy - y^2 + 32 = 0$$

Solution

First, we find dy/dx by implicit differentiation.

$$2x + \left(2x\frac{dy}{dx} + 2y\right) - 2y\frac{dy}{dx} = 0$$

$$\frac{dy}{dx}(2x - 2y) + (2x + 2y) = 0$$

$$\frac{dy}{dx} = \frac{y + x}{y - x}, \qquad y \neq x$$

Differentiating both sides of this equation, we obtain

$$\frac{d^2y}{dx^2} = \frac{d}{dx}\left(\frac{y + x}{y - x}\right) = \frac{(y - x)\dfrac{d}{dx}(y + x) - (y + x)\dfrac{d}{dx}(y - x)}{(y - x)^2}$$

$$= \frac{(y - x)\left(\dfrac{dy}{dx} + 1\right) - (y + x)\left(\dfrac{dy}{dx} - 1\right)}{(y - x)^2}$$

$$= \frac{\left(y\dfrac{dy}{dx} + y - x\dfrac{dy}{dx} - x\right) - \left(y\dfrac{dy}{dx} - y + x\dfrac{dy}{dx} - x\right)}{(y - x)^2}$$

which on simplification yields

$$\frac{d^2y}{dx^2} = \frac{2y - 2x\dfrac{dy}{dx}}{(y - x)^2}$$

Substituting for dy/dx, we get

$$\frac{d^2y}{dx^2} = \frac{2y - 2x\left(\dfrac{y + x}{y - x}\right)}{(y - x)^2}$$

$$= \frac{2(y^2 - 2xy - x^2)}{(y - x)^3}$$

Using $y^2 - 2xy - x^2 = 32$ from the original equation, we conclude that

$$\frac{d^2y}{dx^2} = \frac{64}{(y - x)^3}, \qquad y \neq x$$

Example 3.33

Find the equation of the tangent to the circle $x^2 + y^2 = 25$ at the point (3, 4).

Solution

Differentiating with respect to x, we have

$$2x + 2y\frac{dy}{dx} = 0$$

or

$$\frac{dy}{dx} = -\frac{x}{y}, \qquad y \neq 0$$

The slope of the tangent line to the curve at (3, 4) is given by

$$m = \left(\frac{dy}{dx}\right)_{(3,4)} = -\frac{3}{4}$$

The equation of the tangent line through (3, 4) is

$$y - 4 = m(x - 3)$$
$$y - 4 = -\tfrac{3}{4}(x - 3)$$

This yields

$$4y - 16 = -3x + 9$$

or

$$4y + 3x = 25$$

Thus, the equation of the tangent line to the circle at (3, 4) is

$$4y + 3x = 25$$

Exercises

Find dy/dx by implicit differentiation problems 1–20.

1. $y^2 = 4x$

2. $xy = 16$

3. $x^2 + 4y^2 = 24$

4. $9x^2 + 4y^2 = 36$

5. $x^2 - 3y^2 = 18$

6. $x^3 + y^3 = 15$

7. $x^3 + y^3 = 18xy$

8. $x^2 + 2xy = 25$

9. $x^2 + xy + y^2 = 4$

10. $2x^2 - xy + 3y^2 = 24$

11. $x + \sqrt{xy} = y^2$

12. $xy^2 - 2x^2y = 12$

13. $y^2(x^2 + y^2) = 64$

14. $y^3 = 4(x^2 + y^2)$

15. $x^4 - x^2y^2 + y^4 = a^4$

16. $\sqrt{x} + \sqrt{y} = \sqrt{a}, \qquad x \geq 0,\ y \geq 0$

17. $\sqrt{x} + \sqrt{xy} + \sqrt{y} = 25; \qquad x, y \geq 0$

18. $x^{1/3} + y^{1/3} = 27$

19. $x^5 - 2x^3y^2 + 4x^2y^4 + y^5 = 10$

20. $x^6 + 5x^4y^2 + 3x^2y^4 + y^6 = 32$

Find dy/dx and d^2y/dx^2 by implicit differentiation in problems 21–26.

21. $x^2 + y^2 = a^2$

22. $3x^2 + 4y^2 = 12$

23. $25x^2 - 16y^2 = 400$

24. $xy + y^2 = 16$

25. $x^{2/3} + y^{2/3} = a^{2/3}$

26. $x^2 + xy + y^2 = 25$

Find the equation of the tangent to each of the following curves at the point specified.

27. $x^2 + 4y^2 = 17; \qquad (1, 2)$

28. $x^2 + y^2 = 25; \qquad (4, 3)$

29. $xy = 8; \qquad (2, 4)$

30. $xy = 1; \qquad (1, 1)$

31. $2x^2 - xy + 3y^2 = 18; \qquad (3, 1)$

32. $x^2 + xy + y^2 = 4; \qquad (2, -2)$

33. $x^3 - 3axy + y^3 = 0; \qquad \left(\frac{3a}{2}, \frac{3a}{2}\right)$

34. $x^3 + y^3 = 18xy; \qquad (8, 4)$

35. $\sqrt{x} + \sqrt{y} = 6; \qquad (4, 16)$

36. $x^2y^3 - 5y^3 - x = 6; \qquad (2, -2)$

37. Show that the equation of the tangent to the circle

$$x^2 + y^2 = a^2$$

at the point (x_0, y_0) is given by

$$xx_0 + yy_0 = a^2$$

38. Show that the equation of the tangent to the ellipse

$$\frac{x^2}{a^2} + \frac{y^2}{b^2} = 1$$

at the point (x_0, y_0) is given by

$$\frac{xx_0}{a^2} + \frac{yy_0}{b^2} = 1$$

39. Show that the equation of the tangent to the parabola $y^2 = 4ax$ at the point (x_0, y_0) is given by

$$yy_0 = 2a(x + x_0)$$

40. Show that the equation of the tangent to the hyperbola

$$\frac{x^2}{a^2} - \frac{y^2}{b^2} = 1$$

at the point (x_0, y_0) is given by

$$\frac{xx_0}{a^2} - \frac{yy_0}{b^2} = 1$$

Exercises for Review

1. Define each of the following concepts:
   a. difference quotient
   b. instantaneous rate of change
   c. function differentiable at a point
   d. slope of the tangent line at a point
   e. marginal demand
   f. rate of change in the marginal demand
2. Classify each of the following statements as true or false.
   a. If a function is continuous at a point, then it is differentiable at that point.
   b. If a function is differentiable at a point, then it is continuous at that point.
3. Using the five-step rule, find the derivative of each of the following functions.

   a. $f(x) = 3x + 4$

   b. $f(x) = x^2 + 3$

   c. $f(x) = \dfrac{1}{x}$

   d. $f(x) = \dfrac{1}{x^2}$

   e. $f(x) = \sqrt{x + 1}$

   f. $f(x) = \dfrac{1}{\sqrt{x + 2}}$

4. Differentiate each of the following functions.

a. $f(x) = 3x^2 - 5x + 7$

b. $f(x) = 4x^3 + 5x^2 - 2x + 1$

c. $f(x) = 4x + \frac{1}{x}$

d. $f(x) = \sqrt{x} + \frac{1}{\sqrt{x}}$

e. $f(t) = 128t - 16t^2$

f. $f(t) = 192t - 16t^2 + 100$

g. $f(t) = (2t + 1)(2t + 3)$

h. $f(t) = (t + 2)\sqrt{t}$

i. $f(x) = (\sqrt{x} + 1)(\sqrt{x} + 2)$

j. $f(t) = \frac{t - 1}{t + 1}$

k. $f(t) = \frac{t^2 - t + 1}{t^2 + t + 1}$

l. $f(t) = \sqrt{t^2 + 1}$

m. $f(x) = (4 - x^2)^3$

n. $f(x) = (4x + 1)^2(3x + 2)^3$

o. $f(x) = \left(\frac{x + 1}{x - 1}\right)^4$

p. $f(x) = \left(\frac{2x + 1}{3x + 4}\right)^{1/2}$

5. Find dy/dx in each of the following problems.

a. $x^2 + y^2 = 144$

b. $xy = 8$

c. $xy^2 - 2x^2y = 12$

d. $\sqrt{x} + \sqrt{xy} + \sqrt{y} = 25$

6. Find the first two derivatives in each of the following problems.

a. $y = (2x + 3)^2$

b. $y = \frac{1}{x + 1}$

c. $y = \frac{x + 2}{x + 3}$

d. $y = x\sqrt{x}$

7. Find $f'(0)$, $f'(1)$, and $f'(2)$ for each of the following functions.

a. $f(x) = 3x^2 - 4x + 1$

b. $f(x) = (2x + 5)^2$

c. $f(x) = \frac{1}{x + 2}$

d. $f(x) = \sqrt{x + 8}$

8. Find the slope and the equation of the line tangent to each of the following curves at the specified value of x.

a. $y = x^2 + 2x + 3; \quad x = 1$

b. $y = \sqrt{x}; \quad x = 4$

c. $y = (x^2 - 3)^5(2x - 1)^3; \quad x = 1$

d. $y = \left(\frac{x + 1}{x - 1}\right)^2; \quad x = 3$

e. $y = \sqrt{25 - x^2}; \quad x = 3$

9. Find the equation of the line tangent to each of the following curves at the point specified.

a. $x^2 + y^2 = 169; \quad (5, 12)$

b. $\sqrt{x} + \sqrt{y} = 4; \quad (4, 4)$

c. $xy = 16; \quad (2, 8)$

10. The number of bacteria, $f(t)$, in a certain culture after t hours is given by

$$f(t) = 2t^3 + 3t^2 + 100$$

Determine the instantaneous rate of change in the number of bacteria at

a. $t = 1$ hour **b.** $t = 2$ hours.

11. The concentration of a drug in the bloodstream of a patient is approximated by the function

$$f(t) = \frac{1}{t + 1}$$

where $f(t)$ is the quantity of the drug (in suitable units) in the bloodstream t hours after it is injected. How fast is the drug losing its effectiveness at

a. $t = 1$ hour **b.** $t = 2$ hours

12. A psychologist has prepared a list of nonsense words for use in a memory experiment. The number of words, $N(t)$, that a person is able to memorize after t hours of practice is given by

$$N(t) = 40\sqrt{t}, \qquad 0 \leq t \leq 9$$

Determine the instantaneous rate of improvement when

a. $t = 1$ hour **b.** $t = 2$ hours **c.** $t = 4$ hours

13. The demand for a certain product is given by the function

$$D(p) = \frac{500}{\sqrt{2p + 1}}$$

where $D(p)$ represents the demand of the product at the price p. Find the marginal demand at $p = 4$.

14. A ball is thrown vertically into the air. Its height s (in feet) at time t (in seconds) is given by the equation

$$s = 96t - 16t^2$$

Find the velocity and the acceleration of the ball at $t = 2$ seconds.

15. Let $f(x)$ be defined as follows:

$$f(x) = \begin{cases} x - 1, & x \geq 1 \\ 1 - x, & x < 1 \end{cases}$$

a. Graph $f(x)$.
b. Is $f(x)$ continuous at $x = 1$?
c. Is $f(x)$ differentiable at $x = 1$?

4 Applications of the Derivative

4.1 Introduction

The goal of maximizing profit is of foremost interest to business firms. The management of a firm attempts to maximize its profit or the rate of growth of its operations and attempts to minimize the cost of its products. These basic assumptions help explain the behavior of economic institutions. Consequently, the mathematical tools that help the management maximize profit and minimize costs are vitally important to business.

Consider these examples:

A. A real estate office handles a 150-unit apartment complex. Past experience shows that all units are occupied if the rent is \$200 per month, but for each \$10 increase in rent, five units become vacant. What rent should the management charge each tenant so as to maximize the revenue?

B. A printed poster is to have a margin of 2 inches on the sides and 4 inches at the top and bottom. What are the most economical dimensions for the poster board if the printed area is to contain 200 square inches of illustrations and instructions?

C. A farmer wishes to fence off a rectangular pasture along a straight river, so that the side along the river requires no fence. She has enough barbed wire to build a fence one mile long. How long should the sides of the rectangle be so as to enclose the largest possible area?

D. Another farmer estimates that if he begins picking his peaches today, he will have a crop of 500 pounds and can make a profit of \$0.30 a pound. He could wait as long as a week before the peaches are overripe; for each day he waits, he will have 25 extra pounds, but his profit will decrease about \$0.01 per pound per day. When should he pick his crop so as to maximize his profit?

The theory of optimization deals with questions of this nature. Optimization theory has a wide range of applications in agriculture, engineering, medicine, physics, marketing, and many other related areas of business and management.

4.2 Maxima and Minima

We begin with some background material necessary to define the maximum and minimum values of a function. We will make a distinction between a *relative* and *absolute* maximum value and a similar distinction between a *relative* and *absolute* minimum value. Recall from Chapter 1 that the closed interval $[a, b]$ is the set of real numbers $\{x \mid a \leq x \leq b\}$, the open interval (a, b) is the set of real numbers $\{x \mid a < x < b\}$, and the semi-closed intervals $(a, b]$ and $[a, b)$ are the sets of real numbers $\{x \mid a < x \leq b\}$ and $\{x \mid a \leq x < b\}$, respectively. These intervals are of finite length and are called *bounded* intervals.

The notation for a point (x, y) in the plane is the same as the notation for an open interval (a, b) of real numbers. You have to determine which meaning is intended by the context.

Definition 4.1

Let f be a continuous function defined on the open interval (a, b) and let $x_0 \in (a, b)$. Then $f(x_0)$ is a relative maximum of f if there exists $\delta > 0$ such that

$$f(x_0) \geq f(x)$$

for all $x \in (x_0 - \delta, x_0 + \delta) \subseteq (a, b)$.

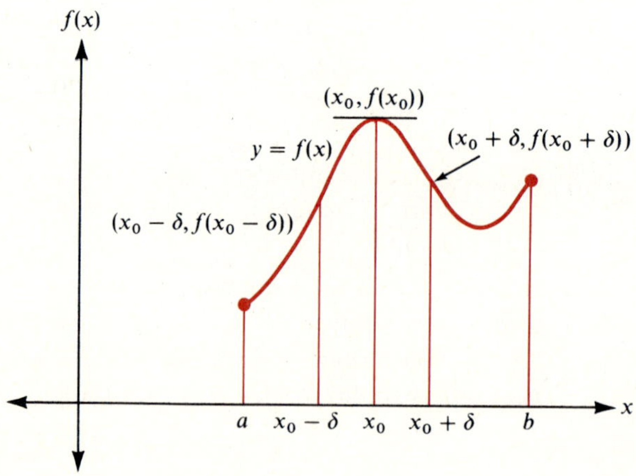

Figure 4.1

This definition states that a function f has a relative maximum at x_0 if the function evaluated at x_0 is greater than or equal to the function evaluated at any other point in the interval $(x_0 - \delta, x_0 + \delta)$. This is shown in Figure 4.1.

Definition 4.2

A continuous function f defined on the open interval (a, b) has a relative minimum at a point x_0 if there exists $\delta > 0$ such that

$$f(x_0) \le f(x)$$

for all $x \in (x_0 - \delta, x_0 + \delta) \subseteq (a, b)$.

This is equivalent to saying that a function f has a relative minimum at x_0 if the value of the function at x_0 is less than or equal to the value of the function at any other point $x \in (x_0 - \delta, x_0 + \delta)$. This is shown in Figure 4.2.

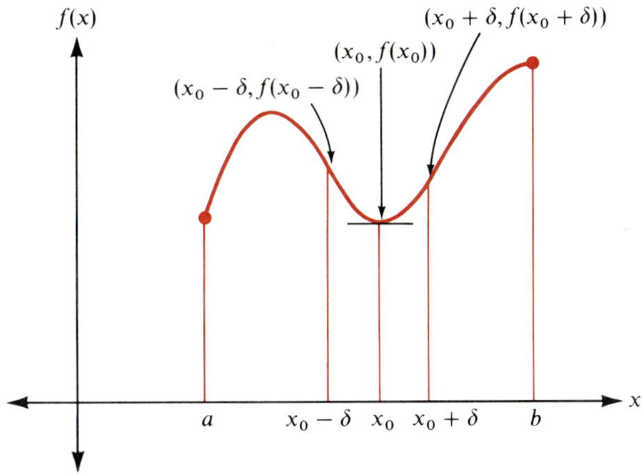

Figure 4.2

The function $y = f(x)$ in Figure 4.1 and Figure 4.2 is differentiable for all $x \in (a, b)$. Remember that the derivative of a function f at $x = x_0$ represents the slope of the tangent to the graph of the function at the point $(x_0, f(x_0))$. Intuitively, it should be clear that if $f(x)$ is to have a relative maximum or relative minimum at $x = x_0$, then the tangent line must be horizontal at $(x_0, f(x_0))$. Since a horizontal line has slope 0, $f'(x) = 0$ at a relative maximum or relative minimum point. It is possible that the derivative may fail to exist at a given point even though the function has a relative maximum or relative minimum at that point. Consider, for example, the graph of

$$f(x) = \begin{cases} 3 - x, & -2 \le x \le 1 \\ x + 1, & 1 < x \le 4 \end{cases}$$

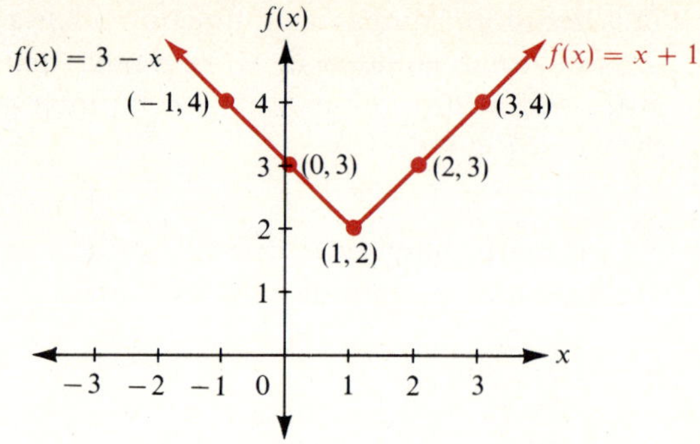

Figure 4.3

in Figure 4.3. Note that even though $f'(1)$ does not exist (why?), f has a relative minimum at $x = 1$. Thus for locating a relative maximum or relative minimum point, it is essential to check any point x_0 in (a, b) at which $f'(x_0) = 0$ and to consider the points in (a, b) where the derivative does not exist.

Definition 4.3

Points in the domain of a function f where $f'(x) = 0$ or where $f'(x)$ does not exist are called *critical points* of the function.

Proposition 4.1

If a function has a relative maximum or relative minimum at a point x_0 in the open interval (a, b), then x_0 is a critical point of the function f.

The converse of Proposition 4.1 is *not* true because the existence of a critical point x_0 does not necessarily imply that the function has a relative maximum or relative minimum at x_0. The function $f(x) = x^3$, for example (see Figure 4.4), has a derivative $f'(x) = 3x^2$ which is zero at $x = 0$, but the point $(0, 0)$ is neither a relative maximum nor a relative minimum of the function f. Thus the knowledge that $f(x)$ has a critical point at x_0 is not enough to conclude that f has a relative maximum or minimum at x_0.

Example 4.1

Discuss the function

$$f(x) = 20 + 60x - 5x^2, \qquad 0 \le x \le 12$$

with regard to relative maxima.

Solution

The function is differentiable everywhere in the open interval $(0, 12)$ and hence the solutions of $f'(x) = 0$ are the only critical points. Differen-

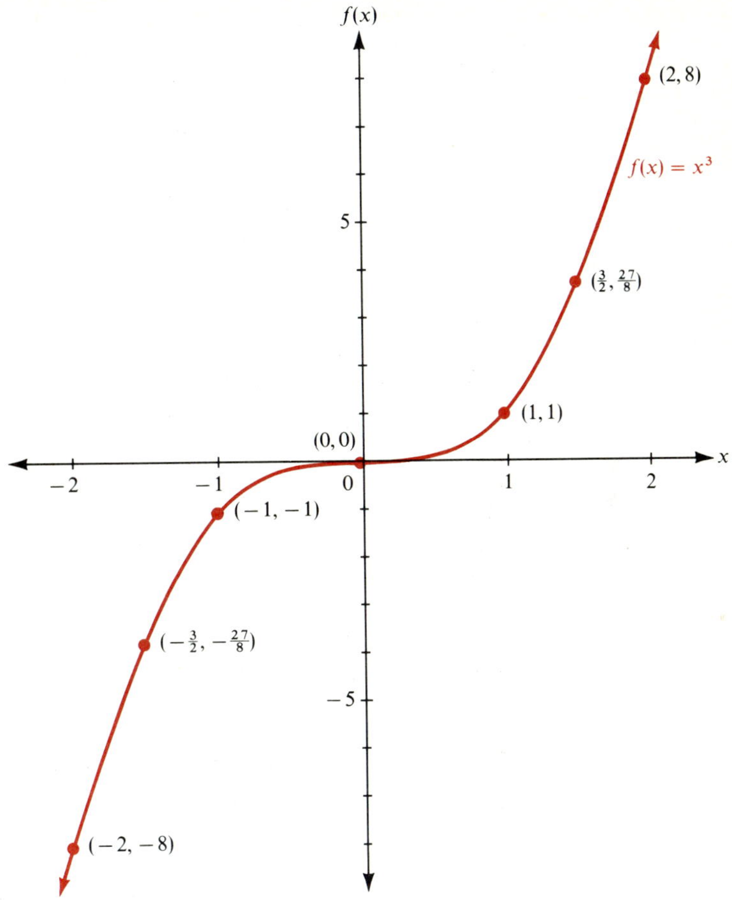

Figure 4.4

tiating $f(x)$, we get

$$f'(x) = 60 - 10x$$

Setting $f'(x) = 0$, we get $x = 6$.

Table 4.1 shows the values of $f(x)$ that correspond to some selected values of $x \in [0, 12]$. Plotting the ordered pairs (0, 20), (1, 75), (2, 120), . . . , (12, 20), we have the graph of f as shown in Figure 4.5.

Table 4.1

x	0	1	2	4	6	8	10	11	12
$f(x)$	20	75	120	180	200	180	120	75	20

The function $f(x)$ has a relative maximum at the critical point $x = 6$; the relative maximum is $f(6) = 200$. Later in the chapter we shall learn how to use the first and second derivatives to verify that this is indeed true.

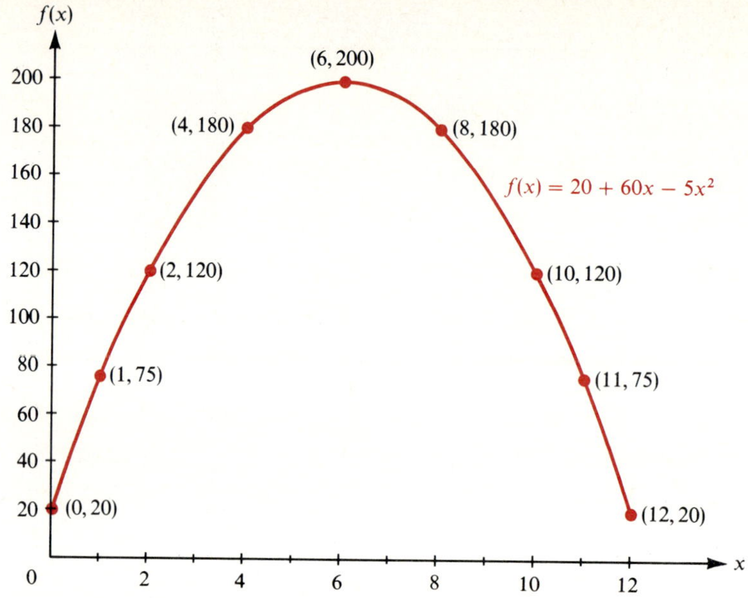

Figure 4.5

Example 4.2

Discuss the function

$$f(x) = x^3 - 3x^2 - 9x + 15, \qquad -4 \leq x \leq 6$$

with regard to relative maxima and minima.

Solution

The derivative $f'(x)$ exists for all values of $x \in (-4, 6)$ and hence the solutions of $f'(x) = 0$ are the only critical points we need to examine for relative maxima and relative minima. Observe that the derivative

$$\begin{aligned} f'(x) &= 3x^2 - 6x - 9 \\ &= 3(x^2 - 2x - 3) \\ &= 3(x - 3)(x + 1), \qquad -4 \leq x \leq 6 \end{aligned}$$

and $f'(x) = 0$ if and only if $x = 3$ or $x = -1$.

Table 4.2 shows the values of $f(x)$ that correspond to some selected values of $x \in [-4, 6]$.

Table 4.2

x	-4	-3	-2	-1	0	1	2	3	4	5	6
$f(x)$	-61	-12	13	20	15	4	-7	-12	-5	20	69

Plotting some of these ordered pairs $(-3, -12)$, $(-2, 13)$, . . . , $(5, 20)$, we have the graph of f as shown in Figure 4.6. The function $f(x)$ has

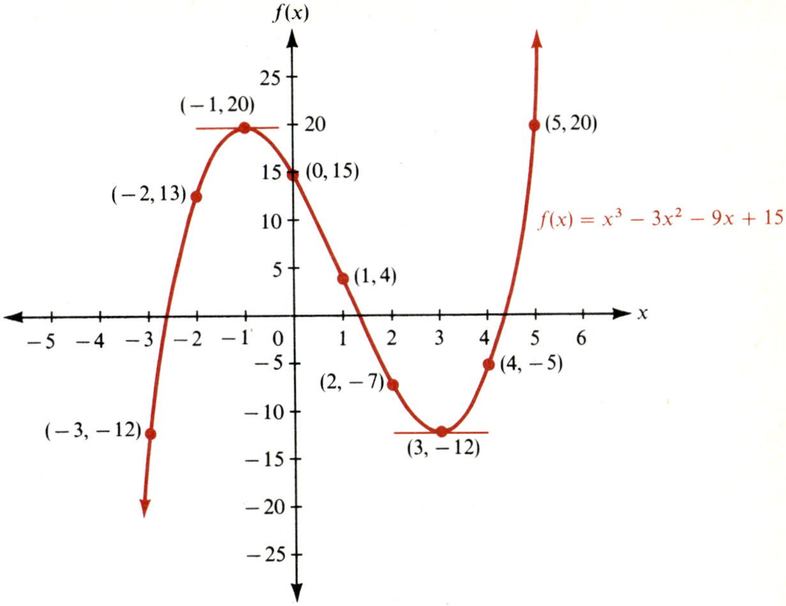

Figure 4.6

a relative maximum at $x = -1$ and a relative minimum at $x = 3$. These values are $f(-1) = 20$ and $f(3) = -12$.

Exercises

Graph each of the following functions and discuss the points of relative maximum and relative minimum.

1. $f(x) = x^2 + 2x + 6, \quad -4 \leq x \leq 4$
2. $f(x) = x^2 - 6x + 13, \quad 0 \leq x \leq 6$
3. $f(x) = 4x - x^2, \quad -2 \leq x \leq 4$
4. $f(x) = 8x - x^2, \quad 0 \leq x \leq 7$
5. $f(x) = x^3 - 3x^2, \quad -3 \leq x \leq 4$
6. $f(x) = x^3 - 12x^2 + 8, \quad -4 \leq x \leq 10$
7. $f(x) = x^3 - 6x^2 + 9x, \quad -1 \leq x \leq 5$
8. $f(x) = x^3 - 9x^2 + 24x + 1, \quad 0 \leq x \leq 5$
9. $f(x) = \begin{cases} x + 1, & 0 \leq x \leq 4 \\ 9 - x, & 4 < x \leq 7 \end{cases}$
10. $f(x) = \begin{cases} 5 - 2x, & -1 \leq x \leq 2 \\ 2x - 3, & 2 < x \leq 5 \\ 22 - 3x, & 5 < x \leq 8 \end{cases}$

4.3 Increasing and Decreasing Functions

In the preceding section, we found relative maxima and relative minima by sketching graphs of functions. Since graphing a function may be time consuming, we must find more efficient methods to determine whether or not a critical point is a point of relative maximum, relative minimum, or neither. We begin with a look at *increasing* and *decreasing* functions.

Definition 4.4

A function $f(x)$ defined on the interval (a, b) is said to be *increasing* on (a, b) if $x_1 < x_2$ implies $f(x_1) < f(x_2)$ for any x_1 and $x_2 \in (a, b)$. A function $f(x)$ defined on the interval (a, b) is said to be *decreasing* on (a, b) if $x_1 < x_2$ implies $f(x_1) > f(x_2)$ for any x_1 and $x_2 \in (a, b)$.

Consider, for example, a demand curve in economics. The demand for a certain product depends largely on its price. Generally speaking, as the price increases, the quantity demanded by consumers decreases. On the other hand, if the price of an item is reduced, the demand for that product increases. Thus the demand of a product is a *decreasing* function of its price. An example of a demand curve is shown in Figure 4.7.

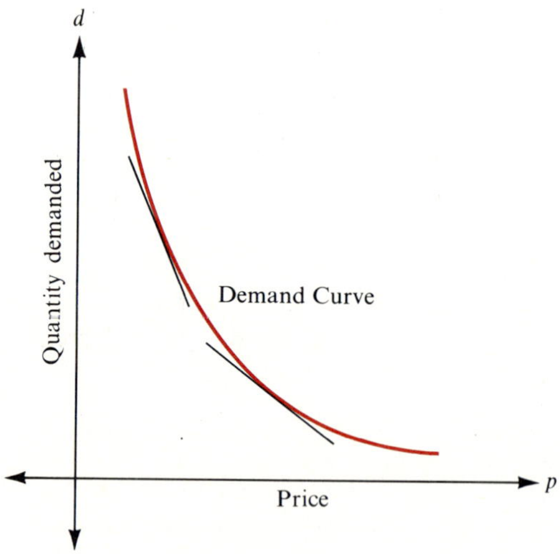

Figure 4.7

Note that the market supply of a product also depends on the product's price. Any increase in the price of a product is an incentive to increase production; consequently, large supplies in the market result in more profit for the suppliers. Thus the supply of a product is an *increasing* function of its price. The graph of a supply curve is shown in Figure 4.8.

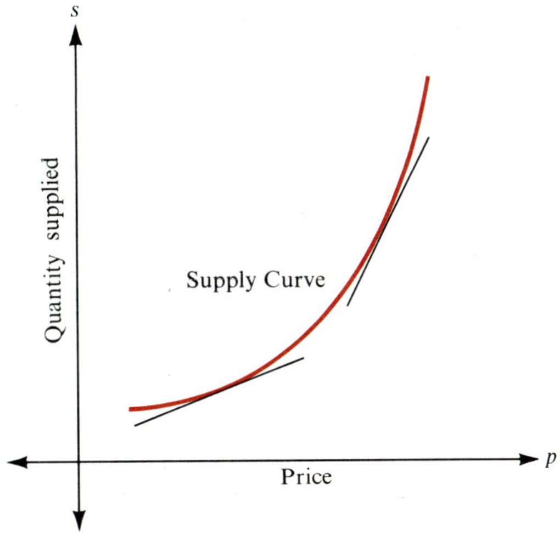

Figure 4.8

It seems reasonable that if a curve is rising as we go from left to right across an interval then the function is increasing on that interval. In such cases the curve has a positive slope at each point in the interval; that is, $f'(x_0) > 0$ for each x_0 in that interval. A curve that is falling from left to right in some interval represents a decreasing function and that curve has a negative slope at each point in the interval; that is, $f'(x_0) < 0$ for each x_0 in that interval.

Proposition 4.2

Let $f(x)$ be differentiable over the interval (a, b). If $f'(x) > 0$ for each $x \in (a, b)$, then f is an increasing function on (a, b); if $f'(x) < 0$ for each $x \in (a, b)$, then f is a decreasing function on (a, b).

This proposition helps us determine the intervals in which a function is increasing and in which it is decreasing. Consider the following examples.

Example 4.3

Find the interval in which the function

$$f(x) = 6x - x^2$$

is increasing and the interval in which it is decreasing.

Solution

Differentiating $f(x)$, we get

$$f'(x) = 6 - 2x$$

Thus $x = 3$ is the only point at which the slope is zero. To examine values of x for which $f'(x) > 0$ and the values of x for which $f'(x) < 0$, we let $f'(x) > 0$ which implies

$$\begin{aligned} 6 - 2x &> 0 \\ 6 &> 2x \\ 3 &> x. \end{aligned}$$

Similarly, $f'(x) < 0$ implies that $x > 3$. Thus $f'(x) > 0$ for $x < 3$ and $f'(x) < 0$ for $x > 3$. Applying Proposition 4.2, we conclude that $f(x)$ is increasing on the interval $(-\infty, 3)$ and decreasing on $(3, \infty)$.

Example 4.4

Find the intervals in which the function

$$f(x) = x^3 - 3x$$

is increasing and the intervals in which $f(x)$ is decreasing.

Solution

First, we find that

$$\begin{aligned} f'(x) &= 3x^2 - 3 \\ &= 3(x^2 - 1) \\ &= 3(x - 1)(x + 1) \end{aligned}$$

Thus, $x = -1$ and $x = 1$ are the only points for which the slope is zero. Figure 4.9 illustrates the values for which the factors $(x - 1)$ and $(x + 1)$ and the function $f'(x)$ are positive and negative.

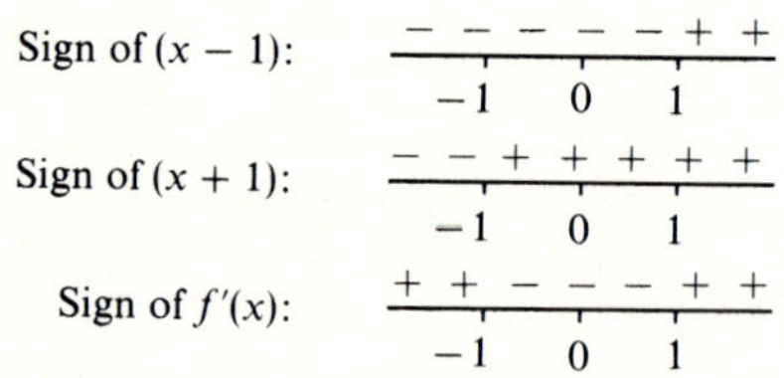

Figure 4.9

Clearly,

$$\begin{aligned} f'(x) &> 0 \quad \text{for} \quad x \in (-\infty, -1) \\ f'(x) &< 0 \quad \text{for} \quad x \in (-1, 1) \\ f'(x) &> 0 \quad \text{for} \quad x \in (1, \infty) \end{aligned}$$

Applying Proposition 4.2 to each of the three intervals, we conclude that $f(x)$ is increasing on the intervals $(-\infty, -1)$ and $(1, \infty)$ and decreasing on the interval $(-1, 1)$.

Example 4.5

Determine the intervals in which the function

$$f(x) = \frac{x}{x^2 + 1}$$

is increasing and the intervals in which the function is decreasing.

Solution

Using the quotient rule, we obtain

$$\begin{aligned} f'(x) &= \frac{(x^2 + 1)\dfrac{d}{dx}(x) - x\dfrac{d}{dx}(x^2 + 1)}{(x^2 + 1)^2} \\ &= \frac{(x^2 + 1) - x(2x)}{(x^2 + 1)^2} \\ &= \frac{1 - x^2}{(x^2 + 1)^2} \\ &= \frac{(1 - x)(1 + x)}{(x^2 + 1)^2} \end{aligned}$$

Since $(x^2 + 1)^2$ is always positive, $f'(x) = 0$ if and only if $x = 1$ or $x = -1$. To study where $f'(x)$ is positive and negative, we observe that the factor $(1 - x)$ is positive for $x < 1$ and negative for $x > 1$. Similarly, the factor $(1 + x)$ is negative for $x < -1$ and positive for $x > -1$. Figure 4.10 shows the sign of each factor along with the sign of the function $f'(x)$.

Sign of $(1 - x)$: + + + + + − −
(axis: −1 0 1)

Sign of $(1 + x)$: − − + + + + +
(axis: −1 0 1)

Sign of $f'(x)$: − − + + + − −
(axis: −1 0 1)

Figure 4.10

Note that

$$\begin{aligned} f'(x) &< 0 \quad \text{for} \quad x < -1 \\ f'(x) &> 0 \quad \text{for} \quad -1 < x < 1 \\ f'(x) &< 0 \quad \text{for} \quad x > 1 \end{aligned}$$

Thus $f(x)$ is decreasing on the intervals $(-\infty, -1)$ and $(1, \infty)$ and increasing on the interval $(-1, 1)$.

Exercises

In problems 1–20, determine the intervals in which the function is increasing and the intervals in which it is decreasing.

1. $f(x) = x^2 - 4x$

2. $f(x) = x^2 - 8x + 12$

3. $f(x) = x^2 + 6x + 1$

4. $f(x) = 6x - x^2$

5. $f(x) = x^3 - 3x + 10$

6. $f(x) = 3x^2 - 2x^3$

7. $f(x) = x^3 - 3x^2$

8. $f(x) = 3 - x - 2x^2$

9. $f(x) = 2x^3 - 9x^2 + 12x$

10. $f(x) = 2x^3 + 3x^2 - 72x$

11. $f(x) = 4x^3 - 3x^2 - 36x + 18$

12. $f(x) = x^3 - 12x + 27$

13. $f(x) = x^4 - 8x^2$

14. $f(x) = x^4 - 4x^3 + 4x^2$

15. $f(x) = x^4 - 2x^2 + 12$

16. $f(x) = x^4 - 2x^2$

17. $f(x) = \dfrac{x}{x^2 + 1}, \qquad x \neq 1$

18. $f(x) = \dfrac{x}{x^2 + 4}$

19. $f(x) = \dfrac{x}{x^2 + 9}$

20. $f(x) = \dfrac{x}{(x^2 + 2)^2}$

21. The cost function for a firm producing a certain commodity is given by

$$f(x) = 100 + 60x - 5x^2$$

where $f(x)$ is the cost in thousands of dollars and x represents the output in hundreds of units per week. Find the production levels for which $f(x)$ is increasing and the levels for which $f(x)$ is decreasing.

22. The profit (in millions of dollars) of a drilling company is determined by the function

$$P(x) = 80x - 10x^2, \qquad 0 \le x \le 8$$

where x is the amount of oil produced (in millions of barrels) and $P(x)$ is the corresponding profit. Determine the production levels for which $P(x)$ is increasing and the levels for which $P(x)$ is decreasing.

23. A psychological study of welders at an industrial plant measured job satisfaction on a scale of 0 to 10 according to the function

$$f(t) = -0.1t^2 + t + 6.5$$

where t was recorded in weeks. Was worker satisfaction increasing or decreasing on the job after 4 weeks?

24. The concentration of a drug x seconds after injection is given by the function

$$f(x) = \frac{x}{x^2 + 900}, \qquad 0 \le x \le 50$$

where x is recorded in minutes. Determine the time intervals for which $f(x)$ is increasing and the time intervals for which $f(x)$ is decreasing.

4.4 The First Derivative Test

The derivative of a function is a powerful tool in analyzing whether a critical point is a point of relative maximum, relative minimum, or neither. Consider, for example, the function in Figure 4.11 that is graphed for $x \in [x_0 - \delta, x_0 + \delta]$, $\delta > 0$.

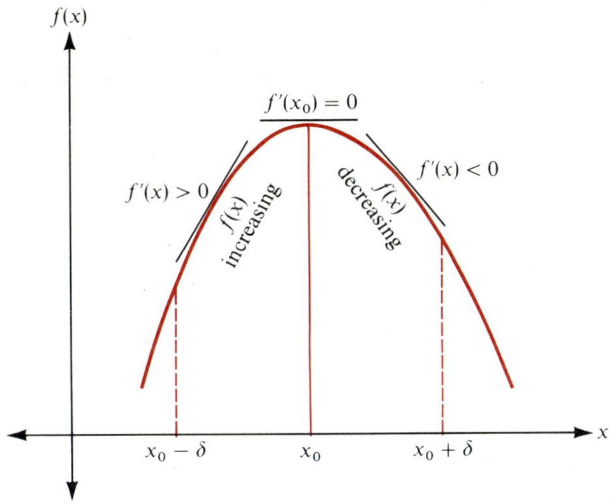

Figure 4.11

At the far left, $f(x)$ increases rapidly, and the first derivative f' is large and positive. As x increases toward x_0, $f(x)$ increases but now more and more slowly until, finally, the graph has a horizontal tangent at x_0. During that interval $[x_0 - \delta, x_0]$, f' becomes smaller and smaller until it reaches zero at $x = x_0$. Thereafter, $f(x)$ decreases, slowly at first, then rapidly. The values of f' in the interval $[x_0, x_0 + \delta]$ start at zero at $x = x_0$ and become more and more negative. In general, $f(x)$ has a relative maximum at x_0 if there exists an interval $(x_0 - \delta, x_0)$ on which $f'(x) > 0$ and an interval $(x_0, x_0 + \delta)$ on which $f'(x) < 0$.

Now let us examine the graph of $f(x)$ in Figure 4.12. At the far left, f is decreasing rapidly and the derivative f' is large and negative. As x increases to x_0, $f(x)$ continues to decrease but at a slower rate until the graph has a horizontal tangent at x_0. Then $f(x)$ increases, slowly at

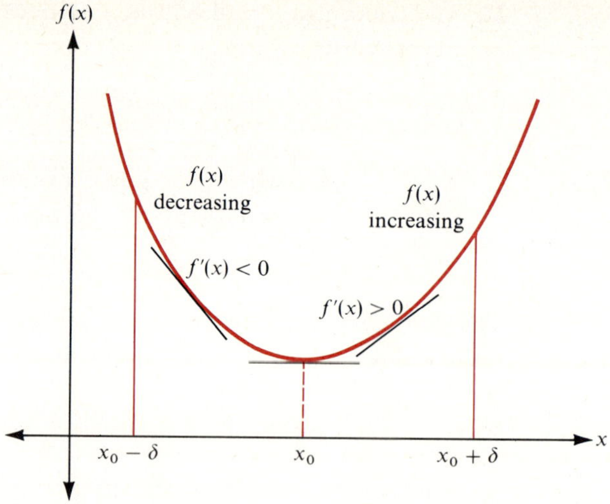

Figure 4.12

first, then rapidly for all values of $x \in (x_0, x_0 + \delta)$, $\delta > 0$. The corresponding values of $f'(x)$ are negative for all values of $x \in (x_0 - \delta, x_0)$. The value of $f'(x)$ is zero at $x = x_0$ and is positive for all values of $x \in (x_0, x_0 + \delta)$. In general, $f(x)$ has a relative minimum at x_0 if there is an interval $(x_0 - \delta, x_0)$ on which $f'(x) < 0$ and an interval $(x_0, x_0 + \delta)$ on which $f'(x) > 0$.

The function $f(x)$ in Figure 4.13 attains neither a maximum nor a minimum in the interval $(x_0 - \delta, x_0 + \delta)$ since $f(x)$ decreases on both sides of the critical point x_0 and $f'(x)$ is negative on both sides of x_0. Note that $f'(x_0) = 0$. These facts lead to the following proposition, which we state without proof.

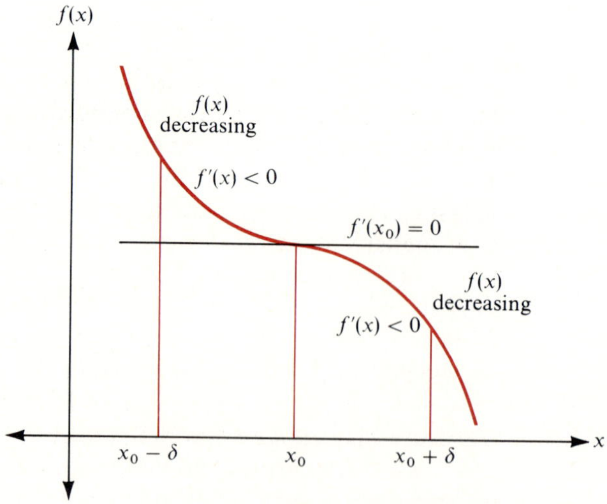

Figure 4.13

Proposition 4.3

Suppose that $f(x)$ is differentiable on the open interval (a, b) and x_0 is a critical point in (a, b). Then the function f has

a. a relative maximum at x_0 if there exists an interval $(x_0 - \delta, x_0)$ on which $f'(x) > 0$ and an interval $(x_0, x_0 + \delta)$ on which $f'(x) < 0$.

b. a relative minimum at x_0 if there exists an interval $(x_0 - \delta, x_0)$ on which $f'(x) < 0$ and an interval $(x_0, x_0 + \delta)$ on which $f'(x) > 0$.

c. neither a relative maximum nor a relative minimum at x_0 if $f'(x)$ has the same sign for both $x \in (x_0 - \delta, x_0)$ and for $x \in (x_0, x_0 + \delta)$.

Example 4.6

Find the points at which

$$f(x) = x^3 - 9x^2 + 24x, \qquad 0 < x < 6$$

has a relative maximum or a relative minimum.

Solution

First, we differentiate $f(x)$ and obtain

$$\begin{aligned} f'(x) &= 3x^2 - 18x + 24 \\ &= 3(x^2 - 6x + 8) \\ &= 3(x - 2)(x - 4) \end{aligned}$$

Thus $f'(x) = 0$ for $x = 2$ or $x = 4$. Further,

$$\begin{aligned} f'(x) &> 0 \quad \text{for} \quad 0 < x < 2 \\ f'(x) &< 0 \quad \text{for} \quad 2 < x < 4 \\ f'(x) &> 0 \quad \text{for} \quad 4 < x < 6 \end{aligned}$$

This means that $f'(x)$ is positive in a neighborhood to the left of $x = 2$, zero at $x = 2$, and negative in a neighborhood to the right of $x = 2$. Consider, for instance, values to the left and to the right of $x = 2$; $x = 1$ and $x = 3$, respectively. (Note that it is extremely important that the values under consideration do not surround another critical point.) For $x = 1$, we have

$$f'(1) = 3(1)^2 - 18(1) + 24 = 9 > 0$$

and for $x = 3$,

$$f'(3) = 3(3)^2 - 18(3) + 24 = -3 < 0$$

Hence by Proposition 4.3, $f(x)$ has a relative maximum at $x = 2$; this relative maximum is $f(2) = 20$. Now observe that

$$f'(5) = 3(5)^2 - 18(5) + 24 = 9 > 0$$

Since $f'(x) < 0$ for $x \in (2, 4)$ and $f'(x) > 0$ for $x \in (4, 6)$, we conclude that $f(x)$ has a relative minimum at $x = 4$; this relative minimum is $f(4) = 16$.

Example 4.7

Discuss the function

$$g(x) = \begin{cases} -x, & -3 < x \leq 0 \\ x^3 - 9x^2 + 24x, & 0 < x < 6 \\ -2x + 48, & 6 \leq x < 9 \end{cases}$$

with regard to relative minima and relative maxima.

Solution

Note that the function $g(x)$ is defined over the open interval $(-3, 9)$ and agrees with the function $f(x)$ of Example 4.6 defined over the open interval $(0, 6)$. The function $g(x)$ is continuous at $x = 0$ and $x = 6$ but is not differentiable at these points (why?). Further,

$$g'(x) = \begin{cases} -1, & -3 < x < 0 \\ 3x^2 - 18x + 24, & 0 < x < 6 \\ -2, & 6 < x < 9 \end{cases}$$

Clearly, $3x^2 - 18x + 24 = 0$ if and only if $x = 2$ or $x = 4$. Thus the critical points are $x = 2$ and $x = 4$ (where $g'(x) = 0$) and $x = 0$ and $x = 6$ (where $g'(x)$ does not exist). Observe that

$$\begin{aligned} g'(x) &< 0 \quad \text{for} \quad x \in (-3, 0) \\ g'(x) &> 0 \quad \text{for} \quad x \in (0, 2) \\ g'(x) &< 0 \quad \text{for} \quad x \in (2, 4) \\ g'(x) &> 0 \quad \text{for} \quad x \in (4, 6) \\ g'(x) &< 0 \quad \text{for} \quad x \in (6, 9) \end{aligned}$$

Hence, $g(x)$ has relative minima at $x = 0$ and $x = 4$ and relative maxima at $x = 2$ and $x = 6$.

Example 4.8

Discuss the function

$$f(x) = x^4 - 18x^2$$

with regard to relative maxima and relative minima.

Solution

Differentiating $f(x)$, we have

$$\begin{aligned} f'(x) &= 4x^3 - 36x \\ &= 4x(x^2 - 9) \\ &= 4x(x - 3)(x + 3) \end{aligned}$$

Clearly, $f'(x) = 0$ if and only if $x = 0$, $x = -3$, or $x = 3$. Note that

$$\begin{aligned} f'(x) &< 0 \quad \text{for} \quad x < -3 \\ f'(x) &> 0 \quad \text{for} \quad -3 < x < 0 \\ f'(x) &< 0 \quad \text{for} \quad 0 < x < 3 \\ f'(x) &> 0 \quad \text{for} \quad x > 3 \end{aligned}$$

This analysis reveals that as x increases from left to right, the derivative $f'(x)$ is negative on the interval $(-\infty, -3)$, zero at $x = -3$, and positive on the interval $(-3, 0)$; by Proposition 4.3, the function has a relative minimum at $x = -3$. Since the derivative $f'(x)$ is positive on the interval $(-3, 0)$ and negative on the interval $(0, 3)$, we conclude that the function has a relative maximum at $x = 0$. Similarly, we observe that the critical point $x = 3$ is a point of relative minimum.

We have studied methods for locating relative maxima and relative minima of a function defined on an open interval. If the function is defined on a closed interval $[a, b]$, it may attain the largest or smallest functional value at an endpoint of the interval. The function whose graph is sketched in Figure 4.14, for example, has a relative maximum at

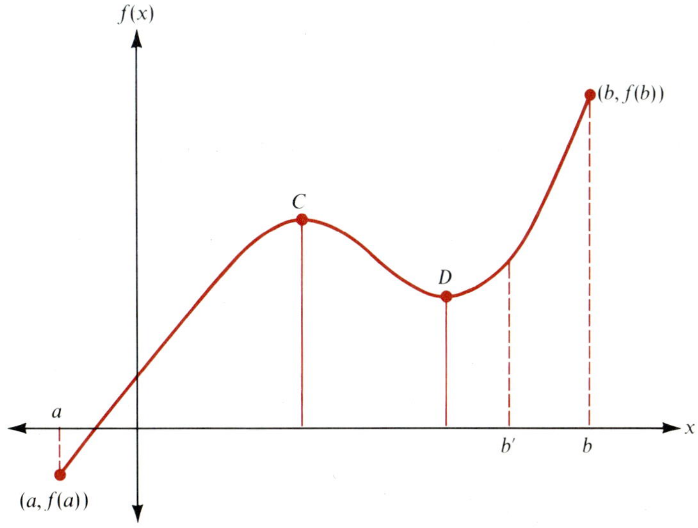

Figure 4.14

C but attains its largest functional value at the endpoint b. Similarly, the function has a relative minimum at D but the smallest functional value occurs at the endpoint a. If $[a, b']$ were the interval of interest, then the largest functional value would occur at the point C and the smallest value at the endpoint a. This discussion enables us to define the *absolute maximum* and *absolute minimum* of a function.

Definition 4.5

A continuous function $f(x)$ has an *absolute maximum* at x_0 if

$$f(x) \le f(x_0)$$

for all x in the domain of the function. The function has an *absolute minimum* at x_0 if

$$f(x) \ge f(x_0)$$

for all x in the domain of the function.

From this definition, we can conclude that if a function is continuous and is defined over $[a, b]$, then an absolute maximum or minimum is attained either at some point in (a, b) or at an endpoint of the interval.

Example 4.9

Determine the absolute maximum and minimum of the function

$$f(x) = x^3 - 12x^2 + 21x + 5$$

where the domain of the function is $[-1, 8]$.

Solution

Note that $f'(x)$ exists for all $x \in (-1, 8)$ and

$$\begin{aligned} f'(x) &= 3x^2 - 24x + 21 \\ &= 3(x^2 - 8x + 7) \\ &= 3(x - 1)(x - 7) \end{aligned}$$

The absolute maximum and minimum values of the function occur either at the endpoints $x = -1$ and $x = 8$ or at the points $x = 1$ and $x = 7$ at which $f'(x) = 0$. Evaluating $f(x)$ at each of these points, we observe that

$$f(-1) = -29, \quad f(1) = 15$$
$$f(7) = -93, \quad f(8) = -83$$

Hence, the absolute maximum occurs at $x = 1$ and the absolute minimum at $x = 7$.

Example 4.10

Determine all maximum and minimum values of the function

$$f(x) = \frac{x^2 + 8}{x - 1}, \qquad x \neq 1$$

on the interval $[-1, 7]$.

Solution

Note that Definition 4.5 does not apply to this example; the function

$$f(x) = \frac{x^2 + 8}{x - 1}, \qquad x \neq 1$$

is not defined at $x = 1$. Thus it is *not* continuous and hence not differentiable on the closed interval $[-1, 7]$. We can study the behavior of the function by graphing it. Table 4.3 shows the values of $f(x)$ that correspond to some values of x in the domain of f.

Table 4.3

x	$f(x)$	x	$f(x)$
−1	−4.5	1.01	902.1
0	−8.0	1.10	92.1
0.50	−16.5	1.50	20.5
0.90	−88.1	2.0	12.0
0.95	−178.05	4.0	8.0
0.99	−898.01	7.0	9.5

Plotting some of the ordered pairs $(-1, -4.5)$, $(0, -8)$, . . . , $(4, 8)$, $(7, 9.5)$ we have the graph of $f(x)$ as shown in Figure 4.15. Note that as x approaches 1 from the right, $f(x)$ approaches positive infinity and as x approaches 1 from the left, $f(x)$ approaches negative infinity. Thus there is no absolute maximum or absolute minimum in the interval. Further, if we differentiate $f(x)$, we obtain

$$\begin{aligned} f'(x) &= \frac{(x-1)\dfrac{d}{dx}(x^2+8) - (x^2+8)\dfrac{d}{dx}(x-1)}{(x-1)^2} \\ &= \frac{(x-1)(2x) - (x^2+8)\cdot 1}{(x-1)^2} \\ &= \frac{x^2 - 2x - 8}{(x-1)^2} \\ &= \frac{(x-4)(x+2)}{(x-1)^2}, \qquad x \neq 1 \end{aligned}$$

Now $f'(x) = 0$ if and only if $x = -2$ or $x = 4$. Since $x = -2$ is not in the domain of the function, the relative maximum or relative minimum

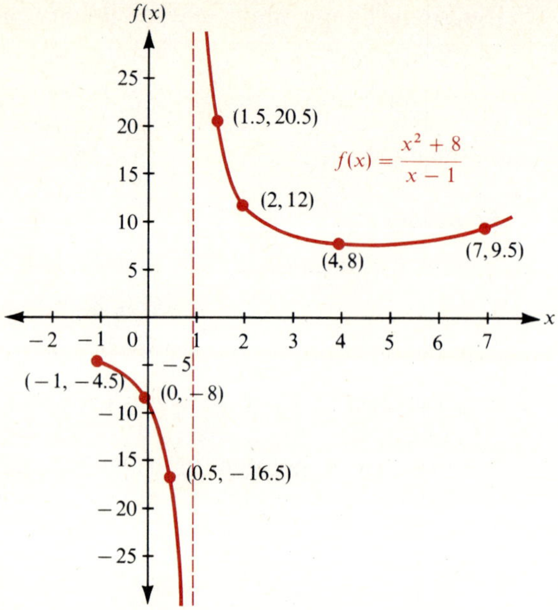

Figure 4.15

for the interval $[-1, 7]$ is attained only at the point $x = 4$. In fact, the point $(4, 8)$ is a relative minimum.

In summary, we use the following methods to locate maximum and minimum points of a *continuous* function defined on a closed interval.

1. Determine the critical points; they are all points for which $f'(x) = 0$, along with the points for which $f'(x)$ does not exist.
2. Find the values of f at the critical points and at the end points of the interval. The largest of the values is the absolute maximum and the smallest is the absolute minimum.

Exercises

Examine each of the following functions for relative maxima and relative minima in the open intervals.

1. $f(x) = x^2 - 6x + 8; \quad (1, 6)$
2. $f(x) = 2x^2 - 8x + 3; \quad (0, 5)$
3. $f(x) = 3x^2 - 6x + 10; \quad (-1, 4)$
4. $f(x) = 4 - 4x - x^2; \quad (-4, 4)$
5. $f(x) = 2x^3 - 9x^2 + 12x + 8; \quad (-4, 4)$
6. $f(x) = x^3 - 3x + 5; \quad (-3, 3)$
7. $f(x) = x^3 + x^2 - x + 1; \quad (-3, 3)$

8. $f(x) = x^3 - 4x^2 + 4x + 5; \qquad (-2, 5)$

9. $f(x) = 3x^4 - 8x^3 - 18x^2 + 9; \qquad (-3, 4)$

10. $f(x) = \begin{cases} x^2 - 1, & x < -1, x > 1 \\ 1 - x^2, & -1 < x < 1 \end{cases}$

11. $f(x) = \begin{cases} 3x + 18, & -6 < x < -3 \\ x^4 - 8x^2, & -3 \le x \le 3 \\ -3x + 18, & 3 < x < 6 \end{cases}$

12. $f(x) = \begin{cases} x + 2, & -5 < x < -1 \\ x^2, & -1 \le x \le 1 \\ -2x + 3, & 1 < x < 6 \end{cases}$

13. $f(x) = \dfrac{x}{x^2 + 1}; \qquad (-2, 2)$

14. $f(x) = \dfrac{x}{x^2 + 4}; \qquad (-3, 4)$

In problems 15 through 28, find the absolute maximum and the absolute minimum of each function on its closed interval.

15. $f(x) = x^2 + 2; \qquad [-2, 3]$

16. $f(x) = x^2 - 4x + 5; \qquad [0, 5]$

17. $f(x) = 3x^2 + 6x - 5; \qquad [-3, 2]$

18. $f(x) = 2x^2 - 8x + 1; \qquad [-1, 4]$

19. $f(x) = 2x^3 + 3x^2 - 12x; \qquad [-3, 4]$

20. $f(x) = x^3 - 2x^2 + x - 1; \qquad [-3, 3]$

21. $f(x) = -x^3 + 12x^2 - 32; \qquad [2, 6]$

22. $f(x) = x^3 - 3x + 8; \qquad [-3, 4]$

23. $f(x) = x^4 - 2x^2 + 1; \qquad [-2, 3]$

24. $f(x) = x^4 + 4x^3 + 6x^2 + 4x + 1; \qquad [-3, 2]$

25. $f(x) = \dfrac{x}{x^2 + 4}; \qquad [-1, 4]$

26. $f(x) = \dfrac{x}{10} + \dfrac{10}{x} + 7, \qquad 1 \le x \le 40$

27. $f(x) = \dfrac{x^2 - 5}{x - 3}, \qquad x \ne 3; [0, 4]$

28. $f(x) = \dfrac{x^2 - 5}{x - 3}, \qquad x \ne 3; [0, 6]$

29. A ball is thrown vertically up into the air. Its height at time t (in seconds) is given by

$$s(t) = 128t - 16t^2$$

where $s(t)$ is the height of the ball (in feet) above the ground at time t. After how many seconds will the ball attain its maximum height?

30. A ball is thrown vertically up into the air. Its height at time t (in seconds) is given by

$$s(t) = 256t - 16t^2$$

where $s(t)$ is the height of the ball (in feet) above the ground at time t. After how many seconds will the ball be at its maximum height?

31. The profit function of a firm producing hair dryers is given by

$$p(x) = 30x - x^2 - 50$$

where x denotes the output of the firm (in hundreds of units per month). Determine the value of x that will maximize profits. What is the total profit of the firm at that level of production?

32. The Alexander Baking Company finds that its profit is directly related to the number of wedding cakes sold in a month. Experience has shown that the profit function is approximated by

$$p(x) = 70x - x^2$$

where x represents the number of cakes sold per month. Determine the number of wedding cakes that must be sold so as to maximize profits.

33. The total profit of a firm that produces a single product is approximated by the function

$$p(x) = 135x + 18x^2 - x^3 - 250$$

where x is the number of units sold. Determine the number of units that must be sold so as to maximize the profit.

34. An average cost function of a product is approximated by

$$f(x) = 10 + \frac{x}{4} + \frac{25}{x}, \qquad 0 < x \leq 40$$

Determine the level of production x for which the average cost is minimum.

35. The number of bacteria (in thousands) present in a certain culture at time t is given approximately by the function

$$f(t) = 4(20t - 2t^2 + 5)$$

where t is recorded in hours. Determine the value of t for which the number of bacteria is at its maximum level.

36. A psychologist has prepared a list of nonsense words for use in a memory experiment. The number of words, $N(t)$, that a person is able to memorize after t hours of practice is given by

$$N(t) = 75t - 15t^2, \qquad 0 < t \leq 4$$

Determine the number of hours a person should practice so as to maximize $N(t)$.

37. The concentration of a drug in the bloodstream of a patient is approximated by the function

$$Q(t) = \frac{t}{t^2 + 625}$$

where $Q(t)$ is the quantity of the drug (in suitable units) in the bloodstream t minutes after it is injected. After how many minutes will the concentration reach its maximum level?

38. The concentration of anesthesia x seconds after injection is given by the function

$$D(x) = \frac{x}{x^2 + 400}, \qquad 0 < x \leq 50$$

After how many seconds will the concentration reach its maximum level?

4.5 The Second Derivative Test

In the preceding sections we have used the first derivative to determine a function's critical points and the intervals over which it is increasing or decreasing. This information enabled us to locate relative maximum and relative minimum points. Now we examine how the second derivative can be used in conjunction with the first derivative to examine critical values. To begin with, let us return to the graph of the function f in Figure 4.11 which has a relative maximum at x_0. Recall that the first derivative, which is associated with the slope of the tangent, is positive for all $x \in (x_0 - \delta, x_0)$, zero for $x = x_0$, and negative for all $x \in (x_0, x_0 + \delta)$. Thus the second derivative, which determines the rate of change of the slope, is negative near a point of relative maximum.

For the function in Figure 4.12, which has a relative minimum at x_0, the first derivative is negative for all $x \in (x_0 - \delta, x_0)$, zero for $x = x_0$, and positive for all $x \in (x_0, x_0 + \delta)$. Thus, the rate of change of the slope, or $f''(x)$, is positive near a point of relative minimum. For the graph of the function in Figure 4.13, we note that the first derivative is negative for all $x \in (x_0 - \delta, x_0)$, zero for $x = x_0$, and negative for all $x \in (x_0, x_0 + \delta)$. Hence there is no change in the rate of change of the slope near x_0 and thus the second derivative at x_0 is zero. These observations lead to the following proposition, which we state without proof.

Proposition 4.4

Let $f(x)$ be a function defined on (a, b). If $x_0 \in (a, b)$ such that $f(x_0)$ is differentiable, $f'(x_0) = 0$, and $f''(x_0)$ exists, then

a. $f(x_0)$ is a relative maximum if $f''(x_0) < 0$

b. $f(x_0)$ is a relative minimum if $f''(x_0) > 0$

The test fails if $f''(x_0) = 0$.

Example 4.11

The function

$$f(x) = x^3 - 9x^2 + 24x$$

has a first derivative

$$\begin{aligned} f'(x) &= 3x^2 - 18x + 24 \\ &= 3(x^2 - 6x + 8) \\ &= 3(x - 2)(x - 4) \end{aligned}$$

Note that the function is differentiable for all x and $f'(x) = 0$ if and only if $x = 2$ or $x = 4$. Thus the critical points are $x = 2$ and $x = 4$. Now we apply the second derivative test. Note that

$$\begin{aligned} f''(x) &= 6x - 18 \\ &= 6(x - 3) \end{aligned}$$

Observe that $f''(2) = -6 < 0$ and $f''(4) = 6 > 0$. Hence $x = 2$ corresponds to a relative maximum and $x = 4$ corresponds to a relative minimum.

Example 4.12

Use the second derivative test to examine the function

$$f(x) = x^4 - 2x^2 + 10$$

for relative maxima and relative minima.

Solution

Differentiating $f(x)$, we have

$$\begin{aligned} f'(x) &= 4x^3 - 4x \\ &= 4x(x^2 - 1) \\ &= 4x(x - 1)(x + 1) \end{aligned}$$

Setting $f'(x) = 0$, we obtain $x = 0$, $x = 1$, and $x = -1$ as critical points. The second derivative of $f(x)$ is

$$f''(x) = 12x^2 - 4 = 4(3x^2 - 1)$$

At the critical points, we have

$$f''(-1) = 8, \quad f''(0) = -4, \quad f''(1) = 8$$

Consequently, $f(0)$ is a relative maximum value and $f(-1)$ and $f(1)$ are relative minimum values of $f(x)$.

Example 4.13

An automobile dealer has determined that local commercial television advertisements have an appreciable impact on new car sales. It is estimated that the number of cars sold, $s(x)$, and the expenditure for advertising, x (in thousands of dollars), are related by the function

$$s(x) = 50 + 36x - 3x^2$$

Determine the amount that must be spent on television commercials so as to maximize sales.

Solution

Differentiating $s(x)$ we obtain

$$\begin{aligned} s'(x) &= 36 - 6x \\ &= 6(6 - x) \end{aligned}$$

The critical points are the solutions to $s'(x) = 0$; that is, $x = 6$ is the only point that could be a relative maximum or minimum point. Observe that since $s''(x) = -6$ for *any* value of x, we conclude that $x = 6$ yields maximum sales.

Example 4.14

A truck uses gasoline at the rate of

$$\frac{1}{10}\left(\frac{25}{x} + \frac{x}{121}\right)$$

gallons per mile while traveling x miles per hour on an interstate highway. If gasoline costs \$1.21 per gallon, determine the speed of the truck that will minimize fuel cost for a 1,500 mile trip. What is the minimum fuel cost?

Solution

The cost of fuel, $c(x)$, is given by

$$\begin{aligned} c(x) &= \text{cost per gallon} \times \text{number of gallons used} \\ &= (1.21) \times 1500\left[\frac{1}{10}\left(\frac{25}{x} + \frac{x}{121}\right)\right] \\ &= 181.5\left(\frac{25}{x} + \frac{x}{121}\right) \\ &= 181.5\left(25x^{-1} + \frac{x}{121}\right) \end{aligned}$$

Thus,

$$c'(x) = 181.5\left(-25x^{-2} + \frac{1}{121}\right)$$

and $c'(x) = 0$ implies that $x^2 = 3025$ or $x = 55$. It is clear that

$$\begin{aligned} c''(x) &= 181.5(-25)(-2)(x^{-3}) \\ &= \frac{9075}{x^3} \end{aligned}$$

and $c''(55) > 0$. Thus the most economical speed for the truck is 55 miles per hour. The minimum fuel cost is

$$c(55) = 181.5\left(\frac{25}{55} + \frac{55}{121}\right) = \$165.00$$

Exercises

In problems 1–15, use the second derivative to identify the relative maxima and relative minima in the open intervals.

1. $f(x) = x^2 - 4x + 3$; $(0, 6)$

2. $f(x) = 2x^2 - 5x + 8$; $(-1, 5)$

3. $f(x) = 40 - 12x - x^2$; $(-8, 8)$

4. $f(x) = 8x - x^2$; $(0, 6)$

5. $f(x) = x^3 - 6x^2$; $(-2, 5)$

6. $f(x) = x^3 - 6x^2 + 9x + 5$; $(-1, 5)$

7. $f(x) = 2x^3 + 3x^2 - 36x + 8$; $(-4, 4)$

8. $f(x) = 2x^3 - 27x^2 + 120x$; $(2, 6)$

9. $f(x) = 2x^3 + 27x^2 + 120x$; $(-6, 6)$

10. $f(x) = x^4 - 2x^2 + 6$; $(-3, 5)$

11. $f(x) = x^2(x^2 - 4x + 4)$; $(-5, 4)$

12. $f(x) = x^4 + \dfrac{16}{3}x^3 + 6x^2$; $(-4, 2)$

13. $f(x) = x + \dfrac{1}{x}$; $(0, 3)$

14. $f(x) = x + \dfrac{4}{x}$; $(1, 5)$

15. $f(x) = x + \dfrac{16}{x}$; $(2, 8)$

In problems 16–20, determine the production level x for which the cost function $c(x)$ attains a minimum value.

16. $c(x) = x^2 - 12x + 25$

17. $c(x) = x^3 - 6x^2 - 15x + 30$

18. $c(x) = 100 - 30x - \frac{9}{2}x^2 + x^3$

19. $c(x) = \frac{1}{3}x^3 - 5x^2 + 24x + 75$

20. $c(x) = 6x^3 - 90x^2 + 1350x + 8000$

21. Find the production level x that maximizes the profit function

$$P(x) = 200x - 0.5x^2$$

22. The total profit made by selling x units of a single product is given by

$$P(x) = 160x - x^2 - 300$$

Determine the number of units of x that will maximize the profit.

23. The number of air conditioners expected to be sold next summer in the ABC Department Store is given by

$$s(x) = 15 + 20x - x^2, \qquad 0 \le x \le 15$$

where x is the number of hot days in the summer. Determine the number of hot days next summer that will maximize the sales of air conditioners.

24. A station wagon uses gasoline at the rate of

$$\frac{1}{12}\left(\frac{25}{x} + \frac{x}{100}\right)$$

gallons per mile while traveling x miles per hour on an interstate highway. If gasoline costs \$1.10 per gallon, determine the speed of the station wagon that will minimize the fuel costs for a 600 mile trip. Find the minimum fuel cost.

25. The total cost function of a product is

$$c(x) = x^3 - 120x^2 + 2100x + 1000$$

where x is the number of units produced. Determine the value of x that would minimize the total cost.

26. The total cost function of a product is

$$c(x) = x^3 - 315x^2 + 27{,}000x + 20{,}000$$

where x is the number of units produced. Determine the number of units that should be produced to minimize the cost.

27. The cost function of a product in a certain industry is

$$c(x) = 0.5x^2 + 10x + 200, \qquad 0 < x \leq 25$$

where x is the number of products the industry produces in one month. Determine the value of x for which the average cost, $c(x)/x$, is minimum.

28. The cost function of a product in an industry is

$$c(x) = 0.01x^2 + 50x + 900, \qquad 0 < x \leq 500$$

Determine the number of units that should be produced so as to minimize the average cost, $c(x)/x$.

29. A baseball player hits a foul ball. Its height at time t (in seconds) is given by

$$s(t) = 80t - 16t^2$$

where $s(t)$ is the height of the ball (in feet) above the ground at time t. After how many seconds will the ball be at its maximum height? What is the maximum height of the ball?

30. The distance d (in miles) between two ships at sea is a function of time t (in hours). If

$$d(t) = 256 - 70t + 5t^2$$

find the time at which the distance between the ships is minimum. Determine the minimum distance between the ships.

31. The population of a certain town is expected to increase according to the formula

$$N(t) = 20 + 4.5t^2 - t^3$$

where $N(t)$ is the number of persons (in thousands) at time t (in years). Determine the value of t at which the population of this town will be at its maximum level.

32. The number of harmful bacteria per cubic centimeter in a reservoir t days after it was treated with bacteriacide is given by

$$N(t) = 1000 - 200t + 20t^2, \qquad 0 \leq t \leq 8$$

After how many days will the concentration of the bacteria, $N(t)$, be at its minimum level?

33. The number of new words in a foreign language that a student is able to memorize after t hours of practice is given by

$$f(t) = 60t - 15t^2, \qquad 0 \leq t \leq 4$$

Determine the number of hours a student should plan to study so as to maximize the number of new words, $f(t)$.

4.6 Optimization Problems

So far we have been given polynomial functions and rational functions and asked to evaluate the functions at their relative maximum, relative minimum and endpoint values. In this section we encounter problems in which we must set up the function before we may proceed to find the lowest cost, the maximum profit, the largest area, and the like.

Example 4.15

Find two positive integers that have a sum of 36 such that their product is as large as possible.

Solution

If x is one of the positive integers, then the other integer is $(36 - x)$. Thus the function to be maximized is

$$\begin{aligned} f(x) &= x(36 - x), \qquad 0 < x < 36 \\ &= 36x - x^2 \end{aligned}$$

The derivative

$$f'(x) = 36 - 2x, \qquad 0 < x < 36$$

is zero if and only if $x = 18$. Since $f''(x) = -2 < 0$, we conclude that $f(x)$ attains its maximum value at $x = 18$. The positive integers are both equal to 18 and their product is 324.

Example 4.16

A farmer has 3000 feet of barbed wire to enclose a rectangular pasture. Her plan includes fencing the entire area and then subdividing it by running a fence across the width as shown in Figure 4.16. Determine the dimensions of the pasture which would enclose the maximum area.

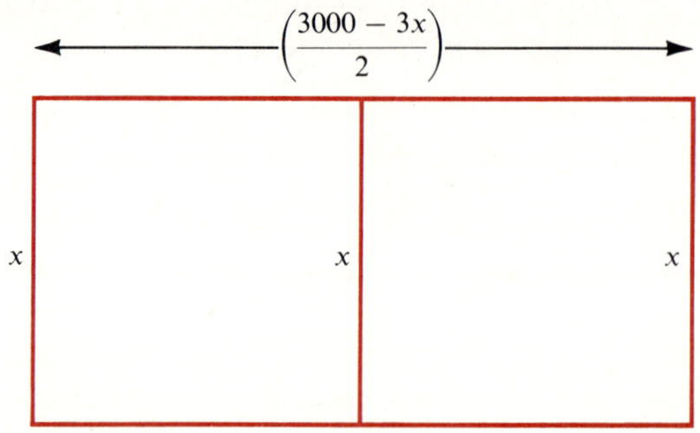

Figure 4.16

Solution

Let x be the width of the pasture (in feet). Then the length of the area to be fenced is $(3000 - 3x)/2$. The function to be maximized is

$$\begin{aligned} A(x) &= x\left(\frac{3000 - 3x}{2}\right), \qquad x \geq 0 \\ &= \tfrac{3}{2}(1000x - x^2) \end{aligned}$$

Then

$$A'(x) = \tfrac{3}{2}(1000 - 2x), \qquad x \geq 0$$

$A'(x) = 0$ if and only if $x = 500$. Since $A''(x) = -3 < 0$, the value $x = 500$ maximizes the function $A(x)$. Thus 3000 feet of fencing for the plan illustrated in Figure 4.16 can enclose a pasture 750 feet long and 500 feet wide.

Example 4.17

An open box is to be made from a rectangular sheet of tin by cutting equal squares out of each corner and folding up the sides. If the sheet is 21 inches long and 16 inches wide and the box is to have the maximum volume, what size squares should be cut out?

Solution

We let x be the side of the square to be cut from each corner of the sheet as shown in Figure 4.17. Then the box obtained by folding the sides up along the dotted lines is $(21 - 2x)$ inches long, $(16 - 2x)$ inches wide, and x inches deep (see Figure 4.18).

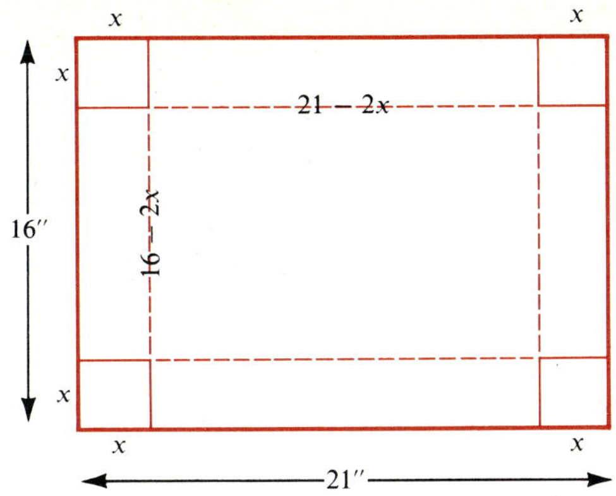

Figure 4.17

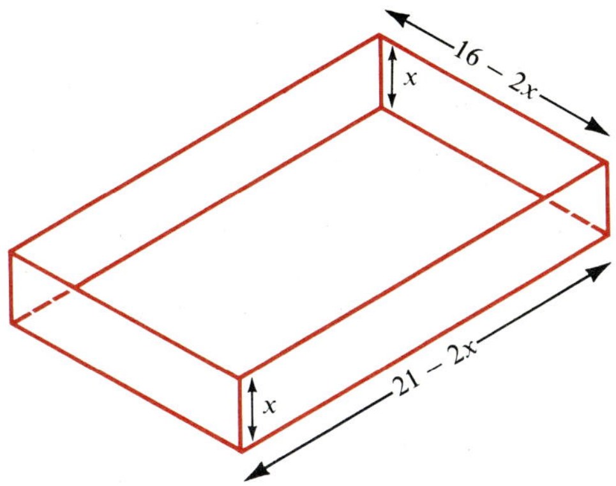

Figure 4.18

Thus the volume of the box is

$$\begin{aligned} V(x) &= x(21 - 2x)(16 - 2x) \\ &= 336x - 74x^2 + 4x^3, \qquad 0 < x < 8 \end{aligned}$$

The restriction $0 < x < 8$ means that the box must have a positive depth ($x > 0$) and a positive width ($x < 8$) for if $x = 8$, the width of the base is zero and we have no box at all. Differentiating $V(x)$ yields

$$\begin{aligned} V'(x) &= 336 - 148x + 12x^2 \\ &= 4(84 - 37x + 3x^2) \\ &= 4(3x - 28)(x - 3) \end{aligned}$$

Setting $V'(x) = 0$, we obtain $x = 3$ and $x = \frac{28}{3}$. The only critical value worth further consideration is $x = 3$ since $x = \frac{28}{3}$ is not in the domain of x. Further,

$$V''(x) = -148 + 24x$$

and $V''(3) = -76 < 0$. Hence the critical value $x = 3$ maximizes the volume of the box; that maximum volume, or $V(3)$, is 450 cubic inches.

Example 4.18

A travel agency has chartered a flight to Paris for 100 high school students. The charge is \$200 with an additional charge of \$2.50 for each cancellation. The plane has a seating capacity of 100 passengers. Determine the total revenue, $R(x)$, as a function of x, the number of cancellations received prior to the departure date. What is the value of x for which $R(x)$ is maximum? What is the maximum revenue?

Solution

If x is the number of cancellations received prior to the departure of the flight, then $(100 - x)$ passengers are on the plane and each pays \$200 plus \$2.50x. Thus the total receipts of the travel agency are

$$R(x) = (100 - x)(200 + 2.50x)$$
$$= 20{,}000 + 50x - 2.50x^2 \qquad 0 \le x \le 100$$

The derivative

$$R'(x) = 50 - 5x$$

is zero if and only if $x = 10$. We need to evaluate $R(x)$ at $x = 10$ and at the endpoint $x = 0$. (The other endpoint is excluded since if 100 cancellations are received, the chartered flight will not take off.) Note that $R''(x) = -5 < 0$ for all x and that

$$R(0) = 20{,}000$$
$$R(10) = 20{,}250$$

Clearly, $x = 10$ maximizes the revenue function. If there are 10 cancellations, each passenger will pay $200 + 10(2.5) = \$225$ and the travel agency will receive \$20250.

Revenue and profit functions typically have maximum points. Consider, for example, a firm that produces a certain quantity x of a single product and generates a sales revenue $R(x)$. If $C(x)$ is the total cost of producing x units, then the profit function is

$$P(x) = R(x) - C(x).$$

Assuming that M is the production capacity of the plant, the management must decide what quantity $(0 \le x \le M)$ it should produce so as to maximize the profit function $P(x)$. If $R(x)$ and $C(x)$ are both differentiable on the closed interval $[0, M]$, then the maximum profit, if it exists, occurs either at the interval's endpoints or at those points for which $P'(x) = 0$.

Example 4.19

The cost and revenue functions for a certain product are

$$C(x) = 5000 + 250x - 0.1x^2$$
$$R(x) = 400x - 0.2x^2$$

Assuming that the plant's production capacity is 1000 units, determine the number x of units that should be produced and sold so as to maximize profit. What is the profit at that level of production?

Solution

The profit function is

$$\begin{aligned} P(x) &= R(x) - C(x) \\ &= (400x - 0.2x^2) - (5000 + 250x - 0.1x^2) \\ &= 150x - 0.1x^2 - 5000 \end{aligned}$$

The derivative

$$P'(x) = 150 - 0.2x$$

is zero if and only if $x = 750$. We need to evaluate $P(x)$ at $x = 750$ and at the endpoints $x = 0$ and $x = 1000$. Note that $P''(x) = -0.2 < 0$ for all x and that

$$\begin{aligned} P(0) &= -5{,}000 \\ P(750) &= 51{,}250 \\ P(1000) &= 45{,}000 \end{aligned}$$

Thus $x = 750$ maximizes the profit function. The maximum profit is \$51,250.

Example 4.20

A manufacturer determines that the price at which a product can be sold is approximated by

$$p(x) = 100 - \frac{x}{2} - \frac{x^2}{12}, \qquad 0 \leq x \leq 20$$

and that the cost function is

$$C(x) = 50 + 20x$$

where x represents the number of units produced (assume that all units produced are sold). Determine the number of units that should be produced to maximize profit. What is the maximum profit?

Solution

The revenue received is

$$\begin{aligned} R(x) &= x \cdot p(x) \\ &= x\left(100 - \frac{x}{2} - \frac{x^2}{12}\right) \\ &= 100x - \tfrac{1}{2}x^2 - \tfrac{1}{12}x^3 \end{aligned}$$

The profit function is

$$\begin{aligned} P(x) &= R(x) - C(x) \\ &= 100x - \tfrac{1}{2}x^2 - \tfrac{1}{12}x^3 - 50 - 20x \\ &= 80x - \tfrac{1}{2}x^2 - \tfrac{1}{12}x^3 - 50 \end{aligned}$$

Differentiating $P(x)$, we get

$$P'(x) = 80 - x - \tfrac{1}{4}x^2$$

Now we need to determine the values of x for which $P'(x) = 0$; that is,

$$80 - x - \tfrac{1}{4}x^2 = 0$$

$$320 - 4x - x^2 = 0$$

$$(20 + x)(16 - x) = 0$$

This means that $x = 16$ or $x = -20$. Since $x = -20$ is not admissible (why?), we evaluate $P(x)$ at $x = 16$ and at the endpoints $x = 0$ and $x = 20$. Observe that $P''(x) = -1 - \frac{1}{2}x$ so that $P''(16) = -9 < 0$. Further, $P(0) = -50$, $P(16) = 760.67$, and $P(20) = 683.33$. We conclude that $x = 16$ maximizes the profit function. At this level of production, the profit is \$760.67.

Examples 4.19 and 4.20 indicate that the maximum profit, if it exists, occurs at a critical point obtained by solving the equation $P'(x) = 0$. If $P'(x) = 0$, then

$$R'(x) - C'(x) = 0$$

which implies that

$$R'(x) = C'(x)$$

Thus if maximum profit is not obtained at 0 or M units, it is obtained when $R'(x) = C'(x)$. This means that for a profit function to attain its maximum value, the production of the plant should be adjusted so that the marginal revenue $R'(x)$ is equal to the marginal cost $C'(x)$. This result is an important statement in economics.

Let us now look at an *inventory management problem* that arises in business. A fundamental objective of management is to develop inventory policies that minimize total operating costs. In effect, two decisions must be made: what quantity to order each time, and when to order that quantity. Infrequently placed orders minimize the ordering costs but involve large shipments which, in turn, involve large storage costs. On the other hand, ordering small quantities reduces storage costs but involves placing orders more frequently, thus increasing the ordering costs. Since either course pushed too far results in reduced profit, we must seek a compromise between these choices. We need to develop an optimal quantity to order so as to minimize the sum of the storing and ordering costs.

Let x be the number of units in an order. We shall assume that stock depletes at a constant rate, that the inventory is at zero when the new order is physically received, and there are no stockouts (see Figure 4.19). The average number of items in storage is one-half of one order; that is, $x/2$.

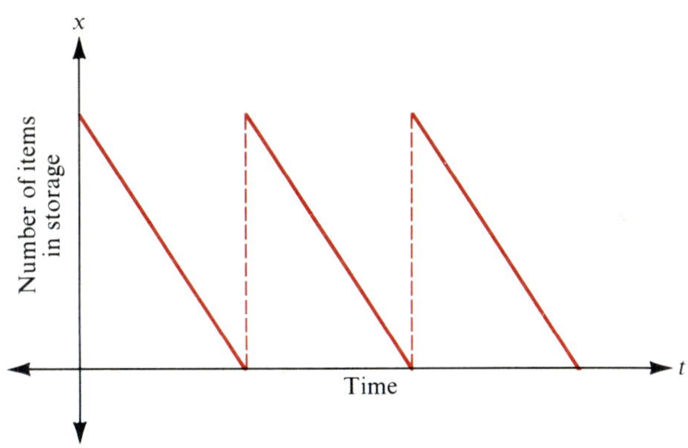

Figure 4.19

If R represents the annual unit requirement, then R/x is the number of orders that will have to be placed during the year. Letting S be the cost of placing an order regardless of its size, then the total ordering costs are

$$\$\left(\frac{R}{x}\right) \cdot S$$

Further, if I is the cost of carrying a unit in stock, then the total annual storage cost for the average inventory $x/2$ is

$$\$\left(\frac{x}{2}\right) \cdot I$$

Let $f(x)$ be the sum of the ordering and storing costs, or the total operating costs. Then

$$f(x) = \frac{RS}{x} + \frac{I}{2} \cdot x$$
$$= RSx^{-1} + \frac{I}{2}x$$

where R, S, and I are assumed to be constants. The graph of $f(x)$ is shown in Figure 4.20.

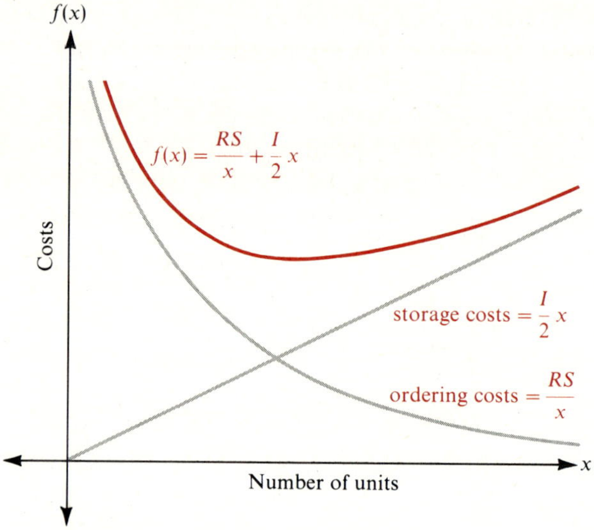

Figure 4.20

The minimum cost occurs at a point where $f'(x) = 0$. Differentiating $f(x)$, we get

$$f'(x) = -RSx^{-2} + \frac{I}{2}$$

Setting $f'(x) = 0$, we find that

$$x = \sqrt{\frac{2RS}{I}}$$

The second derivative of $f(x)$ is

$$f''(x) = -(-2)RSx^{-3} = \frac{2RS}{x^3}$$

Since R, S, and x all have positive values, it follows that $f''(x)$ is positive. Hence, the critical value

$$x = \sqrt{\frac{2RS}{I}}$$

minimizes the total operating costs.

Example 4.21

A manufacturing company expects to buy 8000 units of a certain product. The annual storage cost is \$0.20 per unit and the cost to place an order with the supplier is \$12.50. How many units should the company order each time so as to minimize the total operating cost? How often should this quantity be ordered?

Solution

Here,

$$R = 8000 \text{ units}$$
$$S = \$12.50$$
$$I = \$0.20$$

The optimal order quantity is

$$x = \sqrt{\frac{2RS}{I}} = \sqrt{\frac{2(8000)(12.50)}{(0.20)}} = 1000 \text{ units.}$$

The number of shipments is

$$\frac{8000}{1000} = 8$$

The company should place 8 orders of 1000 units each during the year so as to minimize its total operating costs.

Exercises

1. Find two positive integers that have a sum of 50 such that their product is as large as possible.
2. Find two positive integers that have a product of 144 such that their sum is minimized.
3. The sum of one positive integer and three times another is 60. What numbers should be chosen so as to maximize their product?
4. The sum of twice one positive integer and five times another is 80. What numbers should be chosen so that their product is maximum?
5. A printed poster is to have a margin of 1 inch on the sides and $1\frac{1}{2}$ inches at the top and bottom. What dimensions of the poster board will minimize the total area if the printed area must be 96 square inches?

6. A printed poster is to have a margin of $1\frac{1}{4}$ inches on the sides and $1\frac{1}{2}$ inches at the top and bottom. What are the most economical dimensions for the poster board if the printed area is to contain 30 square inches of illustrations and instructions?

7. A tray is to be made of a rectangular piece of sheet metal by cutting out an equal square from each corner and turning up the sides. If the sheet is 32 inches long and 12 inches wide, determine the dimensions of the tray of largest possible volume.

8. An open box is to be made from a piece of cardboard 15 inches square by cutting a square from each corner and folding up the edges. Determine the dimensions of the box that would have the largest volume.

9. A rancher has 2800 feet of fence to enclose a rectangular pasture. One side of his land is bordered by river and needs no fence. Determine the dimensions of the fence that will enclose the maximum area.

10. A high school is planning to build a rectangular playground that must contain 423,200 square feet. Find the minimum length of fence required and the dimensions of the playground if one side is bordered by a building and needs no fence.

11. A farmer wants to fence in 1,280,000 square feet of land in a rectangular plot and then subdivide it into three equal plots with the fence running across the width as shown in Figure 4.21. What is the least amount of fence required to enclose this area?

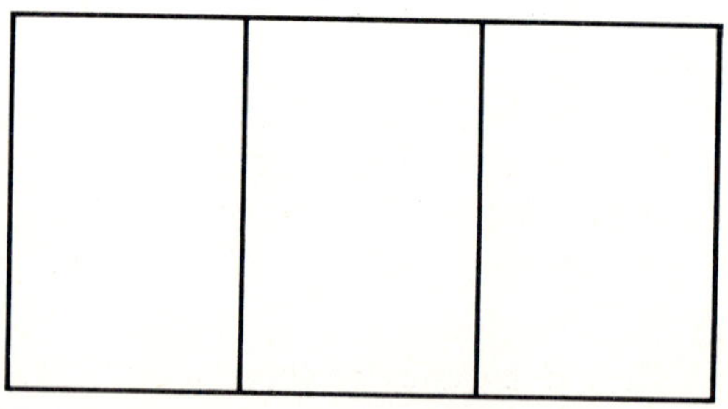

Figure 4.21

12. A rectangular building must contain 405,600 square feet of floor space. The front is to be glass and the other three walls are to be cement block. Assuming that glass construction costs twice as much per foot as cement, find the dimensions of the building that would minimize the cost.

13. A rental agency needs to determine what rent to charge for 200 apartments in order to attain the maximum revenue. Experience shows that if the rent is set at \$150 a month, all units are occupied, but for each \$5 per month increase in rent, five units become vacant. What rent should be charged so as to maximize the revenue? Determine also the maximum revenue.

14. A real estate office handles an apartment complex with 150 units. Past experience shows that all units are occupied if the rent is \$200 per month. For each \$10 increase in rent, five units become vacant. Assuming that the cost of each rented apartment is \$40 per month, what rent should the management charge so as to maximize the profit? Determine also the maximum profit.

15. A high school has arranged a dance for the senior class. The charge is \$5.00 per couple if 40 couples attend and decreases by \$0.05 for each additional couple above the minimum of 40. Determine the number of couples that would maximize the school's revenue. What is the maximum revenue?

16. A resort hotel will provide dinner, dance, and drinks for a local organization. The charge is \$40 per couple with the understanding that the price will be reduced by \$0.20 for each additional couple above 100. Determine the number of couples that will generate the maximum revenue for the hotel.

17. A travel agency plans a student trip to New York City. The fare is \$60 per student if 30 students sign up, and the fare decreases by \$1 for each additional student who goes. Determine the number of students that will generate the maximum revenue for the travel agency.

18. A travel agency plans an excursion trip to West Berlin for 100 members of a college faculty. The plane fare is \$500 with an additional charge of \$10 for each cancellation. Determine the number of cancellations that will generate maximum revenue for the agency.

19. An alumni group makes plans for a charter flight to Key West, Florida. The plane fare is \$100 per person if at least 40 persons go and will decrease by \$0.50 for every additional person above the minimum of 40. Find the number of passengers that will maximize the revenue for the alumni group. What is the maximum revenue?

20. A department store estimates that if it charges x dollars for an electric shaver, it can sell $(380 - 10x)$ shavers. What selling price should the store charge in order to maximize its revenue?

21. A farmer estimates that if he begins picking his peaches today, he will have a crop of 500 pounds and can make a profit of \$0.30 a pound. He could wait as long as a week before the peaches are overripe; for each day he waits, he will have 25 extra pounds of peaches, but his profit will decrease by \$0.01 per pound per day. When should he pick his crop so as to maximize his profit? Determine also his maximum profit.

22. A restaurant owner can have a seating capacity between 90 and 125 in her new restaurant. She has estimated that her profit will average \$12 per week per chair if the seating capacity is 90, then decrease by \$0.10 per chair for each additional seat in excess of 90. Determine the number of seats that must be planned so as to maximize the profit. What is the maximum profit per week?

23. The total cost and revenue functions for a certain product are

$$C(x) = 0.05x^2 + 29x + 1000$$
$$R(x) = 50x - x^2$$

Determine the value of x for which the profit is maximum.

24. A manufacturer has recently determined that the total cost and revenue functions for his product are

$$C(x) = 0.004x^2 + 25x + 2000$$
$$R(x) = 60x - 0.01x^2$$

Find the value of x for which the profit is maximum.

25. A product can be manufactured at a total cost of

$$C(x) = 1000 + 20x$$

where x is the number of units produced. If the price at which each unit is sold is given by

$$p(x) = 50 - 0.1x$$

what production level will maximize the profit? What is the total profit and the price per unit at this level of production?

26. A manufacturer finds that her product can be assembled at a total cost of

$$C(x) = 200 + 30x$$

where x is the number of units produced. Assume that the price at which she can sell each unit is given by

$$p(x) = 150 - 0.2x$$

What level of production will maximize the total profit? What is the corresponding selling price per unit? What is the total profit at this level of production?

27. A certain product can be manufactured at a total cost of

$$C(x) = 0.02x^2 + 50x + 4000$$

where x is the number of units produced. The price at which each unit can be sold is given by

$$p(x) = 100 - 0.005x$$

Determine the production level x at which the profit is maximum. What is the price per unit and the total profit at this level of production?

28. A manufacturer finds that he can sell x units per week of his product at p dollars per unit where

$$p(x) = 127 - \frac{x}{3}$$

and the cost of producing x units per week is

$$C(x) = 200 + 15x + 0.2x^2$$

What level of production would maximize the profit? What is the corresponding price? What is the total profit?

29. A restaurant expects to use 8000 packages of a particular frozen food next month. The cost of placing an order with the supplier is \$6.25, and the cost of maintaining a package in storage for a month is \$0.40. Assuming that the frozen food is consumed at a uniform rate throughout the month and that each shipment arrives on the day that the previous shipment has been used, determine the number of packages the restaurant should order to minimize its cost. How often should that quantity be ordered?

30. A calculator manufacturing company expects to sell 40,000 units of its new model CI 535 next year. The cost of setting up a production run is \$250, and the cost of carrying a unit in inventory is \$1.25 per year. Determine the optimal number of units the company should produce in one production run to minimize its cost. How often should the company schedule its production run? Assume that the demand for the calculators is fairly constant throughout the year and that a production run makes calculators available just when the previous supply is exhausted.

4.7 Curve Sketching

We used the first derivative to determine the critical points of a function and the intervals over which the function is increasing or decreasing. This helped us find relative maxima and relative minima. The second derivative provided another test for relative maxima and minima. We prefer the second derivative test; we generally use the first derivative test when the second derivative test fails or when it is laborious and time consuming to find the second derivative. In this section, we shall use derivatives as an aid in graphing functions. Actually, polynomial and rational functions are graphed more efficiently and accurately by using derivative tests than by plotting and connecting a "sufficient" number of points. We begin with some definitions.

Definition 4.6

Let $f(x)$ be differentiable at each point of an open interval (a, b). Then the graph of the function f is said to be

a. *concave upward* over (a, b) if the tangent line at each $x \in (a, b)$ is below the graph of f.

b. *concave downward* over (a, b) if the tangent line at each $x \in (a, b)$ is above the graph of f.

Consider, for example, the graph of $y = f(x)$ in Figure 4.22. The graph of the function is concave upward. The first derivative $f'(x)$ is negative for all $x < x_0$, zero for $x = x_0$ and positive for all $x > x_0$. This means that the second derivative, $f''(x)$, which determines the rate of change of the slope of the tangent, is positive. In general, whenever the graph of a function f is concave upward on an interval, $f''(x) > 0$. On the other hand, if the graph of f is concave downward on an interval

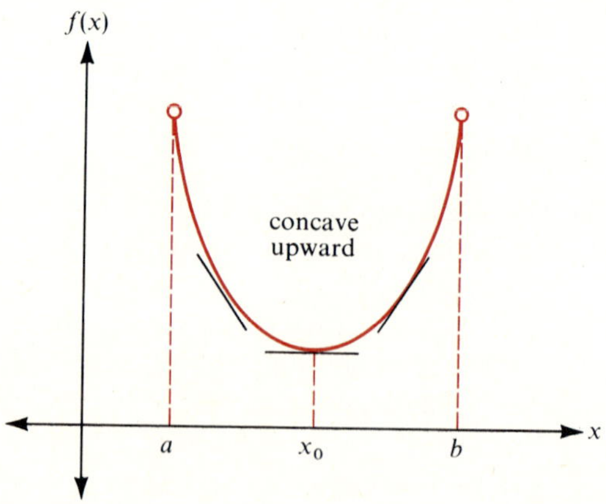

Figure 4.22

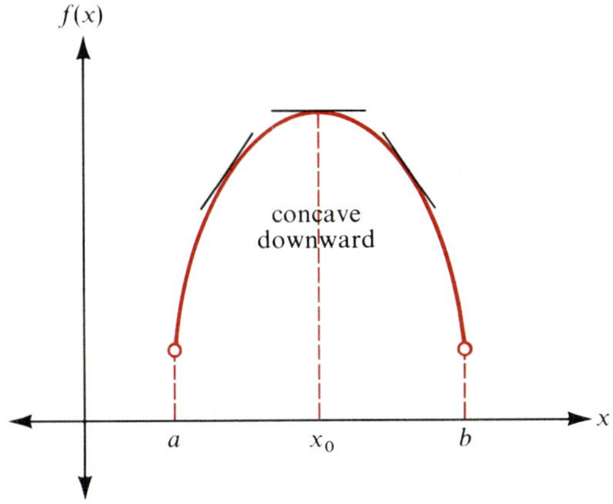

Figure 4.23

(see Figure 4.23), then $f'(x)$ is positive for all $x < x_0$, zero for $x = x_0$, and negative for all $x > x_0$, and the second derivative $f''(x) < 0$. In general, a function that is concave downward on an interval has a second derivative that is negative on that interval. These important results are summarized in the following proposition which we state without proof.

Proposition 4.5

If $f''(x) > 0$ for all $x \in (a, b)$, then the graph of f is concave upward over (a, b). If $f''(x) < 0$ for all $x \in (a, b)$, then the graph of f is concave downward over (a, b).

Example 4.22

The function

$$f(x) = x^2 - 4x$$

is differentiable for all x. Observe that

$$f'(x) = 2x - 4$$
$$f''(x) = 2$$

Since $f''(x) > 0$, we conclude that the curve is concave upward as shown in Figure 4.24.

Example 4.23

The function $f(x) = 4 - x^2$ is differentiable for all x and

$$f'(x) = -2x$$
$$f''(x) = -2$$

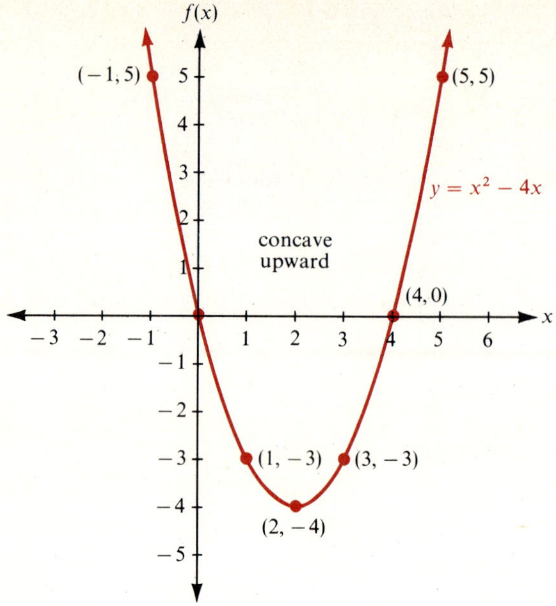

Figure 4.24

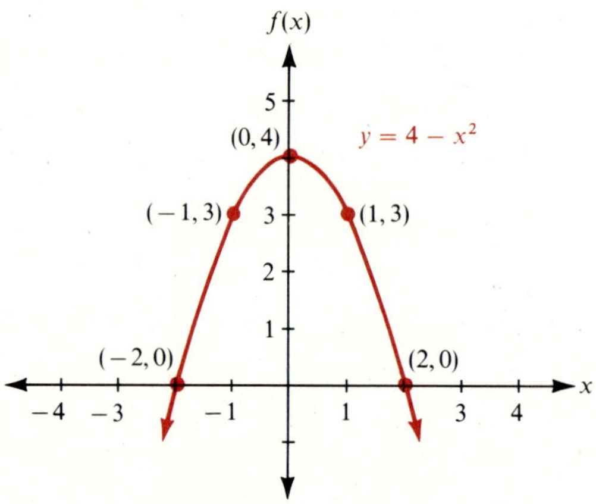

Figure 4.25

Because $f''(x) < 0$, it follows that the curve is concave downward. The graph of $f(x)$ is displayed in Figure 4.25.

The graphs of many functions have some portions that are concave upward and some portions that are concave downward. The point or points where the concavity changes its direction is also of interest.

Definition 4.7

A point $(x_0, f(x_0))$ on the graph of the function f where the concavity changes its direction is called a *point of inflection*.

Consider the graph of $y = f(x)$ in Figure 4.26. The graph of f is concave downward on the interval (a, x_0) and concave upward on the interval (x_0, b). Notice that $f'(x)$ is positive for all $x \in (a, x_0)$, zero for $x = x_0$, and positive for all $x \in (x_0, b)$. In other words, $f'(x)$ is decreasing on both sides of x_0 so there is no change in the *rate of change* of the slope in the immediate vicinity of x_0. Thus the second derivative $f''(x)$ is zero at $x = x_0$.

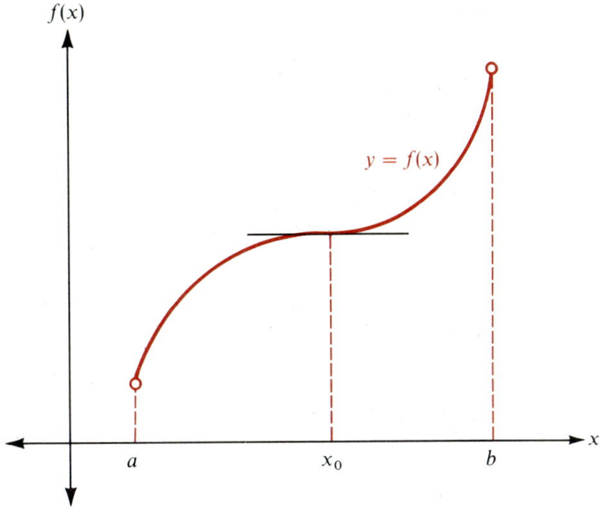

Figure 4.26

Proposition 4.6 If $(x_0, f(x_0))$ is a point of inflection for a function f, then $f''(x_0) = 0$.

The converse of Proposition 4.6 is not true. If $f''(x_0) = 0$, then it does not necessarily follow that x_0 is a point of inflection. For example, for the function $f(x) = x^4$, the second derivative is $f''(x) = 12x^2$ which is zero at $x = 0$. However, the graph of $f(x) = x^4$, as shown in Figure 4.27, is concave upward and has no inflection point.

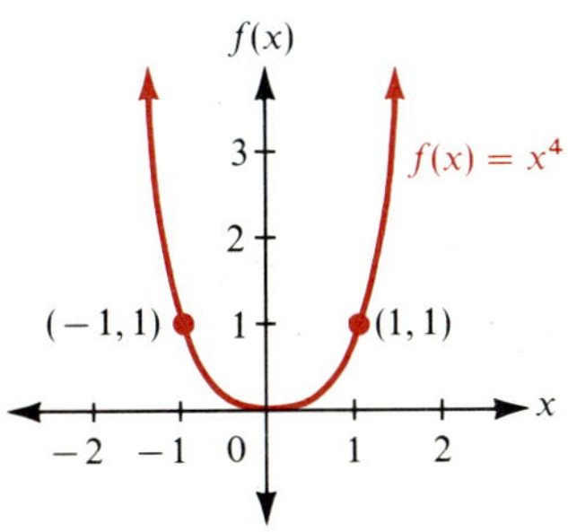

Figure 4.27

We shall now combine the analyses of intervals on which functions are increasing and decreasing with points of relative maximum and minimum, concavity, and points of inflection to obtain an accurate graph of a function.

Example 4.24

Sketch the graph of

$$f(x) = x^3 - 6x^2 + 9x + 5, \qquad -2 < x < 6$$

Solution

Differentiating $f(x)$, we have

$$\begin{aligned} f'(x) &= 3x^2 - 12x + 9 \\ &= 3(x^2 - 4x + 3) \\ &= 3(x - 1)(x - 3) \end{aligned}$$

Thus $x = 1$ and $x = 3$ are the only points for which $f'(x) = 0$. Observe that

$$\begin{aligned} f'(x) > 0 &\quad \text{for} \quad -2 < x < 1 \\ f'(x) < 0 &\quad \text{for} \quad 1 < x < 3 \\ f'(x) > 0 &\quad \text{for} \quad 3 < x < 6 \end{aligned}$$

This means that as x increases from left to right, $f(x)$ is increasing on the open intervals $(-2, 1)$ and $(3, 6)$ and decreasing on the interval $(1, 3)$. Hence, the function has a relative maximum at $x = 1$ and a relative minimum at $x = 3$. These values are $f(1) = 9$ and $f(3) = 5$, respectively.

The second derivative is

$$\begin{aligned} f''(x) &= 6x - 12 \\ &= 6(x - 2) \end{aligned}$$

Clearly, $f''(x) = 0$ if and only if $x = 2$. For $x < 2$, $f''(x) < 0$; consequently, f is concave downward. For $x > 2$, $f''(x) > 0$ and f is concave upward. Since concavity changes its direction at $x = 2$, the point $(2, f(2)) = (2, 7)$ is a point of inflection. Further

$$\begin{aligned} &f(-1) = -11, \quad f(0) = 5, \quad f(1) = 9, \quad f(2) = 7 \\ &f(3) = 5, \qquad f(4) = 9, \quad f(5) = 25 \end{aligned}$$

Putting all this information together, we have the graph of f as shown in Figure 4.28.

Example 4.25

Sketch the graph of

$$f(x) = \tfrac{1}{4}(x^4 - 6x^2 + 8x + 12)$$

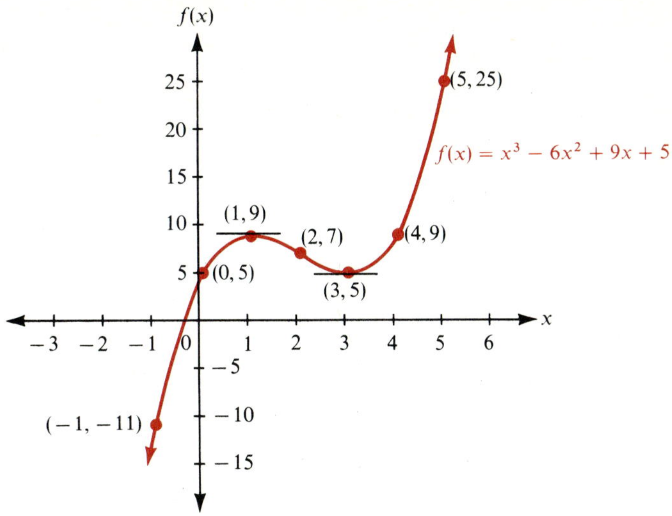

Figure 4.28

Solution

The first derivative is

$$\begin{aligned} f'(x) &= \tfrac{1}{4}(4x^3 - 12x + 8) \\ &= x^3 - 3x + 2 \\ &= (x + 2)(x - 1)^2 \end{aligned}$$

Clearly, $f'(x) = 0$ if and only if $x = -2$ or $x = 1$. Thus $x = -2$ and $x = 1$ are the only candidates which we need to examine for relative maxima and relative minima. Since

$$\begin{aligned} f'(x) &< 0 \quad \text{for} \quad x < -2 \\ f'(x) &> 0 \quad \text{for} \quad x > -2 \end{aligned}$$

it follows that $f(x)$ is decreasing on the interval $(-\infty, -2)$ and increasing on the interval $(-2, \infty)$. Hence $f(x)$ has a relative minimum at $x = -2$; this value is $f(-2) = -3$.

The second derivative is

$$\begin{aligned} f''(x) &= 3x^2 - 3 \\ &= 3(x^2 - 1) \\ &= 3(x - 1)(x + 1) \end{aligned}$$

To determine the intervals over which $f''(x) > 0$ and the intervals over which $f''(x) < 0$, Figure 4.29 illustrates the signs of the factors $(x - 1)$ and $(x + 1)$ and the sign of the function $f''(x)$. Because $f''(x) > 0$ for $x \in (-\infty, -1)$ and $x \in (1, \infty)$, we conclude that f is concave upward

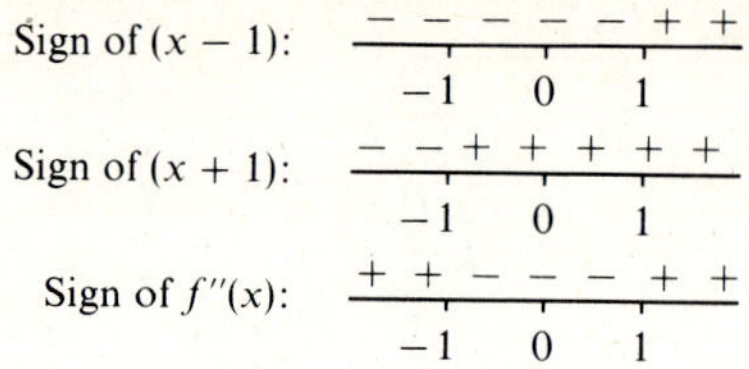

Figure 4.29

over these open intervals. Similarly, since $f''(x) < 0$ for $x \in (-1, 1)$, f is concave downward for $x \in (-1, 1)$. Since concavity changes its direction at $x = -1$ and $x = 1$, the points $(-1, f(-1))$ and $(1, f(1))$ are points of inflection. Further,

$$f(-3) = \tfrac{15}{4}, \quad f(-2) = -3, \quad f(-1) = -\tfrac{1}{4}$$
$$f(0) = 3, \qquad f(1) = \tfrac{15}{4}, \qquad f(2) = 5$$

Putting all the information together, we sketch the graph of f as shown in Figure 4.30.

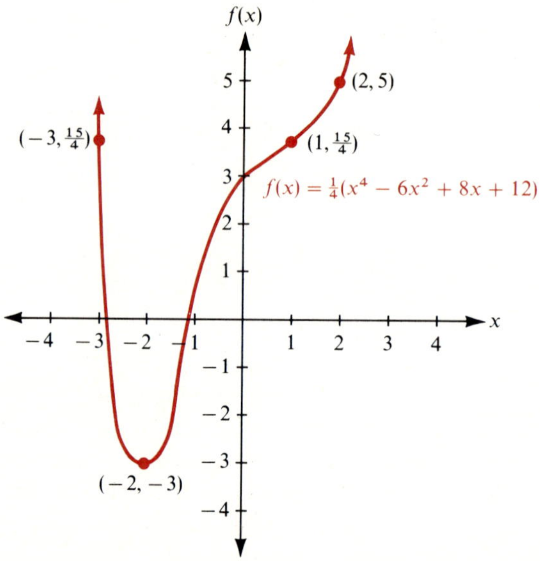

Figure 4.30

Next, we plot the graph of a rational function.

Example 4.26 Sketch the graph of

$$f(x) = \frac{x - 1}{x + 1}$$

Solution

The function $f(x)$ has as its domain the set of real numbers except $x = -1$. At $x = -1$, the denominator $(x + 1)$ is zero and the function $f(x)$ is not defined. The first derivative $f'(x)$ is

$$f'(x) = \frac{d}{dx}\left(\frac{x-1}{x+1}\right) = \frac{(x+1)\cdot 1 - (x-1)\cdot 1}{(x+1)^2} = \frac{2}{(x+1)^2}$$

Thus $f(x)$ is an increasing function for all x except $x = -1$. The second derivative is

$$f''(x) = \frac{d}{dx}\left[2(x+1)^{-2}\right] = 2(-2)(x+1)^{-3} = \frac{-4}{(x+1)^3}$$

For $x < -1$,

$$(x + 1) < 0$$

$$(x + 1)^3 < 0$$

and

$$\frac{-4}{(x+1)^3} > 0$$

Therefore, $f''(x) > 0$ for $x < -1$. Similarly, if $x > -1$, then

$$(x + 1) > 0$$

$$(x + 1)^3 > 0$$

$$\frac{-4}{(x+1)^3} < 0$$

and $f''(x) < 0$ for $x > -1$. Thus f is concave upward for $x < -1$ and concave downward for $x > -1$. Since the concavity does not change at any point $x \in D_f$, the curve has no inflection point.

Let us now examine the behavior of $f(x)$ for values of x sufficiently close to -1. We shall leave it for the reader to see that as x approaches -1, $(x + 1)$ approaches zero and $|f(x)|$ becomes large. Hence, $x = -1$ is a *vertical asymptote* for the graph.

Next, we consider the limit of $f(x)$ as x approaches positive infinity.

$$\begin{aligned}
\lim_{x\to+\infty} f(x) &= \lim_{x\to+\infty}\left(\frac{x-1}{x+1}\right) \\
&= \lim_{x\to+\infty}\left[\frac{1-\frac{1}{x}}{1+\frac{1}{x}}\right] \\
&= \frac{1-0}{1+0} \qquad \left(\lim_{x\to+\infty}\frac{1}{x} = 0\right) \\
&= 1
\end{aligned}$$

Similarly,

$$\lim_{x \to -\infty} f(x) = \lim_{x \to -\infty} \left(\frac{x-1}{x+1}\right) = 1$$

Because $f(x)$ approaches 1 as x approaches positive or negative infinity, we conclude that $f(x) = 1$ is a *horizontal asymptote* for the graph of f. Further,

$$\begin{aligned} f(-4) &= \tfrac{5}{3} & f(-\tfrac{3}{4}) &= -7 \\ f(-3) &= 2 & f(-\tfrac{1}{2}) &= -3 \\ f(-2) &= 3 & f(0) &= -1 \\ f(-\tfrac{3}{2}) &= 5 & f(1) &= 0 \\ f(-\tfrac{5}{4}) &= 9 & f(4) &= \tfrac{3}{5} \end{aligned}$$

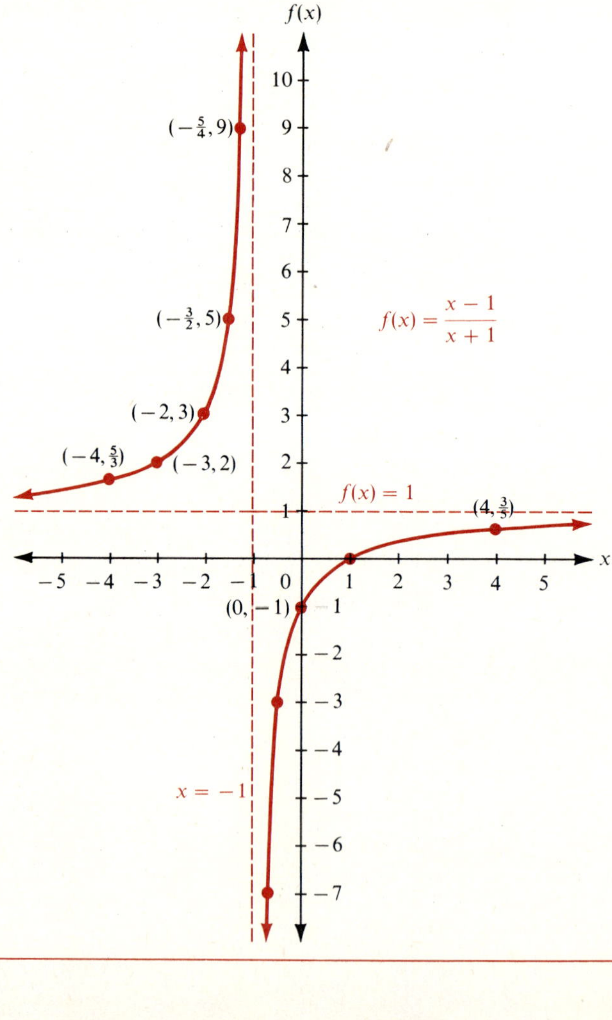

Figure 4.31

Putting all this information together, we display the graph of $f(x)$ in Figure 4.31.

Exercises

Determine the intervals over which each of the following functions is increasing, decreasing, concave upward, and concave downward. Graph each function $f(x)$ showing its inflection points, if any.

1. $f(x) = x^2 - 6x + 15$
2. $f(x) = x^2 + 4x - 4$
3. $f(x) = 12x - x^2$
4. $f(x) = 8x - x^2$
5. $f(x) = x^3 - 3x^2$
6. $f(x) = x^3 + 3x^2 + 4$
7. $f(x) = x^3 - 9x^2 + 24x$
8. $f(x) = x^3 + x^2 - x + 4$
9. $f(x) = x^3 + 3x^2 + 3x - 3$
10. $f(x) = x^3 - 3x^2 + 3x$
11. $f(x) = x^3 - 12x^2 + 36x + 16$
12. $f(x) = x^3 - 3x^2 - 9x + 10$
13. $f(x) = 2x^3 - 9x^2 + 12x - 3$
14. $f(x) = 2x^3 + 3x^2 - 12x$
15. $f(x) = x^4 + 8x^3 + 18x^2 - 5$
16. $f(x) = x^4 - 8x^2$
17. $f(x) = \dfrac{x}{x + 1}$
18. $f(x) = \dfrac{x - 1}{x - 2}$
19. $f(x) = \dfrac{x - 3}{x + 2}$
20. $f(x) = x + \dfrac{1}{x}$

Exercises for Review

1. Define each of the following concepts.
   - **a.** relative minimum
   - **b.** relative maximum
   - **c.** critical point
   - **d.** absolute maximum
   - **e.** absolute minimum
   - **f.** increasing function
   - **g.** decreasing function
   - **h.** concave downward
   - **i.** concave upward
   - **j.** inflection point
2. Determine the intervals on which each of the following functions is increasing and the intervals on which it is decreasing.
   - **a.** $f(x) = x^2 + 6x - 8$
   - **b.** $f(x) = 12x - x^2$
   - **c.** $f(x) = x^3 + x^2 - x + 10$
   - **d.** $f(x) = x^3 + 3x^2 + 12$
   - **e.** $f(x) = x^4 + 8x^3 + 18x^2$
3. Examine each of the following functions for relative maxima and relative minima on the open intervals indicated.
   - **a.** $f(x) = x^2 + x - 30; \quad (-1, 6)$
   - **b.** $f(x) = 8x - x^2 - 6; \quad (0, 5)$

c. $f(x) = x^3 - 3x + 5; \qquad (-3, 3)$
d. $f(x) = x^3 + 3x^2 - 9x - 10; \qquad (-6, 4)$
e. $f(x) = x^4 - 4x^3 + 4x^2 + 1; \qquad (-1, 3)$

4. Find the absolute maximum and the absolute minimum for each of the following functions on the closed intervals indicated.

a. $f(x) = x^3 - 12x^2 + 21x + 5; \qquad [-1, 8]$
b. $f(x) = 2x^3 + 9x^2 - 24x + 3; \qquad [-2, 5]$
c. $f(x) = x^4 + 4x^3 + 6x^2 + 4x + 1; \qquad [-3, 2]$
d. $f(x) = \frac{x}{x^2 + 4}; \qquad [-1, 4]$

5. Determine the intervals over which the function

$$f(x) = 2x^3 - 3x^2 - 12x + 5$$

is increasing, decreasing, concave upward, and concave downward. Use this information and the inflection points, if any, to graph $f(x)$.

6. A ball is thrown vertically into the air. Its height at time t (in seconds) is given by

$$h(t) = 96t - 16t^2$$

where $h(t)$ is the height (in feet) of the ball above the ground. After how many seconds will the ball be at its maximum height?

7. The profit function of a firm is

$$P(x) = 5x - 0.01x^2 - 200$$

where x denotes the output of the firm (in hundreds of units per week). Determine the value of x which will maximize profit. What is the profit of the firm at that level of production?

8. A grocery chain store has a price list which the store expects its new checkers to memorize. The number of items, $N(t)$, whose prices a checker is able to memorize after t hours of practice is approximated by the function

$$N(t) = 60t - 12t^2 \qquad 0 \leq t \leq 4$$

Determine the number of hours a checker should practice to maximize $N(t)$.

9. The concentration of sugar in the bloodstream of a patient is approximated by the function

$$Q(t) = 5.2 + 0.1t - 0.1t^2$$

where $Q(t)$ is the quantity of sugar (in suitable units) in the bloodstream t hours after a sugar metabolism test is performed. After how many hours will the concentration reach its maximum level? What is the maximum amount of sugar in the bloodstream?

10. The number of bacteria (in thousands) present in a certain culture is approximated by the function

$$f(t) = 1000 + 96t - 18t^2 + t^3$$

where t is recorded in hours. Determine the value of t after which the number of bacteria is at its maximum level.

11. The population of a certain species is approximated by the function

$$f(t) = 1000 + 192t + 36t^2 - 4t^3$$

where t is recorded in weeks. Determine the value of t at which the population is at its maximum level.

12. A photographer has a thin piece of wood 40 inches long. How should she cut the wood to make a picture frame that would enclose the maximum area?

13. An open box with a square base is to be made from a square piece of cardboard by cutting equal squares out of each corner and then folding up the edges. If a side of the cardboard is 10 cm, what size squares should be cut out so that the box has the maximum volume?

14. A farmer has 4800 feet of fence to enclose a rectangular pasture. One side of the plot is to be along the bank of a river and needs no fence. Determine the dimensions of this plot so as to enclose the maximum area.

15. Suppose that the farmer in Problem 14 wishes to enclose the plot away from the river and needs, therefore, to fence all four sides of the plot. What are the dimensions of the plot that would enclose the maximum area?

16. A car leasing agency needs to determine the rate it should charge for renting a car. The agency owns a fleet of 120 cars. Experience has shown that all the cars can be rented at a rate of \$30 per day per car. For each \$1 increase in rent, three fewer cars are rented. How much should the agency charge to maximize the total revenue?

17. A closed box with a square top is to have a volume of 31,250 cubic feet. The material for the top and the bottom of the box costs \$4 per square foot while the material for the sides costs \$2 per square foot. What are the dimensions of the box that would minimize the total cost?

18. A certain product can be manufactured at a total cost of

$$C(x) = 200 + 9x$$

where x is the number of units produced. The price at which each unit can be sold is given by

$$p(x) = 25 - 0.2x$$

Determine the production level x that would maximize the profit. What is the price per unit and what is the total profit at this level of production?

19. An oil pipeline is to be built from a source on the shore to a small community on an island 6 kilometers off the shore. The pipeline starts from the source at the point P, runs along the shore to the point Q and then runs underwater to the island at the point T (see Figure 4.32). It costs 1.25 times as much to run the pipeline underwater as it does on the shore. Determine the location of the point Q that minimizes the total cost.

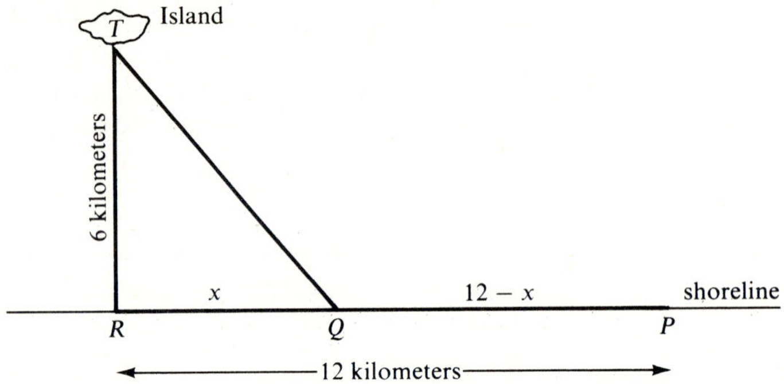

Figure 4.32

20. A manufacturing company expects to use 10,000 units of a certain product each year. The cost of placing an order with the supplier is $35 and the annual storage cost is 17.5 cents per unit. Determine the number of units the company should order each time. How often should that quantity be ordered so as to minimize the total operating cost? Assume that the products are used at a uniform rate and that each shipment arrives on the day the previous supply has been used up.

5 Exponential and Logarithmic Functions

5.1 Introduction

Although linear functions are important for mathematical models, exponential functions and logarithmic functions may have even more widespread applications. These two functions, which are closely related, are particularly useful in describing growth and decay in biology, the social sciences, marketing, medicine, and management. In this chapter we shall study how exponential functions provide important models in these fields.

5.2 Exponential Functions

So far we have used expressions of the type x^p where the base x is a variable and the exponent p is a real number*. Now we turn our attention to situations where the base is constant and the exponent is the independent variable. Consider, for example, an equation of the form $y = b^x$ where b is any positive number but not equal to one and the variable x is a real number. This equation is referred to as an *exponential function.*

Definition 5.1

A function of the form $f(x) = b^x$, where $b > 0$, $b \neq 1$, $x \in R$, is called an *exponential function.*

We wish to emphasize that the base b must be a positive real number; if b were negative, b^x would not be a real number for $x = \frac{1}{2}, \frac{1}{4}$, or any

* See section A.4 of Appendix A.

other fraction with an even denominator. In the case $b = 1$, the function b^x is a constant function and equals 1 for all values of x. The domain of $f(x) = b^x$ is the set of all real numbers and its range, if $b > 0, b \neq 1$, is the set of all positive real numbers.

The following examples illustrate several properties of exponential functions.

Example 5.1

Graph the function

$$f(x) = 2^x$$

Solution

Table 5.1 shows the values of $f(x)$ corresponding to some real values of x.

Table 5.1

x	-4	-3	-2	-1	0	1	2	3
$f(x)$	$\frac{1}{16}$	$\frac{1}{8}$	$\frac{1}{4}$	$\frac{1}{2}$	1	2	4	8

Plotting these ordered pairs and connecting them, we have the graph of $f(x) = 2^x$ in Figure 5.1.

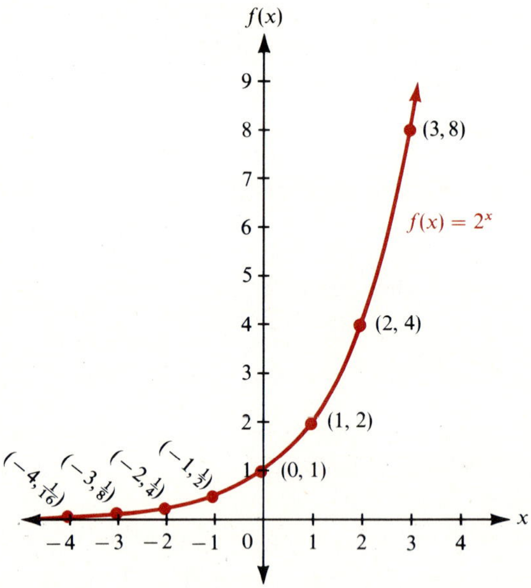

Figure 5.1

Example 5.2

Graph the function

$$f(x) = 3^x$$

Solution

Plotting the ordered pairs $(x, f(x))$ as given in Table 5.2 and connecting them with a smooth curve, we have the graph of the function in Figure 5.2.

Table 5.2

x	-3	-2	-1	0	1	2
$f(x)$	$\frac{1}{27}$	$\frac{1}{9}$	$\frac{1}{3}$	1	3	9

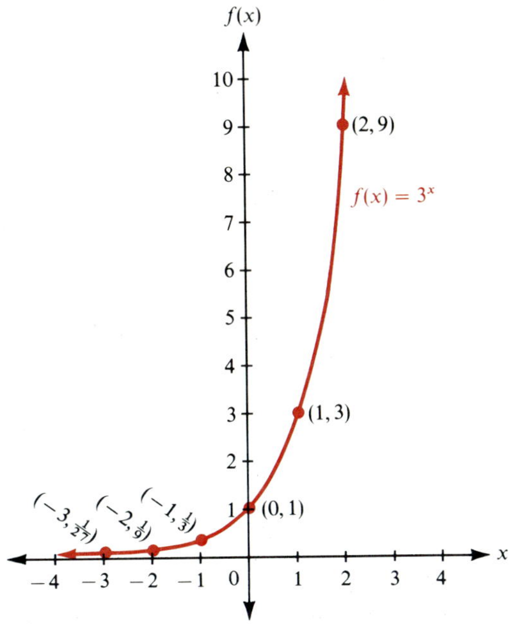

Figure 5.2

An irrational number e, approximately equal to 2.71828, plays an important role in calculus. Specifically, the function e^x is used in applications that require an exponential-type function to describe a physical phenomenon. As a result, e^x is called the *natural* exponential function. Table 1 in the Appendix lists the values of e^x and e^{-x} from $x = 0$ to $x = 9.9$ in steps of 0.1. Alternatively, we may use a scientific calculator to obtain values of e^x and e^{-x}.

We shall take an intuitive look at e in the next section.

Example 5.3 Assuming that $e = 2.71828$, graph $f(x) = e^x$.

Solution

Table 5.3 shows the values of $f(x)$ corresponding to some values of x. Plotting the ordered pairs $(x, f(x))$, we have the graph of e^x in Figure 5.3.

Table 5.3

x	-4	-3	-2	-1	0	1	1.5	2	2.5	3
$f(x)$	0.02	0.05	0.14	0.37	1.00	2.72	4.48	7.39	12.18	20.09

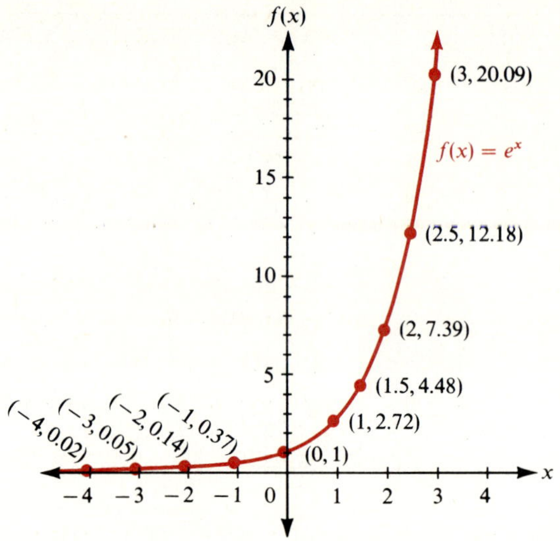

Figure 5.3

The previous examples suggest the following observations about the graph of b^x when $b > 1$.

1. $b^0 = 1$ so the graph of b^x passes through the point $(0, 1)$.
2. $b^x > 0$ for all values of x; hence the graph lies above the x-axis.
3. As x increases, the value of b^x increases. Hence the function $f(x) = b^x$ is an increasing function of x.
4. As x decreases along the negative x-axis, the value of b^x decreases and approaches zero; that is, $\lim_{x \to -\infty} b^x = 0$.
5. For $b \neq 1$, $b^x = b^y$ if and only if $x = y$; thus $f(x) = b^x$ is a one-to-one function.

Next we examine the graph of $y = b^x$ for $0 < b < 1$. Here, the graph has a different appearance. The following examples provide an illustration.

Example 5.4

Graph

$$g(x) = (\tfrac{1}{2})^x$$

Solution

Table 5.4 shows the values of $g(x)$ corresponding to some values of x. Plotting the ordered pairs $(x, g(x))$ from the table, we have the graph in Figure 5.4. Notice that the graph of $g(x) = (\frac{1}{2})^x = 2^{-x}$ is a mirror image of $f(x) = 2^x$ in the y-axis.

Table 5.4

x	-3	-2	-1	0	1	2	3	4
$g(x)$	8	4	2	1	$\frac{1}{2}$	$\frac{1}{4}$	$\frac{1}{8}$	$\frac{1}{16}$

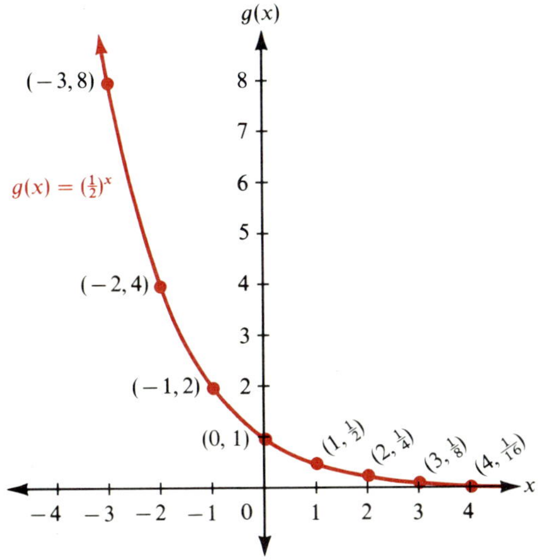

Figure 5.4

In summary, the following are general properties of the exponential function $f(x) = b^x$, $b > 0$, $b \neq 1$.

1. $b^x = 1$ for $x = 0$

2. $b^x > 0$ for all values of x.

3. As x increases, the value of b^x increases when $b > 1$ and decreases when $0 < b < 1$.

4. As x decreases, the value of b^x decreases when $b > 1$ and increases when $0 < b < 1$.

5. $f(x) = b^x$ is a one-to-one function.

Definition 5.2

An equation that contains a variable in an exponent is referred to as an *exponential equation.*

Most exponential equations can be solved by using the fundamental property that if $b^x = b^y$, then $x = y$. The following examples illustrate applications of that fundamental property.

Example 5.5

Solve $2^x = 64$ for x.

Solution

To represent both sides of the equation in the same base, we express 64 as a power of 2. Thus we have

$$2^x = 2^6$$

from which it follows that $x = 6$.

Example 5.6

Solve $5^{-2x} = 625$ for x.

Solution

Since $625 = 5^4$, we may rewrite the given equation as

$$5^{-2x} = 5^4$$

Thus $-2x = 4$ or $x = -2$.

Example 5.7

Solve $9^x = 243$ for x.

Solution

We must rewrite both sides of the equation so that they have the same base. We know that

$$9 = 3^2 \quad \text{and} \quad 243 = 3^5$$

so our original equation can be rewritten as

$$(3^2)^x = 3^5$$
$$3^{2x} = 3^5$$

which implies that $2x = 5$ or $x = \frac{5}{2}$.

Example 5.8

Solve $\left(\frac{3}{4}\right)^{-x} = \frac{243}{1{,}024}$ for x.

Solution

The equation

$$\left(\frac{3}{4}\right)^{-x} = \frac{243}{1{,}024}$$

may also be written as

$$\left(\frac{3}{4}\right)^{-x} = \left(\frac{3}{4}\right)^5$$

from which it follows that $x = -5$.

Exercises

Graph each of the following functions.

1. $y = 2^{x+1}$
2. $y = 3^{x+1}$
3. $y = 2^{x-1}$
4. $y = 3^{x-2}$
5. $y = 4^x$
6. $y = 4^{-x}$
7. $y = 2^x + 1$
8. $y = 2^x - 1$
9. $y = \left(\frac{1}{3}\right)^x$
10. $y = \left(\frac{1}{3}\right)^{x+1}$
11. $y = 2(3^x)$
12. $y = 3(2^x)$
13. $y = 3\left(\frac{1}{2}\right)^x$
14. $y = 2\left(\frac{1}{3}\right)^x$

Solve for x in each of the following equations.

15. $2^x = 1$
16. $2^x = 4$
17. $3^x = 27$
18. $5^x = 125$
19. $(25)^x = 125$
20. $(16)^x = 128$
21. $\left(\frac{1}{3}\right)^x = 81$
22. $5^x = 0.008$
23. $(10)^x = 0.0001$
24. $7^x = \frac{1}{343}$
25. $4^{-2x} = \frac{1}{256}$
26. $\left(\frac{2}{3}\right)^x = \frac{16}{81}$
27. $\left(\frac{4}{5}\right)^{-x} = \frac{125}{64}$
28. $\left(\frac{3}{4}\right)^{-x} = \frac{1{,}024}{243}$
29. $\left(\frac{2}{3}\right)^{-x} = \frac{729}{64}$
30. $\left(\frac{5}{6}\right)^{-x} = \frac{1{,}296}{625}$
31. $\left(\frac{6}{7}\right)^{-x} = \frac{343}{216}$
32. $\left(\frac{3}{5}\right)^{-x} = \frac{3{,}125}{243}$
33. $\left(\frac{8}{9}\right)^{-x} = \frac{729}{512}$
34. $\left(\frac{3}{4}\right)^{-2x} = \frac{256}{81}$

35. Use values of e^x and e^{-x} in Table 1 to graph the following functions.
a. $f(x) = e^{2x}$ **b.** $f(x) = 2e^{-x}$

5.3 Compound Interest

Suppose that Ms. Roberts opens a savings account with \$600 and for two years the bank pays her simple interest at the annual rate of 6 percent. If she withdraws the interest earned each year, she will have an annual income of

$$\$600(0.06) = \$36$$

and the balance in the account will remain at \$600. If she does not withdraw the interest, then the interest for the second year would be computed on the \$636 in the account at the beginning of the second year. The interest for the second year would then be

$$\$636(0.06) = \$38.16$$

instead of \$36. The practice of reinvesting the interest to earn additional interest is called *compounding*. The theory of *compound interest* assumes that the principal and the interest earned thereon continue to be reinvested each year.

To develop a basic formula for compound interest, consider an investment of P dollars at the interest rate of i per year. Since the interest at the end of the first year is Pi, the balance at the end of the year is

$$P + Pi = P(1 + i)$$

This balance is considered as principal at the beginning of the second year and earns

$$P(1 + i)i$$

interest at the end of the second year. The balance at the end of the second year is then

$$P(1 + i) + P(1 + i)i = P(1 + i)(1 + i) = P(1 + i)^2$$

Now $P(1 + i)^2$ is considered as principal at the beginning of the third year. This principal earns interest

$$P(1 + i)^2 i$$

at the end of the third year, which brings the total on deposit to

$$P(1 + i)^2 + P(1 + i)^2 i = P(1 + i)^2(1 + i) = P(1 + i)^3$$

Thus, the general formula for the continued growth of P dollars at yearly interest rate i can be shown to be

5.1 $$A(t) = P(1 + i)^t$$

where $A(t)$ is the accumulated value at the end of t years. The values of $(1 + i)^t$ may be computed by using an electronic calculator. However, values of $(1 + i)^t$ for some values of i and t are given in Table 2 at the end of the book.

Example 5.9

If \$600 is deposited at 6 percent, compounded annually, what will be the total on deposit at the end of 5 years? How much of that will be interest?

Solution

Here,

$$P = \$600, \quad i = 0.06, \quad \text{and} \quad t = 5$$

Using (5.1), we have

$$\begin{aligned} A(5) &= \$600(1 + 0.06)^5 \\ &= \$600(1.06)^5 \\ &= \$600(1.338226) \qquad \textbf{(Table 2)} \\ &= \$802.94 \end{aligned}$$

The accumulated value will be \$802.94 and the interest will be

$$\$802.94 - \$600 = \$202.94$$

We have shown that an investment of P dollars will accumulate to $P(1 + i)^t$ at the end of t years at an annual interest rate i. Sometimes, it may be important to determine how much money a person must invest so as to have a specified sum at the end of t years. In that case, we solve (5.1) for P and obtain

5.2 $$P = A(t)(1 + i)^{-t}$$

This value of P is called the *present value* of the amount $A(t)$. The values of $(1 + i)^{-t}$ for some values of i and t are available in Table 3 at the end of the book. Alternatively, one may use a hand-held calculator to compute these values.

Example 5.10

Mr. Talbot wishes to have \$4,000 available in a bank account for his daughter's first-year college expenses. What sum of money must he invest now at 6.5 percent compounded annually if his daughter is to start college four years hence?

Solution

Using (5.2), we have

$$\begin{aligned} P &= \$4{,}000(1.065)^{-4} \\ &= \$4{,}000(0.777323) \qquad \textbf{(Table 3)} \\ &= \$3{,}109.29 \end{aligned}$$

He should invest \$3,109.29.

In recent years, it has become an established practice among commercial banks, savings and loan institutions, credit unions, and other approved financial companies to pay interest more frequently than once a year. Generally, the rate of interest is quoted on an annual basis but is, in practice, compounded quarterly, monthly, or even daily. Conceivably, this compounding frequency could progress to every hour, every minute, every second, and so on. The frequency with which interest is compounded and paid or reinvested to earn additional interest is called the *conversion period.*

The term "effective rate of interest" is the rate that, compounded annually, would produce the same dollar amount as the quoted or published rate compounded several times a year. An interest rate of 6 percent per year compounded monthly, for instance, does not mean that 6 percent is credited to an account after each month but rather that an interest of $\frac{1}{12}(6) = 0.5$ percent is credited monthly to the existing balance. The 6 percent is called the "nominal" rate of interest. The effective rate of interest is 6.168 percent; that is because each \$1 of principal compounded monthly at 6 percent yields

$$\$(1 + 0.005)^{12} = \$1.061678$$

so that the annual interest earned is

$$\$1.061678 - \$1 = \$0.061678$$

or approximately 6.168 percent. As another illustration, a savings bank advertises that \$1,000 deposited with the bank for a period of four or more years at 12.00 percent compounded monthly results in an actual annual yield of 12.68 percent. The *effective* rate of interest is thus 12.68 percent as compared to the *nominal* rate of 12.00 percent.

Let i_m denote a nominal rate of interest that is compounded m times a year. Clearly, (i_m/m) is the rate used to compute the interest at the end of each conversion period. By an argument similar to the one used in developing (5.1), it can be shown that the accumulated value of P dollars at the end of t years is given by

5.3 $$P\left(1 + \frac{i_m}{m}\right)^{mt}$$

What happens to the expression (5.3) *as m gets larger and larger?* The answer to this question depends on the value of

$$\lim_{n \to +\infty} \left(1 + \frac{1}{n}\right)^n$$

While it is not obvious that this limit exists, we can show intuitively that it does exit. Let us consider values of

$$f(n) = \left(1 + \frac{1}{n}\right)^n$$

for sufficiently large values of n.

$$\begin{aligned} f(1) &= 2 & f(100) &= 2.70481 \\ f(2) &= 2.25 & f(1{,}000) &= 2.71692 \\ f(4) &= 2.44141 & f(10{,}000) &= 2.71815 \\ f(10) &= 2.59374 & f(100{,}000) &= 2.71827 \\ f(50) &= 2.69159 & f(1{,}000{,}000) &= 2.71828 \end{aligned}$$

Thus as n gets larger and larger, $f(n)$ gets closer and closer to an irrational number whose value, correct to five decimal places, is 2.71828. We use the letter e to represent that number. In other words,

5.4 $$\lim_{n \to +\infty} f(n) = \lim_{n \to +\infty} \left(1 + \frac{1}{n}\right)^n = e$$

To evaluate

$$\lim_{m \to +\infty} P\left(1 + \frac{i_m}{m}\right)^{mt}$$

we observe that if we let $n = m/i_m$ then

$$P\left(1 + \frac{i_m}{m}\right)^{mt} = P\left[\left(1 + \frac{1}{n}\right)^n\right]^{i_m t}$$

As n approaches positive infinity, so does m. Thus.

$$\begin{aligned} \lim_{m \to +\infty} P\left(1 + \frac{i_m}{m}\right)^{mt} &= P \lim_{n \to +\infty} \left[\left(1 + \frac{1}{n}\right)^n\right]^{i_m t} \\ &= P\left[\lim_{n \to +\infty} \left(1 + \frac{1}{n}\right)^n\right]^{i_m t} \\ &= Pe^{i_m t} \end{aligned}$$

Thus,

5.5 $$\lim_{m \to +\infty} P\left(1 + \frac{i_m}{m}\right)^{mt} = Pe^{i_m t}$$

Example 5.11

Find the accumulated amount after 10 years if \$2,000 is invested at 6 percent compounded

a. annually **c.** quarterly **e.** daily
b. semi-annually **d.** monthly

Solution

a. Here, $P = \$2{,}000$, $i = 0.06$, $t = 10$. Using (5.1), we obtain

$$\begin{aligned} A &= \$2{,}000(1.06)^{10} \\ &= \$2{,}000(1.790848) \qquad \textbf{(Table 2)} \\ &= \$3{,}581.70 \end{aligned}$$

b. $P = \$2{,}000$, $i_m = 0.06$, $m = 2$; $i_m/m = 0.06/2 = 0.03$, and $t = 10$. Using (5.3), we have

$$\begin{aligned} A &= \$2{,}000(1.03)^{20} \\ &= \$2{,}000(1.806111) \qquad \textbf{(Table 2)} \\ &= \$3{,}612.22 \end{aligned}$$

c. $P = \$2{,}000$, $i_m = 0.06$, $m = 4$; $i_m/m = 0.06/4 = 0.015$, and $t = 10$. Hence, using (5.3), we obtain

$$\begin{aligned} A &= \$2{,}000(1.015)^{40} \\ &= \$2{,}000(1.814018) \qquad \textbf{(Table 2)} \\ &= \$3{,}628.04 \end{aligned}$$

d. $P = \$2{,}000$, $i_m = 0.06$, $m = 12$; $i_m/m = 0.06/12 = 0.005$, $t = 10$. Hence,

$$A = \$2{,}000(1.005)^{120}$$

The value of $(1.005)^{120}$ is not available in Table 2. However, observe that

$$(1.005)^{60} = 1.348850 \qquad \textbf{(Table 2)}$$

and

$$\begin{aligned} (1.005)^{120} &= [(1.005)^{60}]^2 \\ &= (1.348850)^2 \\ &= 1.819396 \end{aligned}$$

Thus

$$\begin{aligned} A &= \$2{,}000(1.005)^{120} \\ &= \$2{,}000(1.819396) \\ &= \$3{,}638.79 \end{aligned}$$

e. Again, $P = \$2{,}000$, $i_m = 0.06$, $m = 365$ and

$$\frac{i_m}{m} = \frac{0.06}{365} = 0.000164384, \quad \text{and} \quad t = 10$$

We are assuming that there are 365 days in a year; thus, there are $10 \times 365 = 3{,}650$ conversion periods. Hence

$$A = \$2{,}000(1.000164384)^{3650}$$

Using an electronic calculator, we observe that

$$(1.000164384)^{3650} = 1.822032$$

Thus,

$$A = \$2{,}000(1.822032)$$
$$= \$3{,}644.06$$

The information obtained in the preceding example is summarized in Table 5.5.

Table 5.5
***P* = $2,000, i_m = 0.06, *t* = 10 years**

Compounded	Value of m	Accumulated Value of P
annually	1	$3,581.70
semi-annually	2	3,612.22
quarterly	4	3,628.04
monthly	12	3,638.79
daily	365	3,644.06

Note that as the interest is compounded more frequently, the accumulated value increases. Further, if interest is compounded continuously, then the accumulated value is given by (5.5). That is,

$$A = Pe^{i_m t}$$

For $P = \$2{,}000$, $t = 10$, and $i_m = 0.06$,

$$A = \$2{,}000e^{(0.06)10}$$
$$= \$2{,}000e^{0.6}$$
$$= \$2{,}000(1.8221) \qquad \textbf{(Table 1)}$$
$$= \$3{,}644.20$$

which is slightly more than the amount obtained by compounding daily.

Example 5.12 Determine how much money should be deposited in a bank if the money is to accumulate to \$2,500 in 5 years at 6 percent interest compounded

a. monthly **b.** daily **c.** continuously

Solution

Observe that equation (5.2) converted to m conversion periods per year becomes

$$P = A(t)\left(1 + \frac{i_m}{m}\right)^{-mt}$$

a. Here, $m = 12$, $i_m = 0.06$, $i_m/m = 0.005$ and $t = 5$. Using the above equation, we obtain

$$\begin{aligned} P &= \$2{,}500(1.005)^{-60} \\ &= \$2{,}500(0.741372) \qquad \textbf{(Table 3)} \\ &= \$1{,}853.43 \end{aligned}$$

b. Here, $m = 365$, $t = 5$, and $i_m/m = 0.06/365 = 0.000164384$. Hence,

$$P = \$2{,}500(1.000164384)^{-1825}$$

Using an electronic calculator, we observe that

$$(1.000164384)^{-1825} = 0.740836$$

Thus,

$$\begin{aligned} P &= \$2{,}500(0.740836) \\ &= \$1{,}852.09 \end{aligned}$$

c. From (5.5), we have

$$A = P \cdot e^{i_m t}$$

or

5.6 $$P = Ae^{-i_m t}$$

Here, $A = \$2{,}500$, $i_m = 0.06$ and $t - 5$ years. Hence,

$$\begin{aligned} P &= \$2{,}500e^{-(0.06)(5)} \\ &= \$2{,}500e^{-0.30} \\ &= \$2{,}500(0.74082) \qquad \textbf{(Table 1)} \\ &= \$1{,}852.05 \end{aligned}$$

Exercises

1. Find the accumulated amount and the compound interest if \$800 is invested for 2 years at 5 percent compounded
a. annually **b.** semi-annually **c.** quarterly

2. Find the accumulated value and the compound interest if \$1,500 is invested for 6 years at 6 percent compounded

a. semi-annually
b. quarterly
c. monthly
d. daily
e. continuously [Use $e^{0.36} = 1.433329$.]

3. Determine the accumulated amount for each of the following.

a. \$700 for 3 years at 5 percent compounded quarterly
b. \$1,200 for 4 years at 6 percent compounded monthly
c. \$2,000 for 3 years at 7 percent compounded continuously [Use $e^{0.21} = 1.233678$.]

4. Dr. Grant deposits \$4,000 in a savings account at 6.5 percent compounded continuously. How much will she have in her account 10 years from now? [Use $e^{0.65} = 1.91554$.]

5. Mr. Smith borrows \$1,500 at 8 percent compounded quarterly. Determine how much he owes if he borrows funds for

a. 6 months
b. 9 months
c. one year
d. two years

6. Find the amount that must be invested at 8 percent interest in order to accumulate \$2,000 at the end of 5 years if interest is compounded

a. annually
b. semi-annually
c. quarterly
d. continuously

7. Find the amount that must be invested so as to accumulate to \$4,500 five years from now at 6 percent compounded

a. semi-annually
b. quarterly
c. monthly
d. daily
e. continuously

8. Find the amount that must be invested so as to accumulate to \$2,500 three years from now if the current rate of interest is 10 percent compounded

a. quarterly
b. continuously

9. Mrs. White wishes to have \$5,000 available for her son who will enter college in 4 years. What sum of money must she invest now if the current rate of interest is 7.5 percent compounded continuously?

10. Mr. Berman wishes to have \$3,500 available for a European trip he is planning to undertake 3 years hence. What sum of money must he invest now in order to accumulate \$3,500 in 3 years if interest is 7.5 percent compounded continuously? [Use $e^{-0.225} = 0.79852$.]

5.4 Applications of Exponential Functions

Many practical situations involving growth or decay can be described by means of exponential functions of the form

$$f(t) = Ae^{kt}$$

where A represents the amount of substance present at time $t = 0$, k is a constant, and $f(t)$ represents the amount of substance present at time t. At time $t = 0$,

$$f(0) = Ae^{k\cdot 0} = Ae^0 = A$$

Thus,

$$f(t) = f(0)e^{kt}$$

When $k > 0$, the function $f(t)$ is an increasing function of time and we speak of "exponential growth." On the other hand, if $k < 0$, the function $f(t)$ is a decreasing function of time and we speak of "exponential decay." These functions are important in learning theory, banking practices, bacteria growth, sales increases, radioactive decay, and so on.

Example 5.13

A company's gross income is given by the formula

$$I(t) = \$10{,}000e^{(0.5)(t-1)}$$

where t is measured in years. Calculate the gross income of the company for $t = 1, 2, 3, 4, 5$ and 6 years. Graph $I(t)$.

Solution

The earnings at time $t = 1$ are

$$I(1) = \$10{,}000e^0 = \$10{,}000$$

Using Table 1, we have

$$\begin{aligned} I(2) &= \$10{,}000e^{0.5} \\ &= \$10{,}000(1.6487) = \$16{,}487 \\ I(3) &= \$10{,}000e^{1} \\ &= \$10{,}000(2.7183) = \$27{,}183 \\ I(4) &= \$10{,}000e^{1.5} \\ &= \$10{,}000(4.4817) = \$44{,}817 \\ I(5) &= \$10{,}000e^{(0.5)(4)} \\ &= \$10{,}000e^{2} \\ &= \$10{,}000(7.3891) = \$73{,}891\text{; and} \\ I(6) &= \$10{,}000^{e(0.5)(5)} \\ &= \$10{,}000e^{2.5} \\ &= \$10{,}000(12.182) = \$121{,}820 \end{aligned}$$

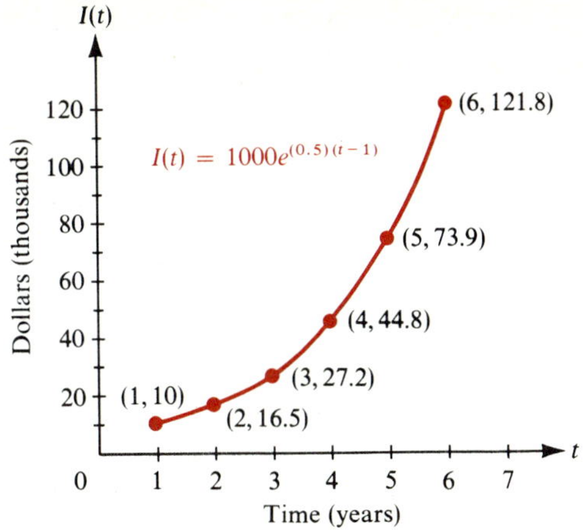

Figure 5.5

Plotting $(t, I(t))$ for $t = 1, 2, 3, 4, 5,$ and 6 and connecting the points with a smooth curve, we have the graph in Figure 5.5.

Example 5.14

The annual profit of a company from the sales of a new product is given by

$$P(t) = \$5{,}000 + 10{,}000e^{-0.4(t-1)}$$

where t is the number of years that the product is on the market. Determine the annual profit at the end of 1, 2, 3, 4, 5, 6, and 10 years. Graph $P(t)$.

Solution

Table 5.6 shows the value of $P(t)$ for different values of t.

Table 5.6

t (in years)	$P(t)$ (in dollars)
1	$5{,}000 + 10{,}000e^{0} = 15{,}000$
2	$5{,}000 + 10{,}000e^{-0.4} = 5{,}000 + 10{,}000(0.67032) = 11{,}703$
3	$5{,}000 + 10{,}000(e^{-0.8}) = 5{,}000 + 10{,}000(0.44933)$ $= 9{,}493$
4	$5{,}000 + 10{,}000(e^{-1.2}) = 5{,}000 + 10{,}000(0.30119)$ $= 8{,}012$
5	$5{,}000 + 10{,}000(e^{-1.6}) = 5{,}000 + 10{,}000(0.20190)$ $= 7{,}019$
6	$5{,}000 + 10{,}000(e^{-2.0}) = 5{,}000 + 10{,}000(0.13534)$ $= 6{,}353$
10	$5{,}000 + 10{,}000(e^{-3.6}) = 5{,}000 + 10{,}000(0.02732)$ $= 5{,}273$

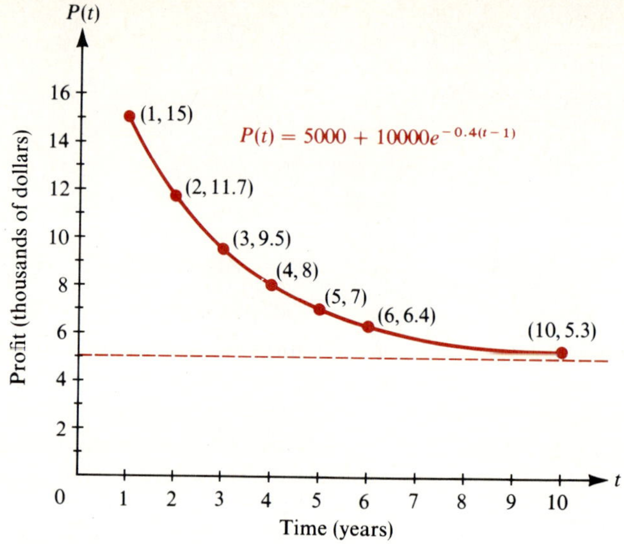

Figure 5.6

Plotting the order pairs $(t, P(t))$, we obtain the graph as shown in Figure 5.6. Notice that profit tends to level off with time and gradually approaches a level of \$15,000.

Example 5.15

The market research organization of a frozen food company estimates that the demand for chicken (in thousands per month) is given by the equation

$$D(t) = (8 - 7e^{-0.8(t-1)})$$

where t is measured in months. Find the estimated sales for $t = 1, 2, 3, 4, 5, 6, 8, 10$, and 12 months. Graph $D(t)$.

Solution

The demand function $D(t)$ for different values of t is given in Table 5.7.

Table 5.7

t (in months)	Demand (in thousands per month)
1	$8 - 7e^0 = 8 - 7 = 1$
2	$8 - 7e^{-0.8} = 8 - 7(0.44933) = 8 - 3.14531 = 4.85469$
3	$8 - 7e^{-1.6} = 8 - 7(0.2019) = 8 - 1.4133 = 6.5867$
4	$8 - 7e^{-2.4} = 8 - 7(0.09072) = 8 - 0.63504 = 7.36496$
5	$8 - 7e^{-3.2} = 8 - 7(0.04076) = 8 - 0.28532 = 7.71468$
6	$8 - 7e^{-4} = 8 - 7(0.01831) = 8 - 0.12824 = 7.87176$
8	$8 - 7e^{-5.6} = 8 - 7(0.00369) = 8 - 0.02583 = 7.97417$
10	$8 - 7e^{-7.2} = 8 - 7(0.00074) = 8 - 0.00518 = 7.99482$
12	$8 - 7e^{-8.8} = 8 - 7(0.00015) = 8 - 0.00105 = 7.99895$

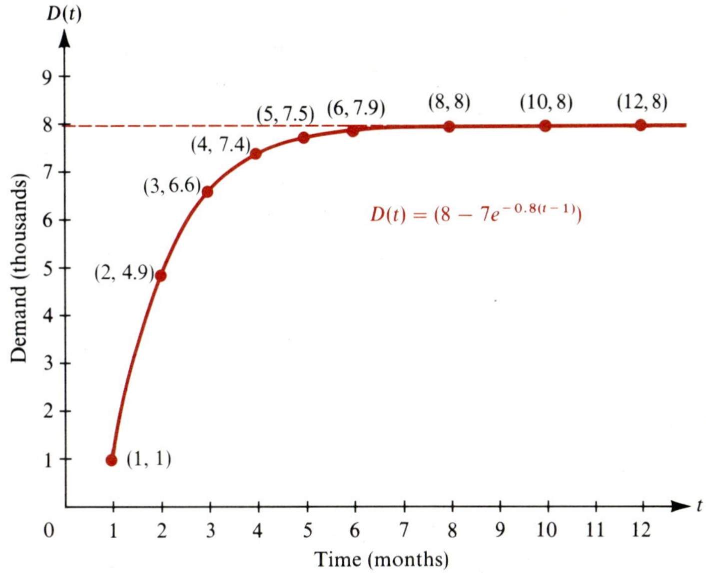

Figure 5.7

Plotting the ordered pairs $(t, D(t))$ and connecting these points with a smooth curve, we have the graph as in Figure 5.7. Note that the demand finally levels off as time passes.

Exercises

1. The number of bacteria, $N(t)$, present in a certain culture at time t is given by $N(t) = 500e^{(0.3)t}$ where t is measured in hours. Find the number of bacteria present at time
   a. $t = 0$ **b.** $t = 2$ **c.** $t = 5$

2. Assume that the number of bacteria, $N(t)$, in a certain culture is given by $N(t) = 1{,}000e^{(0.2)t}$, where t is measured in hours. Find the number of bacteria present when
   a. $t = 0$ **b.** $t = 10$ **c.** $t = 20$

3. The size of a population of a certain city at time t is given by $P(t) = 10^5e^{(0.05)t}$, where t is measured in years. Determine the size of the population when
   a. $t = 2$ **b.** $t = 6$ **c.** $t = 10$ **d.** $t = 20$

4. The quantity $Q(t)$ of radium (a substance subject to radioactive decay), measured in milligrams, at time t is given by the equation
$$Q(t) = 1{,}000e^{(-0.04)t}$$
   where t is measured in centuries. Determine the quantity present when
   a. $t = 5$ **b.** $t = 10$ **c.** $t = 15$ **d.** $t = 20$

5. The concentration of a drug in the bloodstream of a patient is approximated by the equation

$$Q(t) = 5e^{-0.3t}$$

where $Q(t)$ is the quantity of the drug (in suitable units) in the bloodstream t hours after it is injected. Determine $Q(t)$ for $t = 1, 2, 3, 4, 5, 6, 8$, and 10 hours.

6. A psychologist has prepared a list of nonsense words for use in a memory experiment. The number of words, $N(t)$, that a person is able to memorize after t hours of practice is approximated by

$$N(t) = 100(1 - e^{-0.2t})$$

Determine $N(t)$ for $t = 1, 2, 3, 4, 5, 6, 8$, and 10 hours.

7. A manufacturer's annual profit from the sales of a certain product is given by

$$P(t) = \$10{,}000(1 + e^{-0.3t})$$

where t is the number of years the product is on the market. Determine $P(t)$ for $t = 1, 2, 3, 4, 5, 6, 8$, and 10 years.

8. The annual salary raise of a university's faculty is determined by the formula

$$S(t) = \$15{,}000e^{-1.2t}$$

where t represents the number of years of service.
a. Determine the raise in salary after $t = 1, 2, 3, 4$, and 5 years.
b. Graph $S(t)$.

9. The sales of a new product of the ABC company in Detroit are approximated by the equation

$$S(t) = 2{,}000(5 + 4e^{-0.5t})$$

where t is the number of months the product is advertised.
a. Calculate the sales for $t = 1, 2, 3, 4, 5, 6, 7, 8, 10$, and 12.
b. Graph $S(t)$.

10. The monthly sales of a new model of automobile in the state of Connecticut are approximated by the equation

$$S(t) = 4{,}000e^{-0.3t}$$

where t is the number of months this model has been in production. Calculate $S(t)$ for $t = 1, 2, 3, \ldots, 12$ and graph $S(t)$.

5.5 Logarithmic Functions

Thus far, we have discussed exponential functions represented by the equation $y = b^x$, where $b > 0$, $b \neq 1$, and x is any real number. Although the exponential equation $y = b^x$ is not usually easy to solve for x when

y and b are known, one would expect that the equation does have a solution. Further, the graphs of $y = b^x$ in Figure 5.1 (for $b > 1$) and in Figure 5.4 (for $0 < b < 1$) each represent one-to-one correspondences between the set of all real numbers x and the set of all positive real values of y. This means that the exponential function $y = b^x$ has an inverse function. We call this inverse the *logarithmic function*, and we say that if $y = b^x$, then $x = \log_b y$. The number x is called the *logarithm* of y to the base b.

Definition 5.3

For $b > 0$, $b \neq 1$ and $y > 0$, $x = \log_b y$ if and only if $y = b^x$.

In words, the logarithm of a positive number to a given base is the exponent to which the base must be raised to yield that number. Thus if we write $3^4 = 81$, then the logarithm of 81 to the base 3 is 4; that is, $\log_3 81 = 4$. Again, if we write $10^2 = 100$, then $\log_{10} 100 = 2$.

For further illustrations, consider the following examples.

Example 5.16

Determine $\log_2 128$.

Solution

Let $x = \log_2 128$. Using the definition of logarithmic function, we obtain the exponential equation

$$2^x = 128$$

Since $128 = 2^7$, we see that $x = 7$ and so

$$\log_2 128 = 7$$

Example 5.17

Determine $\log_3 (\frac{1}{27})$.

Solution

Let $x = \log_3 (\frac{1}{27})$. Then

$$3^x = \frac{1}{27}$$

Expressing $\frac{1}{27}$ as a power of base 3, we have

$$\frac{1}{27} = \left(\frac{1}{3}\right)^3 = 3^{-3}$$

Thus,

$$3^x = 3^{-3}$$

which implies that $x = -3$. Therefore, $\log_3 (\frac{1}{27}) = -3$.

Example 5.18

Find $\log_8 32$.

Solution

If $x = \log_8 32$, then $8^x = 32$. Since $8 = 2^3$ and $32 = 2^5$, we have

$$2^{3x} = 2^5$$

which implies that $3x = 5$ and $x = \frac{5}{3}$. Hence, $\log_8 32 = \frac{5}{3}$.

We consider it worthwhile to reinforce the fact that the functions $y = \log_b x$ and $y = b^x$ are inverses of each other. Thus if we want to graph $y = \log_b x$ for $b > 1$, we need only graph its inverse function $y = b^x$ and then observe its mirror image in the line $y = x$. See Figure 5.8. From the graph, we observe that

1. $\log_b x$ is defined only for $x > 0$
2. $\log_b x < 0$ for $0 < x < 1$
3. $\log_b x = 0$ for $x = 1$
4. $\log_b x > 0$ for $x > 1$
5. $\log_b x = 1$ if $x = b$
6. $\log_b x$ is an increasing function of x.

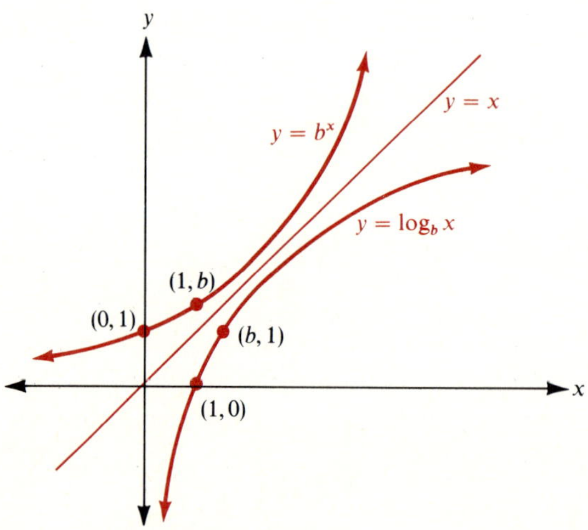

Figure 5.8

There is a definitive relationship between exponential and logarithmic functions. The following rules are immediate consequences of the laws of exponents*.

* See Appendix A for a brief review.

Proposition 5.1

If m and n are positive numbers and b is any base (that is, $b > 0$ and $b \neq 1$), then

a. $\log_b (mn) = \log_b m + \log_b n$

b. $\log_b \left(\dfrac{m}{n}\right) = \log_b m - \log_b n$

Proof

Let $\log_b m = p$ and $\log_b n = q$. Using Definition 5.3, we have

$$m = b^p \quad \text{and} \quad n = b^q$$

a. Observe that

$$mn = b^p b^q = b^{p+q}$$

Hence by Definition 5.3, it follows that

$$\log_b (mn) = p + q = \log_b m + \log_b n$$

b. Also

$$\frac{m}{n} = \frac{b^p}{b^q} = b^{p-q}$$

and using Definition 5.3, we obtain

$$\log_b \left(\frac{m}{n}\right) = p - q = \log_b m - \log_b n$$

which proves our assertion.

Example 5.19

Given that $\log_b 2 = 0.69$, $\log_b 3 = 1.10$, $\log_b 5 = 1.61$, find the following.

a. $\log_b 6$ **b.** $\log_b 15$ **c.** $\log_b (\frac{3}{2})$ **d.** $\log_b (\frac{10}{3})$

Solution

a. $\log_b 6 = \log_b (2 \cdot 3) = \log_b 2 + \log_b 3 = 0.69 + 1.10 = 1.79$

b. $\log_b 15 = \log_b (3 \cdot 5) = \log_b 3 + \log_b 5 = 1.10 + 1.61 = 2.71$

c. $\log_b (\frac{3}{2}) = \log_b 3 - \log_b 2 = 1.10 - 0.69 = 0.41$

d. $\log_b \left(\dfrac{10}{3}\right) = \log_b \left(\dfrac{5 \cdot 2}{3}\right) = \log_b 5 + \log_b 2 - \log_b 3$
$= 1.61 + 0.69 - 1.10 = 1.20$

Proposition 5.2

If m is any positive real number, r is any real number, and b is a base, then

$$\log_b (m^r) = r \log_b m$$

Proof

Let $\log_b m = p$. Then $m = b^p$ and $m^r = b^{pr}$. Hence,

$$\log_b (m^r) = \log_b (b^{pr}) = pr = rp = r \log_b (m)$$

Example 5.20

Given that $\log_b 2 = 0.69$ and $\log_b 3 = 1.10$, determine

a. $\log_b \sqrt{2}$ **b.** $\log_b 72$

Solution

a. $\log_b \sqrt{2} - \log_b (2)^{1/2} = \frac{1}{2} \log_b 2 = \frac{1}{2}(0.69) = 0.345$

b. $\log_b 72 = \log_b (2^3 \cdot 3^2) = \log_b (2^3) + \log_b (3^2)$ **(Proposition 5.1)**

$= 3 \log_b 2 + 2 \log_b 3$ **(Proposition 5.2)**

$= 3(0.69) + 2(1.10)$

$= 4.27$

Proposition 5.3

If a and b are two bases and m is any positive number, then

$$\log_a m = \frac{\log_b m}{\log_b a}$$

Proof

Let $\log_a m = p$ which implies that $m = a^p$. Similarly, if $\log_b a = n$, then $a = b^n$. Thus,

$$m = a^p = (b^n)^p = b^{np}$$

This statement implies that

$$\log_b m = np = (\log_b a)(\log_a m)$$

and since $\log_b a$ cannot equal zero (why?), our assertion follows.

Example 5.21

If $\log_b 2 = 0.69$, $\log_b 3 = 1.10$ and $\log_b 5 = 1.61$, determine

a. $\log_3 5$ **b.** $\log_2 3$

Solution

a. $\log_3 5 = \dfrac{\log_b 5}{\log_b 3} = \dfrac{1.61}{1.10} = 1.46$

b. $\log_2 3 = \dfrac{\log_b 3}{\log_b 2} = \dfrac{1.10}{0.69} = 1.59$

Exercises

1. Express each of the following exponential equations in logarithmic form.

a. $2^3 = 8$
b. $3^2 = 9$
c. $5^4 = 625$
d. $(16)^{1/2} = 4$
e. $(32)^{3/5} = 8$
f. $(216)^{2/3} = 36$

2. Express each of the following logarithmic equations in exponential form.

a. $\log_3 27 = 3$
b. $\log_2 64 = 6$
c. $\log_4 256 = 4$
d. $\log_8 4 = \frac{2}{3}$
e. $\log_{32} 2 = \frac{1}{5}$
f. $\log_6 1 = 0$

3. Solve for x in each of the following equations.

a. $\log_3 x = 4$
b. $\log_2 x = 7$
c. $\log_8 512 = x$
d. $\log_2 (\frac{1}{8}) = x$
e. $\log_5 x = -2$
f. $\log_5 125 = x$
g. $\log_x 16 = 4$
h. $\log_9 3 = x$
i. $\log_x 25 = -2$
j. $\log_x 343 = 3$

4. Solve for x in each of the following equations.

a. $\log_2 (2^4) = x$
b. $\log_3 (x^5) = 5$
c. $\log_x (3^5) = 5$
d. $\log_5 (3x) = 2$

5. Given that $\log_b 2 = 0.69$, $\log_b 3 = 1.10$, $\log_b 5 = 1.61$, $\log_b 7 = 1.95$, evaluate the following expressions.

a. $\log_b (\frac{2}{7})$
b. $\log_b (21)$
c. $\log_b (35)$
d. $\log_b (25)$
e. $\log_b (128)$
f. $\log_b (343)$

6. If $\log_{10} 2 = 0.3010$, $\log_{10} 3 = 0.4771$, $\log_{10} 7 = 0.8451$, evaluate the following expressions.

a. $\log_{10} \sqrt{2}$
b. $\log_{10} (\frac{21}{2})$
c. $\log_{10} (\frac{14}{3})$
d. $\log_3 7$
e. $\log_2 3$
f. $\log_{10} 42$
g. $\log_3 2$
h. $\log_{10} (21)^{1/3}$
i. $\log_{10} (\frac{9}{4})$
j. $\log_{10} (12)^{1/3}$
k. $\log_{10} (3\sqrt{8})$
l. $\log_{10} \sqrt{63}$

7. Given that $\log_{10} 2 = 0.3010$ and $\log_{10} 3 = 0.4771$, find the logarithms of 4, 6, 8, 9, 12, and 16.

8. Prove that

$$(\log_a b)(\log_b a) = 1$$

9. Prove that if m, n, and p are positive numbers and b is any base, then

$$\log_b(mnp) = \log_b m + \log_b n + \log_b p$$

10. Prove that

$$b^{\log_b x} = x$$

for $x > 0$.

5.6 Common and Natural Logarithms

Logarithms have been used in mathematics as an effective tool for simplifying numerical work. Since the number system is in base 10, logarithms to base 10 are the most convenient for numerical calculations. Such logarithms are called *common logarithms*. In other words, the common logarithm of a number is the exponent to which 10 must be raised to yield that number. Since

$$10^1 = 10, \quad 10^2 = 100, \quad 10^3 = 1{,}000, \quad 10^4 = 10{,}000, \quad 10^5 = 100{,}000$$

it follows that the common logarithm of 10 is 1, the common logarithm of 100 is 2, the common logarithm of 1,000 is 3, and so on. Thus the following statements are equivalent.

$$\begin{array}{lcl} 10^1 = 10 & \text{and} & \log_{10} 10 = 1 \\ 10^2 = 100 & \text{and} & \log_{10} 100 = 2 \\ 10^3 = 1{,}000 & \text{and} & \log_{10} 1{,}000 = 3 \\ 10^4 = 10{,}000 & \text{and} & \log_{10} 10{,}000 = 4 \end{array}$$

For brevity, we shall write $\log x$ to mean the logarithm of x to the base 10. Further, because

$$\begin{array}{lcl} 10^{-1} = 0.1, & \text{then} & \log 0.1 = -1 \\ 10^{-2} = 0.01, & \text{then} & \log 0.01 = -2 \\ 10^{-3} = 0.001, & \text{then} & \log 0.001 = -3 \end{array}$$

Thus the logarithms of the integral powers of 10 are precisely those integers. Evidently, if

$$10^x = 75$$

then x must be a number between 1 and 2; that is, x is the integer one plus a decimal fraction. In fact, every logarithm may be expressed as a sum of an integer and a decimal fraction. In practice, the decimal is rounded off to a few places.

Definition 5.4 The integral part of a common logarithm is called the *characteristic*. The decimal part of a common logarithm is called the *mantissa*.

Any positive number x can be expressed in scientific notation as

$$x = n \cdot 10^c$$

where c is an integer (which may be positive, zero, or negative) and n satisfies the inequality $1 \leq n < 10$. Thus for that number x,

$$\begin{aligned} \log x &= \log n + \log 10^c \qquad \textbf{(Proposition 5.1)} \\ &= \log n + c \end{aligned}$$

where c is the characteristic and $\log n$ is the mantissa.

The mantissa, $\log n$, is obtained from Table 4 which lists the logarithms of numbers from 1.00 to 9.99 in steps of 0.01. Consider, for example, log 2.35. In Table 4, we locate the row for 2.3, and then move across this row until we are under the column headed 5, where we find the entry 0.3711. Again, we may use this table to determine that

$$\log 7.82 = 0.8932; \quad \log 5.47 = 0.7380; \quad \log 3.59 = 0.5551$$

and so on.

Example 5.22

Find log 4350.

Solution

Since $4350 = 4.350 \times 10^3$, it follows that $c = 3$ and $n = 4.350$. Using Table 4, we find that $\log 4.350 = 0.6385$. Thus,

$$\begin{aligned} \log 4350 &= 0.6385 + 3 \\ &= 3.6385 \end{aligned}$$

Example 5.23

Find log 0.0728.

Solution

Since

$$0.0728 = 7.28 \times 10^{-2}$$

we have the characteristic $c = -2$ and $n = 7.28$. Thus,

$$\begin{aligned} \log 0.0728 &= (\log 7.28) + (-2) \\ &= 0.8621 - 2 \\ &= -1.1379 \end{aligned}$$

We know how to find $\log_{10} x$ for a given positive number x, but if $\log_{10} x$ is given, how do we determine x? First, we write $\log x$ in the form of

$$\log x = (\log n) + c$$

where c is the characteristic and $\log n$ is the mantissa. Since this statement is equivalent to

$$x = n \cdot 10^c$$

we determine x by finding n and then multiplying it by 10^c. Note that $\log n$ is a number between 0 and 1 and is listed in the body of Table 4.

Example 5.24

Find x if $\log x = 1.8021$

Solution

Since

$$\log x = 0.8021 + 1$$

it follows that $c = 1$ and $\log n = 0.8021$. Referring to Table 4, we find that mantissa 0.8021 corresponds to the number $n = 6.34$. Thus,

$$\begin{aligned} x &= n \times 10^1 \\ &= 6.34 \times 10^1 \\ &= 63.4 \end{aligned}$$

Example 5.25

Find x if $\log x = -1.1232$

Solution

We must express $\log x$ in the form of $\log n + c$, where $\log n$ is a positive decimal fraction. Since Table 4 contains only positive values, we make the mantissa positive by a simple algebraic step. We add and subtract 2 thus obtaining

$$\begin{aligned} \log x &= \underbrace{-1.1232 + 2} - 2 \\ &= 0.8768 + (-2) \end{aligned}$$

which implies that $c = -2$ and $\log n = 0.8768$. Referring to the body of Table 4, we find that mantissa 0.8768 corresponds to the number $n = 7.53$. Thus,

$$\begin{aligned} x &= n \times 10^c \\ &= 7.53 \times 10^{-2} \\ &= 0.0753 \end{aligned}$$

We will now use logarithms for simplifying numerical work. Let us look at some illustrative examples.

Example 5.26

Find $x = (1.19)^{20}$

Solution

$$\begin{aligned} \log x &= \log(1.19)^{20} \\ &= 20 \log 1.19 \qquad \textbf{(Proposition 5.2)} \\ &= 20(0.0755) \\ &= 1.51 \end{aligned}$$

To determine x, we note that $c = 1$ and $\log n = 0.51$. Referring to Table 4, we have $n = 3.24$ approximately and

$$\begin{aligned} x &= n \times 10^c \\ &= 3.24 \times 10^1 \\ &= 32.40 \end{aligned}$$

Example 5.27

Find the value of

$$(215)^{1/3}(80)^{1/5}(0.0025)^{1/4}$$

Solution

Let $x = (215)^{1/3}(80)^{1/5}(0.0025)^{1/4}$. Using Proposition 5.1, we have

$$\begin{aligned} \log x &= \log(215)^{1/3} + \log(80)^{1/5} + \log(0.0025)^{1/4} \qquad \textbf{(Proposition 5.1)} \\ &= \tfrac{1}{3}\log 215 + \tfrac{1}{5}\log 80 + \tfrac{1}{4}\log(0.0025) \qquad \textbf{(Proposition 5.2)} \end{aligned}$$

From Table 4, we have

$$\begin{aligned} \log 215 &= \log(2.15 \times 10^2) = (\log 2.15) + 2 = 2.3324 \\ \log 80 &= \log(8.00 \times 10^1) = (\log 8.00) + 1 = 1.9031 \\ \log 0.0025 &= \log(2.5 \times 10^{-3}) \\ &= (\log 2.5) - 3 \\ &= 0.3979 - 3 \\ &= -2.6021 \end{aligned}$$

Thus,

$$\begin{aligned} \log x &= \tfrac{1}{3}(2.3324) + \tfrac{1}{5}(1.9031) + \tfrac{1}{4}(-2.6021) \\ &= 0.7775 + 0.3806 - 0.6505 \\ &= 0.5076 \end{aligned}$$

To determine x, note that $c = 0$ and $\log n = 0.5076$. From Table 4, we find that $n = 3.22$ approximately. Thus,

$$x = n \cdot 10^c = 3.22 \times 10^0$$
$$= 3.22$$

Example 5.28

Find the value of $\frac{(32.5)(40.8)^{1/3}}{(4.72)}$.

Solution

Let $x = \frac{(32.5)(40.8)^{1/3}}{(4.72)}$. Applying Proposition 5.1, we have

$$\log x = \log 32.5 + \log(40.8)^{1/3} - \log 4.72 \qquad \textbf{(Proposition 5.1)}$$
$$= \log 32.5 + \tfrac{1}{3}\log 40.8 - \log 4.72 \qquad \textbf{(Proposition 5.2)}$$

From Table 4, we find that

$$\log 32.5 = \log(3.25 \times 10^1) = (\log 3.25) + 1 = 1.5119$$
$$\log 40.8 = \log(4.08 \times 10^1) = (\log 4.08) + 1 = 1.6107$$
$$\log 4.72 = 0.6739$$

Thus,

$$\log x = 1.5119 + \tfrac{1}{3}(1.6107) - 0.6739 = 1.3749$$

To determine x, note that $c = 1$ and $\log n = 0.3749$. Referring to Table 4, we have $n = 2.37$ approximately and

$$x = n \cdot 10^c = (2.37) \times 10^1 = 23.70$$

The *natural logarithm* of a number is the exponent to which the base e must be raised to yield that number. We will adhere to the notation that $\ln x$ is to be interpreted as the natural logarithm of x while $\log x$ will continue to be interpreted as logarithm of x to the base 10. Table 5 lists the value of natural logarithms from 1.00 to 9.99 in steps of 0.01. Consider, for example, In 5.24. Referring to Table 5, we locate the row for 5.2, move across until we are under the column headed 0.04, and read the entry 1.6563. Again, we may use this table to determine that

$$\ln 2.02 = 0.7031; \quad \ln 6.38 = 1.8532; \quad \ln 9.99 = 2.3016$$

and so on. To determine ln 10, we use the fact that

$$\log_e 10 = \frac{\log_{10} 10}{\log_{10} e} \qquad \textbf{(Proposition 5.3)}$$

Assuming that $e = 2.71828$, it can be shown that $\log_{10} e = 0.43429$. Thus

$$\log_e 10 = \frac{1}{0.43429} = 2.3026 \qquad (\log_{10} 10 = 1)$$

Observe that

$$\ln 100 = \ln 10^2 = 2 \ln 10 = 2(2.3026) = 4.6052$$
$$\ln 1000 = \ln 10^3 = 3 \ln 10 = 3(2.3026) = 6.9078$$
$$\ln 10000 = \ln 10^4 = 4 \ln 10 = 4(2.3026) = 9.2104$$

and so on.

Example 5.29

Find

a. $\ln 456$ **b.** $\ln 2480$ **c.** $\ln 0.567$

Solution

a. Note that $456 = 4.56 \times 10^2$. Using Proposition 5.1, we have

$$\ln 456 = \ln 4.56 + \ln 10^2 = \ln 4.56 + 2 \ln 10$$

From Table 5, we find that $\ln 4.56 = 1.5173$. Thus

$$\ln 456 = 1.5173 + 4.6052 = 6.1225$$

b. Since $2480 = 2.48 \times 10^3$, we apply Proposition 5.1 and obtain

$$\begin{aligned} \ln 2480 &= \ln 2.48 + \ln 10^3 \\ &= \ln 2.48 + 3 \ln 10 \qquad \textbf{(Proposition 5.2)} \\ &= 0.9083 + 3(2.3026) \\ &= 7.8161 \end{aligned}$$

c. Note that $0.567 = 5.67 \times 10^{-1}$.

$$\begin{aligned} \ln 0.567 &= \ln 5.67 + \ln 10^{-1} \\ &= \ln 5.67 - \ln 10 = 1.7352 - 2.3026 = -0.5674 \end{aligned}$$

Example 5.30

Given $e^{2x} = 17$, determine x.

Solution

Taking natural logarithms, we have

$$\begin{aligned} \ln(e^{2x}) = \ln 17 &= \ln(1.7 \times 10^1) \\ &= \ln 1.7 + \ln 10 \qquad \textbf{(Proposition 5.1)} \\ &= 0.5306 + 2.3026 \\ &= 2.8332 \end{aligned}$$

That is,

$$2x \ln e = 2.8332 \qquad \textbf{(Proposition 5.2)}$$
$$2x = 2.8332 \qquad (\ln e = 1)$$

or

$$x = 1.4166$$

Recall that an exponential function is of the form $f(x) = b^x$ where $b > 0$, $b \neq 1$ and x is any real number. Because exponential functions with base e are easier to work with, we shall indicate briefly how to change an exponential function with base b (such as $f(x) = b^x$) to an exponential function with base e. In other words, we need to determine z such that

$$b^x = e^z$$

Taking natural logarithms of both sides, we have

$$\ln b^x = \ln e^z$$

or

$$x \ln b = z \ln e$$

Because $\ln e = 1$, we conclude that

$$z = x \ln b$$

Hence, we can write

5.7 $$b^x = e^{x \ln b}$$

Example 5.31

The number of milligrams of radium, $Q(t)$, at time t is given by $Q(t) = 1{,}000(0.95)^t$, where t is measured in centuries. Express $Q(t)$ as an exponential function with base e and then determine the quantity present at $t = 10$. [Use $e^{-0.513} = 0.5987$.]

Solution

Using (5.7), the equation

$$Q(t) = 1{,}000(0.95)^t$$

can be expressed as

$$Q(t) = 1{,}000e^{t(\ln 0.95)}$$

Note that

$$\begin{aligned} \ln 0.95 &= \ln 9.5 + \ln 10^{-1} \\ &= \ln 9.5 - \ln 10 \qquad \textbf{(Proposition 5.2)} \\ &= 2.2513 - 2.3026 \\ &= -0.0513 \end{aligned}$$

Thus,

$$Q(t) = 1{,}000e^{-0.0513t}$$

and

$$\begin{aligned} Q(10) &= 1{,}000e^{-0.513} \\ &= 1{,}000(0.5987) \\ &= 598.7 \end{aligned}$$

Exercises

1. Find each of the following logarithms.
   a. $\log 4230$
   b. $\log 423$
   c. $\log 42.3$
   d. $\log 4.23$
   e. $\log 0.423$
   f. $\log 0.0423$
   g. $\ln 4230$
   h. $\ln 423$
   i. $\ln 42.3$
   j. $\ln 4.23$
   k. $\ln 0.423$
   l. $\ln 0.0423$
2. Solve for x.
   a. $10^x = 84$
   b. $10^x = 34.6$
   c. $10^x = 3.46$
   d. $e^x = 10$
   e. $e^x = 24.3$
   f. $2e^x = 130$
3. Use Table 4 to find the value of x in each of the following problems.
   a. $\log x = 0.4048$
   b. $\log x = 1.6972$
   c. $\log x = 3.1614$
   d. $\log x = -1.0752$
   e. $\log x = -2.3420$
   f. $\log x = 1.4564$
   g. $\log x = 3.9284$
   h. $\log x = -1.6180$
4. Use Table 5 to find the value of x in each of the following problems.
   a. $\ln x = 2.3026$
   b. $\ln x = 4.2697$
   c. $\ln x = 2.9704$
   d. $\ln x = 1.2698$
   e. $\ln x = 1.7029$
   f. $\ln x = 2.2279$
5. Use properties of common logarithms to find the value of each of the following expressions.
   a. 9.23×0.914
   b. $\dfrac{24.9 \times 16.2}{37.3}$
   c. $\dfrac{0.0074 \times 0.0137}{0.0252}$
6. Use properties of common logarithms to evaluate the following expressions.
   a. $(20)^{1/4}$
   b. $(0.2)^{1/5}$
   c. $(0.003)^{1/2}$
7. Solve for x.
   a. $2^x = 15$
   b. $3^x = 4$
   c. $5^x = 27$
   d. $(0.01)^x = 8$
   e. $e^{2x} = 121$
   f. $2e^{2x} = 50$
   g. $e^{3x} = 64$

8. Solve for x.

a. $10^x = \frac{123}{231}$ **c.** $3^{2x-1} = 26$ **e.** $5^{x+2} = 8^{2x-3}$

b. $10^{-x} = \frac{123}{231}$ **d.** $3^{1-2x} = 49$

9. The monthly sales of a new product are approximated by the equation

$$s(t) = 20{,}000(0.75)^t$$

where t is the number of months the product has been on the market. Express $s(t)$ as an exponential function with base e and then determine the sales at $t = 8$. [Use $e^{-2.3016} = 0.1001$]

10. The decline in sales of a product is approximated by the equation

$$s(t) = 10{,}000(0.82)^t$$

where t is the number of months the product is not advertised. Express $s(t)$ as an exponential function with base e and then determine $s(t)$ for $t = 5$ and $t = 10$. [Use $e^{-0.99225} = 0.3707$; $e^{-1.9845} = 0.1374$]

5.7 Derivatives of Exponential Functions

We are ready to differentiate

$$y = e^x$$

To do so, we will use the limit definition of the derivative. In the next section, we shall make use of this result to differentiate logarithmic functions.

Proposition 5.4 If $y = e^x$, then $dy/dx = e^x$.

Proof

Using the five-step rule, we have

Step 1. $y = e^x$

Step 2. $y + \Delta y = e^{x+\Delta x} = e^x \cdot e^{\Delta x}$

Step 3. $(y + \Delta y) - y = e^x \cdot e^{\Delta x} - e^x$

That is,

$$\Delta y = e^x(e^{\Delta x} - 1)$$

Step 4. $\dfrac{\Delta y}{\Delta x} = e^x\left(\dfrac{e^{\Delta x} - 1}{\Delta x}\right)$

Step 5. $$\frac{dy}{dx} = \lim_{\Delta x \to 0} \left(\frac{\Delta y}{\Delta x}\right)$$
$$= \lim_{\Delta x \to 0} \left[e^x\left(\frac{e^{\Delta x} - 1}{\Delta x}\right)\right]$$
$$= e^x \lim_{\Delta x \to 0} \left(\frac{e^{\Delta x} - 1}{\Delta x}\right)$$

The evaluation of the expression

$$\lim_{\Delta x \to 0} \left(\frac{e^{\Delta x} - 1}{\Delta x}\right)$$

is beyond our present scope. We state, without proof, that

$$\lim_{\Delta x \to 0} \left(\frac{e^{\Delta x} - 1}{\Delta x}\right) = 1$$

Thus, if $y = e^x$, then

5.8 $$\frac{dy}{dx} = e^x$$

Corollary

If $y = e^{u(x)}$, then

5.9 $$\frac{dy}{dx} = e^{u(x)}u'(x)$$

This follows from Proposition 5.4 and the Chain Rule.

Proposition 5.5

If $y = a^x$, then $dy/dx = a^x \ln a$, $a > 0$.

Proof

Using (5.7), we can write

$$a^x = e^{x \ln a}$$

Thus,

$$y = e^{u(x)} \quad \text{where} \quad u(x) = x \ln a$$

Applying (5.9), we have

$$\frac{dy}{dx} = e^{x \ln a} \frac{d}{dx}(x \ln a)$$
$$= (e^{x \ln a})(\ln a)$$

Thus,

5.10 $$\frac{dy}{dx} = a^x \ln a$$

Corollary

If $y = a^{u(x)}$, $a > 0$, then

5.11
$$\frac{dy}{dx} = a^{u(x)}u'(x) \ln a$$

This follows from Proposition 5.5 and the Chain Rule.

Example 5.32

Find dy/dx if $y = e^{\sqrt{x}}$.

Solution

Here, $u(x) = x^{1/2}$. Differentiating $u(x)$, we have

$$u'(x) = \tfrac{1}{2}x^{-1/2}$$

From (5.9), we get

$$\begin{aligned}\frac{dy}{dx} &= e^{u(x)} \cdot u'(x) \\ &= e^{\sqrt{x}}\, \tfrac{1}{2}x^{-1/2} \\ &= e^{\sqrt{x}}\, \frac{1}{2}\frac{1}{\sqrt{x}} \\ &= \frac{e^{\sqrt{x}}}{2\sqrt{x}}\end{aligned}$$

Example 5.33

Find dy/dx if $y = x^2e^{2x}$.

Solution

Using the product rule, we obtain

$$\begin{aligned}\frac{dy}{dx} &= x^2\,\frac{d}{dx}(e^{2x}) + e^{2x}\,\frac{d}{dx}(x^2) \\ &= x^2e^{2x}\,\frac{d}{dx}(2x) + 2xe^{2x} \\ &= x^2e^{2x}2 + 2xe^{2x} \\ &= 2xe^{2x}(x + 1)\end{aligned}$$

Example 5.34

The concentration of a drug in the bloodstream of a patient is approximated by the equation

$$Q(t) = 10e^{-0.8t}$$

where $Q(t)$ is the quantity of the drug (in suitable units) in the bloodstream t hours after it is injected. Determine the instantaneous rate of change in the concentration when

a. $t = 1$ **b.** $t = 3$

Solution

$$Q(t) = 10e^{-0.8t}$$

$$Q'(t) = \frac{d}{dt}(10e^{-0.8t})$$

$$= 10\frac{d}{dt}(e^{-0.8t})$$

$$= 10(-0.8)e^{-0.8t}$$

$$= -8.0e^{-0.8t}$$

Hence,

a. $Q'(1) = -8.0e^{-0.8} = -8(0.44933) = -3.59464$, and

b. $Q'(3) = -8.0e^{-2.4} = -8(0.09072) = -0.72576$

The rate of change in the concentration continues to decrease as t gets larger.

Example 5.35

The demand function of a certain product is

$$d(p) = 1{,}500e^{-0.2p}$$

where p is price in dollars per unit. Determine the value of p for which the sales revenue is maximum.

Solution

The sales revenue is the product of the price p and the demand $d(p)$. Hence, the function we want to maximize is

$$f(p) = p \cdot d(p)$$

$$= 1{,}500pe^{-0.2p}$$

The function $f(p)$ is differentiable and

$$f'(p) = 1{,}500\left[p\frac{d}{dp}(e^{-0.2p}) + e^{-0.2p}\frac{d}{dp}(p)\right]$$

$$= 1{,}500[p(-0.2e^{-0.2p}) + e^{-0.2p}]$$

$$= 1{,}500e^{-0.2p}(1 - 0.2p)$$

Note that $e^{-0.2p}$ is never zero. Thus $f'(p) = 0$ if and only if $1 - 0.2p = 0$; that is, when $p = 5$. Now we apply the second derivative test.

$$f''(p) = 1{,}500\left[e^{-0.2p}\frac{d}{dp}(1 - 0.2p) + (1 - 0.2p)\frac{d}{dp}(e^{-0.2p})\right]$$

$$= 1{,}500[e^{-0.2p}(-0.2) + (1 - 0.2p)(-0.2)e^{-0.2p}]$$

$$= 300e^{-0.2p}(0.2p - 2)$$

Thus,

$$f''(5) = -300e^{-1} < 0$$

Hence, the sales revenue is maximum when the price is \$5.00 per unit.

Exercises

Find dy/dx in each of the following problems.

1. $y = e^{3x}$

2. $y = 4e^{5x}$

3. $y = e^{x^2}$

4. $y = e^{x^2+2x}$

5. $y = e^{2x^2+3x+1}$

6. $y = e^{3x^2-4x}$

7. $y = e^{\sqrt{x+1}}, \qquad x \geq -1$

8. $y = e^{\sqrt{2x+3}}, \qquad x \geq \dfrac{-3}{2}$

9. $y = e^{x\sqrt{x-1}}, \qquad x \geq 1$

10. $y = e^{x\sqrt{x+1}}, \qquad x \geq -1$

11. $y = e^{\sqrt{x^2-1}}, \qquad |x| \geq 1$

12. $y = e^{\sqrt{2x^2+1}}$

13. $y = x(10)^x$

14. $y = (x+2)e^{-x}$

15. $y = x^2e^{\sqrt{x}}, \qquad x \geq 0$

16. $y = x^2(8)^{3x}$

17. $y = (x^2+1)a^{4x}$

18. $y = (x+1)^2(10)^{4x}$

19. $y = \dfrac{a^x}{1+a^x}$

20. $y = \dfrac{e^{3x}}{1+e^{3x}}$

21. $y = \dfrac{200}{1+10e^{(0.3)x}}$

22. $y = \dfrac{5{,}000}{1+30e^{(0.5)x}}$

23. A psychologist has prepared a list of nonsense words for use in a memory experiment. The number of words, $N(t)$, that a person is able to memorize after t hours of practice is given by

$$N(t) = 100(1 - e^{-0.2t})$$

Determine the instantaneous rate of improvement in $N(t)$ when
a. $t = 1$ **b.** $t = 2$ **c.** $t = 5$

24. The performance record of a particular student learning a foreign language is given approximately by

$$N(t) = 120(1 - e^{-0.1t}), \qquad 0 < t \leq 5$$

where $N(t)$ is the number of words the student learns after t hours of continuous study. Determine the instantaneous rate of improvement after
a. $t = 1$ hour **b.** $t = 3$ hours **c.** $t = 5$ hours

25. The number of bacteria, $N(t)$, present in a certain culture at time t is given by the equation

$$N(t) = 1{,}000e^{0.3t}$$

where t is measured in hours. Determine the rate of change in the number of bacteria after

a. $t = 1$ hour **b.** $t = 2$ hours **c.** $t = 4$ hours

26. The quantity of radium, $Q(t)$, measured in milligrams, at time t is given by the equation

$$Q(t) = 500e^{-0.03t}$$

where t is measured in centuries. Determine the instantaneous rate of change in the quantity present when

a. $t = 2$ **b.** $t = 4$ **c.** $t = 10$

27. The annual profit of a publisher from the sales of a certain book is given by

$$P(t) = \$6{,}000 + 8{,}000e^{-0.5t}$$

where t is the number of years the book has been on the market. Find the instantaneous rate of change in profit when

a. $t = 2$ years **b.** $t = 4$ years **c.** $t = 5$ years

28. The annual sales of a new edition of a text are approximated by

$$S(x) = 60{,}000(1 - e^{-0.4x})$$

where x is the number of years the textbook remains on the market. Determine the instantaneous rate of change in sales when

a. $x = 2$ years **b.** $x = 3$ years **c.** $x = 4$ years

29. The profit (in millions of dollars) from the sales of a certain product is approximated by

$$P(x) = 2.4 - e^{-0.25x}$$

where x is the number of units sold (in thousands). Determine the marginal profit at

a. $x = 2$ **b.** $x = 3$ **c.** $x = 4$

30. The number of monthly sales of a new model of an automobile in the New England states is approximated by

$$S(t) = 5{,}000e^{-0.2t}$$

where t is the number of months that the model has been in production. Determine the instantaneous rate of change in the sales after

a. $t = 1$ month **b.** $t = 3$ months **c.** $t = 6$ months

31. The salvage value of an aircraft after t years of use is approximated by the equation

$$S(t) = \$400{,}000e^{-0.15t}$$

Determine the rate of depreciation after
a. 2 years **b.** 4 years **c.** 10 years

32. The concentration of a drug in the bloodstream of a patient is approximated by the equation

$$Q(t) = 15e^{-0.6t}$$

where $Q(t)$ is the quantity of the drug in the bloodstream t hours after it is injected. Determine the rate of change in the concentration when
a. $t = 1$ hour **b.** $t = 4$ hours **c.** $t = 8$ hours

33. The atmospheric pressure, $P(h)$, measured in pounds per square inch, is given by the formula

$$P(h) = 15e^{-0.2h}$$

where h is the altitude above sea level in miles. Determine the instantaneous rate of change in the pressure when altitude h is
a. 1 mile **b.** 2 miles **c.** 5 miles

34. The demand function for a certain product is

$$d(p) = 2{,}000e^{-0.25p}$$

where p is the price in dollars per unit. Determine the value of p for which the sales revenue is maximum. What is the demand at that price?

35. The management of ABC Company has determined that the demand function for its product is

$$d(p) = 5{,}000e^{-0.04p}$$

where p is the price in cents per unit. Determine the value of p which will maximize the total sales revenue. What will the demand be at that price?

5.8 Derivatives of Logarithmic Functions

We are ready to use Proposition 5.5 to find the derivative of the logarithmic function.

Proposition 5.6 If $y = \ln x$, then $dy/dx = 1/x$

Proof

Note that

$$y = \ln x \quad \text{and} \quad x = e^y$$

are equivalent equations. Using (5.9), we differentiate both sides of

$$e^y = x$$

with respect to x and obtain

$$e^y \frac{dy}{dx} = 1$$

Therefore,

5.12 $$\frac{dy}{dx} = \frac{1}{e^y} = \frac{1}{x} \qquad (x = e^y)$$

as asserted.

Corollary

If $y = \ln u(x)$, $u(x) > 0$, then

5.13 $$\frac{dy}{dx} = \frac{u'(x)}{u(x)}$$

This assertion follows directly from Proposition 5.6 and the Chain Rule.

Example 5.36

Find dy/dx if $y = \ln(1 + x^2)$.

Solution

This function is of the form $\ln u(x)$ where

$$u(x) = (1 + x^2)$$

Differentiating $u(x)$, we get

$$u'(x) = 2x$$

Now applying (5.13), we have

$$\frac{dy}{dx} = \frac{u'(x)}{u(x)} = \frac{2x}{1 + x^2}$$

Example 5.37

Find dy/dx if $y = \ln(1 + e^{3x})$.

Solution

Here,

$$u(x) = 1 + e^{3x}$$
$$u'(x) = \frac{d}{dx}(1 + e^{3x}) = 3e^{3x}$$

Using (5.13), we obtain

$$\frac{dy}{dx} = \frac{u'(x)}{u(x)} = \frac{3e^{3x}}{1 + e^{3x}}$$

Example 5.38

Find dy/dx if $y = \ln \sqrt{x}$, $x > 0$.

Solution

First we apply Proposition 5.2 and simplify the original expression.

$$\begin{aligned} y &= \ln \sqrt{x} \\ &= \ln x^{1/2} \\ &= \tfrac{1}{2} \ln x \end{aligned}$$

Now differentiating y, we obtain

$$\frac{dy}{dx} = \frac{1}{2}\frac{d}{dx}(\ln x) = \frac{1}{2} \cdot \frac{1}{x} \qquad \textbf{(Proposition 5.6)}$$
$$= \frac{1}{2x}, \qquad x > 0$$

Example 5.39

Find dy/dx if $y = (\ln x)/x$, $x > 0$.

Solution

Using the quotient rule, we have

$$\frac{dy}{dx} = \frac{x \cdot \dfrac{d}{dx}(\ln x) - (\ln x) \cdot \dfrac{d}{dx}(x)}{x^2}$$
$$= \frac{x \cdot \left(\dfrac{1}{x}\right) - \ln x}{x^2}$$
$$= \frac{1 - \ln x}{x^2}, \qquad x > 0$$

Example 5.40

Find dy/dx if $y = e^{2x}$. In x, $x > 0$.

Solution

Using the product rule, we get

$$\begin{aligned}\frac{dy}{dx} &= e^{2x}\frac{d}{dx}(\ln x) + \ln x \frac{d}{dx}(e^{2x}) \\ &= e^{2x}\frac{1}{x} + \ln x \cdot e^{2x}\frac{d}{dx}(2x) \\ &= e^{2x}\frac{1}{x} + (\ln x)2e^{2x} \\ &= e^{2x}\left(\frac{1}{x} + 2\ln x\right)\end{aligned}$$

Example 5.41

Find dy/dx if $y = \ln[x(x+1)/(x+2)^2]$, $x > 0$.

Solution

$$\begin{aligned}y &= \ln\left[\frac{x(x+1)}{(x+2)^2}\right] \\ &= \ln(x) + \ln(x+1) - \ln(x+2)^2 && \textbf{(Proposition 5.1)} \\ &= \ln(x) + \ln(x+1) - 2\ln(x+2) && \textbf{(Proposition 5.2)}\end{aligned}$$

Now we differentiate y with respect to x. Thus,

$$\begin{aligned}\frac{dy}{dx} &= \frac{1}{x} + \frac{1}{x+1} - \frac{2}{x+2} \\ &= \frac{3x+2}{x(x+1)(x+2)}\end{aligned}$$

We will now show how easy it is to find the derivative of a product or quotient by taking logarithms of both sides of an equation.

Example 5.42

Find dy/dx if $y = x\sqrt{x+1}/\sqrt{x+3}$, $x > 0$.

Solution

First we take the logarithm of each side and obtain

$$\begin{aligned}\ln y &= \ln\left[\frac{x\sqrt{x+1}}{\sqrt{x+3}}\right] \\ &= \ln x + \ln\sqrt{x+1} - \ln\sqrt{x+3} && \textbf{(Proposition 5.1)} \\ &= \ln x + \tfrac{1}{2}\ln(x+1) - \tfrac{1}{2}\ln(x+3) && \textbf{(Proposition 5.2)}\end{aligned}$$

Now we differentiate both sides with respect to x. Note that the derivative of $\ln y$ is $(1/y)(dy/dx)$. Thus,

$$\frac{1}{y}\frac{dy}{dx} = \frac{1}{x} + \frac{1}{2}\left(\frac{1}{x+1}\right) - \frac{1}{2}\left(\frac{1}{x+3}\right) = \frac{x^2+5x+3}{x(x+1)(x+3)}$$

Hence,

$$\frac{dy}{dx} = y\frac{(x^2+5x+3)}{x(x+1)(x+3)} = \frac{x\sqrt{x+1}}{\sqrt{x+3}}\,\frac{(x^2+5x+3)}{x(x+1)(x+3)} = \frac{(x^2+5x+3)}{(x+1)^{1/2}(x+3)^{3/2}}$$

Example 5.43

Find dy/dx if $y = x^x$, $x > 0$.

Solution

Taking the logarithm of each side, we get

$$\ln y = \ln(x^x) = x \ln x \qquad \textbf{(Proposition 5.2)}$$

Differentiating both sides with respect to x, we obtain

$$\frac{1}{y}\frac{dy}{dx} = x\frac{d}{dx}(\ln x) + \ln x \frac{d}{dx}(x) = x\frac{1}{x} + \ln x = 1 + \ln x$$

Hence,

$$\frac{dy}{dx} = y(1+\ln x) = x^x(1+\ln x)$$

Exercises

Find dy/dx in each of the following cases.

1. $y = \ln(x+1)$

2. $y = \ln(2x+3)$

3. $y = \ln\sqrt{x+1}$

4. $y = \ln\sqrt{2x+3}$

5. $y = \ln(x\sqrt{x - 1})$

6. $y = \ln\sqrt{x^2 + 1}$

7. $y = \ln(\ln x)$

8. $y = \ln(\ln\sqrt{x + 1})$

9. $y = x^2 \ln 3x$

10. $y = x^3 \ln \sqrt{x + 2}$

11. $y = \ln(xe^x)$

12. $y = \ln(x^2e^{2x})$

13. $y = \ln(x^x)$

14. $y = \ln(1 + e^{2x})$

15. $y = \ln\left(\dfrac{e^x}{1 + e^x}\right)$

16. $y = \ln\left(\dfrac{e^{2x}}{1 + x}\right)$

17. $y = \dfrac{e^x - 1}{2 \ln x}$

18. $y = \dfrac{a}{1 + \ln kx}$

19. $y = \dfrac{200}{1 + 10e^{0.3x}}$

20. $y = \dfrac{5{,}000}{1 + 30e^{0.5x}}$

21. $y = \dfrac{\ln x}{x^2}$

22. $y = \dfrac{\ln\sqrt{x + 1}}{x + 3}$

23. $y = \ln\left(\dfrac{x + 1}{x - 1}\right)$

24. $y = \ln\sqrt{\dfrac{x + 3}{x - 4}}$

25. $y = \ln\sqrt{\dfrac{x + 4}{x - 1}}$

26. $y = \ln\left[\dfrac{(x + 1)(2x - 1)}{3x + 4}\right]$

27. $y = \dfrac{x(x + 2)}{\sqrt{x + 1}}$

28. $y = \dfrac{x\sqrt{x + 2}}{x + 3}$

29. $y = \dfrac{x(x + 2)^2}{\sqrt{x + 3}}$

30. $y = \dfrac{(x + 1)^2(x + 3)^3}{(2x + 1)}$

31. $y = \dfrac{\sqrt{x + 1}\sqrt{x + 2}}{(3x + 2)^2}$

32. $y = \dfrac{\sqrt{3x + 2}(x + 4)}{(4x + 5)^2}$

33. $y = \dfrac{\sqrt{x + 2}}{\sqrt{x^2 + 1}}$

34. $y = \sqrt{\dfrac{x + 1}{x + 2}}$

35. $y = (x + 1)^{x + 1}$

36. $y = (x + 2)^x$

37. $y = (2x + 3)^x$

38. $y = (\sqrt{x})^{\sqrt{x}}$

39. $y = (\ln x)^{\ln x}$

40. $y = (\sqrt{x + 1})^{x + 2}$

Discuss each of the following functions with regard to relative maxima and relative minima.

41. $y = x - \ln x, \quad x > 0$

42. $y = \ln(x + e^{-x})$

43. $y = xe^{-x}$

44. $y = x^2e^{-x}$

45. $y = e^{x^2-x}$

46. $y = xe^{-x^2/2}$

Exercises for Review

1. Define each of the following concepts.

a. exponential function
b. exponential equation
c. logarithmic function
d. common logarithms
e. natural logarithms
f. effective rate of interest
g. nominal rate of interest
h. conversion period

2. Solve for x the following equations.

a. $\log_2 x = 5$
b. $\log_x 125 = 3$
c. $\log_4 x = 1$
d. $\log_5 (2x) = -1$
e. $\log_{16} 4 = x$
f. $\log_3 243 = x$

3. Given that

$$\log_{10} 2 = 0.3010, \quad \log_{10} 3 = 0.4771, \quad \log_{10} 5 = 0.6990$$

evaluate each of the following logarithms.

a. $\log_{10} 6$
b. $\log_{10} 30$
c. $\log_{10} (\frac{15}{2})$
d. $\log_{10} (\sqrt{2})$
e. $\log_5 3$
f. $\log_3 5$

4. Find each of the following logarithms.

a. $\log 5140$
b. $\log 326$
c. $\log 24.3$
d. $\log 0.0243$
e. $\ln 567$
f. $\ln 31.8$
g. $\ln 0.234$
h. $\ln 0.00234$

5. Find the value of x in each of the following expressions.

a. $\log x = 1.8476$
b. $\log x = 0.5527$
c. $\log x = -1.2692$
d. $\ln x = 1.6658$
e. $\ln x = 3.8199$
f. $\ln x = -0.5674$

6. Solve for x the following equations.

a. $10^x = 27$
b. $10^x = 3.7$
c. $10^{-x} = \frac{341}{563}$
d. $e^x = 8.30$
e. $e^{-x} = 0.548$
f. $3^{x+2} = 5^{2x-1}$

7. The number of bacteria, $N(t)$, in a certain culture is given by

$$N(t) = 20{,}000e^{-0.01t}$$

where t is recorded in hours. Find the number of bacteria present at time

a. $t = 0$ **b.** $t = 10$ **c.** $t = 20$ **d.** $t = 40$

8. The annual sales of a new edition of a textbook are approximated by the equation

$$S(t) = 50{,}000(1 - e^{-0.4t})$$

where t is the number of years the textbook remains on the market. Calculate $S(t)$ for $t = 1, 2, 3, 4,$ and 5.

9. Find the accumulated amount and the interest if \$5,000 is invested for 3 years at 8 percent compounded

a. quarterly

b. daily

c. continuously. [Use $e^{0.24} = 1.2712$]

10. The Jacksons wish to have \$12,000 available for their son who will enter college in 8 years. What sum of money should be invested now if the current rate of interest is 10 percent compounded continuously?

11. How long will it take for a principal amount to double itself if it is invested at 6 percent annual interest compounded continuously?

12. Differentiate each of the following.

a. $y = e^{4x}$

b. $y = e^{\sqrt{x}}$

c. $y = xe^{2x}$

d. $y = \dfrac{e^x}{1 + e^{2x}}$

e. $y = x^2e^{3x} + xe^{-x}$

f. $y = \ln(x^2 + 4)$

g. $y = \ln(x + e^x)$

h. $y = \ln(\sqrt{3x + 2})$

i. $y = \ln\left(\dfrac{e^{2x}}{1 + e^{2x}}\right)$

j. $y = \ln(\ln x)$

k. $y = (\ln x)^2$

l. $y = \ln(x + 1)^x$

m. $y = (x + 1)^x$

n. $y = \sqrt{\dfrac{x + 1}{x + 2}}$

o. $y = \dfrac{x\sqrt{x + 2}}{2x + 3}$

p. $y = \dfrac{(2x + 1)^2(x - 1)^3}{\sqrt{3x + 4}}$

13. The population, $P(t)$, of a certain county is given approximately by the equation

$$P(t) = 1{,}500e^{0.03t}$$

where t is recorded in years. Determine the instantaneous rate of change in the population when

a. $t = 10$ **b.** $t = 20$

14. The number of words, $N(t)$, that a particular student learning a foreign language is able to memorize after t hours of practice is approximated by the equation

$$N(t) = 100(1 - e^{-0.15t})$$

Determine the rate of improvement after

a. $t = 2$ hours **b.** $t = 4$ hours

15. The demand function for a certain product is

$$d(p) = 1{,}000e^{-0.05p}$$

where p is the price per unit (in cents). Determine the value of p for which the sales revenue will be maximum. What will the demand be at that price?

6 Integral Calculus

6.1 Introduction

Up to this point, our work has centered on finding the derivative $f'(x)$ of a given function $f(x)$. Now we attempt to find a function $f(x)$ when its derivative $f'(x)$ is given. This brings us to the study of integral calculus, a subject we illustrate by computing the area bounded by a given set of curves. Later in the chapter we discuss some applications of integral calculus in the fields of business, biology, the social sciences, and economics.

6.2 Antiderivatives

Given any function $f(x)$, does there exist another function $F(x)$ whose derivative is $f(x)$? Consider, for example, the function $f(x) = 2x$. Can we find a function $F(x)$ such that its derivative $F'(x) = 2x$? Clearly $F(x) = x^2$ is one answer because $(d/dx)(x^2) = 2x$. Similarly, if $F(x) = x^2 + 15$ then $F'(x) = 2x$, and if $F(x) = x^2 - 56$ then also $F'(x) = 2x$. This means that there is more than one function $F(x)$ such that $F'(x) = 2x$, and all such functions are of the form

$$F(x) = x^2 + c$$

where c is a constant.

Definition 6.1

A function $F(x)$ is called an *antiderivative* of the function $f(x)$ on an interval $[a, b]$ if and only if $F'(x) = f(x)$ for all x in $[a, b]$.

Example 6.1

Let $f(x) = x^2$. Then the function $F(x) = x^3/3$ is an antiderivative of $f(x)$ because

$$\frac{d}{dx}\left(\frac{x^3}{3}\right) = \frac{1}{3}\frac{d}{dx}(x^3) = \frac{1}{3}(3x^2) = x^2$$

Similarly, if c is any constant, then $F(x) = (x^3/3) + c$ is also an antiderivative of x^2 since

$$\frac{d}{dx}\left(\frac{x^3}{3} + c\right) = x^2$$

Example 6.2

Let $f(x) = e^x$. Then $F(x) = e^x$ is an antiderivative of $f(x)$ since $F'(x) = e^x = f(x)$. If c is any constant, then $F(x) = e^x + c$ is also an antiderivative of $f(x)$.

Example 6.3

If $f(x) = 1/x$, then $F(x) = \ln|x|$ is an antiderivative of $f(x)$ since

$$\frac{d}{dx}(\ln|x|) = \frac{1}{x}$$

Clearly $\ln|x| + c$, where c is a constant, is also an antiderivative of $f(x)$.

We have observed that $F(x)$ is not unique since a function $f(x)$ may have more than one antiderivative; in fact, infinitely many functions $F(x)$ exist. Is there any relation between these antiderivatives of function $f(x)$? For example, if $f(x) = x$, then the functions $F_1(x) = x^2/2$, $F_2(x) = (x^2/2) + 70$, $F_3(x) = (x^2/2) - 50$, and $F_4(x) = (x^2/2) + \pi$ are all antiderivatives of $f(x)$. Note that all these antiderivatives differ only by a constant.

Proposition 6.1

If $F_1(x)$ and $F_2(x)$ are antiderivatives of the same function $f(x)$ in the interval $[a, b]$, then

$$F_1(x) = F_2(x) + c$$

where c is an arbitrary constant.

Let us now introduce a notation for antiderivatives.

Definition 6.2

If $F(x)$ is an antiderivative of the function $f(x)$, then we write

$$\int f(x)\,dx = F(x) + c$$

The symbol $\int f(x)\,dx$ represents the set of *all* antiderivatives of $f(x)$.

The function $f(x)$ is called the *integrand* and $F(x) + c$ is called the *indefinite integral*, or simply the *integral*. The adjective *indefinite* refers to the presence of the unspecified constant c, which is called the *constant of integration.*

Example 6.4

a. $\int x\,dx = \frac{x^2}{2} + c$

b. $\int t^2\,dt = \frac{t^3}{3} + c$

The symbols dx and dt indicate which variable, x or t, is the variable of interest.

Because integration is the inverse process of differentiation, the first requirement for developing skill in integration is a mastery of the differentiation formulas. Recall that the derivative of a constant is zero; therefore, the antiderivative of zero is a constant. In other words,

$$\int 0\,dx = c$$

where c is a constant.

We now list integral formulas of certain well-known functions.

1. $\int x^n\,dx = \frac{x^{n+1}}{n+1} + c, \qquad n \neq -1$

2. $\int e^x\,dx = e^x + c$

3. $\int a^x\,dx = \frac{a^x}{\ln a} + c$

4. $\int \frac{dx}{x} = (\ln|x|) + c$

Example 6.5

a. $\int x^2\,dx = \frac{x^3}{3} + c$

b. $\int \frac{dt}{t^3} = \int t^{-3}\,dt = \frac{t^{-3+1}}{-3+1} + c = -\frac{1}{2t^2} + c$

c. $\int \sqrt{x}\,dx = \int x^{1/2}\,dx = \frac{x^{1/2+1}}{\frac{1}{2}+1} + c = \frac{2}{3}x^{3/2} + c, \qquad x > 0$

Recall from Chapter 3 that the derivative of a constant times a function is that constant times the derivative of the function. There is a similar property for integration, the inverse process of differentiation. That is,

6.1
$$\int kf(x)\,dx = k\int f(x)\,dx$$

Example 6.6

a. $\int 5x^3\,dx = 5\int x^3\,dx = \frac{5}{4}x^4 + c$

b. $\int \frac{dx}{3x} = \frac{1}{3}\int \frac{dx}{x} = \frac{1}{3}(\ln|x|) + c$

c. $\int 2e^x\,dx = 2\int e^x\,dx = 2e^x + c$

Example 6.7

a. $\int x^2 t\,dt = x^2\int t\,dt = \frac{x^2t^2}{2} + c$

b. $\int x^2 t\,dx = t\int x^2 dx = \frac{tx^3}{3} + c$

Note that the symbol dt or dx indicates the variable of interest.

Because the derivative of the sum or difference of two functions is the sum or difference of their derivatives, we have

6.2
$$\begin{aligned}\int [f(x) \pm g(x)]\,dx &= \int f(x)\,dx \pm \int g(x)\,dx\\ &= F(x) \pm G(x) + c\end{aligned}$$

where $F'(x) = f(x)$, $G'(x) = g(x)$, and c is an arbitrary constant.

Example 6.8

Find

$$\int \left(x + \frac{4}{x}\right) dx$$

Solution

By (6.2), we have

$$\begin{aligned}\int \left(x + \frac{4}{x}\right) dx &= \int x\,dx + \int \frac{4}{x} dx\\ &= \int x\,dx + 4\int \frac{dx}{x} \qquad \textbf{(6.1)}\\ &= \frac{x^2}{2} + 4(\ln|x|) + c\end{aligned}$$

where c is a constant.

Example 6.9

Find

$$\int (3x^2 + 2x + 1)\,dx$$

Solution

By (6.2),

$$\int (3x^2 + 2x + 1)\,dx = \int 3x^2\,dx + \int 2x\,dx + \int dx$$

which, by (6.1), equals

$$\begin{aligned} &= 3\int x^2\,dx + 2\int x\,dx + \int dx \\ &= 3\left(\frac{x^3}{3}\right) + 2\left(\frac{x^2}{2}\right) + x + c \\ &= x^3 + x^2 + x + c \end{aligned}$$

Example 6.10

Find

$$\int \left(5x^4 + 2x^2 + 3e^x + \frac{2}{x}\right)dx$$

Solution

$$\begin{aligned} \int \left(5x^4 + 2x^2 + 3e^x + \frac{2}{x}\right)dx &= \int 5x^4\,dx + \int 2x^2\,dx + \int 3e^x\,dx + \int \frac{2}{x}\,dx \\ &= 5\int x^4\,dx + 2\int x^2\,dx + 3\int e^x\,dx + 2\int \frac{dx}{x} \\ &= 5\left(\frac{x^5}{5}\right) + 2\left(\frac{x^3}{3}\right) + 3e^x + 2\ln|x| + c \\ &= x^5 + \tfrac{2}{3}x^3 + 3e^x + 2\ln|x| + c \end{aligned}$$

Example 6.11

Find

$$\int \left(2a^x + \frac{1}{3x}\right)dx, \qquad a > 0,\ x > 0$$

Solution

$$\begin{aligned} \int \left(2a^x + \frac{1}{3x}\right)dx &= \int 2a^x\,dx + \int \frac{dx}{3x} \\ &= 2\int a^x\,dx + \frac{1}{3}\int \frac{dx}{x} \\ &= 2\left(\frac{a^x}{\ln a}\right) + \frac{1}{3}\ln|x| + c \end{aligned}$$

Next we discuss a few applications of integration. Besides familiarizing you with the basic usefulness of integral calculus, these applications are intended to emphasize the relationship between integration and differentiation.

Example 6.12

Find the equation of the curve $y = f(x)$ which passes through the point $(2, 16)$ and which satisfies the condition that the slope of the curve at $(x, f(x))$ is $4x^3 - 2x$.

Solution

Since the slope $dy/dx = f'(x) = 4x^3 - 2x$, the curve $y = f(x)$ must be of the form

$$\int (4x^3 - 2x)\,dx = \int 4x^3\,dx - \int 2x\,dx$$

that is,

$$f(x) = x^4 - x^2 + c$$

To evaluate the constant c, note that $f(x)$ passes through the point $(2, 16)$. Substituting $x = 2$ and $f(x) = 16$ in the previous equation,

$$16 = 2^4 - 2^2 + c$$

$$c = 4$$

The required curve is

$$y = x^4 - x^2 + 4$$

Example 6.13

The velocity of an automobile at time t (in seconds) is given by $v(t) = 4t + 5$ where $v(t)$ is measured in feet per second. Determine the distance function $s(t)$ assuming that the automobile is at rest when $t = 0$.

Solution

The velocity $v(t)$ is the derivative of the distance function $s(t)$. Therefore, the distance function $s(t)$ is the antiderivative or the integral of the velocity. That is,

$$\begin{aligned} s(t) &= \int v(t)\,dt \\ &= \int (4t + 5)\,dt \\ &= 2t^2 + 5t + c \end{aligned}$$

where c is a constant to be determined. We know that $s(t) = 0$ when $t = 0$. This means that

$$0 = 2(0)^2 + 5(0) + c$$

Hence $c = 0$ and the distance function is

$$s(t) = 2t^2 + 5t$$

Example 6.14

The instantaneous rate of change in the population $P(t)$ of a colony of bacteria is given by $P'(t) = 50t - 3t^2$. Determine the population function $P(t)$, given that $P(t) = 1{,}000$ when $t = 0$.

Solution

Integrating $P'(t)$, we obtain

$$\begin{aligned} P(t) &= \int P'(t)\,dt \\ &= \int (50t - 3t^2)\,dt \\ &= 25t^2 - t^3 + c \end{aligned}$$

where c is the constant to be determined from the fact that $P(0) = 1{,}000$. Since

$$P(0) = 25(0)^2 - (0)^3 + c$$

$c = P(0) = 1{,}000$ and

$$P(t) = 25t^2 - t^3 + 1{,}000$$

Example 6.15

The management of a publishing company has estimated that the marginal cost function of publishing books is

$$C'(x) = 20x + \frac{100}{\sqrt{x}}$$

where x denotes the number (in thousands) of books published per year. Determine the cost function $C(x)$ and the average cost function $A(x)$, assuming that the fixed cost of production is \$50,000.

Solution

The marginal cost function $C'(x)$ is the derivative of the total cost function $C(x)$. Therefore, the total cost function $C(x)$ is the antiderivative or the integral of the marginal cost function $C'(x)$. Integrating $C'(x)$, we obtain

$$\begin{aligned} C(x) &= \int C'(x)\,dx \\ &= \int \left(20x + \frac{100}{\sqrt{x}}\right) dx \\ &= 10x^2 + 50\sqrt{x} + k \end{aligned}$$

where k is a constant to be determined. The fixed cost, or the cost incurred at zero level of production, is $C(0)$, and

$$C(0) = 10(0)^2 + 50\sqrt{0} + k$$

Thus, $k = C(0) = \$50{,}000$ and

$$C(x) = 50{,}000 + 10x^2 + 50\sqrt{x}$$

The average cost function $A(x)$ is the total cost function divided by the output. Therefore,

$$\begin{aligned} A(x) &= \frac{C(x)}{x} \\ &= 10x + \frac{50{,}000}{x} + \frac{50}{\sqrt{x}} \end{aligned}$$

The same type of reasoning can be applied to the revenue function. Since the marginal revenue $R'(x)$ is the derivative of the revenue function $R(x)$, the revenue function can be obtained by integrating the marginal revenue function. Thus,

$$R(x) = \int R'(x)\,dx$$

The average revenue is then calculated by dividing the total revenue $R(x)$ by the output x.

How do we determine the maximum profit for a firm when the information is given in terms of marginal cost and marginal revenue? Consider the following example.

Example 6.16

The marginal cost of production of a new product and its marginal revenue function are

$$C'(x) = 20 + \frac{x}{20}$$

and

$$R'(x) = 30$$

where x is the number of units produced. Determine the maximum profit assuming that the fixed cost to the firm is \$200.

Solution

The profit is maximum at the level of production for which the marginal cost equals the marginal revenue. Thus solving $C'(x) = R'(x)$ for x, we have

$$20 + \frac{x}{20} = 30$$
$$x = 200$$

Recall that the profit function $P(x)$ is

$$P(x) = R(x) - C(x)$$

Integrating the marginal revenue function $R'(x)$, we obtain

$$\begin{aligned} R(x) &= \int R'(x)\,dx \\ &= \int 30\,dx \\ &= 30x + k_1 \end{aligned}$$

Note that $R(x) = 0$ when there is no production. Thus,

$$R(0) = 0 = 30(0) + k_1$$

and the constant k_1 must equal zero.

Integrating the marginal cost function, we have

$$\begin{aligned} C(x) &= \int C'(x)\,dx \\ &= \int \left(20 + \frac{x}{20}\right) dx \\ &= 20x + \frac{x^2}{40} + k_2 \end{aligned}$$

Given that the cost to the firm is \$200 when $x = 0$,

$$C(0) = 20(0) + \tfrac{1}{40}(0)^2 + k_2$$
$$k_2 = C(0) = 200$$

Thus,

$$C(x) = 200 + 20x + \frac{x^2}{40}$$

Hence,

$$\begin{aligned} P(x) &= R(x) - C(x) \\ &= 30x - \left(20x + \frac{x^2}{40} + 200\right) \\ &= 10x - \frac{x^2}{40} - 200 \end{aligned}$$

When the output is 200 units, the profit is maximized, and

$$P(200) = 10(200) - \frac{(200)^2}{40} - 200$$
$$= 800$$

Exercises

Find each of the following indefinite integrals and verify your results by differentiation.

1. $\int 4x^3\,dx$
2. $\int \frac{dt}{t^2}, \quad t \neq 0$
3. $\int \frac{2}{x^4}\,dx, \quad x \neq 0$
4. $\int x^{2/3}\,dx$
5. $\int \frac{x+1}{x^2}\,dx, \quad x \neq 0$
6. $\int \frac{dt}{t\sqrt{t}}, \quad t \neq 0$
7. $\int t(t+1)\,dt$
8. $\int (x+1)(x+2)\,dx$
9. $\int (x+2)^2\,dx$
10. $\int (2x+3)^2\,dx$
11. $\int \left(x^2 + \frac{1}{x}\right)dx, \quad x > 0$
12. $\int \left(x + \frac{1}{2}\right)^2 dx$
13. $\int \left(x - \frac{1}{2}\right)^2 dx$
14. $\int \left(x^2 + \frac{1}{x^2}\right)^2 dx, \quad x \neq 0$
15. $\int \left(x^2 + \frac{1}{x^2} + e^x\right)dx, \quad x \neq 0$
16. $\int (2+x)\sqrt{x}\,dx, \quad x \geq 0$
17. $\int \left(\frac{e^x + e^{-x}}{2}\right)dx$
18. $\int \left(\frac{e^x - e^{-x}}{2}\right)dx$
19. $\int e^{\ln t}\,dt, \quad t > 0$
20. $\int \log e^t\,dt$
21. $\int xt\,dt$
22. $\int xt\,dx$
23. $\int x^2t^2\,dt$
24. $\int x^2t^2\,dx$
25. $\int x^3t^2\,dx$
26. $\int x^3t^2\,dt$
27. $\int \left(t + \frac{1}{t}\right)^2 dt, \quad t > 0$
28. $\int \left(t - \frac{1}{t}\right)^2 dt, \quad t > 0$

Find the equation of the curve $y = f(x)$ in problems 29–32.

29. $f'(x) = 2x - 3; \qquad f(2) = -2$

30. $f'(x) = x + 5; \qquad f(4) = 32$

31. $f'(x) = 3(x + 2)^2; \qquad f(0) = 9$

32. $f'(x) = e^x + x^2; \qquad f(0) = 4$

33. Find the equation of the curve $y = f(x)$ which passes through the point $(1, \frac{7}{2})$ and which satisfies the condition that the slope of the curve at $(x, f(x))$ is $x^2 - 3x + 2$.

34. Find the equation of the curve $y = f(x)$ which passes through the point $(2, 9)$ and which satisfies the condition that the slope of the curve at $(x, f(x))$ is $3x^2 - 4x + 1/x^2$.

35. A ball is thrown vertically into the air. The velocity of the ball at time t (in seconds) is given by

$$v(t) = 128 - 32t$$

Determine the distance function $s(t)$, where $s(t)$ represents the height of the ball (in feet) above the ground at time t.

36. The velocity of a particle moving along a certain path at time t (in seconds) is given by

$$v(t) = 12t^2 + 24t$$

where $v(t)$ is measured in feet per second. Determine the distance function $s(t)$ assuming that the particle starts from rest at time $t = 0$.

37. The instantaneous population growth rate in a certain town is given by

$$P'(t) = 100 + 12\sqrt{t}, \qquad t \geq 0$$

Determine the population $P(t)$ in t years assuming that the current population of the town is 7,000.

38. The instantaneous population growth rate in a certain town is given by

$$P'(t) = 36t^2 - 60t^{3/2}$$

Determine the population $P(t)$ in t years assuming that the current population of the town is 12,000.

39. The marginal cost function of a product is

$$C'(x) = 500 - 3x$$

where $C'(x)$ is measured in dollars and x denotes the output. Determine the total cost function $C(x)$ and the average cost function $A(x)$ if the fixed cost is \$3,000.

40. The marginal cost of production to a firm is given by

$$C'(x) = 6 - 0.1x + 0.04x^2$$

where x represents the quantity produced. Determine the total cost function $C(x)$ assuming that the fixed cost is \$200.

41. The marginal revenue function of a monopolist is

$$R'(x) = 200 - 0.0004x + 0.01x^2$$

where x denotes the quantity produced and sold. Assuming that the total revenue is zero when $x = 0$, determine the revenue function $R(x)$.

42. The marginal revenue function of a daily newspaper is

$$R'(x) = 100 + 0.02x + 0.03x^2$$

where x is the number (in thousands) of copies published. Assuming that the total revenue is zero when $x = 0$, determine the revenue function $R(x)$.

43. Suppose that $f(t)$ denotes the temperature in an apartment t hours after turning on the airconditioner. The instantaneous rate of change in temperature in the apartment is

$$f'(t) = -10 - 8(t + 1)^{-2}, \qquad 0 \le t \le 3$$

Determine $f(t)$ assuming that it was 80°F in the apartment before the airconditioner unit was turned on.

44. A fixed quantity of a drug is injected in the bloodstream of a patient and the percentage of concentration $f(t)$ of the drug is recorded continuously for t hours. The instantaneous rate of change in the percentage of concentration of the drug is given by

$$f'(t) = t^{-1/2}, \qquad t \ge 0$$

Determine the function $f(t)$ assuming that none of the drug was initially in the bloodstream.

45. A legionnaire's disease hits a certain town. Let $N(t)$ be the number of persons sick with the disease at time t, where t is measured in days from the outbreak of the disease. Given that $N(0) = 5$ and that the sickness is spreading at the rate

$$N'(t) = 36 - 12t^{-1/2}, \qquad 0 \le t \le 25,$$

determine $N(t)$.

46. The instantaneous rate of improvement of a particular student in memorizing a list of nonsense words is approximated by

$$N'(t) = \frac{60}{\sqrt{t}}, \qquad t \geq 0$$

where $N(t)$ is the number of words the student learns after t hours of continuous practice. Determine $N(t)$ if $N(0) = 0$.

47. The marginal cost of a product and its marginal revenue function are

$$C'(x) = 40 + \frac{x}{8}$$

and

$$R'(x) = 50 + \frac{x}{10}$$

where x is the number of units produced. Determine the maximum profit assuming that the fixed cost to the firm is \$500.

48. The marginal cost of a new product and its marginal revenue function are

$$C'(x) = 300 + x$$

and

$$R'(x) = 400 + \frac{x}{5}$$

where x is the number of units produced. What is the maximum profit if the fixed cost of the plant is \$1,250?

6.3 The Definite Integral

The concept of a definite integral arises from the problem of finding the area bounded by the lines $x = a$, $x = b$, the x-axis, and the graph of the curve $y = f(x)$ where $f(x)$ is continuous on the closed interval $[a, b]$. We shall assume in this section that $f(x)$ is non-negative for all $x \in [a, b]$. (See Figure 6.1.)

The fact that three sides of the region are straight lines facilitates an intuitive approach to the problem. We can approximate the area under the curve $y = f(x)$ with the areas of rectangles. The bases of these rectangles will be subintervals of the closed interval $[a, b]$, while heights of the rectangles are determined by points chosen arbitrarily in each subinterval. As the number of rectangles gets larger and larger (with their bases getting smaller and smaller), the limit of the sum of the rectangular areas will give us the area under the curve.

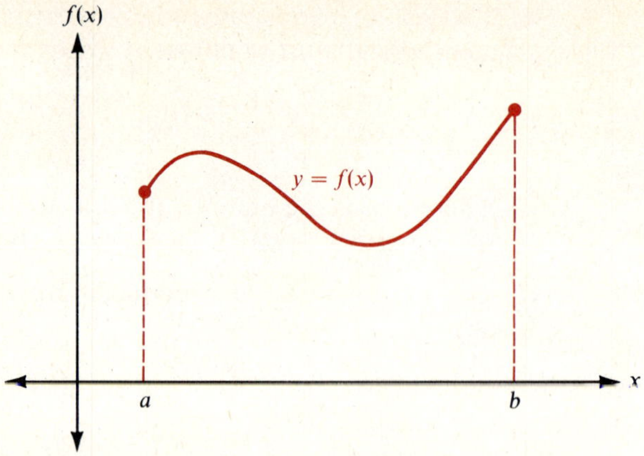

Figure 6.1

In Figure 6.2, the closed interval $[a, b]$ is divided into n (not necessarily equal) parts, so that $x_1, x_2, \ldots, x_{n-1}$ are the points of subdivision with the stipulation that $a = x_0$ and $b = x_n$. Thus we have n closed subintervals

$$[x_0, x_1], [x_1, x_2], [x_2, x_3], \ldots, [x_{n-1}, x_n]$$

which serve as the bases of our rectangles. The height of each rectangle is the value of the function evaluated at a point x_i^* chosen arbitrarily in the subinterval $[x_{i-1}, x_i]$. If

$$\Delta x_i = x_i - x_{i-1}, i = 1, 2, \ldots, n$$

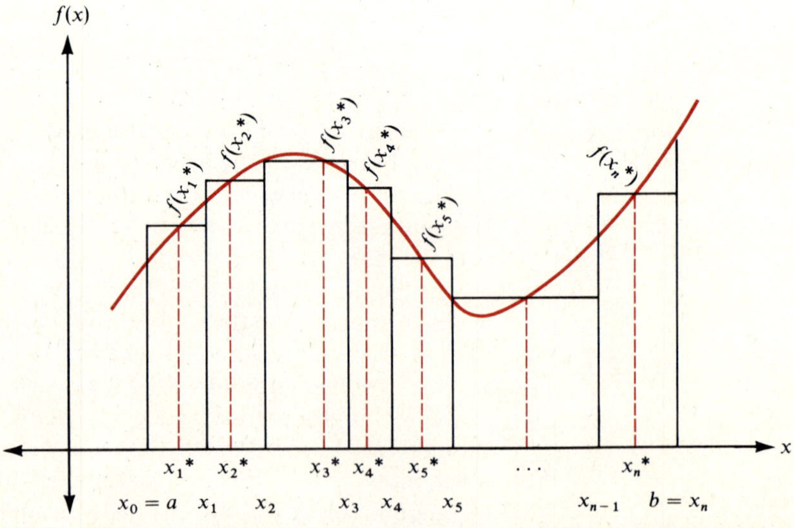

Figure 6.2

then Δx_i is the length of the i-th subinterval, and we have

$$\text{area of first rectangle} = f(x_1^*)\,\Delta x_1$$
$$\text{area of second rectangle} = f(x_2^*)\,\Delta x_2$$
$$\text{area of third rectangle} = f(x_3^*)\,\Delta x_3$$
$$\vdots$$
$$\text{area of last rectangle} = f(x_n^*)\,\Delta x_n$$

Adding the areas of these rectangles, the sum

$$S_n = f(x_1^*)\,\Delta x_1 + f(x_2^*)\,\Delta x_2 + f(x_3^*)\,\Delta x_3 + \cdots + f(x_n^*)\,\Delta x_n$$

is an approximation to the area under the curve. Finally, we take the limit of S_n as n increases indefinitely in such a way that the length of each subinterval approaches zero. This can be described by representing the length of the longest subinterval by Δx so that as $\Delta x \to 0$, the length of every subinterval approaches zero. Thus we have

6.3 $$\lim_{n\to\infty} S_n = \lim_{\Delta x\to 0} [f(x_1^*)\,\Delta x_1 + f(x_2^*)\,\Delta x_2 + f(x_3^*)\,\Delta x_3 + \cdots + f(x_n^*)\,\Delta x_n]$$

Consider, for example, the function $f(x) = x^2$ over the closed interval $[0, 2]$. If we draw the dotted horizontal line in Figure 6.3 to complete

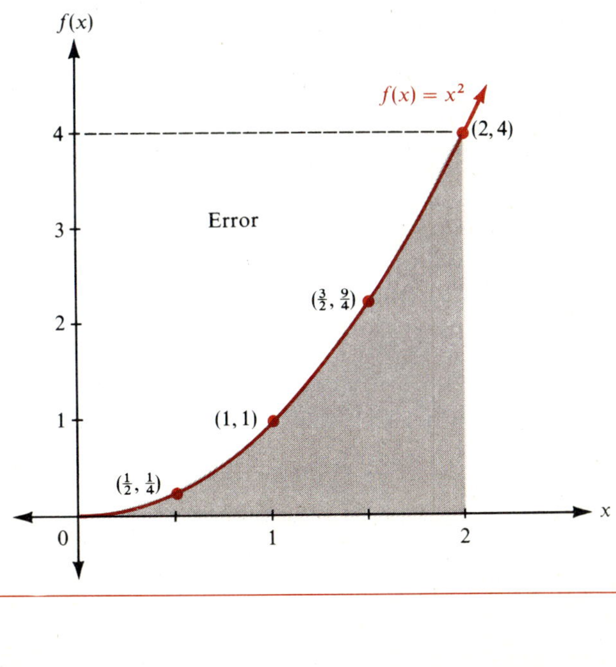

Figure 6.3

the rectangle, we find that the area of this rectangle is

$$\text{base} \times \text{height} = 2 \times 4 = 8$$

which exceeds the desired shaded area by the amount of error indicated in the figure. This rectangular area is the first approximation to the shaded area under the curve. Next, we can divide the interval $[0, 2]$ into 4 parts and choose the height of each rectangle as the value of the function evaluated at the right endpoint of each subinterval. The sum of the four rectangular areas is our second approximation to the shaded area. Note that the error is less than that for our first approximation by the amount shown as "error reduction" in Figure 6.4. It seems

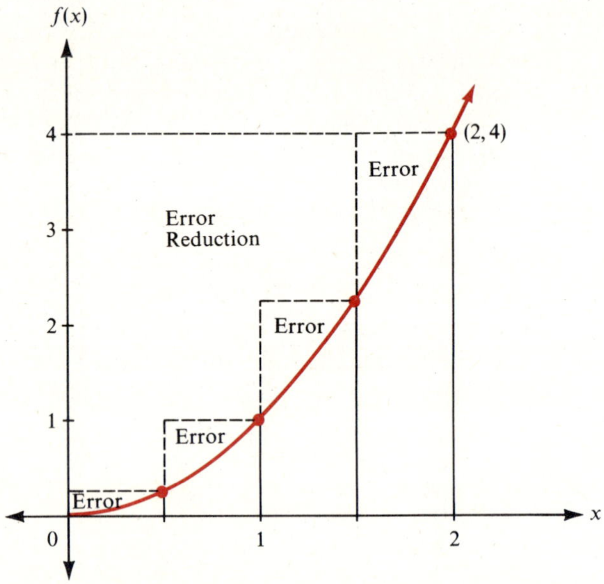

Figure 6.4

reasonable to expect that as we divide the interval $[0, 2]$ into more but smaller subintervals, each time we add the areas of the rectangles we get a better approximation of the area of the shaded region. This introduces the concept of a limit because we are concerned with the behavior of S_n as the subdivision process is continued indefinitely and the length of each subinterval approaches zero.

Suppose that the interval $[0, 2]$ is divided into n equal subintervals and that x_i^* is chosen to be the right-hand endpoint of the subinterval $[x_{i-1}, x_i]$ as shown in Figure 6.5. The length of each subinterval is $2/n$. Thus,

$$\text{area of first rectangle} = \frac{2}{n} \cdot f\left(\frac{2}{n}\right) = \frac{2}{n} \cdot \left(\frac{2}{n}\right)^2$$

$$\text{area of second rectangle} = \frac{2}{n} \cdot f\left(\frac{4}{n}\right) = \frac{2}{n} \cdot \left(\frac{4}{n}\right)^2$$

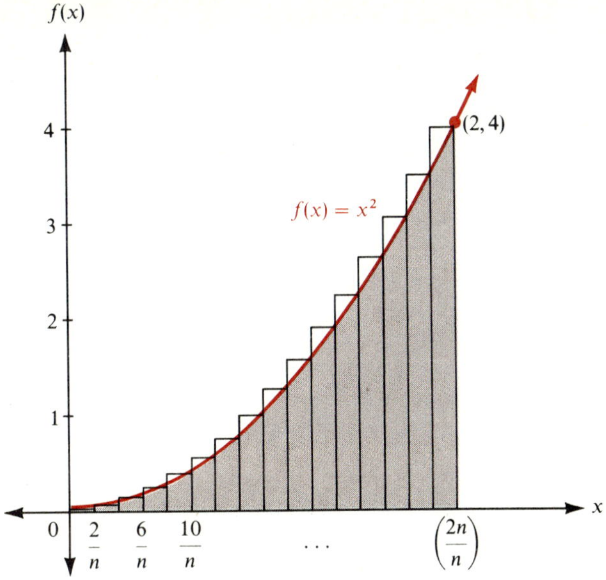

Figure 6.5

$$\text{area of third rectangle} = \frac{2}{n} \cdot f\left(\frac{6}{n}\right) = \frac{2}{n}\left(\frac{6}{n}\right)^2$$

$$\vdots$$

$$\text{area of last rectangle} = \frac{2}{n} \cdot f\left(\frac{2n}{n}\right) = \frac{2}{n} \cdot \left(\frac{2n}{n}\right)^2$$

Adding the areas of these rectangles, we have

$$\begin{aligned} S_n &= \left(\frac{2}{n}\right)\left(\frac{2}{n}\right)^2 + \left(\frac{2}{n}\right)\left(\frac{4}{n}\right)^2 + \left(\frac{2}{n}\right)\left(\frac{6}{n}\right)^2 + \cdots + \left(\frac{2}{n}\right)\left(\frac{2n}{n}\right)^2 \\ &= \left(\frac{2}{n}\right)^3 (1^2 + 2^2 + 3^2 + \cdots + n^2) \end{aligned}$$

Using the formula that

$$1^2 + 2^2 + 3^2 + \cdots + n^2 = \frac{n(n+1)(2n+1)}{6}$$

we have

$$\begin{aligned} S_n &= \left(\frac{2}{n}\right)^3 \frac{n(n+1)(2n+1)}{6} \\ &= \frac{4}{3} \frac{(n+1)(2n+1)}{n^2} \\ &= \frac{4}{3}\left(1 + \frac{1}{n}\right)\left(2 + \frac{1}{n}\right) \end{aligned}$$

Remember that S_n is an approximation to the shaded area under the curve $f(x) = x^2$ and that the error decreases as n, the number of rectangles, increases. Consequently, the area under the curve is $\lim_{n\to\infty} S_n$. As $n \to \infty$, $1/n \to 0$. Thus

$$\begin{aligned}\lim_{n\to\infty} S_n &= \lim_{n\to\infty}\left[\frac{4}{3}\left(1+\frac{1}{n}\right)\left(2+\frac{1}{n}\right)\right]\\ &= \tfrac{4}{3}(1+0)(2+0)\\ &= \tfrac{8}{3}\end{aligned}$$

Let us now examine the use of inscribed rectangles to approximate the area under the same curve $f(x) = x^2$. As before, the interval $[0, 2]$ is divided into n equal parts, but now the height of each rectangle is the value of the function at the left endpoint of the corresponding subinterval as illustrated in Figure 6.6.

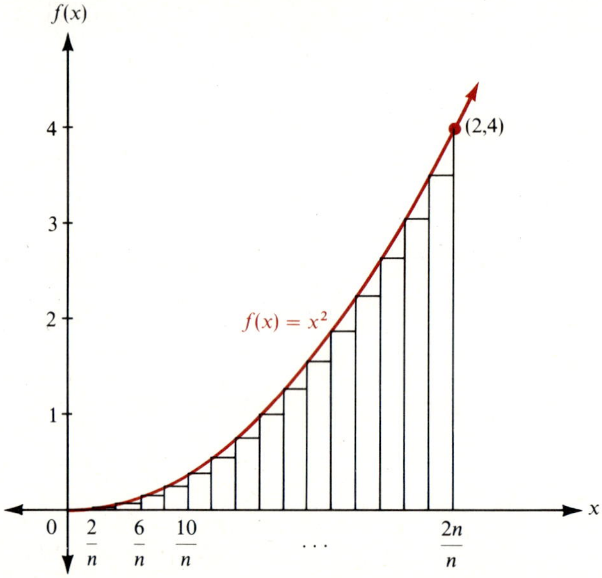

Figure 6.6

In this case,

$$\text{area of first rectangle} = \frac{2}{n}\cdot f(0) = 0$$

$$\text{area of second rectangle} = \frac{2}{n}\cdot f\left(\frac{2}{n}\right) = \frac{2}{n}\cdot\left(\frac{2}{n}\right)^2$$

$$\text{area of third rectangle} = \frac{2}{n}\cdot f\left(\frac{4}{n}\right) = \frac{2}{n}\cdot\left(\frac{4}{n}\right)^2$$

$$\vdots$$

$$\text{area of last rectangle} = \frac{2}{n}\cdot f\left(\frac{2(n-1)}{n}\right) = \frac{2}{n}\left(\frac{2(n-1)}{n}\right)^2$$

Adding the areas of these rectangles, we get

$$S_n = 0 + \frac{2}{n}\cdot\left(\frac{2}{n}\right)^2 + \left(\frac{2}{n}\right)\cdot\left(\frac{4}{n}\right)^2 + \cdots + \left(\frac{2}{n}\right)\cdot\left(\frac{2(n-1)}{n}\right)^2$$
$$= \left(\frac{2}{n}\right)^3 [1^2 + 2^2 + 3^2 + \cdots + (n-1)^2]$$

In order to simplify even further, we use the formula

$$1^2 + 2^2 + \cdots + k^2 = \frac{k(k+1)(2k+1)}{6}$$

For $k = n - 1$, we get

$$1^2 + 2^2 + 3^2 + \cdots + (n-1)^2 = \frac{(n-1)n(2n-1)}{6}$$

Thus,

$$S_n = \left(\frac{2}{n}\right)^3 \frac{(n-1)n(2n-1)}{6}$$
$$= \frac{4}{3}\frac{(n-1)(2n-1)}{n^2}$$
$$= \frac{4}{3}\left(1 - \frac{1}{n}\right)\left(2 - \frac{1}{n}\right)$$

and

$$\lim_{n\to\infty} S_n = \lim_{n\to\infty}\left[\frac{4}{3}\left(1 - \frac{1}{n}\right)\left(2 - \frac{1}{n}\right)\right]$$
$$= \tfrac{4}{3}(1-0)(2-0)$$
$$= \tfrac{8}{3}$$

which agrees with the result we established before. It seems reasonable to expect that if we get the same area in both of the extreme cases, we should get the same result in every intermediate case.

The result is independent of the choice of x_i^* in the i-th subinterval $[x_{i-1}, x_i]$. Furthermore, the result is also independent of the choice of subdivisions of the closed interval $[a, b]$.

Now we return to equation (6.3) and generalize how the previous procedure is used, at least in theory, to obtain the area bounded by the lines $x = a$, $x = b$, $y = 0$ (the x-axis), and $y = f(x)$ where f is an arbitrary continuous function and $f(x) \geq 0$ over the closed interval $[a, b]$. We have observed that as the subdivisions become finer and finer, the number of rectangles increases, the base of each rectangle approaches zero, and the sum of the areas of the rectangles approaches a limit. This limiting value is called the *definite integral* of $f(x)$ from $x = a$ to $x = b$. In symbols, we write

$$\textbf{6.4} \qquad \int_a^b f(x)\,dx = \lim_{\Delta x \to 0} [f(x_1^*)\Delta x_1 + f(x_2^*)\Delta x_2 + \cdots + f(x_n^*)\Delta x_n]$$

where Δx is the length of the largest subinterval, that is, $\Delta x = \max(\Delta x_1, \Delta x_2, \ldots, \Delta x_n)$, a and b are the lower and upper limits of integration, and $f(x)$ is referred to as the integrand. Remember that the letter x in (6.4) merely identifies the variable being used in evaluating the area under the curve $y = f(x)$ and

$$\int_a^b f(x)\,dx, \qquad \int_a^b f(t)\,dt, \qquad \int_a^b f(u)\,du, \ldots$$

have exactly the same meaning.

Let us now look at a few examples.

Example 6.17

Find the area bounded by the lines $x = 1$, $x = 2$, the x-axis, and the curve $f(x) = x(2 - x)$.

Solution

Since the result we get is independent of both the subdivisions and the choice of x_i^* in the interval $[x_{i-1}, x_i]$, we shall divide the interval $[1, 2]$ into n equal parts and choose each x_i^* to be the right-hand endpoint in each subinterval as shown in Figure 6.7.

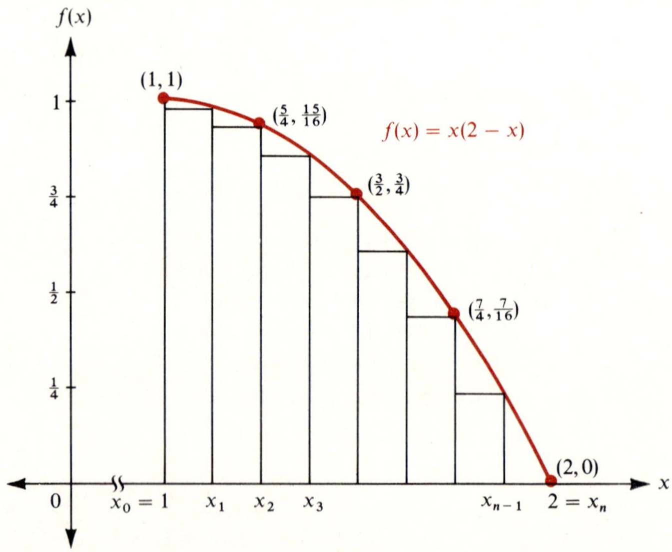

Figure 6.7

The length of each subinterval is $1/n$. The inscribed rectangles are of height

$$f\left(1+\frac{1}{n}\right)=\left(1+\frac{1}{n}\right)\left(2-1-\frac{1}{n}\right)=1-\left(\frac{1}{n}\right)^2$$

$$f\left(1+\frac{2}{n}\right)=\left(1+\frac{2}{n}\right)\left(2-1-\frac{2}{n}\right)=1-\left(\frac{2}{n}\right)^2$$

$$f\left(1+\frac{3}{n}\right)=\left(1+\frac{3}{n}\right)\left(2-1-\frac{3}{n}\right)=1-\left(\frac{3}{n}\right)^2$$

$$f\left(1+\frac{4}{n}\right)=\left(1+\frac{4}{n}\right)\left(2-1-\frac{4}{n}\right)=1-\left(\frac{4}{n}\right)^2$$

$$\vdots$$

$$f\left(1+\frac{n-1}{n}\right)=\left(1+\frac{n-1}{n}\right)\left(2-1-\frac{n-1}{n}\right)=1-\left(\frac{n-1}{n}\right)^2$$

The sum of the areas of the rectangles is

$$S_n=\frac{1}{n}\underbrace{(1+1+\cdots+1)}_{(n-1)\text{ terms}}-\frac{1}{n^3}[1^2+2^2+\cdots+(n-1)^2]$$

Now we need the formula

$$1^2+2^2+\cdots+k^2=\frac{k(k+1)(2k+1)}{6}$$

which, for $k=n-1$, becomes

$$1^2+2^2+\cdots+(n-1)^2=\frac{(n-1)n(2n-1)}{6}$$

Thus

$$S_n=\frac{1}{n}\cdot(n-1)-\frac{1}{n^3}\cdot\frac{n(n-1)(2n-1)}{6}$$

$$=\left(1-\frac{1}{n}\right)-\frac{1}{6}\left(1-\frac{1}{n}\right)\left(2-\frac{1}{n}\right)$$

and

$$\lim_{n\to\infty} S_n=1-\tfrac{1}{3}=\tfrac{2}{3}$$

Hence,

$$\int_1^2 x(2-x)\,dx=\tfrac{2}{3}$$

Example 6.18 Find the area bounded by the lines $x=0$, $x=1$, the x-axis, and the curve $f(x)=x^3$.

Solution

As before, we divide the interval $[0, 1]$ into n equal parts and choose each x_i^* to be the right-hand endpoint in each subinterval as illustrated in Figure 6.8. The length of each subinterval is $1/n$.

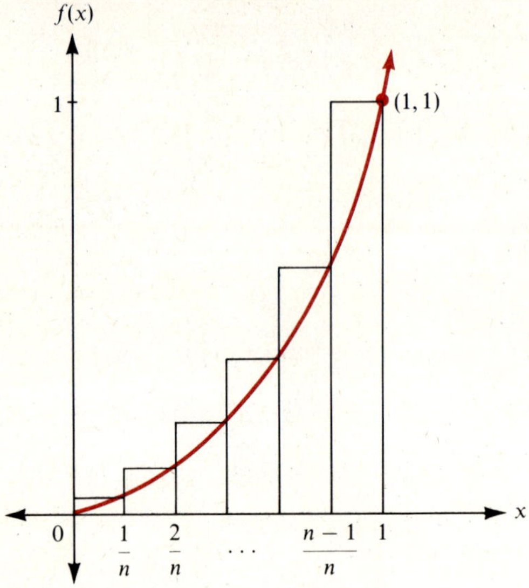

Figure 6.8

The heights of the rectangles are

$$f\left(\frac{1}{n}\right) = \left(\frac{1}{n}\right)^3$$

$$f\left(\frac{2}{n}\right) = \left(\frac{2}{n}\right)^3$$

$$f\left(\frac{3}{n}\right) = \left(\frac{3}{n}\right)^3$$

$$\vdots$$

$$f\left(\frac{n}{n}\right) = \left(\frac{n}{n}\right)^3$$

The sum of the areas of the rectangles is

$$S_n = \frac{1}{n}\left[\left(\frac{1}{n}\right)^3 + \left(\frac{2}{n}\right)^3 + \cdots + \left(\frac{n}{n}\right)^3\right]$$

$$= \frac{1}{n^4}[1^3 + 2^3 + \cdots + n^3]$$

To simplify further, we use the formula

$$1^3 + 2^3 + \cdots + n^3 = \frac{n^2(n+1)^2}{4}$$

and obtain

$$S_n = \frac{1}{n^4}\frac{n^2(n+1)^2}{4}$$
$$= \frac{1}{4}\left(1 + \frac{1}{n}\right)^2$$

As $n \to \infty$, $1/n \to 0$. Thus

$$\lim_{n\to\infty} S_n = \tfrac{1}{4}$$

Hence,

$$\int_0^1 x^3\,dx = \tfrac{1}{4}$$

Exercises

Evaluate the following integrals by dividing the closed interval $[a, b]$ into n subintervals of equal length.

1. $\int_2^4 2x\,dx$

2. $\int_1^3 (x + 2)\,dx$

3. $\int_0^2 (2t + 3)\,dt$

4. $\int_0^5 (3t + 4)\,dt$

5. $\int_0^3 x^2\,dx$

6. $\int_1^3 (x^2 + 2)\,dx$

7. $\int_1^2 (x^2 + 2x)\,dx$

8. $\int_2^4 (x^2 + 3x + 2)\,dx$

9. $\int_2^3 (x + 1)^2\,dx$

10. $\int_1^3 (x^2 + x + 3)\,dx$

11. $\int_1^3 (2x^2 + 3x + 4)\,dx$

12. $\int_1^2 x^2(x + 1)\,dx$

6.4 The Fundamental Theorem of Calculus

It is apparent that the concept of the definite integral is closely related to the idea of area. Specifically, if $f(x)$ is continuous and non-negative over a closed interval $[a, b]$, then

$$\int_a^b f(x)\,dx$$

represents the area bounded by the lines $x = a$, $x = b$, the x-axis, and the curve $y = f(x)$.

We now state an important theorem, known as *The Fundamental Theorem of Calculus*, which serves as the link between integration and differentiation.

Proposition 6.2

If $f(x)$ is a continuous function on the closed interval $[a, b]$ and $F(x)$ is an antiderivative of $f(x)$, then

$$\int_a^b f(x)\,dx = F(b) - F(a)$$

The fundamental theorem states that we can compute the definite integral of a continuous function $f(x)$ over $[a, b]$ by finding an antiderivative $F(x)$, evaluating $F(x)$ at $x = b$ and at $x = a$, and then subtracting. It is customary to use the notation

$$F(x)\Big|_a^b = F(b) - F(a)$$

Using this notation, we formalize the fundamental theorem of calculus as

$$\int_a^b f(x) = F(x)\Big|_a^b \quad \text{where} \quad F'(x) = f(x)$$

In section 6.3, we evaluated the area under the curve $f(x) = x^2$ bounded by $x = 0$, $x = 2$, and the x-axis. This was accomplished by summing the areas of rectangles and then finding the limit of the sum. *The fundamental theorem of calculus provides a remarkably easier procedure*, as illustrated in the following example.

Example 6.19

Evaluate

$$\int_0^2 x^2\,dx$$

Solution

An antiderivative of x^2 is

$$F(x) = \frac{x^3}{3}$$

Evaluating $F(x)$ at $x = 2$ and $x = 0$ we have

$$F(2) = \tfrac{8}{3}$$
$$F(0) = 0$$

Applying Proposition 6.2, we have

$$\int_0^2 x^2\,dx = F(2) - F(0) = \tfrac{8}{3}$$

Let us look at few more examples.

Example 6.20

Evaluate

$$\int_0^1 x^3\,dx$$

Solution

An antiderivative of x^3 is $F(x) = x^4/4$. Using the fundamental theorem, we obtain

$$\begin{aligned}\int_0^1 x^3\,dx = F(x)\Big|_0^1 &= F(1) - F(0)\\ &= \tfrac{1}{4} - 0\\ &= \tfrac{1}{4}\end{aligned}$$

and this agrees with the result we obtained in Example 6.18 using rectangles.

Example 6.21

Evaluate

$$\int_1^2 e^x\,dx$$

Solution

An antiderivative of $f(x) = e^x$ is

$$F(x) = e^x$$

Thus,

$$\begin{aligned}\int_1^2 e^x\,dx = F(x)\Big|_1^2 &= F(2) - F(1)\\ &= e^2 - e^1 = 4.6708\end{aligned}$$

In evaluating the definite integral

$$\int_a^b f(x)\,dx$$

we have assumed that a, the lower limit of integration, is less than b, the upper limit. What happens if a and b are reversed?

Proposition 6.3

If $f(x)$ is a continuous function and $f(x) \geq 0$ over the closed interval $[a, b]$, then

$$\int_a^b f(x)\,dx = -\int_b^a f(x)\,dx$$

The sign of integral is changed if the limits of integration are reversed.

Example 6.22

Evaluate

$$\int_2^0 x^2\,dx$$

Solution

$$\int_2^0 x^2\,dx = -\int_0^2 x^2\,dx = -\tfrac{8}{3}$$

We also agree that if the lower and the upper limits of integration are equal, then

$$\int_a^a f(x)\,dx = 0$$

Further, it seems reasonable that if $f(x)$ is a continuous function, $f(x) \geq 0$ over the closed interval $[a, b]$, and $a < c < b$ (as shown in Figure 6.9), then we may add the area under $f(x)$ from a to c to the area under $f(x)$ from c to b and obtain the area under $f(x)$ over the entire interval $[a, b]$.

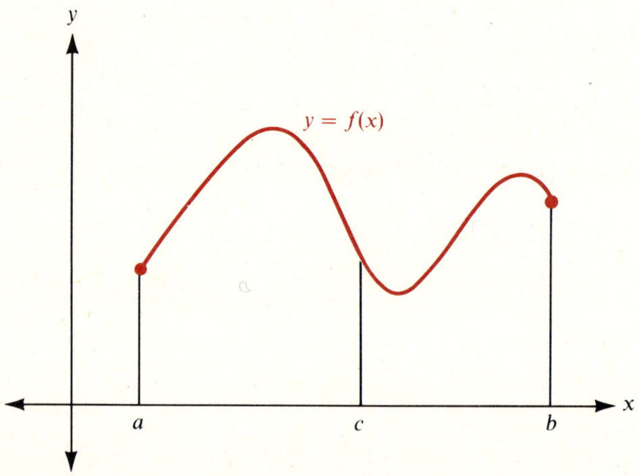

Figure 6.9

Proposition 6.4

If $f(x)$ is a continuous function, $f(x) \geq 0$ in the interval $[a, b]$, and $a < c < b$, then

$$\int_a^b f(x)\,dx = \int_a^c f(x)\,dx + \int_c^b f(x)\,dx$$

We consider it worthwhile to restate the results (6.1) and (6.2) for indefinite integrals. These results can be used to evaluate definite integrals of polynomials and other functions.

Proposition 6.5

If $f(x)$ is a continuous function, $f(x) \geq 0$ in the interval $[a, b]$, and k is any constant, then

$$\int_a^b kf(x)\,dx = k\int_a^b f(x)\,dx$$

Example 6.23

Evaluate

$$\int_1^2 4x^2\,dx$$

Solution

$$\int_1^2 4x^2\,dx = 4\int_1^2 x^2\,dx$$

Using the fundamental theorem, we observe that

$$\int_1^2 x^2\,dx = \frac{x^3}{3}\bigg|_1^2 = \tfrac{8}{3} - \tfrac{1}{3} = \tfrac{7}{3}$$

Thus,

$$\int_1^2 4x^2\,dx = 4\int_1^2 x^2\,dx = 4(\tfrac{7}{3}) = \tfrac{28}{3}$$

Proposition 6.6

If $f(x)$ and $g(x)$ are continuous functions such that $f(x) \geq 0$ and $g(x) \geq 0$ over the interval $[a, b]$, then

$$\int_a^b [f(x) \pm g(x)]\,dx = \int_a^b f(x)\,dx \pm \int_a^b g(x)\,dx$$

Example 6.24

Evaluate

$$\int_1^4 (x^3 + 3x^2 + 4)\,dx$$

Solution

$$\int_1^4 (x^3 + 3x^2 + 4)\,dx = \int_1^4 x^3\,dx + \int_1^4 3x^2\,dx + \int_1^4 4\,dx$$

(Proposition 6.6)

$$= \int_1^4 x^3\,dx + 3\int_1^4 x^2\,dx + 4\int_1^4 dx$$

(Proposition 6.5)

Recognizing that

$$\frac{d}{dx}\left(\frac{x^4}{4}\right) = x^3, \quad \frac{d}{dx}\left(\frac{x^3}{3}\right) = x^2, \quad \text{and} \quad \frac{d}{dx}(x) = 1$$

we may write

$$\int_1^4 x^3\,dx = \tfrac{1}{4}x^4\Big|_1^4 = \tfrac{256}{4} - \tfrac{1}{4} = \tfrac{255}{4}$$

$$3\int_1^4 x^2\,dx = 3(\tfrac{1}{3}x^3)\Big|_1^4 = x^3\Big|_1^4 = 64 - 1 = 63$$

and

$$4\int_1^4 dx = 4\left(x\Big|_1^4\right) = 4(4-1) = 12$$

Thus,

$$\int_1^4 (x^3 + 3x^2 + 4)\,dx = \tfrac{255}{4} + 63 + 12 = \tfrac{555}{4}$$

Let us now consider some applications.

Example 6.25

The population of a species of insects is increasing at a rate of

$$\frac{2{,}000}{\sqrt{1+t}}$$

individuals per week where t is measured in weeks. How many insects are added between the 8th and the 35th week?

Solution

The total increase in the number of insects between the 8th and the 35th week is given by the definite integral

$$\begin{aligned}\int_8^{35} \frac{2{,}000\,dt}{\sqrt{1+t}} &= 2{,}000\int_8^{35} (1+t)^{-1/2}\,dt \\ &= 2{,}000\left[\frac{(1+t)^{-(1/2)+1}}{-\frac{1}{2}+1}\Big|_8^{35}\right] \\ &= 4{,}000\left[(1+t)^{1/2}\Big|_8^{35}\right] \\ &= 4{,}000(\sqrt{36} - \sqrt{9}) \\ &= 4{,}000(6-3) \\ &= 12{,}000\end{aligned}$$

There is an increase of 12,000 insects.

Example 6.26

A calculator company has recently introduced a service contract for its customers. The company has estimated that the rate of maintenance charge of a machine is

$$C(x) = 20 + 0.2e^x$$

where x is the number of years the calculator has been in use. Determine the company's annual charge for each calculator for a five-year contract.

Solution

The total maintenance charge for a 5-year contract is given by

$$\begin{aligned}\int_0^5 C(x)\,dx &= \int_0^5 (20 + 0.2e^x)\,dx \\ &= 20x + 0.2e^x\Big|_0^5 \\ &= [100 + (0.2)(148.4)] - [0 + (0.2)\cdot 1] \\ &= 129.68 - 0.20 \\ &= 129.48\end{aligned}$$

The annual charge would be \$25.90.

Example 6.27

A corporation contemplates opening a regional distribution center at an initial cost of \$174,000. The management believes that increased sales will offset the cost after a short period. The rate of increased sales over 10 years is estimated to be

$$I(x) = 5{,}000 + 8{,}000x$$

where x represents the number of years and $I(x)$ is the increased sales (in dollars) at any given time. Would this regional center pay for itself in 5 years? If not, after how many years of operation would the center pay for itself?

Solution

Because the rate of increased sales is $I(x)$, the actual increase after t years is given by

$$S(t) = \int_0^t I(x)\,dx$$

Thus, the total increase in the first five years is

$$\begin{aligned}S(5) = \int_0^5 I(x)\,dx &= \int_0^5 (5{,}000 + 8{,}000x)\,dx \\ &= 5{,}000x + 4{,}000x^2\Big|_0^5 \\ &= 105{,}000\end{aligned}$$

Obviously, these sales are not enough to offset the cost of \$174,000. To find how long it would take for the regional center to pay for itself, we need to find t such that

$$\int_0^t (5{,}000 + 8{,}000x)\,dx = 174{,}000$$

Thus,

$$5{,}000x + 4{,}000x^2\Big|_0^t = 174{,}000$$

or

$$5{,}000t + 4{,}000t^2 = 174{,}000$$

$$4t^2 + 5t - 174 = 0$$

$$(t - 6)(4t + 29) = 0$$

$$t = 6 \quad \text{or} \quad t = -\tfrac{29}{4}$$

Because t cannot be negative, we conclude that the distribution center would pay for itself in 6 years.

Exercises

Use the fundamental theorem of calculus to evaluate each of the following integrals.

1. $\int_1^2 x\,dx$
2. $\int_2^4 3x\,dx$
3. $\int_1^3 (2x + 3)\,dx$
4. $\int_2^4 (3t + 5)\,dt$
5. $\int_0^3 3t^2\,dt$
6. $\int_1^2 (t^2 + 1)\,dt$
7. $\int_1^4 (x^2 - 3x + 4)\,dx$
8. $\int_0^3 (3x^2 + 2x + 1)\,dx$
9. $\int_3^5 (x^2 + x + 1)\,dx$
10. $\int_2^4 (5x^2 - 2x + 4)\,dx$
11. $\int_0^3 x(x + 1)\,dx$
12. $\int_1^4 2x(x + 4)\,dx$
13. $\int_2^3 (x + 1)(x - 1)\,dx$
14. $\int_1^2 (x + 2)(x + 3)\,dx$
15. $\int_0^2 3t^3\,dt$
16. $\int_1^4 (x^3 + x^2 + 1)\,dx$
17. $\int_2^5 (4x^3 + 3x^2 + 2x + 1)\,dx$
18. $\int_1^3 x(x^2 + 1)\,dx$
19. $\int_2^4 x(x + 1)^2\,dx$
20. $\int_4^6 x(x + 1)(x + 2)\,dx$
21. $\int_0^2 3e^x\,dx$
22. $\int_1^3 (2e^x + 3)\,dx$
23. $\int_{-1}^2 (4e^x + 1)\,dx$
24. $\int_0^1 (4x^3 + 3x^2 + 2e^x)\,dx$

25. Use the fundamental theorem of calculus to show that

$$\int_1^3 (2x+3)\,dx = -\int_3^1 (2x+3)\,dx$$

Verify each of the following results in problems 26–30 using the fundamental theorem of calculus.

26. $\int_0^1 x^2\,dx + \int_1^3 x^2\,dx = \int_0^3 x^2\,dx$

27. $\int_1^4 (2x^2+3x)\,dx = 2\int_1^4 x^2\,dx + 3\int_1^4 x\,dx$

28. $\int_2^5 (x+1)^2\,dx = \int_2^5 x^2\,dx + 2\int_2^5 x\,dx + \int_2^5 dx$

29. $\int_0^2 x^3\,dx = \int_0^1 x^3\,dx + \int_1^2 x^3\,dx$

30. $\int_0^3 x^4\,dx = \int_0^1 x^4\,dx + \int_1^2 x^4\,dx + \int_2^3 x^4\,dx$

Determine the area bounded by $x = a$, $x = b$, the horizontal axis, and the curve $y = f(x)$ in each of the problems 31–36.

31. $f(x) = x^2$; $a = 3$, $b = 5$

32. $f(x) = 2x^2 + 3$; $a = 1$, $b = 4$

33. $f(x) = 2e^x$; $a = 1$, $b = 3$

34. $f(x) = (e^x + 1)$; $a = 2$, $b = 5$

35. $f(x) = \sqrt{x}$; $a = 1$, $b = 4$

36. $f(x) = \sqrt{x}$; $a = 4$, $b = 9$

37. The velocity of a particle moving along a certain path after t seconds is

$$v(t) = 12t^2 + 24t + 5$$

where $v(t)$ is measured in feet per second. Determine how far the particle has traveled

a. in the first 4 seconds **b.** in the next 4 seconds

38. A ball is dropped from a height of 400 feet. Its velocity after t seconds is given by

$$v(t) = 32t$$

a. How far has the ball dropped in the first 4 seconds?
b. After how many seconds will the ball hit the ground? Assume that time is measured from the instant the ball is dropped.

39. The instantaneous rate of growth of the population in a certain town is given by

$$P'(t) = 36t^2 + 60t^{3/2}$$

a. Determine the population $P(t)$ in t years assuming that the current population of the town is 10,000.

b. Determine the projected increase in the population after 4 years.

40. The student enrollment in professional courses in a certain college is expected to increase at a rate of

$$N'(t) = 100 + 32t$$

where $N(t)$ is the number of students enrolled at time t. Determine the expected increase in enrollment in the next 5 years.

41. An influenza hits a certain town. Let $N(t)$ be the number of persons sick at time t, where t is measured from the day the town was hit with flu. The sickness is spreading at the rate of

$$N'(t) = 36 - 12t^{-1/2}, \qquad 0 \leq t \leq 25$$

Determine the increase in the number of persons sick with flu between $t = 4$ and $t = 9$ days.

42. The instantaneous rate of improvement of a particular student in memorizing a list of nonsense words is given by

$$N'(t) = 60t^{-1/2}, \qquad t \geq 0$$

where $N(t)$ is the number of 'words' the student learns after t hours of continuous practice. Determine the increase in the number of words memorized in the first four hours of continuous practice.

43. The marginal revenue function of a daily newspaper is

$$R'(x) = 100 + 0.02x + 0.03x^2$$

where x is the number (in thousands) of copies published. How much additional revenue is generated by an increase in sales from $x = 50$ to $x = 60$?

44. A publishing company is contemplating the purchase of a new computer at a cost of $28,000. The company believes that savings in efficiency will offset the cost after a short period. The rate of cost savings over a period of 10 years is estimated to be

$$C(x) = 500 + 1{,}000x$$

where x represents the number of years and $C(x)$ is the savings (in dollars) per year at any given time. Would the machine pay for itself in 5 years? If not, after how many years of operation would the machine pay for itself?

45. Chicos Corporation is considering the purchase of a regional distribution center at a cost of \$65,000. The management believes that after a short period of adjustment cost savings from the increased sales will offset the cost. The rate of cost savings is estimated to be

$$C(x) = 3{,}000 + 4{,}000x$$

where x represents the number of years and $C(x)$ is the savings (in dollars) per year at any given time. Would the new center pay for itself in three years? If not, after how many years of operation would the regional center pay for itself?

46. Gummos Department Store has developed a service contract for their color televisions. The rate of maintenance cost is estimated to be

$$C(x) = 3x^2 + 2x + 20, \qquad 0 < x \leq 5$$

where x is the number of years the contract remains in force. Determine the amount the store must charge per year for

a. a two-year contract
b. a three-year contract
c. a five-year contract

47. An automobile leasing company has developed a service contract for the new cars it leases. The management has determined that the rate of maintenance charge is approximated by

$$C(x) = 6x^2 + 12x + 10, \qquad 0 < x \leq 3$$

How much should a customer pay per year for

a. a two-year contract
b. a three-year contract

48. In Exercise 47, how much should a customer pay if she leases the car for three years but only buys the maintenance contract for the third year?

6.5 The Area between Two Curves

We have already discussed the area of continuous functions $f(x)$ for which $f(x) \geq 0$ on the closed interval $[a, b]$. We now extend this concept to include functions which are negative over part or all of the closed interval $[a, b]$.

Consider, for example, the area of the curve

$$f(x) = x^2 - 4$$

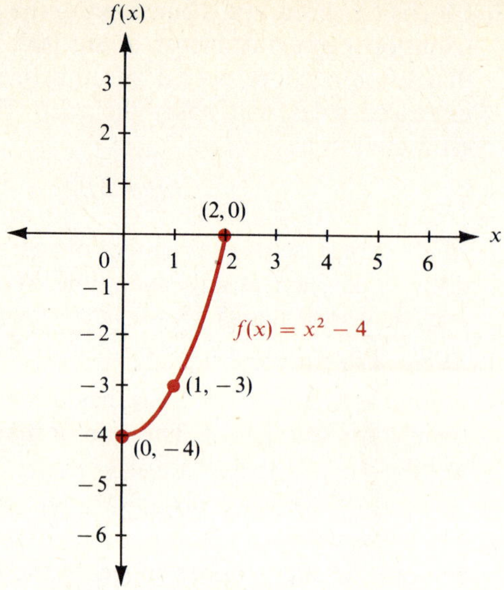

Figure 6.10

bounded by $x = 0$, $x = 2$, and the x-axis. The graph of $f(x)$, as shown in Figure 6.10, makes it obvious that the curve lies below the x-axis in the closed interval $[0, 2]$. If we were to represent the area by the definite integral

$$\int_0^2 f(x)\,dx$$

we would get

$$\begin{aligned}\int_0^2 f(x)\,dx &= \int_0^2 (x^2 - 4)\,dx \\ &= \tfrac{1}{3}x^3 - 4x\Big|_0^2 \\ &= \tfrac{8}{3} - 8 \\ &= -\tfrac{16}{3}\end{aligned}$$

Since area is defined to be a positive number, we must multiply both sides of the above equation by (-1). In general, if $f(x) < 0$ over the closed interval $[a, b]$, we represent the area under the curve $f(x)$ by

6.5 $$A = -\int_a^b f(x)\,dx$$

As another illustration, consider the area of the curve $f(x) = x^3$ bounded by the lines $x = -2$, $x = 2$, and the x-axis. The graph of $f(x)$ (see Figure 6.11) lies partly above and partly below the x-axis. The area below the x-axis is represented by the definite integral $-\int_{-2}^{0} f(x)\,dx$ while the area above the x-axis is $\int_0^2 f(x)\,dx$. Adding these, we get the

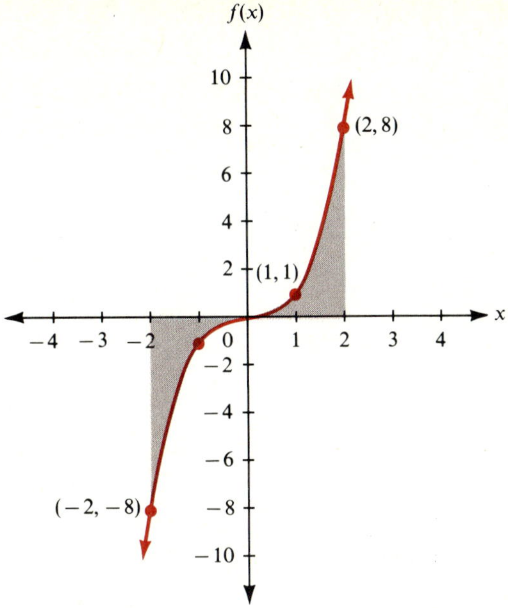

Figure 6.11

total area

$$-\int_{-2}^{0} f(x)\,dx + \int_{0}^{2} f(x)\,dx = -\int_{-2}^{0} x^3\,dx + \int_{0}^{2} x^3\,dx$$
$$= \frac{-1}{4}x^4\Big|_{-2}^{0} + \frac{1}{4}x^4\Big|_{0}^{2}$$
$$= 4 + 4$$
$$= 8$$

Next consider the problem of determining the area of the region bounded by the lines $x = a$ and $x = b$ and two continuous functions $f(x)$ and $g(x)$. Suppose that

$$f(x) \geq g(x)$$

for all x in the closed interval $[a, b]$. Figure 6.12 represents a typical situation. The area of the shaded region is given by

6.6 $$\int_{a}^{b} f(x)\,dx - \int_{a}^{b} g(x)\,dx = \int_{a}^{b} [f(x) - g(x)]\,dx$$

The formula (6.6) holds whether the curves $f(x)$ and $g(x)$ are above or below the x-axis. Consider, for example, the graph of $f(x)$ and $g(x)$ in Figure 6.13. The definite integral $\int_a^b f(x)\,dx$ represents the shaded area *above* the x-axis and $-\int_a^b g(x)$ represents the shaded area *below* the x-axis. Adding these, we have the total area:

$$\int_{a}^{b} [f(x) - g(x)]\,dx$$

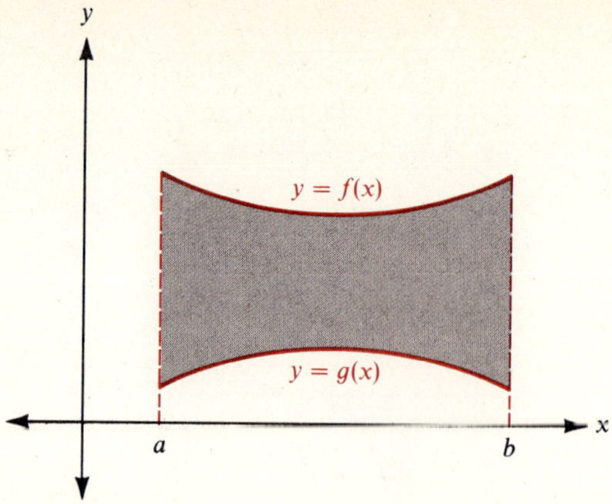

Figure 6.12

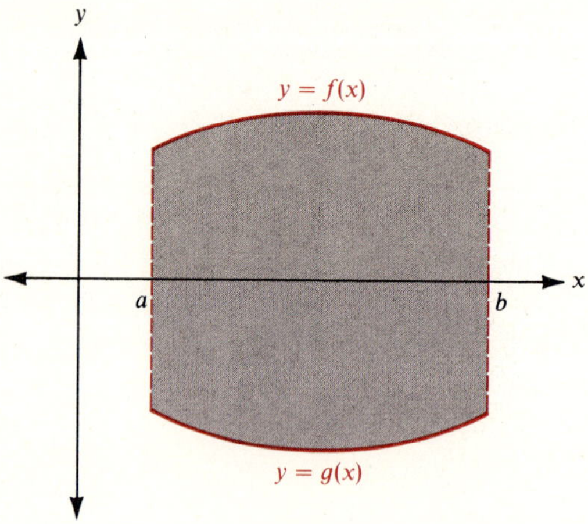

Figure 6.13

Example 6.28

Find the area of the region bounded by the lines $x = \frac{1}{2}$, $x = \frac{3}{2}$, $y = 2x$, and the curve $y = x^2$.

Solution

Since it is not clear from the statement of the problem what the region is, we draw a sketch. Figure 6.14 makes it obvious that $y = 2x$ is above $y = x^2$ in the closed interval $[1, 2]$. If we use (6.6) with $f(x) = 2x$ and $g(x) = x^2$, we find that the area is

$$\int_{1/2}^{3/2} (2x - x^2)\,dx = x^2 - \frac{x^3}{3}\Bigg|_{1/2}^{3/2}$$

$$= (\tfrac{9}{4} - \tfrac{27}{24}) - (\tfrac{1}{4} - \tfrac{1}{24})$$

$$= \tfrac{11}{12}$$

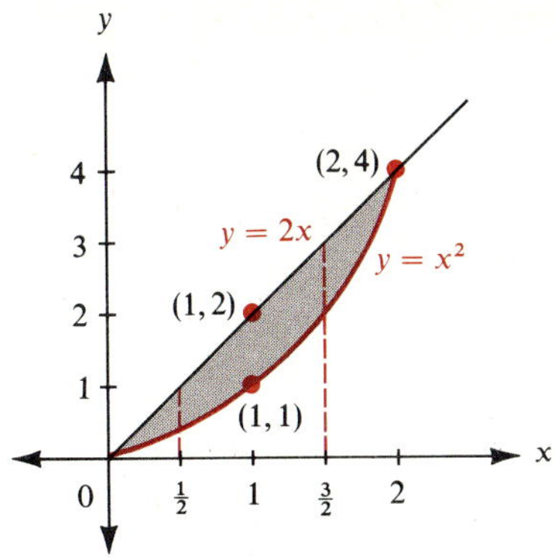

Figure 6.14

Example 6.29

Find the area of the region bounded by the curves $y = x^2 - 4$ and $y = x + 2$.

Solution

To determine the region bounded by the curves, we draw the graphs of the two functions. Figure 6.15 indicates that the curves meet at the points $A(-2, 0)$ and $B(3, 5)$. Note that $y = x + 2$ is above $y = x^2 - 4$ in the closed interval $[-2, 3]$.

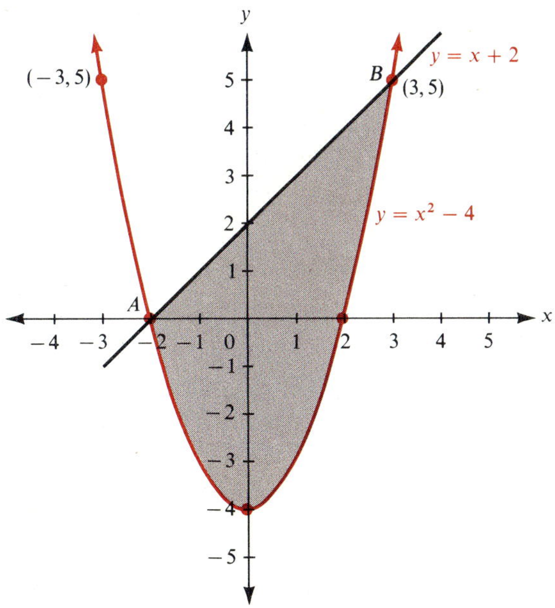

Figure 6.15

Therefore, the bounded area is

$$\int_{-2}^{3} [(x + 2) - (x^2 - 4)]\,dx = \int_{-2}^{3} (6 + x - x^2)\,dx$$

$$= 6x + \frac{x^2}{2} - \frac{x^3}{3}\bigg|_{-2}^{3}$$

$$= (18 + \tfrac{9}{2} - 9) - (-12 + \tfrac{4}{2} + \tfrac{8}{3})$$

$$= \tfrac{125}{6}$$

Example 6.30

Find the area bounded by the curves $y = x$ and $y = x^3$.

Solution

Obviously, a graph is indispensable. Solving the equations $y = x$ and $y = x^3$ by eliminating y, we have

$$x^3 - x = 0$$

$$x(x^2 - 1) = 0$$

$$x(x - 1)(x + 1) = 0$$

Thus $x = 0, 1,$ or -1; the curves intersect at $(0, 0)$, $(1, 1)$, and $(-1, -1)$. After plotting a few points, we obtain the graphs as shown in Figure 6.16. There are two regions, one in the first quadrant and one in the third quadrant. Since the line $y = x$ is above the curve $y = x^3$ in the first

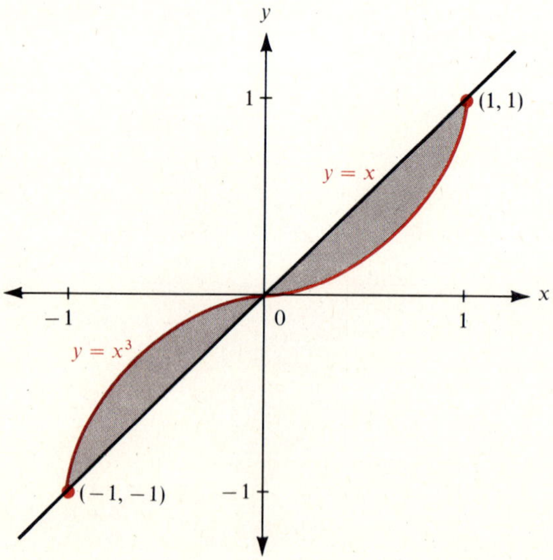

Figure 6.16

quadrant, the area in the first quadrant is

$$\int_0^1 (x - x^3)\,dx = \tfrac{1}{2}x^2 - \tfrac{1}{4}x^4\Big|_0^1$$
$$= \tfrac{1}{2} - \tfrac{1}{4}$$
$$= \tfrac{1}{4}$$

The area *below* the x-axis in the third quadrant is

$$\int_{-1}^0 (x^3 - x)\,dx = \tfrac{1}{4}x^4 - \tfrac{1}{2}x^2\Big|_{-1}^0$$
$$= 0 - (\tfrac{1}{4} - \tfrac{1}{2})$$
$$= \tfrac{1}{4}$$

The total area between the curves is $\frac{1}{4} + \frac{1}{4} = \frac{1}{2}$.

Consumer and Producer Surpluses

An interesting problem in economics that illustrates an application of the area between two curves is finding the quantities called the *consumer's surplus* and the *producer's surplus*. For instance, suppose that the demand for a certain commodity is given by the curve

$$p = D(x)$$

where x represents the number of units demanded by the consumers at the unit price of p dollars. Suppose that the supply curve $S(x)$ for the same product has the equation

$$p = S(x)$$

where x is the number of units produced for sale at p dollars per unit. The demand and supply curves are both shown in Figure 6.17. Market equilibrium is said to occur if the demand equals the supply. Under pure competition, the equilibrium supply x_0 and the equilibrium price p_0 can be determined by solving the two equations simultaneously.

Some consumers who would have paid more than the equilibrium price p_0 have obviously gained by the fixed price p_0 and thus have saved money. The total consumer gain, as a result of the price fixed at p_0, is called the *consumer surplus* and is represented by the area above the line $p = p_0$ and below the demand curve $p = D(x)$ in Figure 6.17. Thus the consumer surplus is given by

6.7 $$\int_0^{x_0} [D(x) - p_0]\,dx = \int_0^{x_0} D(x)\,dx - p_0 x_0$$

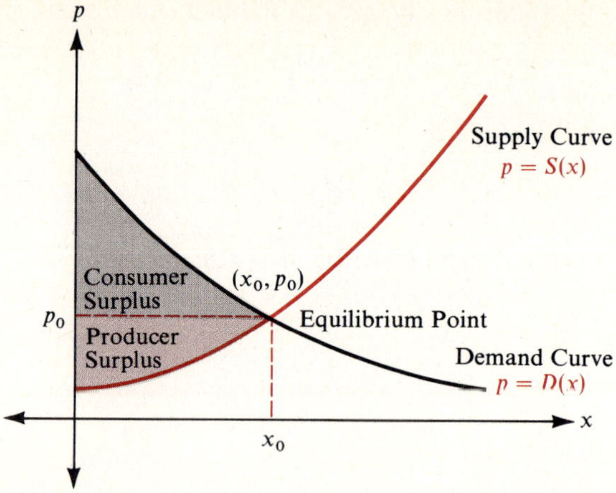

Figure 6.17

Producers of the commodity also realize some gain in the sense that those producers who would have supplied their products at a price below the fixed price price p_0 have made money. The total producer's gain, as a consequence of the price control, is called the *producer surplus*; it is represented by the area below the line $p = p_0$ and above the supply curve $S(x)$. From Figure 6.17, the producer's surplus is

6.8 $$PS = \int_0^{x_0} [p_0 - S(x)]\,dx = p_0 x_0 - \int_0^{x_0} S(x)\,dx$$

Example 6.31

Under pure competition, the quantity demanded and the corresponding price are determined by the demand and supply curves

$$p = 100 - \frac{x^2}{10} \quad \text{and} \quad p = 50 + \frac{x^2}{40}$$

respectively. Determine the corresponding consumer and producer surpluses.

Solution

The demand and supply curves are shown in Figure 6.18.

The point of intersection (20, 60) can also be found by solving the two equations simultaneously. Since the demand equals the supply under pure competition, we have

$$50 + \frac{x^2}{40} = 100 - \frac{x^2}{10}$$

which on simplification yields $x = 20$. This amount determines market equilibrium; the corresponding price is $p = 50 + (20)^2/40 = 60$. Thus

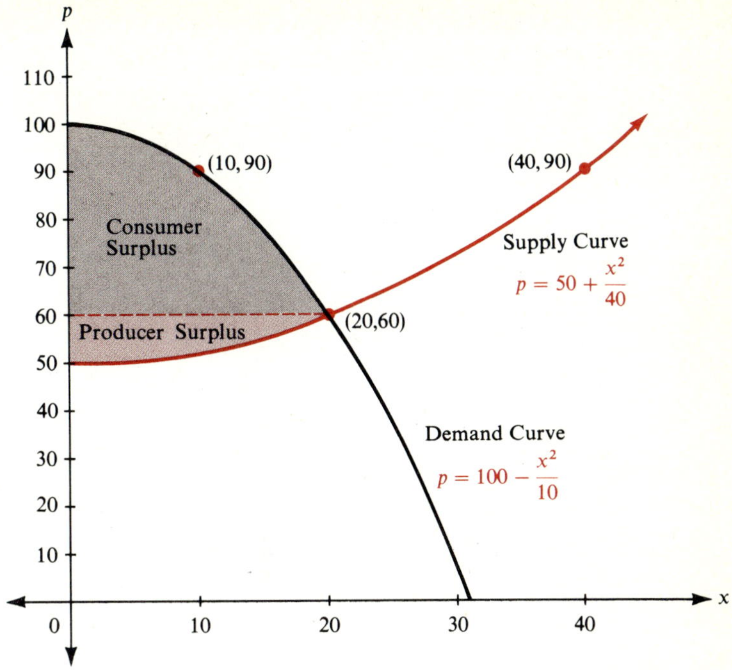

Figure 6.18

the consumer surplus is

$$\begin{aligned} CS &= \int_0^{20} \left[\left(100 - \frac{x^2}{10}\right) - 60\right] dx \\ &= \int_0^{20} \left(40 - \frac{x^2}{10}\right) dx \\ &= 40x - \frac{x^3}{30}\bigg|_0^{20} \\ &= \tfrac{1600}{3} \end{aligned}$$

and the producer surplus is

$$\begin{aligned} PS &= \int_0^{20} \left[60 - \left(50 + \frac{x^2}{40}\right)\right] dx \\ &= \int_0^{20} \left(10 - \frac{x^2}{40}\right) dx \\ &= 10x - \frac{x^3}{120}\bigg|_0^{20} \\ &= \tfrac{400}{3} \end{aligned}$$

Hence the consumer surplus is \$533.33 and the producer surplus is \$133.33.

Exercises

Sketch the region bounded by each set of curves and find its area.

1. $y = 3x + 5, x = -2, x = 3, y = 0$
2. $y = x^3, x = -2, x = 2, y = 0$
3. $y = x^2 - 1, x = -1, x = 3, y = 0$
4. $y = x^2 - 4, x = -2, x = 5, y = 0$
5. $y = x^2, y = x, x = 0, x = 1$
6. $y = x^2, y = 4x, x = 1, x = 3$
7. $y = x^2, y = 6x, x = 1, x = 2$
8. $y = x^2, y = -x, x = -1, x = 0$
9. $y = x^2, y = 3x, x = 1, x = 2$
10. $y = x^2 - 2x, y = \frac{x}{2}, x = 0, x = 2$
11. $y = x^2 - 4, y = \frac{x}{2} + 1$
12. $y = x^2 - 3, y = 2x$
13. $y = 1 - x^2, y = x^2 - 1$
14. $y = 4 - x^2, y = x^2 - 4$
15. $y = x^2, y = 8 - x^2, x = 1, x = -1$
16. $y = 6x - x^2, y = x^2 - 2x$
17. $y = x^2 - x, y = x - x^2$
18. $y = x^2, y = x^3, x = 1, x = 4$
19. $y = e^x, y = 2x + 1, x = -\frac{1}{2}$
20. $y = e^x, y = \frac{1}{x^2}, x = 2, x = 3$

In problems 21–26, the equations of demand and supply are given by $p = D(x)$ and $p = S(x)$, respectively. Determine the point at which the demand equals the supply. Find the consumer and producer surpluses under pure competition.

21. $D(x) = 300 - 40x, S(x) = 30x + 90$
22. $D(x) = 98 - 6x, S(x) = 7x + 20$
23. $D(x) = 10 - \frac{3}{5}x, S(x) = \frac{7}{5}x$
24. $D(x) = (6 - x)^2, S(x) = 14 + x, \qquad 0 \leq x \leq 6$

25. $D(x) = 300 - \frac{5}{6}x$, $S(x) = \frac{x^2}{200} + 128$

26. $D(x) = 900 - 4x - x^2$, $S(x) = x^2 + 10x$

6.6 Integration by Substitution

Success in evaluating the indefinite integral $\int f(x)\,dx$ or the definite integral $\int_a^b f(x)\,dx$ hinges on the ability to find an antiderivative $F(x)$. There are several functions,

$$\int (x^3 + 2)^8\,dx, \quad \int \frac{x\,dx}{2x^2 + 5}, \quad \int xe^{-x^2/2}\,dx$$

to name a few, that cannot be easily identified as derivatives of well-known functions. Thus, we need some additional methods of integration. One such method is called *integration by substitution* or *change of variables*. In the next section, we shall explain *integration by parts*, which is used to transform a complicated integral into a simpler one.

First we introduce the idea of a differential. If $y = f(x)$, where $f(x)$ is differentiable, we may write

$$dy = f'(x)\,dx$$

and say that dy is the *differential* of f.

Example 6.32

a. If $y = x^4$, then $dy = 4x^3\,dx$

b. If $y = e^{5x} + 3x^2 + 6$, then $dy = (5e^{5x} + 6x)\,dx$

c. If $y = a^x + \ln x$, then $dy = \left(a^x \ln a + \frac{1}{x}\right)dx, \qquad x > 0$

d. If $y = \sqrt{1 + x^2}$, then $dy = \frac{x\,dx}{\sqrt{1 + x^2}}$

Differentials, which are a disguised form of the chain rule, are very useful for finding antiderivatives. We are now in a position to study *integration by substitution*. Let us consider some examples.

Example 6.33

Find

$$\int 3x^2(x^3 + 2)^8\,dx$$

Solution

If we consider the substitution $u = x^3 + 2$, then $du/dx = 3x^2$. Taking the differential, we have

$$du = 3x^2\,dx$$

Thus,

$$\int 3x^2(x^3+2)^8\,dx = \int u^8\,du = \frac{u^9}{9} + c = \tfrac{1}{9}(x^3+2)^9 + c$$

To check whether we have obtained the right answer, we must show that the derivative of the integral equals the integrand; that is

$$\frac{d}{dx}\left[\tfrac{1}{9}(x^3+2)^9 + c\right] = \tfrac{1}{9}\left[9(x^3+2)^8(3x^2) + 0\right]$$

$$= (x^3+2)^8(3x^2)$$

Hence, our answer is correct.

Example 6.34

Find

$$\int \frac{2x\,dx}{x^2+1}$$

Solution

If we define the function $u = x^2 + 1$, then $du/dx = 2x$. Taking differentials, we have $du = 2x\,dx$. Thus,

$$\int \frac{2x\,dx}{x^2+1} = \int \frac{du}{u} = \ln|u| + c = \ln|x^2+1| + c$$

where c is an arbitrary constant.

Example 6.35

Find

$$\int \frac{(\ln x)^2}{x}\,dx$$

Solution

Note that dx/x is the differential of $\ln x$. If we substitute $u = \ln x$, then the integral becomes

$$\int u^2\,du$$

which we can handle easily. Thus

$$\int \frac{(\ln x)^2}{x} = \int u^2\,du = \frac{u^3}{3} + c = \frac{1}{3}(\ln x)^3 + c, \qquad x > 0$$

If we differentiate the right-hand side of the equation, we find that our answer is correct.

Example 6.36

Find

$$\int e^{-5x}\,dx$$

Solution

If we substitute $u = -5x$, then

$$du = -5\,dx$$

We can multiply and divide by (-5) to obtain

$$\int e^{-5x}\,dx = -\frac{1}{5}\int e^{-5x}(-5)\,dx = -\frac{1}{5}\int e^{u}\,du = -\frac{1}{5}e^{u} + c$$

$$= -\tfrac{1}{5}e^{-5x} + c$$

where c is an arbitrary constant.

In the previous examples, we looked for a function $u = g(x)$ whose differential $du = g'(x)$ either appeared in the integrand or was closely related to it. The reason that this method of integration works is a consequence of the following proposition, which we state without proof.

Proposition 6.7

Let $f(u)$ be a continuous function where $u = g(x)$ and $du/dx = g'(x)$ exists. Then

$$\int f[g(x)]g'(x)\,dx = \int f(u)\,du = F(u) + c$$

where $F'(u) = f(u)$ and $du = g'(x)\,dx$.

Let us consider more examples.

Example 6.37

Find

$$\int \frac{(e^{x} + 2x)\,dx}{e^{x} + x^{2} + 1}$$

Solution

Notice that the expression $(e^x + 2x)\,dx$ in the numerator is the differential of the expression $(e^x + x^2 + 1)$ in the denominator. If we set $u = e^x + x^2 + 1$, then $du = (e^x + 2x)\,dx$ and we have

$$\int \frac{(e^x + 2x)\,dx}{e^x + x^2 + 1} = \int \frac{du}{u} = (\ln u) + c = \ln(e^x + x^2 + 1) + c$$

Example 6.38

Evaluate

$$\int_0^1 (2x + 3)e^{x^2 + 3x + 1}\,dx$$

Solution

Since $(2x + 3)\,dx$ is the differential of $(x^2 + 3x + 1)$, we substitute $u = x^2 + 3x + 1$. Then $du = (2x + 3)\,dx$ and we have

$$\int (2x + 3)e^{x^2 + 3x + 1}\,dx = \int e^u\,du = e^u + c = e^{x^2 + 3x + 1} + c$$

Thus an antiderivative of $(2x + 3)e^{x^2 + 3x + 1}$ is

$$F(x) = e^{x^2 + 3x + 1}$$

Evaluating $F(x)$ at $x = 1$ and $x = 0$ we have

$$F(1) = e^5$$
$$F(0) = e^1$$

Applying Proposition 6.2, we have

$$\int_0^1 (2x + 3)e^{x^2 + 3x + 1}\,dx = F(1) - F(0) = e^5 - e^1 = 145.6948$$

Let us now look at an application.

Example 6.39

A large rubber manufacturing plant on the eastern seacoast discharges pollutants into the sea at an approximate rate of

$$N'(x) = 3x(x^2 + 16)^{1/2}$$

where $N(x)$ is the number (in thousands) of tons of pollutants discharged into the sea after x years of operation. Determine the quantity of pollutants discharged into the sea during the first three years of the plant's operation.

Solution

The problem is to evaluate

$$\int_0^3 3x(x^2 + 16)^{1/2}\,dx$$

First, we need to find an antiderivative $F(x)$ of $f(x) = 3x(x^2 + 16)^{1/2}$ such that $F'(x) = f(x)$.

Observe that $3x\,dx$ in the integrand is closely related to the differential of $(x^2 + 16)$. If we substitute $u = x^2 + 16$, so that $du = 2x\,dx$, we can multiply and divide the integrand by $(\frac{3}{2})$ and obtain

$$\begin{aligned}\int 3x(x^2 + 16)^{1/2}\,dx &= \frac{3}{2}\int \frac{2}{3}(3x)(x^2 + 16)^{1/2}\,dx\\ &= \frac{3}{2}\int (2x)(x^2 + 16)^{1/2}\,dx\\ &= \frac{3}{2}\int u^{1/2}\,du \qquad [u = x^2 + 16,\ du = 2x\,dx]\\ &= \frac{3}{2}\left[\frac{u^{3/2}}{(3/2)}\right] + c\\ &= u^{3/2} + c\\ &= (x^2 + 16)^{3/2} + c\end{aligned}$$

Thus an antiderivative of $f(x) = 3x(x^2 + 16)^{1/2}$ is

$$F(x) = (x^2 + 16)^{3/2}$$

Evaluating $F(x)$ at $x = 3$ and $x = 0$ we have

$$\begin{aligned}F(3) &= (25)^{3/2} = 125\\ F(0) &= (16)^{3/2} = 64\end{aligned}$$

Applying Proposition 6.2, we obtain

$$\int_0^3 3x(x^2 + 1)^{1/2}\,dx = F(3) - F(0) = 125 - 64 = 61$$

Thus the plant has discharged 61 thousand tons of pollutants into the sea during its first three years of operation.

We wish to remind the reader that in each of the preceding examples we looked for a function $u(x)$ whose differential du either appeared in the integrand or was closely related to it. This enabled us, with the use of differentials, to simplify the integrand so that we were finding antiderivatives of the types we have discussed earlier. The choice of substitution is best made by experience; it takes patience and practice to become acquainted with a suitable substitution in each case.

Exercises

Find each of the following indefinite integrals.

1. $\int \sqrt{x+4}\,dx, \quad x \geq -4$
2. $\int \sqrt{4x+4}\,dx, \quad x \geq -1$
3. $\int \sqrt{4x+3}\,dx, \quad x \geq -\frac{3}{4}$
4. $\int \sqrt{3x+2}\,dx, \quad x \geq -\frac{2}{3}$
5. $\int \frac{dx}{\sqrt{5x+6}}, \quad x \geq -\frac{6}{5}$
6. $\int x(x^2+1)^8\,dx$
7. $\int x^2(x^3+5)^6\,dx$
8. $\int (2x+3)^{-1/2}\,dx, \quad x \geq -\frac{3}{2}$
9. $\int x(1-x^2)^{1/2}\,dx$
10. $\int x^2(x^3+1)^{1/2}\,dx, \quad x \geq -1$
11. $\int 4x^3(x^4+2)^{1/2}\,dx$
12. $\int \frac{3x^2\,dx}{x^3+1}, \quad x \neq -1$
13. $\int \frac{x^3\,dx}{x^4+2}$
14. $\int \frac{x\,dx}{x^2+1}$
15. $\int xe^{x^2}\,dx$
16. $\int x^2e^{x^3}\,dx$
17. $\int e^{3x}\,dx$
18. $\int e^{-3x}\,dx$
19. $\int 3^x\,dx$
20. $\int 10^{2x}\,dx$
21. $\int \frac{e^{\sqrt{x+1}}}{\sqrt{x+1}}\,dx, \quad x \geq -1$
22. $\int \frac{dx}{\sqrt{xe^{\sqrt{x}}}}, \quad x > 0$
23. $\int x(x^2+1)^{1/2}\,dx$
24. $\int (e^x+x)^2(e^x+1)\,dx$
25. $\int \frac{dx}{x(\ln x)^2}, \quad x > 0$
26. $\int \frac{dx}{x \ln x}, \quad x > 0$
27. $\int \frac{e^x\,dx}{1+e^x}$
28. $\int \frac{e^{2x}\,dx}{1+e^{2x}}$
29. $\int \frac{(e^x+e^{-x})\,dx}{e^x-e^{-x}}$
30. $\int \frac{(e^x-e^{-x})\,dx}{(e^x+e^{-x})^2}$

Evaluate each of the following definite integrals.

31. $\int_0^1 x^2(x^3 + 2)^3\, dx$

32. $\int_0^1 \frac{3x^2\, dx}{(x^3 + 1)^2}$

33. $\int_0^4 x\sqrt{x^2 + 9}\, dx$

34. $\int_0^{12} \frac{x\, dx}{\sqrt{x^2 + 25}}$

35. $\int_0^8 e^{0.25t}\, dt$

36. $\int_0^1 e^{2t}\, dt$

37. $\int_0^1 e^{-4t}\, dt$

38. $\int_1^2 te^{t^2}\, dt$

39. $\int_{-1}^0 e^t(e^t + 1)^3\, dt$

40. $\int_1^4 \frac{e^{\sqrt{t}}\, dt}{2\sqrt{t}}$

41. The marginal cost of production of microwave ovens is given by

$$C'(x) = 2x\sqrt{x^2 + 100}$$

where x is the number of units produced. If fixed costs are \$25,000, determine the total cost function.

42. The marginal cost of producing an item is given by

$$C'(x) = \frac{x}{\sqrt{x^2 + 100}}$$

where x represents the quantity produced. Assuming that the fixed cost is \$1,000, determine the total cost function and the average cost function.

43. The instantaneous growth rate in the number of bacteria is

$$N'(t) = 300e^{0.3t}$$

where t is the time measured in hours. Determine $N(t)$, the number of bacteria, assuming that the current number of bacteria is 1,000. What will be the number of bacteria 5 hours from now? What will be the number after the following 5 hours?

44. The instantaneous growth rate in the population of a certain city is given by

$$N'(t) = e^{0.05t}$$

where t is recorded in years. Determine $N(t)$, the size of the population, assuming that the current population of the city is 20 million. What will be the population five years from now?

45. Automobiles depreciate in value at a rate that changes with time t (in years). The annual rate at which the value of a particular model car is dropping is

$$V'(t) = -3{,}000e^{-0.375t}$$

where $V(t)$ is the value of the car t years after it was sold. If a car is sold new for \$8,200, what will be its resale value five years later?

46. The instantaneous rate of decay in the quantity of radium, $Q(t)$, measured in milligrams, is given by the equation

$$Q'(t) = 15e^{-0.03t}$$

where t is measured in centuries. Determine the amount of radium that decayed between the second and fourth century.

6.7 Integration by Parts

The method of *integration by parts* transforms one integration problem into another that is easier to handle. The technique is based on the product rule

$$\frac{d}{dx}[u(x)v(x)] = u(x)v'(x) + v(x)u'(x)$$

Integrating both sides, we have

$$u(x)v(x) = \int u(x)v'(x)\,dx + \int v(x)u'(x)\,dx$$

or

6.9
$$\int u(x)v'(x)\,dx = u(x)v(x) - \int v(x)u'(x)\,dx$$

This is known as the *formula for integration by parts.* The success in evaluating

$$\int u(x)v'(x)\,dx$$

lies in the identification of $u(x)$ and $v'(x)$ and their manipulation into (6.9) so that the integral on the right-hand side of (6.9) is simpler than the one we started with. If a definite integral is involved, then we have

6.10
$$\int_a^b u(x)v'(x)\,dx = u(x)v(x)\Big|_a^b - \int_a^b v(x)u'(x)\,dx$$

The following examples apply this method.

Example 6.40

Find

$$\int xe^x\,dx$$

Solution

Let $u(x) = x$ and $v'(x) = e^x$. Then

$$u'(x) = 1 \quad \text{and} \quad v(x) = e^x + k$$

where k is a constant. Applying (6.9) we obtain

$$\begin{aligned}\int xe^x\,dx &= x(e^x + k) - \int (e^x + k)\,dx \\ &= xe^x + kx - \int e^x\,dx - \int k\,dx + c \\ &= xe^x + kx - e^x - kx + c \\ &= xe^x - e^x + c\end{aligned}$$

Note that k, the first constant, does not appear in the final answer. Because this is always the case, we shall drop the first constant when evaluating $\int v'(x)\,dx$.

It is worthwhile to examine what would happen if we choose $u(x)$ and $v'(x)$ differently. Suppose that

$$u(x) = e^x \quad \text{and} \quad v'(x) = x$$

Then,

$$u'(x) = e^x \quad \text{and} \quad v(x) = \tfrac{1}{2}x^2$$

Applying (6.9) we have

6.11 $$\int xe^x\,dx = e^x(\tfrac{1}{2}x^2) - \int (\tfrac{1}{2}x^2)e^x\,dx$$

which is correct but does not help in the evaluation of $\int xe^x\,dx$. In fact, the integral on the right-hand side of (6.11) is more difficult than the original one. In other words, the choice of $u(x)$ and $v'(x)$ is very important; the trick is to rewrite the expression so that the integral on the right-hand side is simpler than the one we were given.

Example 6.41

Find

$$\int x^2e^x\,dx$$

Solution

Let $u(x) = x^2$ and $v'(x) = e^x$. Then

$$u'(x) = 2x \quad \text{and} \quad v(x) = e^x$$

Using (6.9), we have

$$\int x^2e^x\,dx = x^2e^x - 2\int xe^x\,dx$$

Using the result from Example 6.40 that

$$\int xe^x \, dx = xe^x - e^x + c$$

we get

$$\int x^2 e^x \, dx = x^2 e^x - 2(xe^x - e^x + c)$$
$$= e^x(x^2 - 2x + 2) + c_1$$

where $c_1 = -2c$.

Example 6.42

Find

$$\int \ln x \, dx, \qquad x > 0$$

Solution

If $u(x) = \ln x$ and $v'(x) = 1$, then

$$u'(x) = \frac{1}{x} \quad \text{and} \quad v(x) = x$$

Thus,

$$\int \ln x \, dx = x \ln x - \int x \cdot \frac{1}{x} dx$$
$$= x \ln x - x + c, \qquad x > 0$$

Example 6.43

Evaluate

$$\int_0^3 x\sqrt{1 + x} \, dx$$

Solution

Let $u(x) = x$ and $v'(x) = (1 + x)^{1/2}$. Then

$$u'(x) = 1 \quad \text{and} \quad v(x) = \tfrac{2}{3}(1 + x)^{3/2}$$

Using (6.10), we obtain

$$\int_0^3 x\sqrt{1 + x} \, dx = \frac{2}{3} x(1 + x)^{3/2} \Big|_0^3 - \frac{2}{3} \int_0^3 (x + 1)^{3/2} \, dx$$
$$= \tfrac{2}{3}(24) - \tfrac{2}{3} \cdot \tfrac{2}{5}(x + 1)^{5/2} \Big|_0^3$$
$$= 16 - \tfrac{4}{15}(31)$$
$$= 16 - \tfrac{124}{15}$$
$$= \tfrac{116}{15}$$

Exercises

Use *integration by parts* in problems 1–17 to find the integral.

1. $\int x \ln x \, dx$
2. $\int x^2 \ln x \, dx$
3. $\int (x+1)e^x \, dx$
4. $\int (2x+3)e^x \, dx$
5. $\int xe^{2x} \, dx$
6. $\int xe^{3x} \, dx$
7. $\int x^2e^{2x} \, dx$
8. $\int x^3e^{3x} \, dx$
9. $\int xe^{-2x} \, dx$
10. $\int x(1+x)^3 \, dx$
11. $\int_0^1 xe^{-7x} \, dx$
12. $\int_0^4 x^2e^{-x/4} \, dx$
13. $\int_1^2 x \ln x \, dx$
14. $\int_2^4 x\sqrt{x-1} \, dx$
15. $\int_5^9 \frac{x \, dx}{\sqrt{x-1}}$
16. $\int_0^5 x\sqrt{3x+1} \, dx$
17. $\int_0^7 x(1+x)^{3/2} \, dx$

18. The velocity of a particle moving along a certain path at time t (in seconds) is given by

$$v(t) = te^{-0.5t}$$

where $v(t)$ is measured in feet per second. Determine the distance function $s(t)$ assuming that the particle starts from rest at time $t = 0$.

19. The demand rate for a new cologne introduced in the holiday season by a large advertising campaign is estimated to be

$$S'(t) = 500te^{-t}$$

where $S(t)$ is the total sales of the product t days after the end of the advertising campaign. Determine the total sales for the cologne for the first five days after the campaign.

20. The daily rate of sales for a new product is given by

$$N'(t) = 100te^{-0.05t}$$

where $N(t)$ is the number of units sold in t days after the product has been on the market. Determine the total sales for

a. the first five days
b. the first ten days

6.8 Numerical Integration*

The familiar antiderivative rule for the evaluation of the definite integral is useful if we can find an antiderivative. Unfortunately, there are many functions for which an antiderivative cannot be found by any of the methods we have discussed so far. Nevertheless, if a function is continuous over a closed interval $[a, b]$, then its definite integral exists and its value may be approximated to any degree of accuracy. Many methods have been devised to approximate the value of a definite integral. In this section we introduce two such methods: the *trapezoidal rule* and *Simpson's rule.*

First we illustrate the trapezoidal rule. Suppose that the definite integral

$$\int_a^b f(x)\,dx$$

is to be evaluated. We divide the closed interval $[a, b]$ into n equal parts so that $x_1, x_2, \ldots, x_{i-1}, x_i, \ldots, x_{n-1}$ are points of subdivision with $a = x_0$ and $b = x_n$ as shown in Figure 6.19. Thus we have n closed subintervals,

$$[x_0, x_1], [x_1, x_2], \ldots, [x_{i-1}, x_i], \ldots, [x_{n-1}, x_n],$$

each of length $\Delta x = (b - a)/n$. The area under the curve $y = f(x)$ is approximated by the sum of the areas of the n trapezoids.

The area of i-th trapezoid is

$$A_i = \frac{\Delta x}{2}[f(x_{i-1}) + f(x_i)], \qquad i = 1, 2, 3, \ldots, n$$

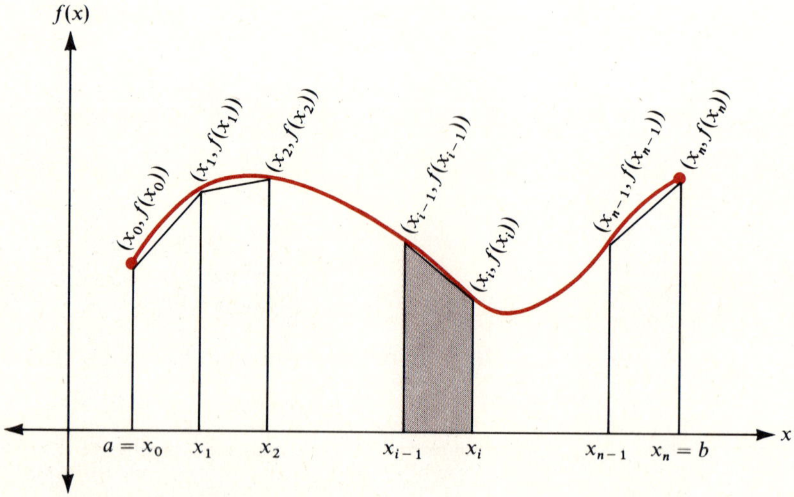

Figure 6.19

* This section may be omitted, if necessary, without any loss of continuity.

and the total area of the n trapezoids is

$$A = \frac{\Delta x}{2}\{[f(x_0) + f(x_1)] + [f(x_1) + f(x_2)] + \cdots + [f(x_{i-1}) + f(x_i)]$$
$$+ \cdots + [f(x_{n-1}) + f(x_n)]\}$$
$$= \frac{\Delta x}{2}[f(x_0) + 2f(x_1) + 2f(x_2) + \cdots + 2f(x_{n-1}) + f(x_n)]$$

Since $\Delta x = (b - a)/n$, we conclude that

6.12 $$\int_a^b f(x)\,dx \approx \frac{b-a}{2n}[f(x_0) + 2f(x_1) + 2f(x_2) + \cdots + 2f(x_{n-1}) + f(x_n)]$$

Example 6.44

Compute

$$\int_0^1 \frac{dx}{1+x}$$

using the trapezoidal rule with $n = 5$. Check the result by integrating and finding the exact value.

Solution

Since $a = 0$, $b = 1$, and $n = 5$, we have

$$\frac{b-a}{n} = 0.2$$

Further,

$$\begin{array}{ll} x_0 = 0 & f(x_0) = 1 \\ x_1 = 0.2 & f(x_1) = 0.833333 \\ x_2 = 0.4 & f(x_2) = 0.714825 \\ x_3 = 0.6 & f(x_3) = 0.625 \\ x_4 = 0.8 & f(x_4) = 0.555555 \\ x_5 = 1.0 & f(x_5) = 0.5 \end{array}$$

Using (6.12), we get

$$\int_0^1 \frac{dx}{x+1} \approx \frac{0.2}{2}[1.0 + 1.66667 + 1.42857 + 1.25 + 1.11111 + 0.5]$$
$$= (0.1)(6.95635)$$
$$= 0.695635$$

The exact value is

$$\int_0^1 \frac{dx}{x+1} = \ln(x+1)\Big|_0^1 = \ln 2 = 0.69314$$

Next, we illustrate Simpson's rule. The interval $[a, b]$ is divided into an *even* number of equal subintervals. Again, $x_0, x_1, x_2, \ldots, x_n$ are the points of subdivision and we have, as before, n closed subintervals

$$[x_0, x_1], [x_1, x_2], [x_2, x_3], [x_3, x_4], \ldots, [x_{n-1}, x_n]$$

each of length $\Delta x = (b - a)/n$. The intervals are considered in pairs so that each portion of the curve covers a subinterval of length $2(\Delta x)$. For the first pair of intervals, $[x_0, x_1]$ and $[x_1, x_2]$, the points on the curve are $(x_0, f(x_0))$, $(x_1, f(x_1))$, and $(x_2, f(x_2))$. The area under the curve $y = f(x)$ from $x = x_0$ to $x = x_2$ is approximated by the parabolic (dotted) arc

$$y = ax^2 + bx + c, \qquad a \neq 0$$

rather than the line segments as in the trapezoidal rule (see Figure 6.20). It can be shown that the area under the parabolic arc from x_0 to x_2 is

$$\int_{x_0}^{x_2} (ax^2 + bx + c)\,dx = \frac{\Delta x}{3}\,[f(x_0) + 4f(x_1) + f(x_2)]$$

Similarly, the area under the parabolic arc from x_2 to x_4 is given by

$$\int_{x_2}^{x_4} (ax^2 + bx + c)\,dx = \frac{\Delta x}{3}\,[f(x_2) + 4f(x_3) + f(x_4)]$$

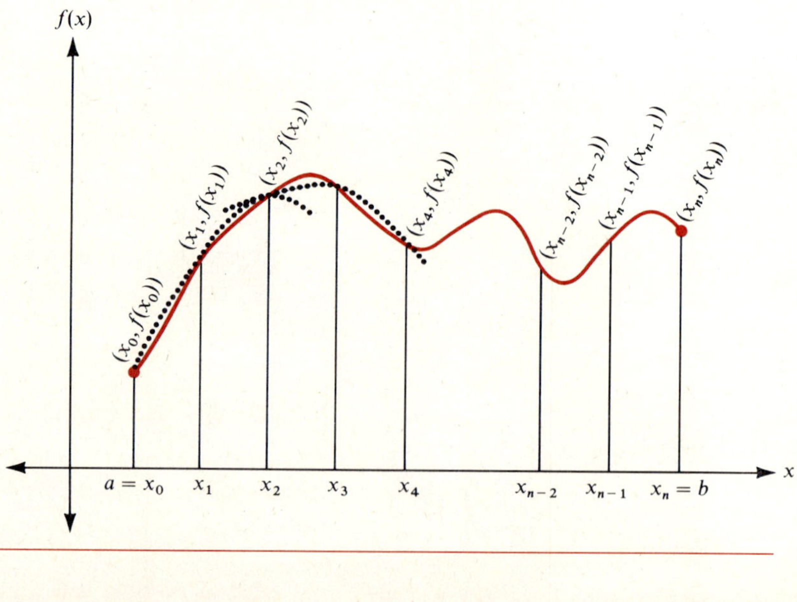

Figure 6.20

and so on. Finally, the area of the last piece of the curve $y = f(x)$ is approximated by the parabolic arc $y = ax^2 + bx + c$ from x_{n-2} to x_n and is given by

$$\int_{x_{n-2}}^{x_n} (ax^2 + bx + c)\,dx = \frac{\Delta x}{3}[f(x_{n-2}) + 4f(x_{n-1}) + f(x_n)]$$

The total area under all the parabolic arcs is

$$\begin{aligned} A &= \frac{\Delta x}{3}[f(x_0) + 4f(x_1) + f(x_2) + f(x_2) + 4f(x_3) + f(x_4) \\ &\quad + \cdots + f(x_{n-2}) + 4f(x_{n-1}) + f(x_n)] \\ &= \frac{\Delta x}{3}\{f(x_0) + 4[f(x_1) + f(x_3) + \cdots + f(x_{n-1})] + 2[f(x_2) + f(x_4) \\ &\quad + \cdots + f(x_{n-2})] + f(x_n)]\} \end{aligned}$$

which is taken as an approximate value of $\int_a^b f(x)\,dx$. Because $\Delta x = (b - a)/n$, it follows that

6.13
$$\begin{aligned} \int_a^b f(x)\,dx \approx \frac{b-a}{3n}[&y_0 + 4(y_1 + y_3 + \cdots + y_{n-1}) \\ &+ 2(y_2 + y_4 + \cdots + y_{n-2}) + y_n] \end{aligned}$$

where $y_0, y_1, \ldots, y_n$ are the ordinates of the curve $y = f(x)$ at the points of subdivision $x_0, x_1, \ldots, x_n$.

This is Simpson's rule. It generally gives a better approximation than the trapezoidal rule.

Example 6.45

Compute

$$\int_0^1 \frac{dx}{x+1}$$

using Simpson's rule with $n = 4$.

Solution

Here $a = 0$, $b = 1$, and $n = 4$; consequently $\Delta x = 0.25$. Further, we have

$$\begin{aligned} x_0 &= 0; & y_0 &= f(x_0) = 1 \\ x_1 &= 0.25; & y_1 &= f(x_1) = 0.8 \\ x_2 &= 0.50; & y_2 &= f(x_2) = 0.666667 \\ x_3 &= 0.75; & y_3 &= f(x_3) = 0.571428 \\ x_4 &= 1.00; & y_4 &= f(x_4) = 0.5 \end{aligned}$$

Using (6.13), we obtain

$$\int_0^1 \frac{dx}{x+1} \approx \frac{0.25}{3}[1 + 4(0.8 + 0.571428) + 2(0.666667) + 0.5]$$

$$= \frac{0.25}{3}(8.319046)$$

$$= 0.69325$$

Exercises

In problems 1–15, use both the (a) trapezoidal rule and (b) Simpson's rule to approximate the given integral to 5 decimal places.

1. $\int_0^1 x\,dx, \quad n = 4$

2. $\int_0^3 x\,dx, \quad n = 6$

3. $\int_0^3 (x+2)\,dx, \quad n = 4$

4. $\int_2^4 (2x+3)\,dx, \quad n = 4$

5. $\int_1^3 x^2\,dx, \quad n = 4$

6. $\int_1^2 (x^2+1)\,dx, \quad n = 4$

7. $\int_0^1 x^3\,dx, \quad n = 4$

8. $\int_0^3 x^2(1+x)\,dx, \quad n = 6$

9. $\int_0^1 \sqrt{x}\,dx, \quad n = 4$

10. $\int_0^1 \sqrt{x}\,dx, \quad n = 8$

11. $\int_4^8 \frac{dx}{x}, \quad n = 4$

12. $\int_2^{10} \frac{dx}{x+1}, \quad n = 8$

13. $\int_0^1 \frac{dx}{1+x^2}, \quad n = 4$

14. $\int_0^1 \frac{dx}{1+x^2}, \quad n = 8$

15. $\int_0^1 2x\sqrt{x^2+1}\,dx, \quad n = 4$

6.9 Differential Equations

Differential equations provide a powerful tool for solving many problems in biology, chemistry, economics, paleontology, physics, and mathematics. The solutions to differential equations also play an important role in the study of the planets and satellites. Since the subject of differential equations can be extremely difficult, in this section we consider only the simplest type of elementary differential equation.

Definition 6.3

A *differential equation* is an equation that involves differentials or derivatives of an unknown function.

Thus

$$\frac{dy}{dx} = 2x + 3$$

$$x\,dx + y\,dy = 0$$

and

$$(x^2 + y^2)\,dx - 2xy\,dy = 0$$

are all examples of differential equations.

Definition 6.4 A function $y = g(x)$ is a *solution of a differential equation* if the equation is satisfied when y and its derivatives are replaced throughout by $g(x)$ and its corresponding derivatives.

Examples 6.46 The function $y = cx^2$ is a solution of the differential equation

6.14 $$x\frac{dy}{dx} = 2y$$

because if we substitute cx^2 for y and $2cx$ for dy/dx in (6.14), we get

$$x(2cx) = 2cx^2 = 2y$$

In computing antiderivatives of functions in Section 6.2, we were actually solving differential equations. For instance, in Example 6.9 we found that

$$\int (3x^2 + 2x + 1)\,dx = x^3 + x^2 + x + c$$

where c is an arbitrary constant. Thus the function

$$y = x^3 + x^2 + x + c$$

is a solution of the differential equation

$$\frac{dy}{dx} = 3x^2 + 2x + 1$$

In general, a solution to the differential equation

6.15 $$\frac{dy}{dx} = f(x)$$

is given by

6.16 $$y = \int f(x)\,dx + c$$

where c is an arbitrary constant. The solution (6.16) is called a *general solution* of the differential equation (6.15). A solution obtained by assigning a particular value to the constant c is called a *particular solution.*

In solving applied problems, we usually have additional information about x and y that allows us to find the value of c.

Example 6.47

The population N of a certain town is increasing according to the relationship

$$\frac{dN}{dt} = 400 + 100e^t$$

where t is measured in years. Determine $N(t)$, the size of the population at time t, assuming $N(0) = 500$.

Solution

First, we find the general solution of the differential equation

$$\frac{dN}{dt} = 400 + 100e^t$$

and then evaluate the constant using the information that $N(0) = 500$. Observe that

$$\begin{aligned} N(t) &= \int (400 + 100e^t)\,dt \\ &= 400t + 100e^t + c \end{aligned}$$

where c is a constant. Since $N(0) = 500$, we have

$$\begin{aligned} 500 &= 400(0) + 100e^0 + c \\ c &= 400 \end{aligned}$$

Hence the particular solution is given by

$$N(t) = 100(4t + e^t + 4)$$

Let us now consider the differential equation

6.17 $$P(x, y) + Q(x, y)\frac{dy}{dx} = 0$$

where $P(x, y)$ and $Q(x, y)$ are functions of two variables. The equation (6.17) may be expressed in the differential form

6.18 $$P(x, y)\,dx + Q(x, y)\,dy = 0$$

Fortunately, a wide class of differential equations can be simplified by collecting all y terms with dy and all x terms with dx. Thus if

$$P(x, y) = A(x)B(y)$$
$$Q(x, y) = C(x)D(y)$$

then we may rewrite (6.18) as

6.19 $$A(x)B(y)\,dx + C(x)D(y)\,dy = 0$$

where $A(x)$ and $C(x)$ are functions of x and $B(y)$ and $D(y)$ are functions of y.

Definition 6.5 A differential equation that can be written in the form (6.19) is said to be a *separable* differential equation.

If we let

$$u(x, y) = \frac{1}{B(y)C(x)}$$

and multiply (6.19) by $u(x, y)$, we obtain

$$\frac{A(x)}{C(x)}dx + \frac{D(y)}{B(y)}dy = 0$$

or

6.20 $$f(x)\,dx + g(y)\,dy = 0$$

where

$$f(x) = \frac{A(x)}{C(x)} \quad \text{and} \quad g(y) = \frac{D(y)}{B(y)}$$

Integrating (6.20) term by term, we get

$$\int f(x)\,dx + \int g(y)\,dy = c$$

where c is some arbitrary constant.

Example 6.48 Solve the differential equation

$$\frac{dy}{dx} = e^{x-y}$$

Solution

Changing to the differential form (6.18) and separating the variables, we get

$$e^y\,dy - e^x\,dx = 0$$

Integrating term by term, we get

$$e^y - e^x = c$$

where c is a constant.

Example 6.49 Solve the differential equation

6.21 $$(y+1)\,dx - 2x\,dy = 0$$

Solution

Comparing (6.21) with (6.19), we see that

$$A(x) = 1, \quad B(y) = y + 1, \quad C(x) = x, \quad D(y) = -2$$

Let

$$u(x, y) = \frac{1}{x(y+1)}$$

Multiplying (6.21) by $u(x, y)$, we have

$$\frac{dx}{x} - \frac{2dy}{y+1} = 0$$

Integrating the above equation term by term, we have

$$\ln x - 2\ln(y+1) = c_1$$

$$\ln x - \ln(y+1)^2 = c_1$$

$$\ln \frac{x}{(y+1)^2} = c_1$$

$$x = c(y+1)^2$$

where $c = \ln c_1$.

There are many situations in which a variable y changes with time t according to the equation

6.22 $$\frac{dy}{dt} = ky$$

where k is a constant. Such situations occur frequently in biology, chemistry, economics, and the social sciences. Separating variables, (6.22) becomes

$$\frac{dy}{y} = k\,dt$$

Integrating both sides, we have

$$\ln y = kt + c$$

If we suppose that $y = y_0$ at time $t = 0$, then

$$\ln y_0 = c$$

and

$$\ln y = kt + \ln y_0$$

$$\ln y - \ln y_0 = kt$$

$$\ln \frac{y}{y_0} = kt$$

This is equivalent to

$$\frac{y}{y_0} = e^{kt}$$

6.23 $$y = y_0 e^{kt}$$

The variable y decreases with increasing time if $k < 0$ and increases if $k > 0$.

Example 6.50

The number of freshmen students who drop out of a certain university is proportional to the number of students enrolled. If there were 3,000 freshmen enrolled in a given semester and 10 percent dropped out after 4 weeks, determine the number of students who remained after 12 weeks. Assume that $\ln 0.9 = -0.10536$ and $e^{-0.31608} = 0.72900$.

Solution

First we must evaluate the constants y_0 and k in (6.23). Using the fact that $y = 3{,}000$ at time $t = 0$, we obtain

$$3{,}000 = y_0 e^0 = y_0$$

Thus

$$y = 3{,}000e^{kt}$$

We now use the information that after 4 weeks, the number of remaining students was

$$(0.90) \cdot (3{,}000) = 2{,}700$$

This means that

$$3{,}000e^{4k} = 2{,}700$$

$$e^{4k} = \frac{2{,}700}{3{,}000} = 0.90$$

$$4k = \ln 0.9$$

$$k = -0.02634$$

Hence,

$$y = 3{,}000e^{-0.02634t}$$

When $t = 12$,

$$\begin{aligned} y &= 3{,}000e^{(-0.02634)12} \\ &= 3{,}000e^{(-0.31608)} \\ &= 3{,}000(0.72900) \\ &= 2259.90 \end{aligned}$$

The number of students remaining after 12 weeks is approximately 2,260.

Example 6.51

A rumor tends to spread through a community of population N at a rate directly proportional to the number x of residents who have heard the rumor and the number $N - x$ of residents who have not yet heard it. In other words,

6.24
$$\frac{dx}{dt} = kx(N - x)$$

a. Solve this differential equation assuming that at $t = 0$ (measured in days), $x = 1$

b. If $N = 500$ and $k = 0.001$, determine the number of residents who have heard the rumor after 10 days.

Solution

a. Separating the variables, equation (6.24) becomes

$$\frac{dx}{x(N - x)} = k\,dt$$

Integrating both sides, we obtain

6.25
$$\int \frac{dx}{x(N - x)} = kt + c_1$$

where c_1 is some constant. To integrate

$$\int \frac{dx}{x(N - x)}$$

we use the following algebraic trick. Observe that

$$\frac{1}{x(N-x)} = \frac{1}{N}\left(\frac{1}{x} + \frac{1}{N-x}\right)$$

Thus,

$$\begin{aligned}\int \frac{dx}{x(N-x)} &= \frac{1}{N}\int \frac{dx}{x} + \frac{1}{N}\int \frac{dx}{N-x}\\ &= \frac{1}{N}\ln x - \frac{1}{N}\ln(N-x) + k_1\\ &= \frac{1}{N}\ln\left(\frac{x}{N-x}\right) + k_1\end{aligned}$$

where k_1 is some constant. Returning to (6.25), we have

$$\begin{aligned}\frac{1}{N}\ln\left(\frac{x}{N-x}\right) + k_1 &= kt + c_1\\ \ln\left(\frac{x}{N-x}\right) &= Nkt + N(c_1 - k_1)\\ \frac{x}{N-x} &= ce^{Nkt} \qquad [c = e^{N(c_1-k_1)}]\end{aligned}$$

To determine the constant c, we use the information that at time $t = 0$, $x = 1$. This yields

$$\frac{1}{N-1} = c$$

Thus we obtain

$$\begin{aligned}\frac{x}{N-x} &= \frac{1}{N-1}e^{Nkt}\\ (N-1)x &= (N-x)e^{Nkt}\\ x[(N-1) + e^{Nkt}] &= Ne^{Nkt}\\ x &= \frac{Ne^{Nkt}}{(N-1) + e^{Nkt}}\end{aligned}$$

Dividing the numerator and denominator by e^{Nkt}, we get

$$x = \frac{N}{1 + (N-1)e^{-Nkt}}$$

b. Setting $N = 500$, $k = 0.001$, and $t = 10$, we have

$$\begin{aligned}x &= \frac{500}{1 + 499e^{-5}}\\ &= 114.62\end{aligned}$$

Approximately 115 residents have heard the rumor after 10 days.

Exercises

In problems 1–4, show that the equation is the solution of the corresponding differential equation.

1. $y = x^2 + x + c$ $\qquad \dfrac{dy}{dx} = 2x + 1$

2. $y = cx^3$ $\qquad x\dfrac{dy}{dx} = 3y$

3. $y = x^3 - x^2 + c$ $\qquad \dfrac{dy}{dx} = 3x^2 - 2x$

4. $\ln y + \dfrac{x}{y} = c$ $\qquad (y - x)\dfrac{dy}{dx} + y = 0$

In problems 5–12, find the solution for each differential equation.

5. $\dfrac{dy}{dx} = 3x^2 + 2x + 5$

6. $\dfrac{dy}{dt} = 4t^3 + 3t^2 + 2t$

7. $\dfrac{dy}{dx} = e^{-x} + x$

8. $\dfrac{dy}{dt} = t + 1$

9. $\dfrac{dy}{dt} = te^t$

10. $\dfrac{dy}{dt} = t^2e^t$

11. $\dfrac{dy}{dt} = 2t\sqrt{t^2 + 1}$

12. $\dfrac{dy}{dt} = 3t\sqrt{t^2 + 100}$

In problems 13–16, find the particular solution for each differential equation.

13. $\dfrac{dy}{dx} = x + 1;$ $\qquad y = 25$ when $x = 8$

14. $\dfrac{dy}{dx} = x^3;$ $\qquad y = 6$ when $x = 2$

15. $\dfrac{dy}{dx} = xe^x;$ $\qquad y = 5$ when $x = 1$

16. $\dfrac{dy}{dx} = \dfrac{2}{x};$ $\qquad y = 1$ when $x = 1$

In problems 17–30, separate the variables and solve the differential equation.

17. $\dfrac{dy}{dx} = \dfrac{y}{x}$

18. $\dfrac{dy}{dx} = x^2y$

19. $\dfrac{dy}{dx} = xy$

20. $\dfrac{dy}{dx} = \dfrac{y+1}{2xy}$

21. $4y\dfrac{dy}{dx} + 3x = 0$

22. $(4 + x)\dfrac{dy}{dx} = y^3$

23. $x\,dx + y\,dy = 0$

24. $x^2\,dx + y^2\,dy = 0$

25. $x\,dy + y\,dx = 0$

26. $x\,dy + (y - 2)\,dx = 0$

27. $e^x\,dy + e^y\,dx = 0$

28. $2(y + 3)\,dx - xy\,dy = 0$

29. $x^2y\,dy - e^y\,dx = 0$

30. $e^x(y-1)\,dx + 2(e^x+4)\,dy = 0$

31. Let A be the amount to which \$100 will grow in t years if it is invested at 6 percent per year compounded continuously. Assume that the rate of growth of A is

$$\frac{dA}{dt} = 0.06A$$

Find what effective rate of interest is earned in one year.

32. Susan deposits \$3,000 in a bank giving 6 percent annual interest compounded continuously. Determine the amount she will have in her account at the end of

a. two years **b.** three years **c.** five years

33. In 1960, the population of a city was 80,000. In 1980, it was 100,000. If the population increases according to

$$\frac{dy}{dt} = ky$$

what will be the population in 2010? Assume that $\ln 1.25 = 0.2231$ and $e^{0.5578} = 1.7468$.

34. The number N of bacteria in a certain culture is increasing according to

$$\frac{dN}{dt} = 0.02N$$

If $N = 5{,}000$ at time $t = 0$, find N when $t = 10$.

35. The number y of bacteria in a culture is increasing according to

$$\frac{dy}{dt} = ky$$

If there are 5,000 bacteria at time $t = 0$ and 10,000 at time $t = 2$, find the number of bacteria at $t = 5$. (Use $\ln 2 = 0.6931$, $e^{1.733} = 5.66$.)

36. The number y of bacteria in a certain culture is increasing according to

$$\frac{dy}{dt} = ky$$

If there are 10^5 bacteria at time $t = 0$ and $3(10^5)$ at $t = 2$, find the number of bacteria at $t = 6$. (Use $\ln 3 = 1.0986$, $e^{3.2958} = 26.999$.)

37. The population of the United States was 150 million in 1950 and 180 million in 1960. Assuming that the population increases at a rate proportional to itself, estimate the population for the year 1990. (Use $\ln 1.2 = 0.1823$, $e^{0.7292} = 2.0734$.)

38. The rate of change of barometric pressure above sea level is proportional to the barometric pressure. If the barometer reads 30 inches at sea level and 25 inches at 4,000 feet, determine the barometric pressure at 20,000 feet. (Use $e^{-0.912} = 0.4017$.)

39. A radioactive substance disintegrates at a rate proportional to the quantity Q of the substance that exists at time t.

a. Determine a formula for the quantity Q remaining after time t assuming that the initial quantity of the substance is Q_0.

b. How long will it take for half of the substance to decay?

40. A radioactive substance decays at a rate proportional to the quantity Q of the substance that exists at time t. Assuming that one-half of the substance decomposes in 2,000 years, determine the quantity remaining after 1,000 years if 50 grams were present at time $t = 0$. (Use $e^{-0.34655} = 0.7071$.)

41. A rumor tends to spread through a community of population 800 at a rate directly proportional to the number x of residents who have heard the rumor and the number $800 - x$ of residents who have not yet heard it. Thus the spreading rumor satisfies the differential equation

$$\frac{dx}{dt} = kx(800 - x)$$

a. Solve this differential equation assuming that at $t = 0$ (measured in days), $x = 1$.

b. Assuming that $k = 0.001$, determine the number of residents who have heard the rumor after 5 days.

42. The number of people in a small community of 500 that are coming down with a certain type of influenza is directly proportional to the number x of residents who are currently infected and the number $500 - x$ of residents who have not been infected. In other words, the influenza infection satisfies the differential equation

$$\frac{dx}{dt} = kx(500 - x)$$

a. Solve this differential equation, given that at time $t = 0$ (measured in days), $x = 1$

b. Determine the number of residents who have been infected after 15 days assuming that the constant of proportionality $k = 0.001$.

Exercises for Review

1. Define each of the following concepts.

a. antiderivative of a function
b. indefinite integral
c. definite integral
d. consumer's surplus
e. producer's surplus
f. differential equation
g. solution of a differential equation
h. separable differential equation

2. State the *Fundamental Theorem of Calculus.*

3. Find each of the following indefinite integrals and verify your results by differentiation.

a. $\int xt\,dt$

b. $\int xt\,dx$

c. $\int \left(x^2 - 4x + \frac{1}{x}\right) dx$

d. $\int (e^x + \sqrt{x})\,dx$

e. $\int \sqrt{2x + 5}\,dx$

f. $\int \frac{dx}{\sqrt{3x + 4}}$

g. $\int \frac{dx}{3x + 4}$

h. $\int (2x + 3)e^{x^2 + 3x}\,dx$

i. $\int x\sqrt{x - 1}\,dx$

j. $\int \frac{e^x\,dx}{1 + e^x}$

k. $\int x\sqrt{x^2 + 4}\,dx$

l. $\int (e^x + x)^2(e^x + 1)\,dx$

m. $\int x \ln x\,dx$

n. $\int xe^x\,dx$

o. $\int xe^{-2x}\,dx$

p. $\int x(1 + x)^4\,dx$

4. Evaluate each of the following definite integrals.

a. $\int_0^1 (x^2 + x + 1)\,dx$

b. $\int_2^4 x(x + 1)\,dx$

c. $\int_0^2 (x^3 + 3x^2 + 2x + 4)\,dx$

d. $\int_2^5 (3e^x + 4)\,dx$

e. $\int_{-2}^3 (6 + x - x^2)\,dx$

f. $\int_8^{24} \frac{dx}{\sqrt{1 + x}}$

g. $\int_0^3 3x(x^2 + 16)^{1/2}\,dx$

h. $\int_{-1}^1 xe^x\,dx$

i. $\int_3^5 x \ln x\,dx$

j. $\int_3^5 x\sqrt{x^2 - 9}\,dx$

5. Evaluate the following integrals by dividing the closed interval $[a, b]$ into n subintervals of equal length.

a. $\int_1^2 (x + 4)\,dx$

b. $\int_1^3 (2x + 3)\,dx$

c. $\int_1^3 x(x - 2)\,dx$

d. $\int_2^4 x^2\,dx$

6. Use (a) the trapezoidal rule and (b) Simpson's rule to approximate each of the following definite integrals, to five decimal places.

a. $\int_1^5 \frac{dx}{x + 2}$, $n = 4$

b. $\int_0^1 \sqrt{x + 1}\,dx$, $n = 8$

c. $\int_0^3 \sqrt{x^2 + 1}\,dx$, $n = 6$

7. Find the general solution for each of the following differential equations.

a. $\frac{dy}{dx} = 3x^2 + 2x + 1$

b. $\frac{dy}{dt} = e^{-t}$

c. $\frac{dy}{dx} = xe^x$

d. $y\frac{dy}{dx} = xe^x + 2$

e. $\frac{dy}{dx} = ky$

f. $\frac{dy}{dx} = x^2y^3$

g. $\frac{dy}{dx} = \frac{2y}{x + 1}$

h. $2x^2y\frac{dy}{dx} - (y^2 + 1) = 0$

i. $2x\,dx - 3y^2\,dy = 0$

j. $x\,dy + y\,dx = 0$

8. Find a particular solution for each of the following differential equations.

a. $\frac{dy}{dx} = 4x^3 + 3x^2 + 2x + 1$; $y = 8$ when $x = 1$

b. $\frac{dy}{dx} = \frac{x}{y^2}$; $y = 3$ when $x = 2$

9. Find the area of the region bounded by each set of given curves.

a. $y = 2x + 3, \quad x = 2, x = 4, y = 0$
b. $y = x^2 - 6x + 5, \quad y = 0$
c. $y = x + 2, \quad y = x^2$
d. $y = 2x + 1, \quad y = x^2 + 1$
e. $y = 2x, \quad y = 6x - x^3$

10. Find the equation of the curve $y = f(x)$ which passes through the point (2, 12) and for which the slope of the curve is $3x^2 - 2x + 1$.

11. The marginal cost of production for a firm is

$$C'(x) = 30 + 4x - 0.01x^2$$

where x represents the quantity produced. Determine the cost function $C(x)$ assuming that the fixed cost is \$500.

12. The velocity of an object after t hours is

$$v(t) = 16 - 10(t + 1)^{-2}$$

where $v(t)$ is recorded in miles per hour. How far does the object travel during the first two hours?

13. The instantaneous population growth rate in a certain town is estimated to be

$$P'(t) = 200 + 300t^{2/3}$$

Determine the population $P(t)$ in t years assuming that the current population of the town is 15,000. What will be the population in 8 years?

14. The marginal cost and the marginal revenue functions for a product are

$$C'(x) = 40 + \frac{x}{12}$$

and

$$R'(x) = 45 + \frac{x}{15}$$

where x is the number of units produced. Determine the maximum profit assuming that the fixed cost to the firm is \$500.

15. Student enrollment in professional courses in a certain state college is expected to increase at a rate of

$$N'(t) = 3{,}000(t + 4)^{-3/2}$$

where $N(t)$ is the number of students enrolled at time t. Determine the projected enrollment 5 years from now if the current college enrollment is 4,000.

16. A factory discharges pollutants into a lake at a rate approximated by

$$P'(t) = (2t - 1)\sqrt{t^2 - t + 1}$$

where $P(t)$ is the number of tons of pollutants discharged into the lake after t years of operation. Determine the quantity of pollutants discharged into the lake during the first four years of the factory's operation.

17. Under pure competition, the quantity demanded and the corresponding price are determined by the demand and supply curves

$$D(x) = 900 - 20x - x^2 \quad \text{and} \quad S(x) = x^2 + 10x$$

Determine the consumer's surplus and the producer's surplus.

18. Kathy deposited \$2,500 in a savings bank at 9.5 percent annual interest, compounded continuously. Determine the amount she will have in her account at the end of

a. three years **b.** four years

19. The number of bacteria in a culture is increasing each day at a rate equal to 10 percent of its population. If there are 15,000 bacteria initially, find the number of bacteria after 10 days.

20. A certain substance in a solution decomposes at a rate proportional to its weight. If 9 grams of the substance are initially present and 1 gram decomposes in the first hour, how long will it take before 5 grams of the substance decompose?

21. A radioactive substance decays at a rate proportional to the amount present. Assuming that its half-life is 5 years, how long will it be until only 1 percent of the original substance remains?

22. A drug injected into the bloodstream of a patient loses its concentration according to the relationship

$$\frac{dQ}{dt} = -0.06Q$$

where t is recorded in hours. Solve this differential equation, given that 0.3 grams of the drug was injected initially. How much drug will remain in the bloodstream after 5 hours?

23. In a small community of 1,200, the number of residents who are catching a certain type of influenza is directly proportional to the number x of residents who are currently infected and the number $1{,}200 - x$ of residents who have not yet been infected. Thus the influenza infection satisfies the differential equation

$$\frac{dx}{dt} = kx(1{,}200 - x)$$

a. Solve this differential equation, given that at time $t = 0$ (measured in days) $x = 1$.

b. Determine the number of residents infected after 10 days assuming that $k = 0.0005$.

7 Functions of Several Variables

7.1 Introduction

Most of the functions we have encountered were expressed by the equation $y = f(x)$. This means that we had an explicit description of a variable y in terms of a single variable x. Practical applications, however, involve functions of several variables more frequently than functions of one variable. For example, the cost of a product may depend on the several factors of raw materials' price, labor wages, overhead expenses, and shipping costs. The sale of a new car may depend on the sticker price, the price of competing automobiles, the amount spent on advertising, the company rebate, and other factors. Dividends paid by a corporation may depend on the corporation's earnings, the potential growth of its products, and other financial conditions. The salary of an employee in a firm may depend on educational background, periodic job ratings, and length of service.

In this chapter we shall examine many of the concepts of calculus that we studied for one variable and extend them to more than one variable. We shall begin by extending the rectangular coordinate system to discuss functions of more than one variable, and then we shall develop the concepts of limits, continuity, partial derivatives, and maxima and minima for functions of several variables. Finally, we shall introduce the use of Lagrange multipliers and the method of least squares to solve some practical problems.

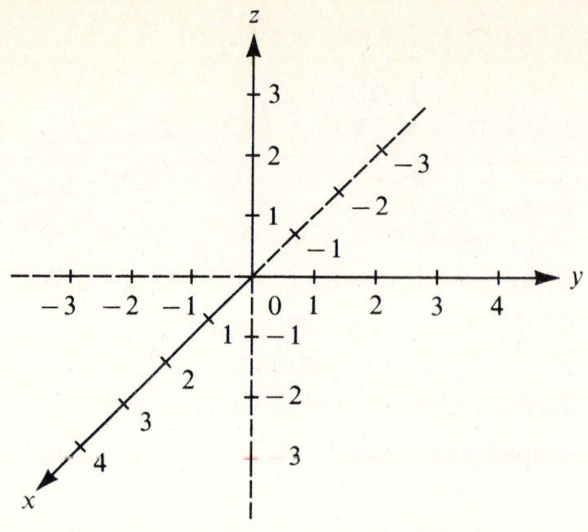

Figure 7.1

7.2 Rectangular Coordinates in Space

We have been using a two-dimensional coordinate system to graph functions of one variable. To graph a function of two variables, we use a three-dimensional coordinate system. The x-, y-, and z-axes (shown in Figure 7.1) are mutually perpendicular and intersect at a point 0, called the *origin*. These axes determine three mutually perpendicular coordinate planes: the xy-plane, determined by the x- and y-axes, the yz-plane, by the y- and z-axes, and the zx-plane, by the z- and x-axes. An arbitrary unit of length is used to mark a number scale on the x-axis, y-axis, and z-axis, as shown in Figure 7.1.

In this system, the three coordinate planes divide space into eight *octants*, analogous to the two coordinate axes which divide the plane into four quadrants. The first octant is the one in which the three coordinates are positive.

We are now ready to set up a one-to-one correspondence between the set of *ordered triples* (x, y, z) of real numbers and the points in space. If P is a point in space, we draw three planes through P, one perpendicular to the x-axis, one perpendicular to the y-axis and one perpendicular to the z-axis. Suppose that the planes intersect the x-axis in the point with coordinate a, the y-axis in the point with coordinate b and the z-axis in the point with coordinate c. Then the ordered triple associated with the point P is (a, b, c), as shown in Figure 7.2. (We are assuming that a, b, and c are all positive.) The number a is the x-coordinate of P, b is the y-coordinate of P and c is the z-coordinate of P. Thus, if $(1, 4, 3)$ represents the coordinates of a point Q, then Q is 1 unit from the yz-plane, 4 units from the zx-plane and 3 units from the xy-plane. One can visualize that any point in the xy-plane has z-coordinate 0, any point in the yz-plane has x-coordinate 0, and any point in zx-plane has y-coordinate 0.

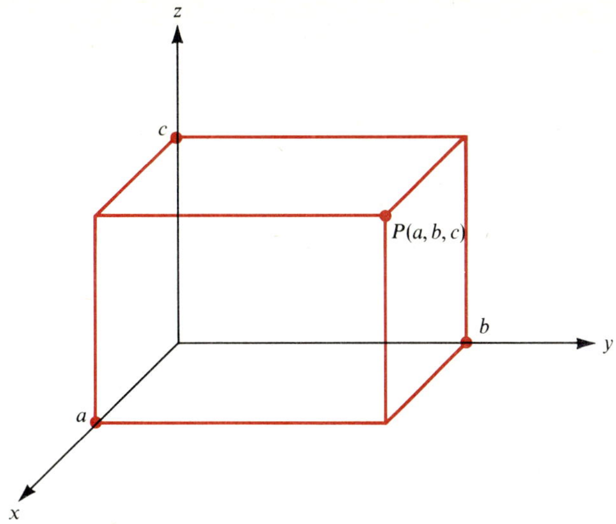

Figure 7.2

Example 7.1

Draw a three-dimensional system and plot the points $P(4, 2, 5)$ and $Q(-2, -3, -4)$.

Solution

To locate point P, we first count off four units on the positive x-axis, then move two units parallel to the positive y-axis, and then measure five units in the direction of positive z-axis. The plotted point P is shown in Figure 7.3.

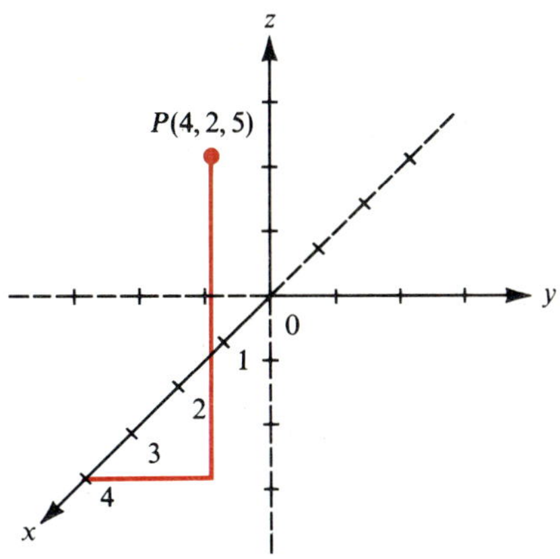

Figure 7.3

To locate the point $Q(-2, -3, -4)$, we measure two units along the negative x-axis, then three units in the direction of the negative y-axis, and finally four units in the negative direction of the z-axis. The point Q is plotted in Figure 7.4.

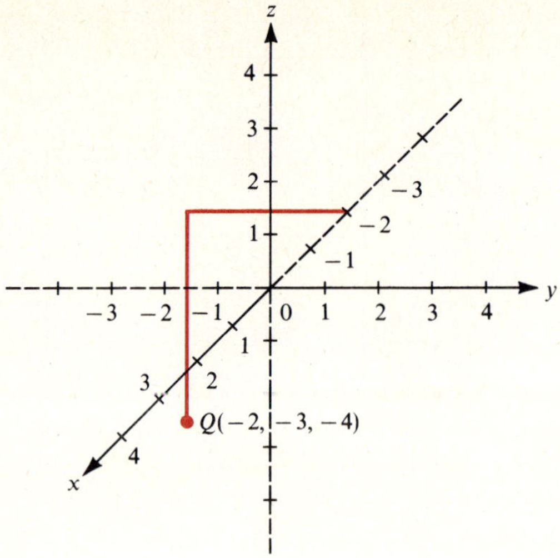

Figure 7.4

The distance between the points (x_1, y_1) and (x_2, y_2) in the xy-plane is

$$\sqrt{(x_2 - x_1)^2 + (y_2 - y_1)^2}$$

The following proposition generalizes this formula to three dimensions.

Proposition 7.1 The distance between the points $P(x_1, y_1, z_1)$ and $Q(x_2, y_2, z_2)$ is

$$\sqrt{(x_2 - x_1)^2 + (y_2 - y_1)^2 + (z_2 - z_1)^2}$$

Proof

Figure 7.5 shows the points P and Q and the rectangular solid they determine with each face parallel to one of the three coordinate planes. Since the line segments PR and QR are perpendicular, the triangle PQR is a right triangle. Using the Pythagorean Theorem, we have

7.1 $$PQ^2 = PR^2 + QR^2$$

Using the Pythagorean Theorem again, we have

7.2 $$PR^2 = PS^2 + RS^2$$

From (7.1) and (7.2), we get

7.3 $$PQ^2 = PS^2 + RS^2 + QR^2$$

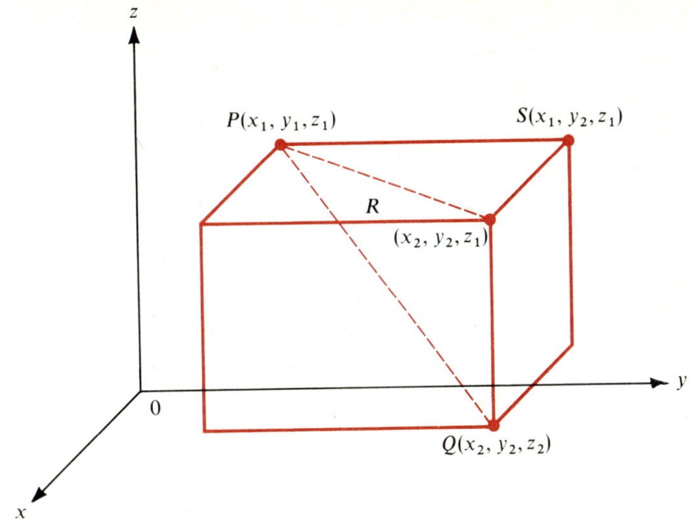

Figure 7.5

Since the x- and z-coordinates of P and S are equal, it follows that

$$PS^2 = (y_2 - y_1)^2$$

Similarly,

$$RS^2 = (x_2 - x_1)^2$$

and

$$QR^2 = (z_2 - z_1)^2$$

Substituting these quantities in (7.3) and rearranging terms, we get

$$PQ^2 = (x_2 - x_1)^2 + (y_2 - y_1)^2 + (z_2 - z_1)^2$$

Hence,

$$PQ = \sqrt{(x_2 - x_1)^2 + (y_2 - y_1)^2 + (z_2 - z_1)^2}$$

Example 7.2

Find the distance between the points $P(4, 3, 1)$ and $Q(7, 5, 2)$.

Solution

Letting $x_1 = 4$, $y_1 = 3$, $z_1 = 1$, and $x_2 = 7$, $y_2 = 5$, and $z_2 = 2$ in Proposition 7.1, we have

$$\begin{aligned} PQ &= \sqrt{(7-4)^2 + (5-3)^2 + (2-1)^2} \\ &= \sqrt{3^2 + 2^2 + 1^2} \\ &= \sqrt{14} \end{aligned}$$

Example 7.3

Find the distance between $P(5, -6, 4)$ and $Q(3, 11, -2)$.

Solution

Letting $x_1 = 5$, $y_1 = -6$, $z_1 = 4$, and $x_2 = 3$, $y_2 = 11$, and $z_2 = -2$, we have

$$\begin{aligned} PQ &= \sqrt{(3-5)^2 + (11-(-6))^2 + (-2-4)^2} \\ &= \sqrt{(-2)^2 + (17)^2 + (-6)^2} \\ &= \sqrt{4 + 289 + 36} \\ &= \sqrt{329} \end{aligned}$$

Definition 7.1

A *sphere* with radius r and center (a, b, c) is the set of all points that are at distance r from the point (a, b, c).

A point (x, y, z) lies *on* the sphere if the distance between it and the center (a, b, c) is r; that is, if

$$\sqrt{(x-a)^2 + (y-b)^2 + (z-c)^2} = r$$

or equivalently,

7.4 $$(x-a)^2 + (y-b)^2 + (z-c)^2 = r^2$$

Equation (7.4) is an algebraic description of a sphere of radius r and center (a, b, c). If the center of the sphere is the origin $(0, 0, 0)$ and the

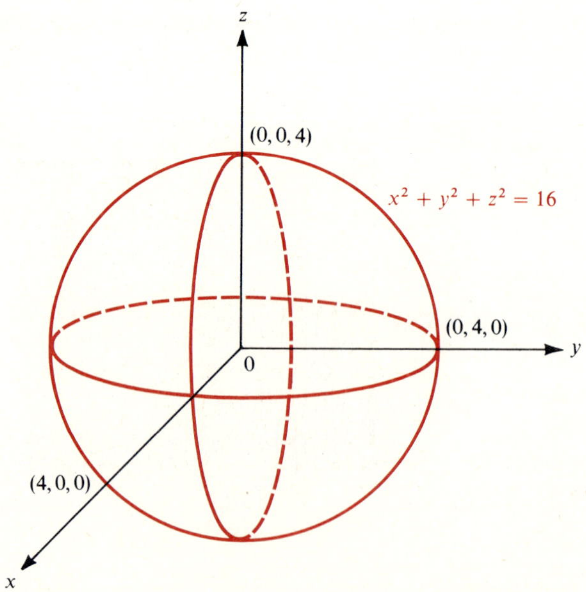

Figure 7.6

radius of the sphere is r, then the corresponding equation is

7.5 $$x^2 + y^2 + z^2 = r^2$$

For instance, the equation

$$x^2 + y^2 + z^2 = 16$$

describes a sphere of radius 4 and center (0, 0, 0). (See Figure 7.6.)

Example 7.4

Find the equation of the sphere with center (2, −3, 4) and radius 6.

Solution

Letting $a = 2$, $b = -3$, and $c = 4$ in (7.4), we obtain

$$(x - 2)^2 + (y + 3)^2 + (z - 4)^2 = 36$$

Definition 7.2

The set of points (x, y, z) that satisfy a given equation in x, y, and z constitutes the *graph* of that equation.

The graph of

$$x^2 + y^2 + z^2 = 9$$

is a sphere of radius 3 and center (0, 0, 0). The graph of $x = 0$ is the yz-plane.

The plane $x = 2$ is the set of all points of the form $(2, y, z)$. This set of points is the plane which is perpendicular to the x-axis and which passes through the point (2, 0, 0) as shown in Figure 7.7. The jagged borders in the diagram suggest that the plane is endless.

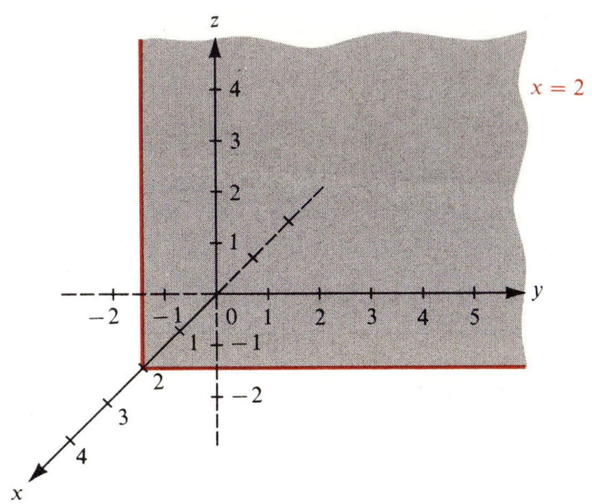

Figure 7.7

Example 7.5

Graph the plane $y = 3$

Solution

The graph consists of all points with y-coordinate equal to 3. The set of points $(x, 3, z)$ is the plane which is perpendicular to the y-axis and which passes through the point $(0, 3, 0)$, as shown in Figure 7.8.

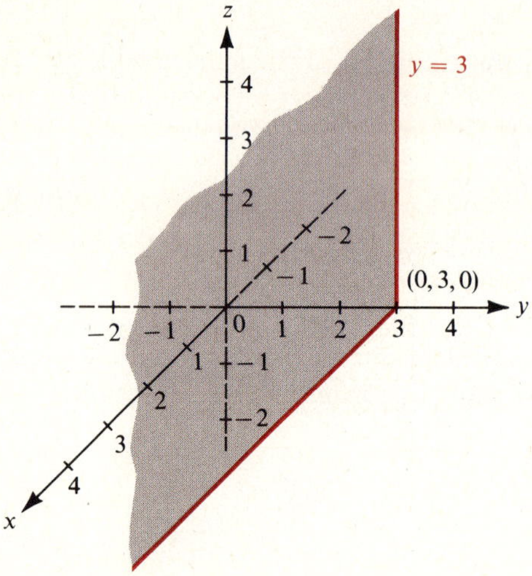

Figure 7.8

Example 7.6

Graph the plane

$$2x + y = 4$$

Solution

If we consider the xy-plane (for which $z = 0$), this equation represents a line. But we get the same line for $z = 1$, $z = 2$, $z = -1$, and so on.

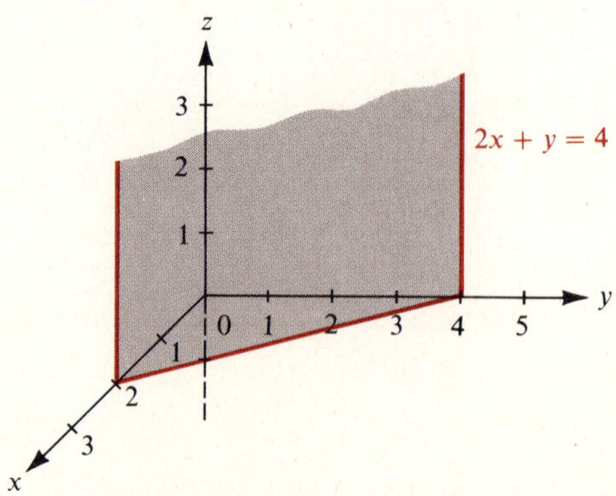

Figure 7.9

Therefore, the graph of this equation is a plane that is parallel to the z-axis and that intersects the xy-plane in the line $2x + y = 4$ (see Figure 7.9).

Example 7.7

Graph the plane

$$6x + 3y + 4z = 12$$

Solution

We need to find some representative points (x, y, z) that satisfy the equation of the plane $6x + 3y + 4z = 12$. The simplest points to find are the intercepts (the points where the plane crosses the coordinate axes), which are $(2, 0, 0)$, $(0, 4, 0)$ and $(0, 0, 3)$. Thus we have the plane shown in Figure 7.10.

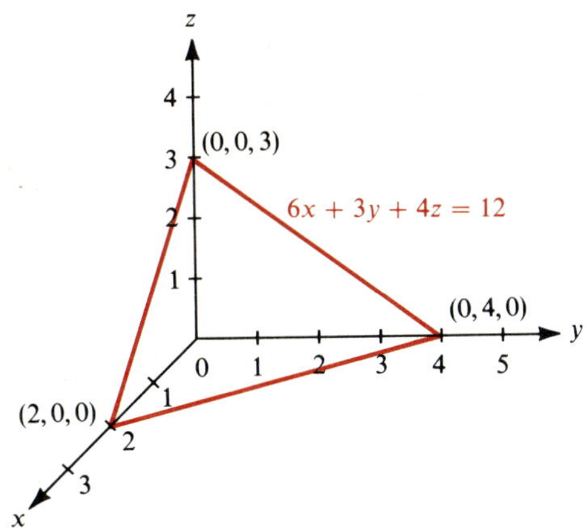

Figure 7.10

Exercises

In problems 1–12, plot the point.

1. $(2, 1, 4)$
2. $(3, -2, 1)$
3. $(-1, -1, 0)$
4. $(-2, -3, 5)$
5. $(-1, 5, 8)$
6. $(1, 0, 2)$
7. $(0, 3, 6)$
8. $(3, 2, -5)$
9. $(5, -2, 3)$
10. $(-2, -1, -3)$
11. $(-1, -4, -6)$
12. $(0, 0, -4)$

In problems 13–20, find the distance between the two points P and Q.

13. $P(1, 2, 3)$, $Q(5, 4, 6)$
14. $P(1, -1, 4)$, $Q(3, 2, -5)$
15. $P(4, -1, 1)$, $Q(-2, -3, 4)$
16. $P(3, -2, 1)$, $Q(-1, -1, 0)$
17. $P(-2, -1, 3)$, $Q(-1, 5, 8)$
18. $P(-1, -2, -3)$, $Q(-6, -8, 10)$
19. $P(-1, -7, -2)$, $Q(-4, 3, 5)$
20. $P(8, -2, -3)$, $Q(-7, -5, 2)$

21. a. Find the equation of the sphere that has center $(1, 2, 3)$ and radius 4.
 b. Is the point $(5, 2, 3)$ on this sphere?
 c. Is the point $(3, 0, 6)$ on this sphere?

22. What are the coordinates of the point where the sphere $x^2 + y^2 + z^2 = 36$ meets
 a. the x-axis b. the y-axis c. the z-axis

23. Find the equation of the sphere with center $(2, -1, 4)$ and radius 5.

24. Find the equation of the sphere with center $(1, 1, -1)$ and radius $\sqrt{2}$.

Graph each of the following planes.

25. $x - 2 = 0$

26. $y + 3 = 0$

27. $z = 4$

28. $x + y = 4$

29. $2x + y = 6$

30. $x + z = 5$

31. $y + z = 3$

32. $x - y = 2$

33. $2y + 3z = 6$

34. $x - 2z = 4$

35. $x + y + z = 1$

36. $2x + 3y + z = 6$

37. $3x - y + 2z = 6$

38. $4x - 2y - 3z = 12$

39. $x - 4y - z = 8$

40. $-2x + 3y - z = 18$

7.3 Functions of Two or More Variables

The volume of a right circular cylinder is given by

$$V = \pi r^2 h$$

where r is the radius of the cylinder and h is its height. We say that the volume V is a function of two variables r and h, and we indicate this using the notation

$$V = f(r, h)$$

As another example, we may consider the equation

$$z = 25x + 30y + 100$$

which describes the cost z of a product (in dollars per unit) in terms of the cost x of raw materials (in dollars per pound) and the cost y of labor (in dollars per hour). Observe that for every pair of values for x and y, there is one and only one value of z. We say that z is a function of x and

y and denote this relationship by the notation

$$z = f(x, y)$$

This leads us to the following definition.

Definition 7.3

Let D be a region in the xy-plane. Then a *function f of two variables* is a rule that associates with every $(x, y) \in D$ exactly one real number. The set D is called the *domain* of f. The set $\{f(x, y): (x, y) \in D\}$, which is the set of values f takes on, is called the *range* of f.

Example 7.8

The equation

$$z = x^2 + xy + y^2$$

defines a function of two variables, x and y, because to each ordered pair of real numbers (x, y) there corresponds one and only one number z such that $z = x^2 + xy + y^2$. The domain D is understood to be the xy-plane. If $x = 3$ and $y = -2$, then

$$\begin{aligned} z &= 3^2 + 3(-2) + (-2)^2 \\ &= 9 - 6 + 4 \\ &= 7 \end{aligned}$$

Thus the function associates the real number 7 with the point $(3, -2)$ in the xy-plane. For $x = -4$ and $y = 5$,

$$\begin{aligned} z &= (-4)^2 + (-4)(5) + 5^2 \\ &= 21 \end{aligned}$$

In function notation, we may express this association as

$$f(-4, 5) = 21$$

Example 7.9

Let D be the xy-plane. For $(x, y) \in D$, we assign the number

$$f(x, y) = x^2 + y^4$$

Since $x^2 + y^4 \geq 0$ for every pair of real numbers (x, y), we see that the range of f is the set of non-negative real numbers.

When the domain D is not specified, we shall take D to be the largest subset of the xy-plane for which $f(x, y)$ is defined. The following two examples illustrate how D may be restricted.

Example 7.10

Find the domain of the function

$$f(x, y) = \sqrt{25 - x^2 - y^2}$$

and find

a. $f(0, 0)$ **b.** $f(4, 0)$ **c.** $f(0, 3)$ **d.** $f(3, 4)$ **e.** $f(6, 0)$

Solution

The function f is defined when the expression under the radical is non-negative. Thus,

$$D = \{(x, y) | x^2 + y^2 \leq 25\}$$

This is a *disc* (a circle and its interior points) in the xy-plane, centered at the origin, with radius 5 (see Figure 7.11).

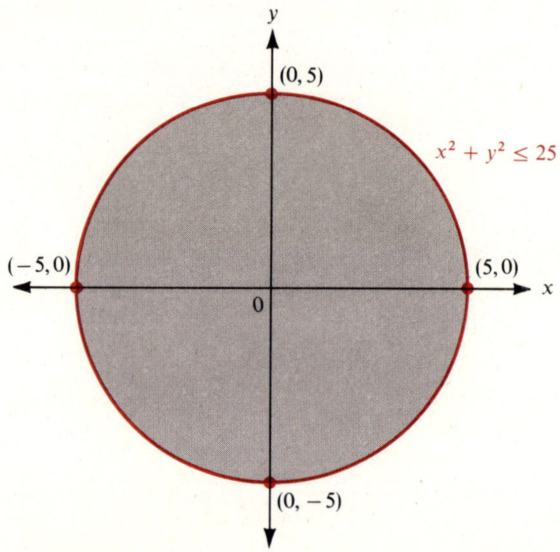

Figure 7.11

Evaluating the function at the specified points, we get

a. $f(0, 0) = \sqrt{25 - 0 - 0} = 5$

b. $f(4, 0) = \sqrt{25 - 16 - 0} = 3$

c. $f(0, 3) = \sqrt{25 - 0 - 9} = 4$

d. $f(3, 4) = \sqrt{25 - 9 - 16} = 0$

e. $f(6, 0)$ is not defined since $(6, 0) \notin D$.

Example 7.11

Find the domain of the function

$$f(x, y) = \frac{x + y}{x - y}$$

and find

a. $f(3,1)$ **b.** $f(4,5)$ **c.** $f(5,8)$ **d.** $f(7,3)$ **e.** $f(1,1)$

Solution

The domain of f is the set of all points in the xy-plane that are off the line $y = x$. (Why?)

a. $f(3,1) = \dfrac{3+1}{3-1} = 2$

b. $f(4,5) = \dfrac{4+5}{4-5} = -9$

c. $f(5,8) = \dfrac{5+8}{5-8} = -\dfrac{13}{3}$

d. $f(7,3) = \dfrac{7+3}{7-3} = \dfrac{5}{2}$

e. $f(1,1)$ is not defined since $(1,1) \notin D$

We can define functions of three variables in a completely analogous manner.

Definition 7.4

Let D be a region in three-dimensional space. A *function f of three variables* is a rule that assigns to each triple $(x, y, z) \in D$ a unique real number, which we denote $f(x, y, z)$. The set D is the *domain* of f and the set $\{f(x, y, z) \mid (x, y, z) \in D\}$, which is the set of values f takes on, is called the *range* of f.

Example 7.12

Find the domain of the function

$$w = f(x, y, z) = \sqrt{16 - x^2 - y^2 - z^2}$$

and calculate

a. $f(4,0,0)$ **b.** $f(2,2,1)$ **c.** $f(3,0,3)$

Solution

The domain

$$D = \{(x, y, z) \mid x^2 + y^2 + z^2 \leq 16\}$$

is the sphere and its interior with center $(0,0,0)$ and radius 4.

a. $f(4,0,0) = \sqrt{16 - 16 - 0 - 0} = 0$

b. $f(2,2,1) = \sqrt{16 - 4 - 4 - 1} = \sqrt{7}$

c. $f(3,0,3)$ is not defined since $(3,0,3) \notin D$

It is natural to ask how one goes about graphing a function of more than one variable. Unfortunately, there is no easy answer to this question. In the discussion that follows, we shall direct our attention to some methods of graphing functions of two variables.

Definition 7.5

The *graph* of a function f of two variables is the representation of the set of all points (x, y, z), where $z = f(x, y)$ for all (x, y) in the domain. In three-dimensional space, this graph is called a *surface.*

Example 7.13

Sketch the surface whose equation is

$$x^2 + z^2 = 1$$

Solution

In the zx-plane, this equation represents a circle with center $(0, 0, 0)$ and radius 1. Since y can be any real number, the equation represents the cylindrical surface shown in Figure 7.12. Observe that the cylinder extends indefinitely along the y-axis.

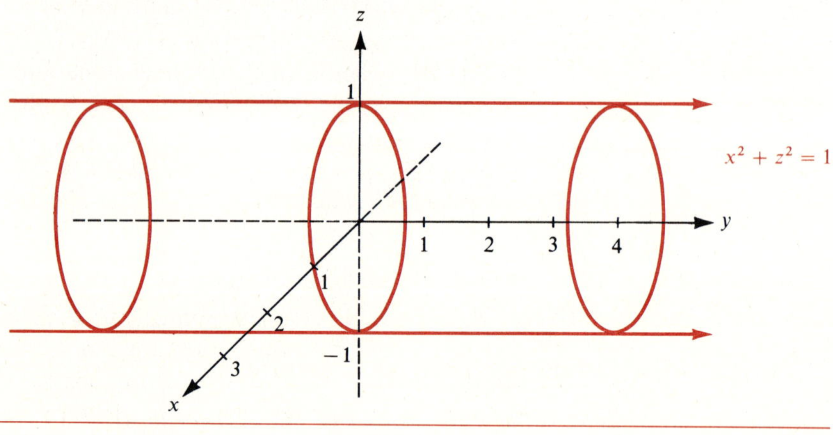

Figure 7.12

Example 7.14

Sketch the surface

$$z = 9 - x^2 - y^2$$

Solution

To help visualize the graph, consider its intersections with the coordinate planes. The plane $z = 0$ meets the surface $z = 9 - x^2 - y^2$ in the circle $x^2 + y^2 = 9$ that has center at the origin and radius 3. The plane $x = 0$ meets the surface $z = 9 - x^2 - y^2$ in the parabola

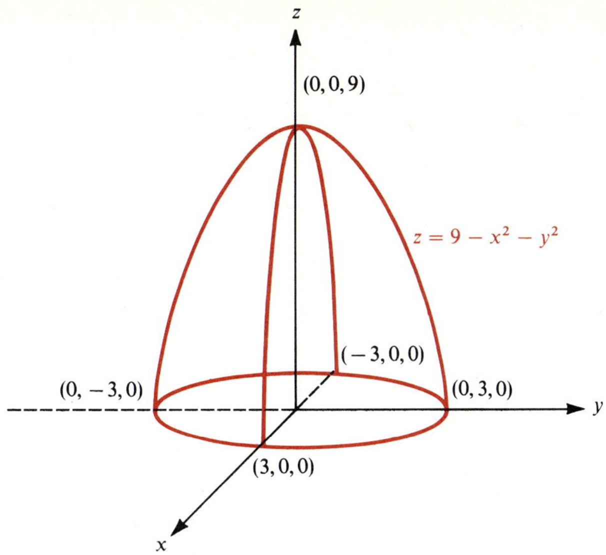

Figure 7.13

$z = 9 - y^2$ in the yz-plane. Similarly, the plane $y = 0$ meets the surface in the parabola $z = 9 - x^2$ in the zx-plane. A sketch incorporating these curves of intersection is shown in Figure 7.13; it is a surface that opens downward.

Example 7.15

Sketch the surface

$$z = x^2 + y^2$$

Solution

Since x^2 and y^2 are both non-negative, $z = x^2 + y^2 \geq 0$, so the graph of the surface lies above the xy-plane. Let us now consider the intersection of the surface with the coordinate planes and with planes parallel to the coordinate planes. When $x = 0$, the yz-plane meets the surface $z = x^2 + y^2$ in the parabola $z = y^2$. Similarly, when $y = 0$ the xz-plane meets the surface in the parabola $z = x^2$. Consider next the surface intersecting a plane parallel to the xy-plane, say $z = 9$. If a point (x, y, z) lies on the surface $z = x^2 + y^2$ and also on the plane $z = 9$, then it must satisfy the equation $x^2 + y^2 = 9$. Thus the point (x, y, z) lies on the circle $x^2 + y^2 = 9$ in the plane $z = 9$ (a circle of radius 3). If we consider a plane further above the xy-plane, z is increased and the radius of the circle is larger. Thus, the surface $z = x^2 + y^2$ is built up of circles whose centers are on the z-axis. The graph, which has the shape of a headlight, is shown in Figure 7.14. The dashed border in the diagram suggests that surface gets wider and wider as z increases.

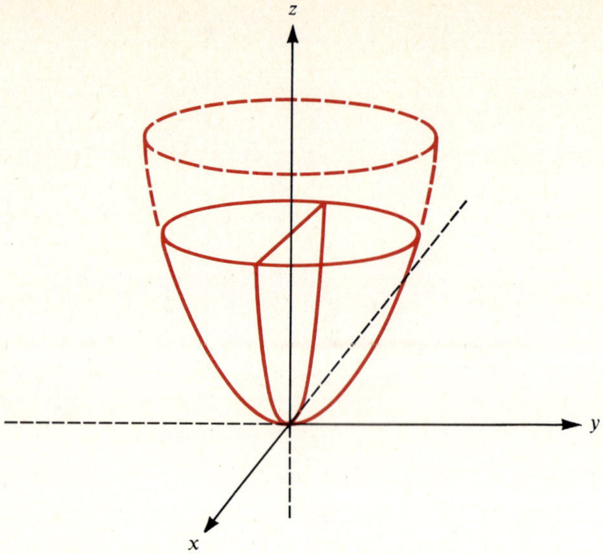

Figure 7.14

Exercises

1. Let $f(x, y) = 2x + 3y - 4$. Find:
   a. $f(0, 0)$ c. $f(2, 1)$ e. $f(-1, -3)$
   b. $f(1, 1)$ d. $f(3, 4)$ f. $f(-2, 4)$

2. Let $f(x, y) = x^2 + y^2 - 9$. Find:
   a. $f(3, 0)$ c. $f(4, -1)$ e. $f(-3, -4)$
   b. $f(2, 3)$ d. $f(-1, 5)$ f. $f(0, 5)$

3. If $f(x, y) = \sqrt{x^2 + y^2}$, find:
   a. $f(0, 0)$ c. $f(2, 1)$ e. $f(0, 5)$
   b. $f(1, -1)$ d. $f(3, 4)$ f. $f(5, 12)$

4. If $g(x, y) = \dfrac{x^2 - y^2}{x - y}$, $x \neq y$, find:
   a. $g(1, 0)$ c. $g(4, 1)$ e. $g(-5, -4)$
   b. $g(0, 2)$ d. $g(3, -2)$ f. $g(-12, 6)$

5. Let $h(x, y, z) = x^2 + 2y^2 + z^2 + 4$. Find:
   a. $h(0, 0, 1)$ c. $h(0, 2, 3)$ e. $h(2, 3, -4)$
   b. $h(1, 1, -1)$ d. $h(-1, 2, -3)$ f. $h(5, -1, 4)$

6. Let $h(x, y, z) = \sqrt{x^2 + y^2 + z^2}$. Find
   a. $h(0, 0, 2)$ c. $h(3, 2, 4)$ e. $h(2, 1, -4)$
   b. $h(1, 0, 3)$ d. $h(-1, 1, 5)$ f. $h(2, 0, 6)$

7. A company spends x dollars on research and development and y dollars on classified advertisements. Assuming that the profit (in

dollars) of the company is given by

$$P(x, y) = 40{,}000 + 50x + 30y + \frac{xy}{100}$$

for all positive integers x and y less than 25,000, determine the company's profit

a. if it spends \$2,000 on research and development and \$5,000 on advertisement.

b. if it spends \$8,000 on research and development and \$10,000 on advertisement.

In problems 8–16, sketch the graph of each equation.

8. $y^2 + z^2 = 1$

9. $x^2 + y^2 = 4$

10. $x^2 + z^2 = 9$

11. $y = x^2 + z^2$

12. $x = y^2 + z^2$

13. $x + 2y - z = 6$

14. $x^2 + y^2 + z^2 = 4$

15. $z = 4 - x^2 - y^2$

16. $z = 16 - x^2 - y^2$

7.4 Limits and Continuity

The ideas we present here are direct generalizations of the material in Sections 5 and 8 of Chapter 2. We suggest that the reader review those sections before proceeding further.

For a function of one variable, we observed that

$$\lim_{x \to a} f(x)$$

does not exist if $f(x)$ approaches two different values as x approaches a from left and from right; that is,

$$\lim_{x \to a^-} f(x) \neq \lim_{x \to a^+} f(x)$$

For a function of two variables, the situation is more complicated because (x, y) can approach (a, b) from an infinite number of directions, not just from two. Some of these directions are illustrated in Figure 7.15. Nevertheless, in order that

$$\lim_{(x,y) \to (a,b)} f(x, y)$$

exists, the function must approach a unique value no matter how (x, y) approaches the fixed point (a, b). As before, the value of the function $f(x, y)$ at $x = a$, $y = b$ has no effect on the existence or nonexistence of the limit.

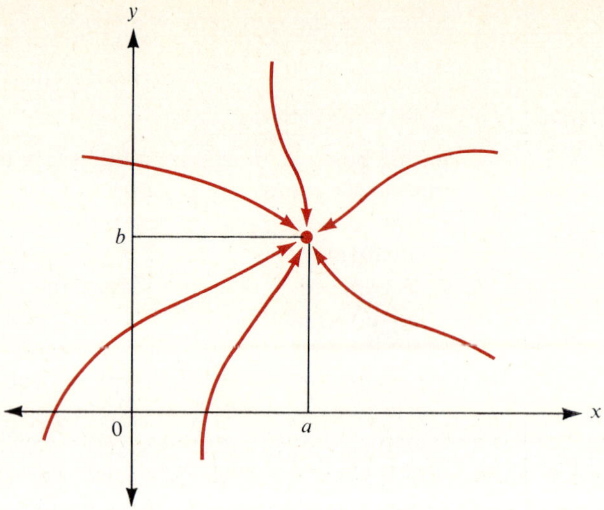

Figure 7.15

We can show that a particular limit does not exist if we get different values when we approach the fixed point (a, b) from different directions. Let us look at some examples.

Example 7.16

Let

$$f(x, y) = \frac{x - y}{x + y}$$

for $(x, y) \neq (0, 0)$. We shall show that

$$\lim_{(x,y)\to(0,0)} f(x, y)$$

does not exist.

If the point $(0, 0)$ is approached along the x-axis, then $y = 0$ and

$$\lim_{(x,y)\to(0,0)} \left(\frac{x - y}{x + y}\right) = \lim_{x\to 0} \left(\frac{x}{x}\right) = \lim_{x\to 0} 1 = 1$$

However, if we approach $(0, 0)$ along the y-axis, then $x = 0$ and

$$\lim_{(x,y)\to(0,0)} \left(\frac{x - y}{x + y}\right) = \lim_{y\to 0} \left(-\frac{y}{y}\right) = \lim_{y\to 0} (-1) = -1$$

Thus, except at the origin, the function $f(x, y) = 1$ at all points on the x-axis and $f(x, y) = -1$ on the y-axis. Further, we could also show that $f(x, y) = 0$ at all points on the line $y = x$ in the xy-plane. Since we get different answers depending on how we approach the point $(0, 0)$, we conclude that the limit in question does not exist.

Example 7.17

Let

$$f(x, y) = \frac{xy^2}{x^2 + y^4}$$

for $(x, y) \neq (0, 0)$. We shall demonstrate that

$$\lim_{(x,y)\to(0,0)} f(x, y)$$

does not exist. First, let us approach the origin along a straight line $y = mx$, $m \neq 0$. Then

$$f(x, y) = \frac{x(m^2x^2)}{x^2 + m^4x^4} = \frac{m^2x}{1 + m^2x^2}$$

which approaches $0/1 = 0$ as $x \to 0$. Thus along every straight line through the origin, $f(x, y) \to 0$ as $(x, y) \to (0, 0)$. Now let us approach $(0, 0)$ along the curve $x = y^2$. For this direction,

$$f(x, y) = \frac{y^2(y^2)}{y^4 + y^4} = \frac{1}{2}$$

so that along the path $x = y^2$, $f(x, y) \to \frac{1}{2}$ as $(x, y) \to (0, 0)$. Hence the limit does not exist.

Example 7.18

Determine

$$\lim_{(x,y)\to(0,0)} f(x, y)$$

where $f(x, y) = \left(\dfrac{x^3 - y^3}{x - y}\right)$

Solution

Observe that f is not defined for any pair (x, y) with $y = x$. Nevertheless,

$$\lim_{(x,y)\to(0,0)} f(x, y)$$

exists, since

$$\begin{aligned}
\lim_{(x,y)\to(0,0)} f(x, y) &= \lim_{(x,y)\to(0,0)} \left(\frac{x^3 - y^3}{x - y}\right) \\
&= \lim_{(x,y)\to(0,0)} \left[\frac{(x - y)(x^2 + xy + y^2)}{(x - y)}\right] \\
&= \lim_{(x,y)\to(0,0)} (x^2 + xy + y^2) \qquad [(x, y) \neq (0, 0)] \\
&= 0
\end{aligned}$$

Hence,

$$\lim_{(x,y)\to(0,0)} f(x, y) = 0$$

If a function of two variables is defined at a point (a, b), if it has a limit at (a, b), and if that limit is the same as the value of the function at (a, b), then the function is said to be continuous at (a, b). This is stated in the following definition.

Definition 7.6

A function $f(x, y)$ of two variables is *continuous* at (a, b) if

a. $f(x, y)$ is defined at (a, b)

b. $\lim_{(x,y)\to(a,b)} f(x, y)$ exists, and

c. $\lim_{(x,y)\to(a,b)} f(x, y) = f(a, b)$

If $f(x, y)$ is not continuous at (a, b), we say that $f(x, y)$ is *discontinuous* at the point (a, b).

Definition 7.7

A function $f(x, y)$ is said to be *continuous in a region* if it is continuous at every point in that region.

Example 7.19

Let

$$f(x, y) = \begin{cases} 3x + 2y, & x \neq 1,\ y \neq 4 \\ 11 & x = 1,\ y = 4 \end{cases}$$

Determine whether or not $f(x, y)$ is continuous at the point $(1, 4)$.

Solution

Observe that

$$\lim_{(x,y)\to(1,4)} (3x + 2y) = 3(1) + 2(4) = 11$$

Since $f(1, 4)$ is defined and also equals 11, we conclude that $f(x, y)$ is continuous at $(1, 4)$.

Example 7.20

Show that the function $f(x, y)$ defined by

$$f(x, y) = \frac{2xy}{x^2 + y^2}$$

is not continuous at the origin.

Solution

Observe that $f(0,0)$ is not defined and this violates the first condition in Definition 7.6. Thus $f(x, y)$ is not continuous at the origin. We could stop here, but it is interesting to see that $\lim_{(x,y)\to(0,0)} f(x, y)$ fails to exist. If the point $(0,0)$ is approached along the path $y = mx$ in the xy-plane, we notice that

$$f(x, y) = \frac{2x(mx)}{x^2 + m^2x^2} = \frac{2m}{1 + m^2}$$

and this yields different values for different values of m. Hence,

$$\lim_{(x,y)\to(0,0)} f(x, y)$$

does not exist.

Example 7.21

Let

$$f(x, y) = \begin{cases} \dfrac{xy^2}{x^2 + y^4}, & (x, y) \neq (0,0) \\ 0 & (x, y) = (0,0) \end{cases}$$

Here $f(x, y)$ is defined at $(0,0)$ but is discontinuous there because

$$\lim_{(x,y)\to(0,0)} f(x, y)$$

does not exist. (See Example 7.17.)

Example 7.22

Let

$$f(x, y) = \begin{cases} \dfrac{x^3 - y^3}{x - y}, & (x, y) \neq (0,0) \\ 15 & (x, y) = (0,0) \end{cases}$$

We have shown in Example 7.18 that $\lim_{(x,y)\to(0,0)} f(x, y)$ exists. Since

$$\lim_{(x,y)\to(0,0)} f(x, y) = 0 \neq f(0,0) = 15$$

condition (c) in Definition 7.6 is violated, and we conclude that $f(x, y)$ is discontinuous at the origin. However, if $f(0,0)$ were originally defined to equal 0, then $f(x, y)$ would be continuous at $(0,0)$.

As in functions of one variable, the continuous functions of two variables have certain properties which we state below.

Proposition 7.2

If $f(x, y) = c$, where c is a constant for all (x, y) in a region, then $f(x, y)$ is continuous in that region.

Proposition 7.3

If $f(x, y)$ is continuous in a certain region, then $c \cdot f(x, y)$ is also continuous in that region.

Proposition 7.4

If $f(x, y)$ and $g(x, y)$ are continuous in a certain region, then the functions

$$H(x, y) = f(x, y) + g(x, y)$$

$$J(x, y) = f(x, y) \cdot g(x, y)$$

$$K(x, y) = \frac{f(x, y)}{g(x, y)}, \qquad g(x, y) \neq 0$$

are also continuous.

This says that, for functions of two variables, the sum and product of continuous functions are continuous, and the quotient of two continuous functions is continuous if the denominator function does not equal zero at any point in the region.

Exercises

Evaluate the limit in each of the problems 1–14.

1. $\lim_{(x,y)\to(0,0)} (4x + 3y + 8)$
2. $\lim_{(x,y)\to(0,1)} (3x^2 + 4y^2 - 2)$
3. $\lim_{(x,y)\to(2,3)} (3x - 4y + 6)$
4. $\lim_{(x,y)\to(3,4)} (x^2 + y^2)$
5. $\lim_{(x,y)\to(-1,2)} (xy + 4y^2x^3)$
6. $\lim_{(x,y)\to(-4,3)} \left(\frac{1 + xy}{1 - xy}\right)$
7. $\lim_{(x,y)\to(1,1)} \left(\frac{x^2 - y^2}{x - y}\right)$
8. $\lim_{(x,y)\to(2,2)} \left(\frac{x^3 - y^3}{x - y}\right)$
9. $\lim_{(x,y)\to(2,1)} \left(\frac{xy}{x^2 - y^2}\right)$
10. $\lim_{(x,y)\to(3,1)} \left(\frac{x^3 - y^3}{x - y}\right)$
11. $\lim_{(x,y)\to(4,3)} \left(\frac{x^2 + y^2}{x^2 - y^2}\right)$
12. $\lim_{(x,y)\to(3,2)} \left(\frac{x^3 + y^3}{x^3 - y^3}\right)$

13. $\lim_{(x,y)\to(-1,2)} \left(\dfrac{4x^3y^2 - 2xy^5 + 7y - 1}{3xy - y^4 + 3x^3}\right)$

14. $\lim_{(x,y)\to(2,2)} \left(\dfrac{x^3 - 2xy + 3x^2 - 2y}{x^2y + 4y^2 - 6x^2 + 24y}\right)$

15. Show that

$$\lim_{(x,y)\to(0,0)} \left(\frac{xy}{x^2 + y^2}\right)$$

does not exist.

16. Show that

$$\lim_{(x,y)\to(0,0)} \left(\frac{x^2y}{x^3 + y^3}\right)$$

does not exist.

In problems 17–23, determine whether or not each function $f(x, y)$ is continuous at all points in its domain.

17. $f(x, y) = \begin{cases} 4x + 3y, & x \neq 1, y \neq 1 \\ 7 & x = 1, y = 1 \end{cases}$

18. $f(x, y) = \begin{cases} x^2 + y^2, & x \neq 3, y \neq 4 \\ 15, & x = 3, y = 4 \end{cases}$

19. $f(x, y) = \begin{cases} \dfrac{xy}{x + y}, & x \neq 0, y \neq 0 \\ 0, & x = 0, y = 0 \end{cases}$

20. $f(x, y) = \begin{cases} x^2 + xy + y^2, & x \neq 3, y \neq 2 \\ 19, & x = 3, y = 2 \end{cases}$

21. $f(x, y) = \begin{cases} \dfrac{x^2 + xy + y^2}{x - y}, & x \neq 3, y \neq 2 \\ 12, & x = 3, y = 2 \end{cases}$

22. $f(x, y) = \begin{cases} \dfrac{x^3 - y^3}{x - y}, & x \neq 4, y \neq 1 \\ 21, & x = 4, y = 1 \end{cases}$

23. $f(x, y) = \begin{cases} \dfrac{x^3 + y^3}{x + y}, & x \neq 3, y \neq 1 \\ 7, & x = 3, y = 1 \end{cases}$

24. Show that the function $f(x, y)$ defined by

$$f(x, y) = \frac{xy}{x^2 - y}$$

is discontinuous for all points on the parabola $y = x^2$.

25. Show that the function $f(x, y)$ defined by

$$f(x, y) = \frac{x^2 - y^2}{x^2 + y^2}$$

is discontinuous at the origin.

26. Show that the function $f(x, y)$ defined by

$$f(x, y) = \frac{x^3 - y^3}{x^3 + y^3}$$

is discontinuous at the origin.

7.5 Partial Differentiation

When studying functions of a single variable, we determined the slope of the curve at several points in its domain. This was done by looking at the derivative of the function. We now generalize this notion to surfaces.

For a function $z = f(x, y)$, if the value of y is held constant, then z becomes a function of one independent variable x. In this situation, z may be differentiated with respect to x, treating y as a constant. Such a derivative is called the *partial derivative* of $f(x, y)$ with respect to x and is commonly denoted by

$$\frac{\partial f}{\partial x}, \quad \frac{\partial z}{\partial x}, \quad \text{or} \quad f_x(x, y)$$

Similarly, the *partial derivative* of $f(x, y)$ with respect to y is usually denoted by

$$\frac{\partial f}{\partial y}, \quad \frac{\partial z}{\partial y}, \quad \text{or} \quad f_y(x, y)$$

It is found by holding x constant and differentiating $f(x, y)$ with respect to y.

Definition 7.8

Let $z = f(x, y)$ be a function of two independent variables. The *partial derivatives* of $f(x, y)$ with respect to x and with respect to y are defined by

$$\frac{\partial z}{\partial x} = f_x(x, y) = \lim_{\Delta x \to 0} \left[\frac{f(x + \Delta x, y) - f(x, y)}{\Delta x}\right]$$

and

$$\frac{\partial z}{\partial y} = f_y(x, y) = \lim_{\Delta y \to 0} \left[\frac{f(x, y + \Delta y) - f(x, y)}{\Delta y}\right]$$

provided that each limit exists.

Example 7.23

Find $\partial z/\partial x$ and $\partial z/\partial y$ for

$$z = x^2 + xy + y^2$$

Solution

Treating y as a constant and differentiating with respect to x, we obtain the partial derivative of z with respect to x:

$$\begin{aligned}\frac{\partial z}{\partial x} &= \frac{\partial}{\partial x}(x^2 + xy + y^2)\\ &= 2x + y\end{aligned}$$

Similarly, the partial derivative of z with respect to y is obtained by treating x as a constant and differentiating z with respect to y:

$$\begin{aligned}\frac{\partial z}{\partial y} &= \frac{\partial}{\partial y}(x^2 + xy + y^2)\\ &= x + 2y\end{aligned}$$

The rules for differentiation that we learned in Chapter 3 are still in force. The only difference is to remember which variable is held constant.

Example 7.24

Find $\partial z/\partial x$ and $\partial z/\partial y$ for

$$z = \sqrt{x^2 + y^2} + x^2e^y$$

Solution

If we hold y constant and differentiate z with respect to x, we obtain

$$\begin{aligned}\frac{\partial z}{\partial x} &= \frac{\partial}{\partial x}(\sqrt{x^2 + y^2} + x^2e^y)\\ &= \frac{\partial}{\partial x}(x^2 + y^2)^{1/2} + e^y\frac{\partial}{\partial x}(x^2)\\ &= \tfrac{1}{2}(x^2 + y^2)^{-1/2}(2x) + e^y 2x\\ &= \frac{x}{\sqrt{x^2 + y^2}} + 2xe^y, \qquad x^2 + y^2 \neq 0\end{aligned}$$

Similarly, treating x as a constant and differentiating z with respect to y, we have

$$\begin{aligned}\frac{\partial z}{\partial y} &= \frac{\partial}{\partial y}(\sqrt{x^2 + y^2} + x^2 e^y)\\ &= \frac{\partial}{\partial y}(x^2 + y^2)^{1/2} + x^2 \frac{\partial}{\partial y}(e^y)\\ &= \tfrac{1}{2}(x^2 + y^2)^{-1/2}(2y) + x^2 e^y\\ &= \frac{y}{\sqrt{x^2 + y^2}} + x^2 e^y\end{aligned}$$

Example 7.25

Find $\partial z/\partial x$ and $\partial z/\partial y$ for

$$z = \frac{x^2 + y}{x + y^2}$$

Solution

Using the quotient rule, we have

$$\begin{aligned}\frac{\partial z}{\partial x} &= \frac{(x + y^2)\dfrac{\partial}{\partial x}(x^2 + y) - (x^2 + y)\dfrac{\partial}{\partial x}(x + y^2)}{(x + y^2)^2}\\ &= \frac{(x + y^2)(2x) - (x^2 + y)\cdot 1}{(x + y^2)^2}\\ &= \frac{2x^2 + 2xy^2 - x^2 - y}{(x + y^2)^2}\\ &= \frac{x^2 + 2xy^2 - y}{(x + y^2)^2}\end{aligned}$$

Similarly,

$$\begin{aligned}\frac{\partial z}{\partial y} &= \frac{(x + y^2)\dfrac{\partial}{\partial y}(x^2 + y) - (x^2 + y)\dfrac{\partial}{\partial y}(x + y^2)}{(x + y^2)^2}\\ &= \frac{(x + y^2)\cdot 1 - (x^2 + y)(2y)}{(x + y^2)^2}\\ &= \frac{x + y^2 - 2x^2 y - 2y^2}{(x + y^2)^2}\\ &= \frac{x - 2x^2 y - y^2}{(x + y^2)^2}\end{aligned}$$

The notation $f_x(a, b)$ or $(\partial z/\partial x)_{(a,b)}$ represents the value of a partial derivative at the point (a, b) in the xy-plane. Thus if

$$f(x, y) = x^3 + 3x^2y + y^3$$

then $f_x(2, 3)$ is obtained by first calculating

$$f_x(x, y) = 3x^2 + 6xy$$

and then substituting $x = 2$, $y = 3$ in $f_x(x, y)$. Thus

$$\begin{aligned} f_x(2, 3) &= 3(2)^2 + 6(2)(3) \\ &= 12 + 36 \\ &= 48 \end{aligned}$$

Similarly, $f_y(1, -4)$ is obtained by first calculating

$$f_y(x, y) = 3x^2 + 3y^2$$

and then substituting $x = 1$, $y = -4$ in $f_y(x, y)$. Thus

$$\begin{aligned} f_y(1, -4) &= 3(1)^2 + 3(-4)^2 \\ &= 3 + 48 \\ &= 51 \end{aligned}$$

Partial derivatives have simple geometric interpretations. Consider a function $z = f(x, y)$ with domain D. The plane $y = c$ intersects this surface in the plane curve APB as shown in Figure 7.16. The partial derivative of $f(x, y)$ with respect to x, as y keeps the constant value

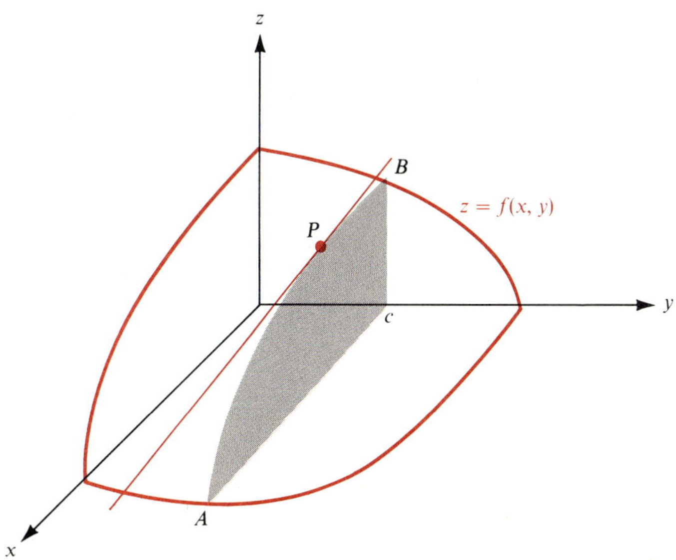

Figure 7.16

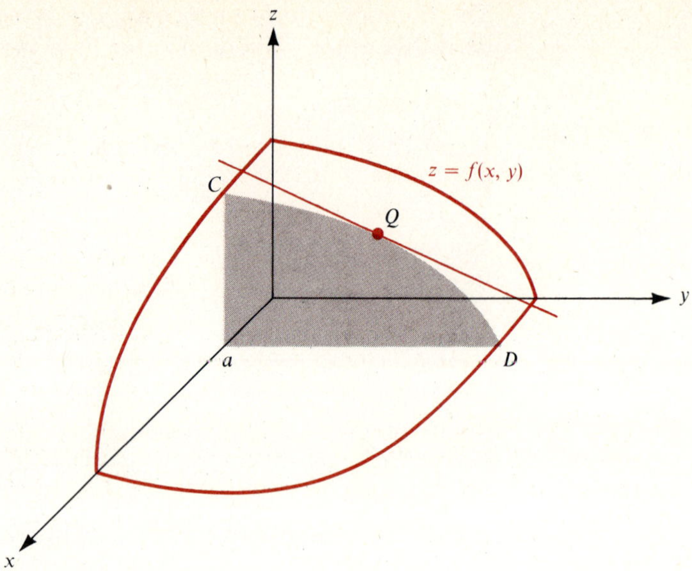

Figure 7.17

c, is the slope of the tangent to the plane curve APB at the point $P(x, c, f(x, c))$. This partial derivative can also be associated with the *rate of change* of $z = f(x, y)$ with respect to x for a fixed value of y. A similar explanation applies to the partial derivative of $f(x, y)$ with respect to y. The plane $x = a$ cuts the surface $z = f(x, y)$ in the plane curve CQD as shown in Figure 7.17. The partial derivative $f_y(a, y)$ is the slope of the tangent to the plane curve CQD at the point $Q(a, y, f(a, y))$. This derivative gives the *rate of change* of $z = f(x, y)$ with respect to y when x is held constant.

In the next few examples, we illustrate how to apply partial derivatives in real problems.

Example 7.26

The volume of a right circular cylinder is given by

$$V = \pi r^2 h$$

where r is the radius of the cylinder and h is its height. Thus the rate of change of V with respect to r (with h fixed) is

$$\frac{\partial V}{\partial r} = \frac{\partial}{\partial r}(\pi r^2 h) = 2\pi r h$$

and the rate of change of V with respect to h (with r fixed) is

$$\frac{\partial V}{\partial h} = \frac{\partial}{\partial h}(\pi r^2 h) = \pi r^2$$

Example 7.27

The cost function $z = C(x, y)$ relates z, the cost (in dollars) of a certain product, to x, the cost of raw material (in dollars per pound), and y, the cost of labor (in dollars per hour). The partial derivatives $\partial z/\partial x$ and $\partial z/\partial y$ are called the *marginal cost functions*. Thus if

$$z = C(x, y) = 50 + 4x + 6y$$

then

$$\frac{\partial z}{\partial x} = 4 \quad \text{and} \quad \frac{\partial z}{\partial y} = 6$$

for all x and y. This means that when the cost of labor is fixed, an increase of \$1.00 per pound in the cost of raw materials causes an increase of \$4.00 in the cost of the product. When the cost of raw materials is fixed, an increase of \$1.00 in the hourly cost of labor incurs an increase of \$6.00 in the cost of the product.

Example 7.28

Let $P(x, y)$ be the profit function of a company when it sells x units of product A and y units of product B. Then $P_x(x, y)$ is the *marginal profit with respect to* x and $P_y(x, y)$ is the *marginal profit with respect to* y. If

$$z = P(x, y) = x^2 + 3y^2 + 5xy$$

then

$$\frac{\partial z}{\partial x} = P_x(x, y) = 2x + 5y$$

and

$$\frac{\partial z}{\partial y} = P_y(x, y) = 6y + 5x$$

for all x and y. Given specific values for x and y, such as $x = 3$ and $y = 4$, the marginal profits with respect to each variable can be calculated. Thus

$$\left(\frac{\partial z}{\partial x}\right)_{(3,4)} = P_x(3, 4) = 2(3) + 5(4) = 26$$

and

$$\left(\frac{\partial z}{\partial y}\right)_{(3,4)} = P_y(3, 4) = 6(4) + 5(3) = 39$$

Exercises

In problems 1–20, calculate $\partial z/\partial x$ and $\partial z/\partial y$.

1. $z = x^2 + y$
2. $z = 3x^2 + 4y^3$
3. $z = x^2y$
4. $z = xy^2$
5. $z = x^2y^2$
6. $z = (x - 2y)^3$
7. $z = (2x - y)^4$
8. $z = (2x + 3y^2)^{3/2}$
9. $z = \dfrac{x}{y}$
10. $z = \dfrac{x}{y} - \dfrac{y}{x}$
11. $z = \sqrt{xy}$
12. $z = \sqrt{x^2 + y^2}$
13. $z = x \log y$
14. $z = x^2 \log(3 + y)$
15. $z = \log(x^2 + y^2)$
16. $z = e^{2xy}$
17. $z = e^{-xy}$
18. $z = e^{xy}(x + y)$
19. $z = xe^y + x^3y^2$
20. $z = 3e^{xy^3}$

In problems 21–30, find $f_x(x, y)$ and $f_y(x, y)$ at the indicated point.

21. $f(x, y) = x^2 - y^2, \quad x = 1, y = 1$
22. $f(x, y) = x^3 - y^4, \quad x = 1, y = -1$
23. $f(x, y) = x^2 + y^2 - 3xy, \quad x = 5, y = 2$
24. $f(x, y) = x + \dfrac{y^2}{x}, \quad x = 1, y = 2$
25. $f(x, y) = 8x^2 - 5xy + y^2, \quad x = 2, y = 1$
26. $f(x, y) = 3x^2 + y^2 - 3xy - 12x - 5y, \quad x = -2, y = 5$
27. $f(x, y) = \log(x^2 + y^4), \quad x = 3, y = 1$
28. $f(x, y) = \log(x^2 + y^2 + xy), \quad x = -1, y = 4$
29. $f(x, y) = \dfrac{x^2 - y^2}{x^2 + y^2}, \quad x = 2, y = -3$
30. $f(x, y) = e^{\sqrt{x^2 + y}}, \quad x = 0, y = 4$
31. Given that

$$z = \tfrac{1}{2} \log(x^2 + y^2)$$

show that

$$\left(\frac{\partial z}{\partial x}\right)^2 + \left(\frac{\partial z}{\partial y}\right)^2 = \frac{1}{x^2 + y^2}$$

32. Given that

$$u = x^3 - 3xy^2 \quad \text{and} \quad v = 3x^2y - y^3$$

show that

a. $\dfrac{\partial u}{\partial x} = \dfrac{\partial v}{\partial y}$ **b.** $\dfrac{\partial u}{\partial y} = -\dfrac{\partial v}{\partial x}$

33. A bank determines the amount of money it is willing to loan for a one-family house by means of the formula

$$z = 3x + 0.02y^2, \qquad x \geq 10{,}000,\ y \geq 1{,}000$$

where x is the down payment and y is the applicant's monthly salary. Assuming that all figures are in dollars, calculate $\partial z/\partial x$ and $\partial z/\partial y$ and describe what each quantity represents.

34. The cost to a manufacturer of producing x units of product A and y units of product B is given by

$$c(x, y) = \frac{50}{2 + x} + \frac{125}{(3 + y)^2}$$

Calculate the marginal cost of each product.

35. A manufacturer's revenue function from the sale of x units of product A and y units of product B is

$$R(x, y) = \log(1 + 50x + 75y) + \sqrt{1 + 40x + 125y}$$

Calculate the *marginal revenue* from each of the two products.

36. If a production function is given in the form $z = f(x, y)$, then $\partial z/\partial x$ and $\partial z/\partial y$ are called the *marginal productivities*. Find the marginal productivities for each of the following production functions.

a. $z = 5xy - 2x^2 - 2y^2, \qquad x = 1, y = 1$
b. $z = 4(4xy - x^2 - 3y^2), \qquad x = 1, y = \frac{1}{2}$

37. A company's profit function from the sale of x units of product A and y units of product B is

$$P(x, y) = -10 + 10x + x^2 - y^2 + 2xy$$

Find the marginal profits $\partial P/\partial x$ and $\partial P/\partial y$ for

a. $x = 4, y = 2$ **c.** $x = 7, y = 3$
b. $x = 5, y = 1$ **d.** $x = 2, y = 6$

38. Suppose that the demand functions for two related products are $D_1 = f(x, y)$ and $D_2 = g(x, y)$ where x and y are the prices of the two products. The two products are said to be *competitive* for all prices (a, b) for which $f_y(a, b)$ and $g_x(a, b)$ are both positive. Determine whether or not the following products are competitive.

a. $f(x, y) = 5 - 2x + y, g(x, y) = 6 + x - y$
b. $f(x, y) = 17 - 2x - y, g(x, y) = 20 - x - 3y$
c. $f(x, y) = 100 - 10x + 2y, g(x, y) = 200 - 25y + 4x$

7.6 Higher Order Partial Derivatives

Recall that if $y = f(x)$, then

$$\frac{dy}{dx} = f'(x) \quad \text{and} \quad \frac{d^2y}{dx^2} = \frac{d}{dx}\left(\frac{dy}{dx}\right) = f''(x)$$

The second derivative of y is the derivative of the first derivative of y. Analogously, if $z = f(x, y)$ then we can differentiate each of the partial derivatives $\partial z/\partial x$ and $\partial z/\partial y$ with respect to either x or y and obtain four second order partial derivatives.

Definition 7.9

Let $z = f(x, y)$ be a function of two independent variables x and y. Then the *second order partial derivatives* of $f(x, y)$ are defined as follows:

$$\frac{\partial}{\partial x}\left(\frac{\partial z}{\partial x}\right) = \frac{\partial^2 z}{\partial x^2} = f_{xx}(x, y)$$

$$\frac{\partial}{\partial y}\left(\frac{\partial z}{\partial x}\right) = \frac{\partial^2 z}{\partial y\, \partial x} = f_{xy}(x, y)$$

$$\frac{\partial}{\partial x}\left(\frac{\partial z}{\partial y}\right) = \frac{\partial^2 z}{\partial x\, \partial y} = f_{yx}(x, y)$$

and

$$\frac{\partial}{\partial y}\left(\frac{\partial z}{\partial y}\right) = \frac{\partial^2 z}{\partial y^2} = f_{yy}(x, y)$$

The distinction between the mixed partial derivatives $\partial^2 z/(\partial y\, \partial x)$ and $\partial^2 z/(\partial x \partial y)$ should be noted carefully. The second order partial derivative $\partial^2 z/(\partial y\, \partial x)$ is obtained by first finding the partial derivative of z with respect to x and then finding the partial derivative of the result with respect to y. On the other hand, $\partial^2 z/\partial x\, \partial y$ is obtained by first differentiating z with respect to y and then with respect to x.

Example 7.29

Find all four second order partial derivatives of z where

$$z = ax^3 + 3bx^2y + 3cxy^2 + ky^3$$

Solution

First we compute $\partial z/\partial x$ and $\partial z/\partial y$.

$$\frac{\partial z}{\partial x} = 3ax^2 + 6bxy + 3cy^2$$

and

$$\frac{\partial z}{\partial y} = 3bx^2 + 6cxy + 3ky^2$$

Then

$$\frac{\partial^2 z}{\partial x^2} = \frac{\partial}{\partial x}\left(\frac{\partial z}{\partial x}\right) = \frac{\partial}{\partial x}(3ax^2 + 6bxy + 3cy^2)$$
$$= 6ax + 6by$$

$$\frac{\partial^2 z}{\partial y^2} = \frac{\partial}{\partial y}\left(\frac{\partial z}{\partial y}\right) = \frac{\partial}{\partial y}(3bx^2 + 6cxy + 3ky^2)$$
$$= 6cy + 6ky$$

$$\frac{\partial^2 z}{\partial y\,\partial x} = \frac{\partial}{\partial y}\left(\frac{\partial z}{\partial x}\right) = \frac{\partial}{\partial y}(3ax^2 + 6bxy + 3cy^2)$$
$$= 6bx + 6cy$$

and

$$\frac{\partial^2 z}{\partial x\,\partial y} = \frac{\partial}{\partial x}\left(\frac{\partial z}{\partial y}\right) = \frac{\partial}{\partial x}(3bx^2 + 6cxy + 3ky^2)$$
$$= 6bx + 6cy$$

Example 7.30

Find the second order partial derivatives $\partial^2 z/(\partial y\,\partial x)$ and $\partial^2 z/(\partial x\,\partial y)$ where $z = \log(e^x + e^y)$.

Solution

Here

$$\frac{\partial z}{\partial x} = \frac{e^x}{e^x + e^y} \quad \text{and} \quad \frac{\partial z}{\partial y} = \frac{e^y}{e^x + e^y}.$$

Then

$$\frac{\partial^2 z}{\partial y\,\partial x} = \frac{\partial}{\partial y}\left(\frac{\partial z}{\partial x}\right) = \frac{\partial}{\partial y}\left(\frac{e^x}{e^x + e^y}\right)$$
$$= \frac{(e^x + e^y)0 - e^x \cdot e^y}{(e^x + e^y)^2}$$
$$= -\frac{e^x e^y}{(e^x + e^y)^2}$$

and

$$\frac{\partial^2 z}{\partial x\, \partial y} = \frac{\partial}{\partial x}\left(\frac{\partial z}{\partial y}\right) = \frac{\partial}{\partial x}\left(\frac{e^y}{e^x + e^y}\right)$$

$$= \frac{(e^x + e^y) \cdot 0 - e^y \cdot e^x}{(e^x + e^y)^2}$$

$$= -\frac{e^y \cdot e^x}{(e^x + e^y)^2}$$

Notice, in both Examples 7.29 and 7.30, that

$$\frac{\partial^2 z}{\partial y\, \partial x} = \frac{\partial^2 z}{\partial x\, \partial y}$$

This is not a coincidence and will happen in most cases, in fact, whenever both second order mixed partial derivatives are continuous. This is stated, without proof, in the following proposition.

Proposition 7.5

If $z = f(x, y)$ is a continuous function and if $f_{xy}(x, y)$ and $f_{yx}(x, y)$ are also continuous, then

$$f_{xy}(x, y) = f_{yx}(x, y)$$

The definition of second order partial derivatives and the proposition on the equality of mixed partials is easily extended to functions of three variables. If $w = f(x, y, z)$, then we have the nine second order partial derivatives (assuming that they exist):

$$\frac{\partial^2 w}{\partial x^2} = f_{xx}(x, y, z), \quad \frac{\partial^2 w}{\partial x\, \partial y} = f_{yx}(x, y, z), \quad \frac{\partial^2 w}{\partial x\, \partial z} = f_{zx}(x, y, z)$$

$$\frac{\partial^2 w}{\partial y^2} = f_{yy}(x, y, z), \quad \frac{\partial^2 w}{\partial y\, \partial x} = f_{xy}(x, y, z), \quad \frac{\partial^2 w}{\partial y\, \partial z} = f_{zy}(x, y, z)$$

$$\frac{\partial^2 w}{\partial z^2} = f_{zz}(x, y, z), \quad \frac{\partial^2 w}{\partial z\, \partial x} = f_{xz}(x, y, z), \quad \frac{\partial^2 w}{\partial z\, \partial y} = f_{yz}(x, y, z)$$

Proposition 7.6

If the function $w = f(x, y, z)$ is continuous, and if the three first order partial derivatives and all six second order mixed partials are also continuous, then

$$f_{xy}(x, y, z) = f_{yx}(x, y, z)$$

$$f_{xz}(x, y, z) = f_{zx}(x, y, z)$$

$$f_{yz}(x, y, z) = f_{zy}(x, y, z)$$

Example 7.31

Let

$$w = f(x, y, z) = xy^3 - zx^5 + x^2yz$$

then

$$\frac{\partial w}{\partial x} = y^3 - 5zx^4 + 2xyz$$

$$\frac{\partial w}{\partial y} = 3xy^2 + x^2z$$

$$\frac{\partial w}{\partial z} = -x^5 + x^2y$$

and

$$\frac{\partial^2 w}{\partial x^2} = \frac{\partial}{\partial x}\left(\frac{\partial w}{\partial x}\right) = -20zx^3 + 2yz$$

$$\frac{\partial^2 w}{\partial y^2} = \frac{\partial}{\partial y}\left(\frac{\partial w}{\partial y}\right) = 6xy$$

$$\frac{\partial^2 w}{\partial z^2} = \frac{\partial}{\partial z}\left(\frac{\partial w}{\partial z}\right) = 0$$

Further,

$$\frac{\partial^2 w}{\partial y\,\partial x} = \frac{\partial}{\partial y}(y^3 - 5zx^4 + 2xyz) = 3y^2 + 2xz$$

$$\frac{\partial^2 w}{\partial x\,\partial y} = \frac{\partial}{\partial x}(3xy^2 + x^2z) = 3y^2 + 2xz$$

$$\frac{\partial^2 w}{\partial z\,\partial x} = \frac{\partial}{\partial z}(y^3 - 5zx^4 + 2xyz) = -5x^4 + 2xy$$

$$\frac{\partial^2 w}{\partial x\,\partial z} = \frac{\partial}{\partial x}(-x^5 + x^2y) = -5x^4 + 2xy$$

$$\frac{\partial^2 w}{\partial y\,\partial z} = \frac{\partial}{\partial y}(-x^5 + x^2y) = x^2$$

and

$$\frac{\partial^2 w}{\partial z\,\partial y} = \frac{\partial}{\partial z}(3xy^2 + x^2z) = x^2$$

We conclude this section by pointing out that we could easily define partial derivatives of order higher than two. For example,

$$\frac{\partial^3 w}{\partial x\,\partial y\,\partial z} = f_{zyx}(x, y, z) = \frac{\partial}{\partial x}\left(\frac{\partial^2 w}{\partial y\,\partial z}\right)$$

Exercises

In problems 1–20, calculate the four second order partial derivatives of z.

1. $z = xy^2$
2. $z = x^2y$
3. $z = x^3y^2$
4. $z = x^2 + xy + y^2$
5. $z = x^3 - 3x^2y + y^3$
6. $z = x^2y^3 + xy^4$
7. $z = x^4 + 3x^2y^2 + y^3$
8. $z = (3x + 2y)^{3/2}$
9. $z = \dfrac{x}{y}$
10. $z = \dfrac{x}{y} + \dfrac{y}{x}$
11. $z = \sqrt{x^2 + y^2}$
12. $z = (x^3 + x^2y^2 + y^3)^{1/2}$
13. $z = \dfrac{x + y}{x - y}$
14. $z = \dfrac{1}{(x + y)^2}$
15. $z = \log(3x + 4y)$
16. $z = \log(x^2 + y^2)$
17. $z = \log(x^3 + y^3)$
18. $z = \log(x^2 + xy + y^2)$
19. $z = e^{xy}$
20. $z = xe^y + x^3y^2$

In problems 21–24, show that $\dfrac{\partial^2 z}{\partial x\, \partial y} = \dfrac{\partial^2 z}{\partial y\, \partial x}$.

21. $z = 2x^2y^3 - x^3y^5$
22. $z = (x^3 + y^2)^5$
23. $z = \sqrt{x^3 + y^3}$
24. $z = 3xy - y^3 + (y^2 - 2x)^{3/2}$

In problems 25–28, calculate the nine second order partial derivatives and show that the three pairs of mixed partials are equal.

25. $f(x, y, z) = xyz$
26. $f(x, y, z) = x^2y^3z^4$
27. $f(x, y, z) = \dfrac{x + y}{z}$
28. $f(x, y, z) = \log(xy + z)$

29. Given that

$$u = x^3 - 3xy^2 \quad \text{and} \quad v = 3x^2y - y^3$$

show that

a. $\dfrac{\partial^2 u}{\partial x^2} + \dfrac{\partial^2 u}{\partial y^2} = 0$

b. $\dfrac{\partial^2 v}{\partial x^2} + \dfrac{\partial^2 v}{\partial y^2} = 0$

30. Given that

$$z = \tfrac{1}{2}\log(x^2 + y^2)$$

show that

$$\frac{\partial^2 z}{\partial x^2} + \frac{\partial^2 z}{\partial y^2} = 0$$

In problems 31–33, show that

$$\frac{\partial^2 u}{\partial x^2} + \frac{\partial^2 u}{\partial y^2} + \frac{\partial^2 u}{\partial z^2} = 0$$

31. $u = x^2 + y^2 - 2z^2$

32. $u = 2z^3 - 3(x^2 + y^2)z$

33. $u = (x^2 + y^2 + z^2)^{-1/2}$

7.7 Maxima and Minima

An important application of functions of one variable is the determination of relative maxima and relative minima. The basic fact we used in Chapter 4 was that if f is continuous on an open interval (a, b) then f takes on a relative maximum or relative minimum value in (a, b). We then defined a critical point to be a number $x \in (a, b)$ at which $f'(x) = 0$ or for which $f(x)$ exists but $f'(x)$ does not. Finally, we gave conditions on first and second derivatives to determine whether a critical point is a relative maximum, relative minimum, or point of inflection.

In this section, we shall develop techniques for finding relative maximum and relative minimum values of a function $z = f(x, y)$. Tests for absolute maximum or absolute minimum are rather complicated. Fortunately, in most applied problems relative maximum and relative minimum values turn out to be absolute maximum and absolute minimum values.

We shall define the relative maximum and the relative minimum for functions of two variables by using the idea of a neighborhood. When we refer to the points in a circular region containing (a, b) as the *neighborhood* of (a, b), we mean the points in a "small" disc centered at (a, b) as shown in Figure 7.18. This is the set of all points (x, y) for

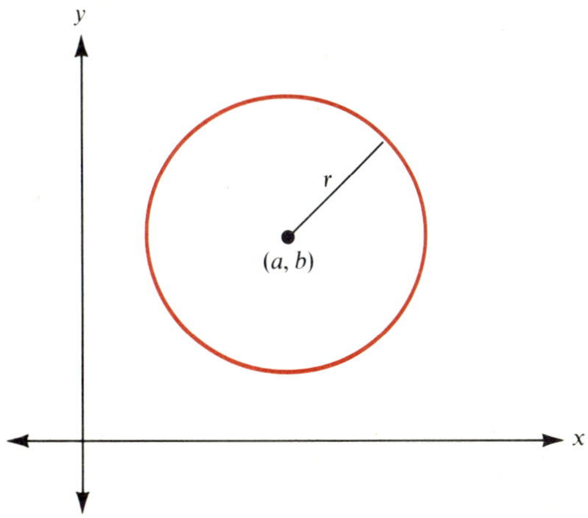

Figure 7.18

which the distance from (a, b) is less than r; that is,

$$\sqrt{(x-a)^2 + (y-b)^2} < r$$

where r is the radius of the disc.

Definition 7.10

Let $z = f(x, y)$ be a function with domain D in the xy-plane. Then f has

a. a *relative maximum* at (a, b) if

$$f(x, y) \leq f(a, b)$$

for all (x, y) in some neighborhood of (a, b) contained in D, or

b. a *relative minimum* at (a, b) if

$$f(x, y) \geq f(a, b)$$

for all points (x, y) in some neighborhood of (a, b) contained in D.

Examples of a relative maximum and a relative minimum are shown in Figures 7.19 and 7.20, respectively. Observe that at a relative maximum the graph of f looks like a mountain peak and at a relative minimum it looks like a bowl.

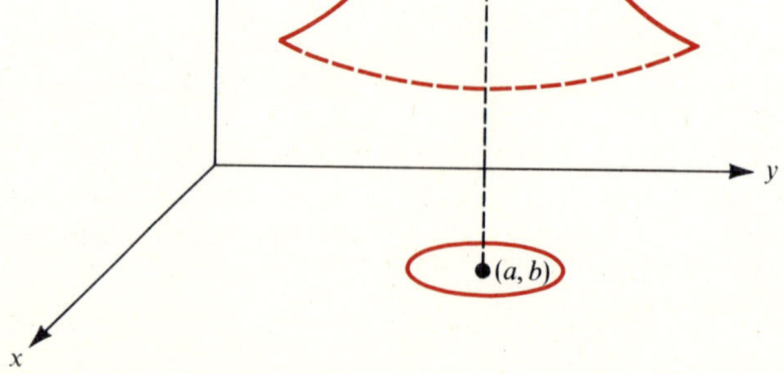

Figure 7.19

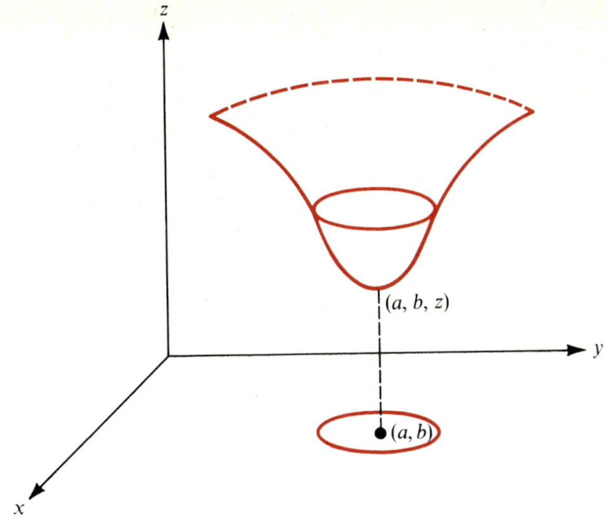

Figure 7.20

Suppose that a function f assumes a relative maximum (minimum) value at a point (a, b) in the domain D. If we fix $y = b$, then $f(x, b)$ is a function of one variable x which has its relative maximum (minimum) at $x = a$, so its derivative must be zero at that point. This means that $f_x(a, b) = 0$. Similarly, $f_y(a, b) = 0$.

Definition 7.11 A point (a, b) on a surface $z = f(x, y)$ for which the partial derivatives $f_x(a, b)$ and $f_y(a, b)$ are both zero, or for which the partial derivatives $f_x(a, b)$ and $f_y(a, b)$ do not exist, is called a *critical point* of the function f.

Proposition 7.7 Let $z = f(x, y)$ be a function defined on a domain D that includes the point (a, b) and all points within some neighborhood of (a, b). If f has a relative maximum or a relative minimum at (a, b), then (a, b) is a critical point of the function f.

The converse of Proposition 7.7 is not true because the existence of a critical point $(a, b) \in D$ does not guarantee that a function has a relative maximum or relative minimum at (a, b). We have encountered this problem for functions of one variable: consider $x = 0$ for $f(x) = x^3$. A typical example for two variables is the function

$$f(x, y) = y^2 - x^2$$

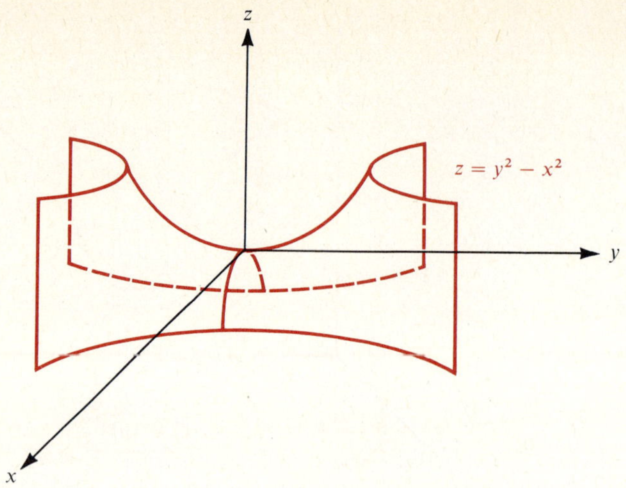

Figure 7.21

Its graph, known as hyperbolic paraboloid, is shown in Figure 7.21. Observe that the surface is saddle-shaped near the origin.

The partial derivatives of f are

$$f_x(x, y) = -2x \quad \text{and} \quad f_y(x, y) = 2y$$

Setting these equal to zero, we find that $(0, 0)$ is the only critical point, yet $f(0, 0)$ is neither a relative maximum nor a relative minimum. To see this, note that the surface goes down as we move along the x-axis and goes up along the y-axis. The critical point $(0, 0)$ in this example, is called a *saddle point*.

Example 7.32

Find all the points where the function

$$f(x, y) = x^2 + 2xy + 2y^2 - 6y$$

may have a relative maximum or relative minimum.

Solution

Taking partial derivatives, we obtain

$$f_x(x, y) = 2x + 2y$$

and

$$f_y(x, y) = 2x + 4y - 6$$

Setting $f_x(x, y)$ and $f_y(x, y)$ equal to zero, we have

$$2x + 2y = 0$$

and

$$2x + 4y - 6 = 0$$

Solving $2x + 2y = 0$ for x, we get $y = -x$. Substituting $-x$ for y in the equation $2x + 4y - 6 = 0$, we have

$$\begin{aligned} 2x - 4x - 6 &= 0 \\ -2x - 6 &= 0 \\ x &= -3 \end{aligned}$$

Since $y = -x$, we get $y = 3$. Thus, the relative maximum or minimum, if it exists, must occur at $(-3, 3)$.

The above example indicates that more is needed to determine whether a critical point is a relative maximum, relative minimum, or neither. One might suspect that the answer has something to do with the signs of the second order partial derivatives of f, analogous to functions of one variable. The following proposition, which we state without proof, gives the answer.

Proposition 7.8

Let f and all its first and second partial derivatives be continuous in some neighborhood of the critical point (a, b). Let

$$A = f_{xx}(a, b), \quad B = f_{xy}(a, b), \quad C = f_{yy}(a, b)$$

Then f has

a. a relative maximum at (a, b) if $AC - B^2 > 0$ and $A < 0$

b. a relative minimum at (a, b) if $AC - B^2 > 0$ and $A > 0$

c. a saddle point at (a, b) if $AC - B^2 < 0$

The test fails if $AC - B^2 = 0$.

Let us look at some examples.

Example 7.33

Find all critical points and test them for relative maxima and minima for the function

$$f(x, y) = 4 + x^2 + 3y^2$$

Solution

Taking first order partial derivatives, we get

$$\begin{aligned} f_x(x, y) &= 2x \\ f_y(x, y) &= 6y \end{aligned}$$

Setting $f_x(x, y) = 0$ and $f_y(x, y) = 0$, we find that $(0, 0)$ is the only critical point of f. To investigate whether $(0, 0)$ is a point of relative maxima, relative minima, or neither, we compute the second order partial derivatives at (0, 0). Thus,

$$A = f_{xx}(0,0) = 2, \quad B = f_{xy}(0,0) = 0, \quad C = f_{yy}(0,0) = 6$$

and

$$AC - B^2 = 12$$

Since $AC - B^2$ and A are both positive, we conclude that f has a relative minimum at (0, 0).

Example 7.34

Find all relative maximum and minimum values of the function

$$f(x, y) = x^2 + xy + y^2 - 3x + 16$$

Solution

Here,

$$f_x(x, y) = 2x + y - 3 \quad \text{and} \quad f_y(x, y) = x + 2y$$

Setting each partial derivative equal to zero, we get

$$2x + y - 3 = 0$$
$$x + 2y = 0$$

Solving $x + 2y = 0$ for x, we have $x = -2y$. Substituting $x = -2y$ in the equation $2x + y - 3 = 0$, we get

$$2(-2y) + y - 3 = 0$$
$$-3y - 3 = 0$$
$$y = -1$$

and $x = -2y = (-2)(-1) = 2$. Thus $(2, -1)$ is the only critical point which needs further investigation. Computing second order partial derivatives, we have

$$A = f_{xx}(2, -1) = 2, \quad B = f_{xy}(2, -1) = 1, \quad C = f_{yy}(2, -1) = 2$$

and

$$AC - B^2 = 2 \cdot 2 - (1)^2 = 3$$

Thus, $AC - B^2 > 0$ and $A > 0$. Hence, f has a relative minimum at $(2, -1)$. This minimum value of f is

$$f(2, -1) = 2^2 + 2(-1) + (-1)^2 - 3(2) + 16$$
$$= 4 - 2 + 1 - 6 + 16$$
$$= 13$$

Example 7.35

Find all relative maximum and minimum values of the function

$$f(x, y) = x^3 + y^3 - 3xy + 4$$

Solution

The first order partial derivatives of f are

$$f_x(x, y) = 3x^2 - 3y \quad \text{and} \quad f_y(x, y) = 3y^2 - 3x$$

Setting $f_x(x, y)$ and $f_y(x, y)$ equal to zero, we have

$$3x^2 - 3y = 0 \quad \text{or} \quad x^2 = y$$

and

$$3y^2 - 3x = 0 \quad \text{or} \quad y^2 = x$$

Substituting $x = y^2$ in the first equation $x^2 = y$, we get

$$y^4 = y$$
$$y^4 - y = 0$$
$$y(y^3 - 1) = 0$$
$$y = 0 \quad \text{or} \quad y = 1$$

If $y = 0$ then $x = 0$, and if $y = 1$ then $x = 1$. Thus $(0, 0)$ and $(1, 1)$ are the two critical points. Computing second order partial derivatives, we obtain

$$f_{xx}(x, y) = 6x, \quad f_{xy}(x, y) = -3, \quad f_{yy}(x, y) = 6y$$

At the critical point $(0, 0)$,

$$A = f_{xx}(0, 0) = 0, \quad B = f_{xy}(0, 0) = -3, \quad C = f_{yy}(0, 0) = 0$$

and

$$AC - B^2 = 0 - (-3)^2 = -9$$

The fact that $AC - B^2 < 0$ implies that $(0, 0)$ is a saddle point. For the critical point $(1, 1)$,

$$A = f_{xx}(1, 1) = 6, \quad B = f_{xy}(1, 1) = -3, \quad C = f_{yy}(1, 1) = 6$$

and

$$AC - B^2 = 36 - 9 = 27$$

Since $AC - B^2$ and A are both positive, it follows that f has a relative minimum at $(1, 1)$. This minimum value of f is

$$f(1, 1) = 1 + 1 - 3 + 4$$
$$= 3$$

Exercises

Find all relative maximum and minimum values of the functions in each of the following problems.

1. $f(x, y) = x^2 + y^2 - 2x + 4y - 2$

2. $f(x, y) = x^2 + y^2 + 2x - 4y$

3. $f(x, y) = x^2 + xy + y^2 + 3x - 4y$

4. $f(x, y) = x^2 + xy + y^2 - x - 2y + 1$

5. $f(x, y) = x^2 + 3xy + 3y^2 - 6x + 3y + 63$

6. $f(x, y) = x^2 - xy + y^2 + 2x + 2y$

7. $f(x, y) = 5xy - 7x^2 - y^2 + 3x - 6y$

8. $f(x, y) = 2xy - 5x^2 - 2y^2 + 4x + 4y$

9. $f(x, y) = 7x^2 - 8xy + 3y^2$

10. $f(x, y) = y^2 + xy - 2x - 2y + 2$

11. $f(x, y) = x^2 + xy + 3x + 2y + 6$

12. $f(x, y) = xy - x^3 - y^2$

13. $f(x, y) = x^3 + y^3 - 9xy$

14. $f(x, y) = 2xy - x^3 - y^2$

7.8 Lagrange Multipliers*

This section is intended to give an intuitive, geometric explanation of the use of *Lagrange multipliers* to maximize or minimize a function subject to a side condition. Suppose, for instance, that we need to determine the minimal surface area of a right cylindrical tin can of radius r and height h that would enclose a volume of 100 cubic inches. In the language of mathematics, the problem is to minimize the surface area

$$S = 2\pi r^2 + 2\pi rh$$

subject to the constraint that

$$\pi r^2 h = 100$$

We will illustrate the method of Lagrange multipliers by using it to solve this problem.

The first step is to introduce a new variable λ and form a function, say u, of the variables r, h, and λ. The new function is

7.6 $$u(r, h, \lambda) = (2\pi r^2 + 2\pi rh) + \lambda(\pi r^2 h - 100)$$

* This section may be omitted, if necessary, without any loss of continuity.

Then we compute the partial derivatives

$$\frac{\partial u}{\partial r}, \quad \frac{\partial u}{\partial h}, \quad \text{and} \quad \frac{\partial u}{\partial \lambda}$$

and obtain

$$\frac{\partial u}{\partial r} = 4\pi r + 2\pi h + 2\pi\lambda rh$$

$$\frac{\partial u}{\partial h} = 2\pi r + \lambda\pi r^2$$

$$\frac{\partial u}{\partial \lambda} = \pi r^2 h - 100$$

Setting these partial derivatives equal to zero, we have three equations in the three unknowns r, h, and λ.

7.7 $$4\pi r + 2\pi h + 2\pi\lambda rh = 0$$

7.8 $$2\pi r + \lambda\pi r^2 = 0$$

7.9 $$\pi r^2 h - 100 = 0$$

From (7.8), we have

$$\pi r(2 + \lambda r) = 0$$

Since πr is not zero, it follows that

7.10 $$\lambda r = -2$$

Substituting $\lambda r = -2$ in (7.7), we get

$$2\pi(2r + h - 2h) = 0$$

$$2\pi(2r - h) = 0$$

7.11 $$h = 2r$$

The equation (7.11) shows that the can of minimal surface area has a diameter equal to the height.

To find r and h explicitly, we need to combine (7.9) and (7.11). This yields

$$\pi r^2(2r) - 100 = 0$$

$$2\pi r^3 = 100$$

$$r^3 = \frac{50}{\pi}$$

Thus

$$r = \left(\frac{50}{\pi}\right)^{1/3} \quad \text{and} \quad h = 2\left(\frac{50}{\pi}\right)^{1/3}$$

Let us now exhibit a general method for the Lagrange multipliers.

Problem

Find the maximum or minimum of the function $v = f(x, y)$ subject to the condition that $g(x, y) = 0$. We suggest the following steps:

Step 1. Introduce a new variable λ and set up the function

$$u(x, y, \lambda) = f(x, y) + \lambda \cdot g(x, y)$$

Step 2. Find the partial derivatives

$$\frac{\partial u}{\partial x}, \quad \frac{\partial u}{\partial y}, \quad \text{and} \quad \frac{\partial u}{\partial \lambda}$$

Step 3. Find all points (x, y) that satisfy the equations

$$\frac{\partial u}{\partial x} = 0, \quad \frac{\partial u}{\partial y} = 0, \quad \text{and} \quad \frac{\partial u}{\partial \lambda} = 0$$

Although the maximum or minimum need not occur at a solution to the equations in Step 3, in many cases it does. Moreover, one can usually decide from the problem itself for which of the solutions the maximum or minimum occurs.

Example 7.36

Maximize $f(x, y) = x^2y$ subject to the condition that $x^2 + y^2 = 12$.

Solution

Here

$$f(x, y) = x^2y \quad \text{and} \quad g(x, y) = x^2 + y^2 - 12$$

The function $u(x, y, \lambda)$ is

7.12 $$u(x, y, \lambda) = x^2y + \lambda \cdot (x^2 + y^2 - 12)$$

The three partial derivatives, set equal to zero, are

7.13 $$\frac{\partial u}{\partial x} = 2xy + 2\lambda x = 0$$

7.14 $$\frac{\partial u}{\partial y} = x^2 + 2\lambda y = 0$$

7.15 $$\frac{\partial u}{\partial \lambda} = x^2 + y^2 - 12 = 0$$

From (7.13), we have

$$2x(y + \lambda) = 0$$

Since x is not zero, it follows that

7.16 $$y = -\lambda$$

Substituting $y = -\lambda$ in (7.14), we get

$$x^2 - 2y^2 = 0$$

7.17 $$x^2 = 2y^2$$

Combining (7.15) and (7.17), we have

$$2y^2 + y^2 = 12$$
$$3y^2 = 12$$
$$y^2 = 4$$
$$y = 2 \quad \text{or} \quad y = -2$$

From (7.17),

$$x = \sqrt{2}y \quad \text{or} \quad x = -\sqrt{2}y$$

Thus the four points to be considered on the circle $x^2 + y^2 = 12$ are

$$(2\sqrt{2}, 2), \quad (-2\sqrt{2}, 2), \quad (2\sqrt{2}, -2) \quad \text{and} \quad (-2\sqrt{2}, -2)$$

Observe that x^2y is positive at the points $(2\sqrt{2}, 2)$ and $(-2\sqrt{2}, 2)$, whereas x^2y is negative at the points $(2\sqrt{2}, -2)$ and $(-2\sqrt{2}, -2)$. The first two points provide the maximum value of x^2y on the circle $x^2 + y^2 = 12$, namely

$$(\pm 2\sqrt{2})^2 \cdot 2 = 16$$

The third and fourth points provide the minimum value of x^2y, namely

$$(\pm 2\sqrt{2})^2(-2) = -16$$

The Lagrange multipliers method can be extended to functions of three variables. In order to find critical values of the function $w = f(x, y, z)$ subject to the constraint $g(x, y, z) = 0$, we introduce, as before, a new variable λ and set up a function of four variables. We go through the same three steps suggested earlier.

Step 1. The new function is

$$u(x, y, z, \lambda) = f(x, y, z) + \lambda \cdot g(x, y, z)$$

Step 2. Find the partial derivatives

$$\frac{\partial u}{\partial x}, \quad \frac{\partial u}{\partial y}, \quad \frac{\partial u}{\partial z}, \quad \text{and} \quad \frac{\partial u}{\partial \lambda}$$

Step 3. Find all points (x, y, z) that satisfy the four equations:

$$\frac{\partial u}{\partial x} = 0, \qquad \frac{\partial u}{\partial y} = 0$$

$$\frac{\partial u}{\partial z} = 0, \qquad \frac{\partial u}{\partial \lambda} = 0$$

We shall not discuss a method for determining whether a critical point is a maximum point, a minimum point, or neither. In many applied problems, this difficulty is overcome by the nature of the problem.

Example 7.37

Find the minimum value of

$$f(x, y, z) = x^2 + y^2 + z^2$$

subject to the condition that $x + y + z = 8$.

Solution

The new function is

$$u(x, y, z, \lambda) = (x^2 + y^2 + z^2) + \lambda(x + y + z - 8)$$

Computing the partial derivatives and setting them equal to zero, we obtain

$$\frac{\partial u}{\partial x} = 2x + \lambda = 0$$

$$\frac{\partial u}{\partial y} = 2y + \lambda = 0$$

$$\frac{\partial u}{\partial z} = 2z + \lambda = 0$$

and

$$\frac{\partial u}{\partial \lambda} = x + y + z - 8$$

From the first three equations, we get

$$x = y = z$$

Substituting this into the last equation, we have

$$x = y = z = \tfrac{8}{3}$$

and so the minimum value of $f(x, y, z) = (\frac{8}{3})^2 + (\frac{8}{3})^2 + (\frac{8}{3})^2 = \frac{64}{3}$.

Example 7.38

A rectangular box is to be made out of 294 square inches of cardboard material. Determine the dimensions of the box that would enclose the maximum volume.

Solution

Let x, y, and z be the dimensions of the rectangular box as shown in Figure 7.22. The volume of the box is

$$f(x, y, z) = xyz$$

and the surface area of the box is

$$2xy + 2yz + 2zx$$

Thus,

$$2xy + 2yz + 2zx = 294$$

The problem is to maximize $f(x, y, z) = xyz$ subject to the constraint that $2xy + 2yz + 2zx = 294$. The new function of four variables is

7.18 $$u(x, y, z, \lambda) = xyz + \lambda(2xy + 2yz + 2zx - 294)$$

Now we find partial derivatives of u with respect to x, y, z, and λ and set these equal to zero. We obtain

7.19 $$\frac{\partial u}{\partial x} = yz + \lambda(2y + 2z) = 0$$

7.20 $$\frac{\partial u}{\partial y} = zx + \lambda(2x + 2z) = 0$$

7.21 $$\frac{\partial u}{\partial z} = xy + \lambda(2y + 2x) = 0$$

and

7.22 $$\frac{\partial u}{\partial \lambda} = 2xy + 2yz + 2zx - 294 = 0$$

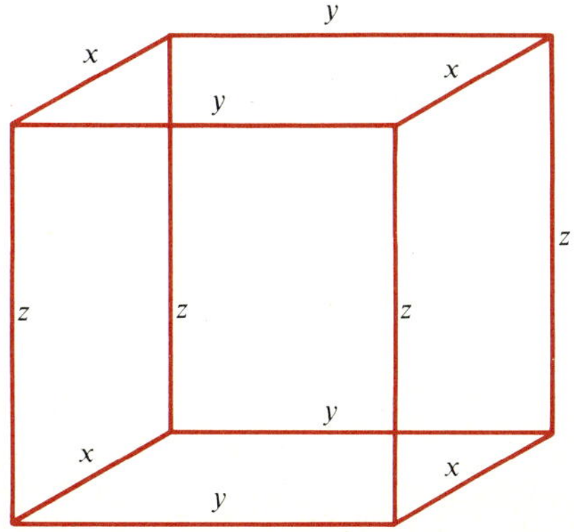

Figure 7.22

From (7.19) and (7.20), we have

$$(x - y)(z + 2\lambda) = 0$$

from which it follows that either $z = -2\lambda$ or $x = y$. If $z = -2\lambda$, then from (7.19), we get

$$\begin{aligned} y(-2\lambda) + \lambda(2y - 4\lambda) &= 0 \\ -4\lambda^2 &= 0 \\ \lambda &= 0 \end{aligned}$$

Setting $\lambda = 0$ in (7.19), (7.20), and (7.21), we obtain

$$yz = zx = xy = 0$$

and this contradicts (7.22). Thus $x = y$. Solving (7.20) and (7.21), we get $y = z$. Thus,

$$x = y = z \tag{7.23}$$

Using this condition in (7.22), we can easily determine that

$$x = y = z = 7 \text{ inches}$$

Exercises

In Problems 1–16, use the method of Lagrange multipliers.

1. Maximize $f(x, y) = xy$ subject to the condition that $4x + 3y - 1200 = 0$.
2. Maximize $f(x, y) = xy$ subject to $x^2 + y^2 = 4$
3. Maximize $f(x, y) = 4xy^2$ subject to $3x - 2y = 5$
4. Minimize $f(x, y) = x^2 + y^2$ subject to $x + 2y = 10$
5. Minimize $f(x, y) = x^2 + 2y^2 - xy$ for points on the line $x + y = 8$
6. Minimize $f(x, y) = x^2 + y^2$ for points on the line $2x + 3y - 6 = 0$
7. Maximize $f(x, y, z) = xyz$ subject to the condition that $x + y + z = 50$
8. Maximize $f(x, y, z) = xyz$ subject to $x + 3y + 4z = 108$
9. Maximize $f(x, y, z) = xyz$ subject to $x^2 + y^2 + z^2 = 1$
10. Maximize $f(x, y, z) = xy^2z^3$ subject to $x + y + z = 50$
11. Minimize $f(x, y, z) = x^2 + y^2 + z^2$ subject to $x + y + z = 50$
12. Find three numbers x, y, and z whose sum is 90 and where product is maximum.
13. A farmer wants to fence three sides of a rectangular field. Assuming that the total area of the plot is 5,000 square feet, find the length and width of the plot that minimizes the number of feet of fencing.

14. A rectangular wooden box with an open top is to contain 108 cubic inches of some material. Ignoring the thickness of the wood, determine the dimensions of the box that minimize the amount of wood needed for the box.

15. Determine the maximum volume of an open-top rectangular box that can be built from 108 square meters of wood.

16. A manufacturing company has three plants, A, B, and C, which produce x, y, and z units, respectively. The annual revenue from the production is given by

$$R(x, y, z) = 6xyz^2 - 400{,}000(x + y + z)$$

If the company is to produce 1,000 units annually, how should the production be allocated so as to maximize profits?

7.9 The Method of Least Squares

In this section we shall consider the problem of estimating or predicting the value of a dependent variable y on the basis of a known value of an independent variable x. For example, in business and economic situations we may like to predict the consumption of gasoline in terms of price per gallon, the potential sales of new cars in terms of auto industry rebates, the import of foreign crude oil in terms of new tax levies, the spending budget of a state in terms of its revenue, the salaries of employees in terms of their experience, the amount a family saves in terms of its annual income, and so on. In education, we may want to predict the grade point average of students at the end of their freshman year in relation to their verbal scores on the Scholastic Aptitude Test. In agriculture, a farmer may like to predict the yield of wheat in terms of the seasonal rainfall.

Of the many mathematical equations used for prediction, the simplest and most widely used is the linear equation

7.24 $$\hat{y} = mx + b$$

where m is the slope of the line and the constant b is the value of $\hat{y}$ (read "y-hat") at the y-axis when $x = 0$.

An example will serve to introduce the concepts and techniques of the method of least squares which can be used to estimate m and b. Six used car dealers were surveyed to determine the relationship between the amount of classified advertising and subsequent car sales. The data in Table 7.1 show the amount of classified advertising (in hundreds of lines) and the numbers of cars sold for each of the six randomly selected dealers. We shall use this data to predict the number of used cars sold on the basis of the classified advertisement.

Table 7.1

Advertising (in hundreds of lines) x	Number of Used Cars Sold y
74	139
45	108
48	98
36	76
27	62
16	57

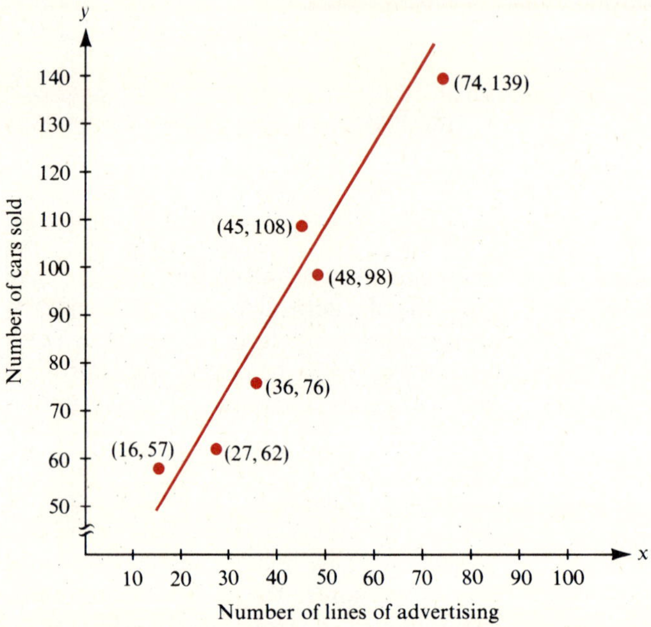

Figure 7.23

An initial approach to the analysis of the information in Table 7.1 is to plot the data as points on a graph (see Figure 7.23). The number of lines of advertising is the independent variable and is represented along the horizontal axis, whereas the number of used cars sold, the dependent variable, is shown along the vertical axis. Plotting these pairs of observations, we note that the points (x, y) on the graph are spread in an irregular pattern called a *scatter diagram*. However, the points are clustered along a straight line which may be used to estimate the relationship between the number of used cars sold and the number of lines of advertising.

We want to determine the equation of the line, $y = mx + b$, that in some sense provides the "best" fit to the set of points (x, y) in Figure 7.23. To do so, we need to determine the constants m and b in such a way that the points in the scatter diagram lie as close to the line as possible. Note that after the line of best fit has been determined, it will yield a y-value for each x-value. This y-value is an *estimate* of the actual value and is denoted by $\hat{y}$. To determine the line (7.24) of best fit, we use the *method of least squares*. According to this method, the meaning of "best fit" is that the sum of the squares of the vertical distances from the plotted points to the line is minimized, and m and b are chosen to minimize that quantity. Thus the problem reduces to finding, for a given set of observations, the constants m and b in the equation (7.24) in such a way that

$$\sum_{i=1}^{6} (y_i - \hat{y}_i)^2$$

is minimized*.

To explain how the method of least squares can be used to determine this *line of best fit*, consider n measurements. Each measurement gives us an x_i, which is a value of a quantity x, and a y_i, which is a value of a quantity y, $i = 1, 2, 3, \ldots, n$. This gives us a set of n pairs, (x_1, y_1), $(x_2, y_2), \ldots, (x_n, y_n)$, which might represent students' verbal scores on the Scholastic Aptitude Test and grade point averages at the end of the freshman year, test scores of job applicants and their production ratings, the number of employees in a department store and the store's monthly sales volume, average heights and weights of men in Southeast Asia, and so on. When plotted on a graph, if the points (x_i, y_i) seem to cluster around a straight line, we must determine the equation of that line which provides the best fit to the data (see Figure 7.24). Note that for each value of x_i we have an *observed* value y_i (the scattered points) and

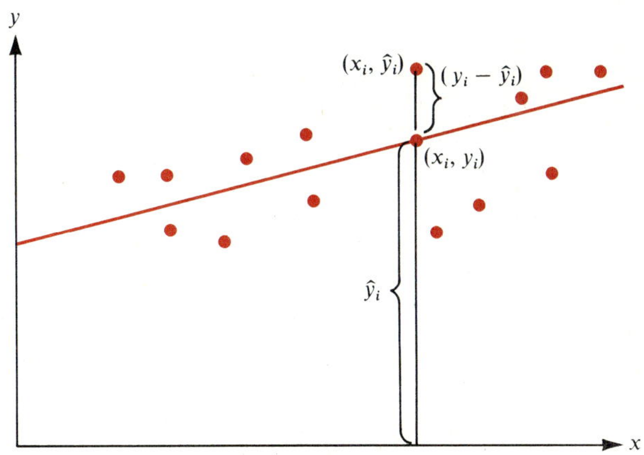

Figure 7.24

* See Appendix B for Summation Notation.

an *estimated* value $\hat{y}_i$ (the points on the line of best fit). The least-squares criterion requires that the constants m and b in the equation of the line be determined so as to minimize

7.25 $$z = \sum_{i=1}^{n} (y_i - \hat{y}_i)^2 = \sum_{i=1}^{n} [y_i - (mx_i + b)]^2 = \sum_{i=1}^{n} (y_i - mx_i - b)^2$$

We find the partial derivatives of z with respect to the unknowns b and m. These are

$$\frac{\partial z}{\partial b} = -\sum_{i=1}^{n} 2(y_i - mx_i - b)$$

and

$$\frac{\partial z}{\partial m} = -\sum_{i=1}^{n} 2x_i(y_i - mx_i - b)$$

Setting $\partial z/\partial b$ and $\partial z/\partial m$ equal to zero and rearranging terms, we obtain

$$\sum_{i=1}^{n} (y_i - mx_i - b) = 0$$

and

$$\sum_{i=1}^{n} x_i(y_i - mx_i - b) = 0$$

This leads us to a system of two equations in two unknowns, b and m. We can manipulate the equations to obtain

7.26 $$\sum_{i=1}^{n} y_i = nb + m \sum_{i=1}^{n} x_i$$

7.27 $$\sum_{i=1}^{n} x_i y_i = b \sum_{i=1}^{n} x_i + m \sum_{i=1}^{n} x_i^2$$

To solve these equations for m and b, we multiply the first equation (7.26) by $\sum_{i=1}^{n} x_i$, the second equation (7.27) by n, subtract the first equation from the second, and solve for m. We obtain

7.28 $$m = \frac{n \sum_{i=1}^{n} x_i y_i - \left(\sum_{i=1}^{n} x_i\right)\left(\sum_{i=1}^{n} y_i\right)}{n \sum_{i=1}^{n} x_i^2 - \left(\sum_{i=1}^{n} x_i\right)^2}$$

Substituting this value of m in (7.26), we solve for b:

7.29 $$b = \frac{\sum_{i=1}^{n} y_i - m \sum_{i=1}^{n} x_i}{n}$$

To compute m and b in the problem of the number of lines advertised and the number of cars sold, we calculate

$$\sum_{i=1}^{6} x_i, \quad \sum_{i=1}^{6} y_i, \quad \sum_{i=1}^{6} x_i^2, \quad \sum_{i=1}^{6} x_i y_i$$

These calculations are shown in Table 7.2.

Table 7.2

x_i	y_i	$x_i y_i$	x_i^2
74	139	10,286	5,476
45	108	4,860	2,025
48	98	4,704	2,304
36	76	2,736	1,296
27	62	1,674	729
16	57	912	256
246	540	25,172	12,086

Thus,

$$\sum_{i=1}^{6} x_i = 246, \qquad \sum_{i=1}^{6} y_i = 540$$

$$\sum_{i=1}^{6} x_i y_i = 25172, \qquad \sum_{i=1}^{6} x_i^2 = 12086$$

Using (7.28), we obtain

$$m = \frac{6(25172) - (246)(540)}{6(12086) - (246)^2} = \frac{18192}{12000} = 1.516$$

and from (7.29) we obtain

$$b = \frac{540 - (1.516)(246)}{6} = \frac{167.064}{6} = 27.844$$

Thus the equation of the line of best fit is

$$\hat{y} = 1.516x + 27.844$$

We must caution the reader that this linear equation gives an *estimated* value of y corresponding to each value of x. These are the "best" predictions of the actual values in a specific sense; the sum of the squares of the deviations of the predicted value of $\hat{y}$ from the actual value of y is minimized.

In order to graph the line of best fit, we need to determine two points on the line. For $x = 45$ we have $\hat{y} = 96.064$, and for $x = 60$ we obtain $\hat{y} = 118.804$. Connecting these points, we obtain the graph of the line of best fit (see Figure 7.23).

Exercises

1. A manufacturer keeps a monthly record of sales per month and the corresponding costs. For the first five months of 1980, the data are as follows:

Advertising Expenditures	1	2	3	4	5
Sales	2	4	7	9	11

Find the equation of the line of best fit for this data. What are the estimated sales for advertising expenditures of 6 units and 10 units?

2. A study is made to relate employee test scores and factory productivity after three months of personnel training. The six pairs of scores shown below were obtained by testing six randomly selected applicants and later measuring each one's productivity.

Aptitude Score, x	9	17	20	19	20	23
Productivity, y	23	35	29	33	43	32

a. Plot the data.
b. Find the equation of the line of best fit which can be used to estimate the productivity from the aptitude score.
c. Determine the estimated productivity for an applicant with an aptitude score of 18.

3. The following data show the advertising costs (x) and the weekly sales (y) in eight randomly selected weeks in 1980.

Advertising costs, x	$ 40	20	25	30	50	40	50	25
Weekly sales, y	$ 490	365	480	475	560	525	510	400

a. Find the line of best fit for the data.
b. What are the expected sales per month for advertising costs of 12 or of 15 units?

4. The following data show the number of automobile registrations and the total consumption of gasoline in seven towns.

Auto Registrations (in thousands)	1.1	3.2	4.1	6.2	7.5	7.9	8.5
Gasoline Consumption (in millions of gallons)	1.0	2.0	2.2	2.5	3.0	3.2	3.5

a. Plot the data.
b. Determine the line of best fit.
c. What is the predicted consumption of gasoline if the auto registrations are set at 5,700?

5. The following experimental data show the amount of water applied (in inches) and the corresponding rice yield (in tons per acre).

Water, x	12	18	24	30	36	42	48
Yield, y	5.27	6.8	6.25	7.21	8.02	8.71	8.42

Obtain the line of best fit for the yield of rice as a function of the amount of water applied. Estimate the most probable yield of rice for 40 inches of water.

6. The following data show the Intelligence Test Score (x) and Mathematics Achievement Score (y) for ten students randomly selected from a high school freshman class.

Intelligence Test Score (x)	45	55	56	58	60	65	68	70	75	80
Mathematics Achievement Score (y)	76	70	68	80	82	84	85	90	94	92

a. Plot the data.
b. Obtain the line of best fit for the data.
c. Estimate the Mathematics Achievement Score for a student who gets an Intelligence Test Score of 50.

7. Ten students randomly selected from a college sophomore class were asked how many hours they had studied for their Statistics final examination. The responses, along with the students' scores on the final examination, are given below.

Number of Hours Studied, x	2	3	4	5	6	7	8	9	10	12
Scores on the Final Examination, y	40	45	60	55	50	65	80	95	90	80

a. Plot the data.
b. Find the line of best fit which may be used to predict the final examination score from the number of hours studied.
c. Estimate the score for a student who studies for 7.5 hours.

8. A college administration collected the following data on the number of applications received by January 1 and the number of new students enrolled in the following fall.

Year	Number of Applications (x)	Number of New Students Registered (y)
1975	2350	1500
1976	2190	1420
1977	2380	1520
1978	2570	1640
1979	2810	1890
1980	2590	1680

a. Determine the line of best fit using the method of least squares.
b. Estimate the number of new students who will register next fall if 2,700 students have applied for admission by January 1.

9. Let x and y represent the stockmarket index and sales in a department store, respectively. The following data show the calculations from a sample of 12 observations.

$$\sum_{i=1}^{12} x_i = 725, \qquad \sum_{i=1}^{12} x_i^2 = 44475$$

$$\sum_{i=1}^{12} y_i = 1011, \qquad \sum_{i=1}^{12} x_i y_i = 61865$$

Determine the line of best fit and estimate the sales for a stockmarket index of 9.2.

10. An automobile dealer collected data for the past three years on the number of inquiries about cars received during a week and the number of cars sold during the following week. The data collected can be summarized as follows:

$$\sum_{i=1}^{156} x_i = 54{,}649 \qquad \sum_{i=1}^{156} y_i = 11{,}492$$

$$\sum_{i=1}^{156} x_i y_i = 4{,}047{,}975 \qquad \sum_{i=1}^{156} x_i^2 = 855{,}766$$

where x = number of inquiries received during a given week
y = number of cars sold during the next week
n = 156 weeks

a. Determine the line of best fit using the method of least squares.
b. Estimate the number of cars that will be sold during the next week if the dealer receives 400 inquiries this week.

Exercises for Review

1. Define each of the following concepts:
   - a. a function of two variables
   - b. the partial derivative of $f(x, y)$ with respect to x
   - c. the partial derivative of $f(x, y)$ with respect to y
   - d. a second order partial derivative
   - e. relative minimum
   - f. relative maximum
   - g. saddle point
   - h. marginal productivities

2. Plot the following points in three-dimensional space:
   - a. $P(2, 3, 4)$
   - b. $Q(0, 2, -3)$

3. Find the distance between the points $P(-3, 5, 6)$ and $Q(-6, 4, -2)$.

4. Find the equation of the sphere with center $C(1, -2, 3)$ and radius 5.

5. Given that

$$f(x, y) = \frac{x^2 - y^2}{x^2 + y^2}$$

   find each of the following:
   - a. $f(1, 0)$
   - b. $f(0, 1)$
   - c. $f(1, 1)$
   - d. $f(2, -1)$
   - e. $f(3, 4)$
   - f. $f(12, 5)$

6. Sketch the graph of each of the following:
   - a. $z = 2$
   - b. $3x + y = 6$
   - c. $2x + 3y + 4z = 12$
   - d. $(x - 3)^2 + (y - 2)^2 + (z - 5)^2 = 9$

7. Evaluate each of the following limits, if it exists.
   - a. $\displaystyle\lim_{(x,y)\to(0,0)} \left(\frac{x^3 - y^3}{x^2 - y^2}\right)$
   - b. $\displaystyle\lim_{(x,y)\to(0,0)} \left(\frac{x^4 - y^4}{x^4 + y^4}\right)$

8. Find the partial derivatives $f_x(x, y)$ and $f_y(x, y)$ for each of the following functions.
   - a. $f(x, y) = x^3y^2$
   - b. $f(x, y) = xe^y$
   - c. $f(x, y) = \sqrt{x^2 + y^2} + xe^{2y}$
   - d. $f(x, y) = \dfrac{x}{y} + \dfrac{y}{x}$

9. Evaluate $f_x(x, y)$ and $f_y(x, y)$ for each of the following functions at the point indicated.
   - a. $f(x, y) = 2x^3y^2, \quad x = 2, y = 3$
   - b. $f(x, y) = x + x^2y, \quad x = 1, y = -1$
   - c. $f(x, y) = x^3e^y, \quad x = 2, y = 0$
   - d. $f(x, y) = \sqrt{x^2 + y^2}, \quad x = 3, y = 4$

10. Find f_{xx}, f_{xy}, f_{yx} and f_{yy} for each of the following functions.

a. $f(x, y) = x^2y + xy^2$ **c.** $f(x, y) = \dfrac{1}{x + y}$

b. $f(x, y) = \dfrac{4x}{y} + y^3$ **d.** $f(x, y) = \log(3x^2 + 4y^2)$

11. Given that

$$f(x, y) = x^3y^3 + 2x^2y$$

show that $f_{xy}(x, y) = f_{yx}(x, y)$.

12. Given that

$$f(x, y) = \frac{x}{x + y}$$

show that $f_{xy}(x, y) = f_{yx}(x, y)$.

13. A company's profit function from the sale of x units of product A and y units of product B is

$$P(x, y) = 4x^3 + 3xy^2 + 2y^3 - 100.$$

Find the marginal profit for $x = 2$ and $y = 3$.

14. Find the marginal productivities for each of the following production functions.
a. $f(x, y) = 300(4x^2 + 3y^2)$, $x = 2$, $y = 3$
b. $f(x, y) = 45x^{2/3}y^{1/3}$, $x = 27$, $y = 8$

15. Find all relative maximum and minimum values for each of the following functions.
a. $f(x, y) = 2xy - 2x^2 - 3y^2 + 900x + 1800y$
b. $f(x, y) = x^2 + y^2 + xy - 9x - 6y + 50$
c. $f(x, y) = 2xy + x^3 - y^3$
d. $f(x, y) = x^2 + y^3 - 2x - 3y - 2xy$

16. Maximize the function

$$f(x, y, z) = xyz$$

subject to the constraint that $x + y + z = 60$.

17. Maximize the function

$$f(x, y, z) = 4xyz$$

subject to the constraint that $x + 2y + 3z = 54$.

18. A company's profit from the sale of x units of product A and y units of product B is given by

$$P(x, y) = 3xy + y^2 - 6x + 100$$

If the company is to produce 50 units, how should the production be allocated so as to maximize profits?

19. A farmer wants to fence three sides of a rectangular plot. Assuming that 60 feet of fencing is available, find the dimensions of the plot that would enclose the maximum area.

20. An open rectangular glass tank is to contain 256 cubic feet of space. Determine the dimensions of the tank so as to minimize the amount of material needed to build the glass tank.

21. An economist collected the following data with respect to bank debits (x) and Federal Reserve discount rate (y) for the period 1972–78.

Year	Bank debits (in trillions of dollars)	Federal Reserve discount rate
1972	3.43	3.00
1973	3.75	3.28
1974	4.52	3.88
1975	5.13	4.38
1976	5.94	4.50
1977	6.35	4.25
1978	7.99	5.25

Determine the line of best fit using the method of least squares.

22. The following data show the undergraduate grade point average (x) and the score on a managerial trainee examination (y) for 12 randomly selected corporate employees.

Grade Point Average	2.2	2.4	3.1	2.5	3.5	3.6	2.5	2.0	2.2	2.6	2.7	3.3
Examination Score	76	89	83	79	91	95	82	69	66	75	80	85

a. Determine the line of best fit using the method of least squares.
b. Estimate the examination score for an employee who has a grade point average of 3.4.

Appendix A Review of Algebra

A.1 Products and Factoring

We know that if a, b, and c are real numbers, then the distributive property of multiplication over addition states that

$$a(b + c) = ab + ac$$

The real numbers a and $(b + c)$ are *factors* of $ab + ac$. Similarly, since $2ab$ is a common factor of the terms $6a^2b$ and $4ab^2$, we can write

$$6a^2b + 4ab^2 = 2ab(3a + 2b)$$

The terms $2ab$ and $(3a + 2b)$ are factors of the product $6a^2b + 4ab^2$.

For the real numbers x, a, and b, the distributive property gives us

$$\begin{aligned}(x + a)(x + b) &= (x + a)x + (x + a)b\\ &= x^2 + ax + bx + ab\\ &= x^2 + (a + b)x + ab\end{aligned}$$

Thus,

A.1 $$(x + a)(x + b) = x^2 + (a + b)x + ab$$

If $b = a$, then (A.1) can be written

A.2 $$(x + a)^2 = x^2 + 2ax + a^2$$

Replacing a by $(-a)$ in (A.2), we have

$$(x - a)^2 = x^2 + 2(-a)x + (-a)^2$$

that is,

A.3 $$(x - a)^2 = x^2 - 2ax + a^2$$

If $b = -a$ in (A.1), we obtain

A.4 $$(x + a)(x - a) = x^2 - a^2$$

Thus $(x + a)$ and $(x - a)$ are factors of $x^2 - a^2$. Let us look at some examples.

Example A.1

a. $(x + 4)(x + 5) = x^2 + (4 + 5)x + (4)(5)$
$= x^2 + 9x + 20$

b. $(x - 2)(x - 3) = x^2 + [(-2) + (-3)]x + (-2)(-3)$
$= x^2 - 5x + 6$

c. $(x + 6)(x - 1) = x^2 + [6 + (-1)]x + (6)(-1)$
$= x^2 + 5x - 6$

d. $(x + 8)^2 = x^2 + 2 \cdot 8 \cdot x + (8)^2$
$= x^2 + 16x + 64$

e. $(x - 5)^2 = x^2 - 2 \cdot 5 \cdot x + (-5)^2$
$= x^2 - 10x + 25$

f. $(x + 7)(x - 7) = x^2 - (7)^2$
$= x^2 - 49$

Example A.2

Factor

$$x^2 + 8x + 12$$

Solution

Comparing this expression with (A.1), we observe that

$$a + b = 8 \quad \text{and} \quad ab = 12$$

The problem now reduces to finding two numbers a and b with sum 8 and product 12. Since $ab = 12$, it follows that a and b are both positive or both negative. Further, since $a + b$ is positive, we conclude that a and b are both positive. There are only three ways to express 12 as the product of two positive numbers.

$$12 = 12 \cdot 1, \quad 12 = 6 \cdot 2, \quad \text{and} \quad 12 = 4 \cdot 3$$

The second set of factors, 6 and 2, sum to 8. Thus

$$x^2 + 8x + 12 = (x + 6)(x + 2)$$

Example A.3

Factor

$$x^2 - 10x + 24$$

Solution

If we compare this expression with (A.1), we note that

$$a + b = -10 \quad \text{and} \quad ab = 24$$

The fact that ab is positive means that a and b are both positive or both negative. We know that $a + b$ is a negative number, so a and b are both negative. There are four ways to express 24 as the product of two negative numbers; they are shown in Table A.1.

Table A.1

ab	a	b	$a + b$
24	-1	-24	-25
24	-2	-12	-14
24	-3	-8	-11
24	-4	-6	-10

The only numbers with sum -10 and product 24 are -4 and -6. Hence,

$$x^2 - 10x + 24 = (x - 4)(x - 6)$$

Example A.4

Factor

$$x^2 + 8x - 33$$

Solution

We proceed as before and compare this expression with (A.1). Observe that

$$a + b = 8 \quad \text{and} \quad ab = -33$$

Since ab is negative, it follows that one factor is negative and the other positive. The ways to express -33 as the product of one negative and one positive number are shown in Table A.2.

Table A.2

ab	a	b	$a + b$
-33	-1	33	$-1 + 33 = 32$
-33	1	-33	$1 + (-33) = -32$
-33	-3	11	$-3 + 11 = 8$
-33	3	-11	$3 + (-11) = -8$

The third set of factors, $a = -3$ and $b = 11$, satisfies the conditions that $a + b = 8$ and $ab = -33$. Thus

$$x^2 + 8x - 33 = (x - 3)(x + 11)$$

We now factor an expression of the form

A.5 $$ax^2 + bx + c, \quad a \neq 1, \quad a \neq 0$$

Consider the following example.

Example A.5

Factor

$$7x^2 - 19x + 10$$

Solution

Comparing this expression with (A.5), we note that $a = 7$, $b = -19$ and $c = 10$. First, we need to find two numbers with product $ac = (7)(10) = 70$, and sum $b = -19$. Since the product is positive, the two numbers are both positive or both negative; but since their sum is negative, the numbers are both negative. Table A.3 lists the ways to express 70 as the product of two negative numbers.

Table A.3

Products	Factors	Sum of Factors
70	$(-1)(-70)$	-71
70	$(-2)(-35)$	-37
70	$(-5)(-14)$	-19
70	$(-10)(-7)$	-17

The third set of factors, -5 and -14, satisfies the conditions. Next, we write

$$7x^2 - 19x + 10 = 7x^2 - 14x - 5x + 10$$

Observe that $7x$ is the greatest common factor in the first two terms $(7x^2)$ and $(-14x)$, and (-5) is a common factor in the last two terms, $(-5x)$ and (10). Thus

$$\begin{aligned} 7x^2 - 14x - 5x + 10 &= 7x(x - 2) - 5(x - 2) \\ &= (7x - 5)(x - 2) \end{aligned}$$

Hence,

$$7x^2 - 19x + 10 = (7x - 5)(x - 2)$$

A.2 Quadratic Equations

In this section we shall discuss the solution of the quadratic equation

A.6 $$ax^2 + bx + c = 0$$

where a, b, and c are real numbers and a is not zero. By *a solution* of an equation we mean a number that satisfies the equation, while *the solution set* means the set of all real numbers that satisfy the given equation. Solving an equation is thus a matter of describing explicitly the solution set of the equation.

In order to solve quadratic equations, it is necessary that we introduce the following property of real numbers.

Proposition A.1

If a and b are real numbers and $ab = 0$, then either $a = 0$ or $b = 0$.

Let us consider some illustrative examples.

Example A.6

Solve the equation

$$x^2 - 5x + 4 = 0$$

Solution

First, factor $x^2 - 5x + 4$ to obtain

$$x^2 - 5x + 4 = (x - 1)(x - 4)$$

Since $(x - 1)(x - 4) = 0$, then by Proposition A.1, either $(x - 1) = 0$ or $(x - 4) = 0$. If $x - 1 = 0$ then $x = 1$ while if $x - 4 = 0$ then $x = 4$. Thus the solution set of the equation $x^2 - 5x + 4 = 0$ is $\{1, 4\}$.

Example A.7

Solve the equation

$$x^2 - x - 12 = 0$$

Solution

Factoring the equation yields

$$(x - 4)(x + 3) = 0$$

By Proposition A.1, either $x - 4 = 0$ or $x + 3 = 0$.

$$x - 4 = 0 \quad \text{implies that} \quad x = 4.$$

$$x + 3 = 0 \quad \text{implies that} \quad x = -3.$$

The solution set of the equation $x^2 - x - 12 = 0$ is $\{-3, 4\}$.

Example A.8

Solve the equation

$$7x^2 - 19x + 10 = 0$$

Solution

In Example A.5 we established that

$$7x^2 - 19x + 10 = (7x - 5)(x - 2)$$

Since $(7x - 5)(x - 2) = 0$ then, by Proposition A.1, either $7x - 5 = 0$ or $x - 2 = 0$. If $7x - 5 = 0$ then $x = 5/7$; if $x - 2 = 0$ then $x = 2$. Hence, the solution set is $\{\frac{5}{7}, 2\}$.

In the above three examples, we were able to determine the solution set for each equation by factoring. Since this method is not always successful, we shall develop a general method for solving quadratic equations. The method we will develop is called *completing the square.*

Observe that expressions such as $x^2 + 4x + 4$ and $x^2 - 6x + 9$ are perfect squares:

$$x^2 + 4x + 4 = (x + 2)^2$$

and

$$x^2 - 6x + 9 = (x - 3)^2$$

Now suppose we are given $x^2 + 10x$. What should we add to $x^2 + 10x$ so that the result is a perfect square? Observe that by adding 25 we obtain $x^2 + 10x + 25$ which equals $(x + 5)^2$. Thus, by adding 25 to the expression $x^2 + 10x$ we have completed the square. In general, if we have an expression of the form $x^2 + kx$, we add the square of one-half the coefficient of x, that is $(k/2)^2$, so that the resulting expression is a perfect square.

Example A.9

The equation

$$x^2 + 6x - 1 = 0$$

cannot be solved by factoring. Therefore, we write the given equation in the form

$$x^2 + 6x = 1$$

and complete the square of the left-hand side. To do that, we add the square of one-half the coefficient of x, that is 9, to both sides of the equation, and obtain

$$x^2 + 6x + 9 = 1 + 9$$

$$(x + 3)^2 = 10$$

Taking the square root of each side of the equation, we get

$$x + 3 = \sqrt{10} \qquad \text{or} \qquad x + 3 = -\sqrt{10}$$
$$x = -3 + \sqrt{10} \quad \text{or} \qquad x = -3 - \sqrt{10}$$

The solution set is $\{-3 - \sqrt{10}, -3 + \sqrt{10}\}$

We shall now apply the method of completing the square to the general quadratic equation

$$ax^2 + bx + c = 0, \qquad a \neq 0$$

Adding $(-c)$ to both sides of the equation, we obtain

A.7 $$ax^2 + bx = -c$$

Dividing the equation (A.7) by a, the coefficient of x^2, we have

A.8 $$x^2 + \frac{b}{a}x = -\frac{c}{a}$$

Adding $(b/2a)^2$, the square of one-half the coefficient of x, to both sides of (A.8), we get

$$x^2 + \frac{b}{a}x + \left(\frac{b}{2a}\right)^2 = \left(\frac{b}{2a}\right)^2 - \frac{c}{a} = \frac{b^2}{4a^2} - \frac{c}{a}$$

which can be written

A.9 $$\left(x + \frac{b}{2a}\right)^2 = \frac{b^2 - 4ac}{4a^2}$$

Taking the square root of both sides, we have

A.10 $$x + \frac{b}{2a} = \frac{\pm\sqrt{b^2 - 4ac}}{2a}$$

Finally, we solve for x and obtain the *quadratic formula*

A.11 $$x = \frac{-b \pm \sqrt{b^2 - 4ac}}{2a}$$

Example A.10

Solve the equation

$$3x^2 + 13x - 10 = 0$$

using the quadratic formula.

Solution

Comparing this equation with (A.6), we observe that

$$a = 3, \quad b = 13, \quad c = -10$$

Then, using the quadratic formula (A.11), we have

$$x = \frac{-13 \pm \sqrt{(13)^2 - 4(3)(-10)}}{(2)(3)}$$

$$= \frac{-13 \pm \sqrt{169 + 120}}{6}$$

$$= \frac{-13 \pm \sqrt{289}}{6}$$

$$= \frac{13 \pm 17}{6}$$

Thus,

$$x = \frac{-13 + 17}{6} = \frac{2}{3}$$

or

$$x = \frac{-13 - 17}{6} = -5$$

The solution set is $\{-5, \frac{2}{3}\}$.

Example A.11

Solve the equation

$$2x^2 - 3x - 4 = 0$$

using the quadratic formula.

Solution

Here $a = 2$, $b = -3$, and $c = -4$. By the quadratic formula, we have

$$x = \frac{-(-3) \pm \sqrt{(-3)^2 - 4(2)(-4)}}{(2)(2)}$$

$$= \frac{3 \pm \sqrt{9 + 32}}{4}$$

$$= \frac{3 \pm \sqrt{41}}{4}$$

The solution set is $\left\{\frac{3 - \sqrt{41}}{4}, \frac{3 + \sqrt{41}}{4}\right\}$

Exercises

In problems 1–8, find the product of each pair of factors.

1. $(x + 1)(x + 2)$

2. $(x - 3)(x + 2)$

3. $(x - 4)(x - 7)$

4. $(x - 4)(x + 6)$

5. $(x + 3)(x - 3)$

6. $(x - 8)(x + 8)$

7. $(x - 8)(x + 6)$

8. $(x - 8)(x - 5)$

In problems 9–22, factor each expression.

9. $x^2 + 5x + 6$

10. $x^2 + 9x + 18$

11. $x^2 - 9x + 14$

12. $x^2 - 11x + 30$

13. $x^2 - 2x - 15$

14. $x^2 - 2x - 35$

15. $x^2 - 13x - 30$

16. $x^2 + 4x - 21$

17. $x^2 + 14x - 32$

18. $x^2 + 8x - 48$

19. $2x^2 + 13x + 15$

20. $6x^2 + 11x - 10$

21. $6x^2 - 7x - 5$

22. $10x^2 - 23x - 5$

In problems 23–32, solve each equation by factoring.

23. $x^2 + 7x + 12 = 0$

24. $x^2 - 5x + 6 = 0$

25. $x^2 + 11x + 28 = 0$

26. $x^2 - x - 2 = 0$

27. $x^2 - 3x - 40 = 0$

28. $x^2 + x - 20 = 0$

29. $x^2 - x - 6 = 0$

30. $x^2 - x - 12 = 0$

31. $6x^2 + 19x + 10 = 0$

32. $5x^2 - 12x + 4 = 0$

In problems 33–44, use the quadratic formula to solve each equation.

33. $x^2 + 5x + 3 = 0$

34. $x^2 - 7x - 8 = 0$

35. $x^2 + 5x + 2 = 0$

36. $3x^2 + 10x - 8 = 0$

37. $3x^2 + 23x + 14 = 0$

38. $5x^2 + 7x - 6 = 0$

39. $2x^2 + 5x + 2 = 0$

40. $x^2 - 4x + 1 = 0$

41. $3x^2 + 2x - 2 = 0$

42. $3x^2 + 7x - 1 = 0$

43. $2x^2 + 5x - 4 = 0$

44. $x^2 + 6x + 6 = 0$

A.3 Rational Expressions

Definition A.1

A fraction which includes algebraic expressions in both the numerator and the denominator is called a *rational expression.* The fractions

$$\frac{1}{x}, \quad \frac{3x+4}{4x+1}, \quad \frac{x^2+3x+1}{x(x+2)}, \quad \text{and} \quad \frac{3x+2}{(4/x)}$$

are examples of rational expressions.

The basic rules that govern the operations with rational expressions are similar to those involving rational numbers. We shall state some of them explicitly. If A, B, C, and D are algebraic expressions, then

A.12 $$\frac{A}{1} = A$$

A.13 $$\frac{A}{B} = \frac{C}{D} \quad \text{if and only if} \quad AD = BC, B \neq 0, D \neq 0$$

A.14 $$\frac{A}{B} + \frac{C}{D} = \frac{AD + BC}{BD}, B \neq 0, D \neq 0$$

A.15 $$\frac{A}{B} \cdot \frac{C}{D} = \frac{AC}{BD}, B \neq 0, D \neq 0$$

A.16 $$\frac{A}{B} \div \frac{C}{D} = \frac{AD}{BC}. B \neq 0, C \neq 0, D \neq 0$$

A.17 $$\frac{AC}{BC} = \frac{CA}{CB} = \frac{A}{B} \text{ for } B \neq 0, C \neq 0$$

Let us illustrate these operations with some examples.

Example A.12

Reduce the rational expression

$$\frac{x^2 - 1}{(x - 1)^2}, \qquad x \neq 1$$

to lowest terms.

Solution

Observe that the numerator $(x^2 - 1)$ can be factored as follows:

$$x^2 - 1 = (x - 1)(x + 1)$$

Thus

$$\frac{x^2 - 1}{(x - 1)^2} = \frac{(x - 1)(x + 1)}{(x - 1)(x - 1)} = \frac{x + 1}{x - 1}, \qquad x \neq 1$$

Remember that to reduce a rational expression we must divide the numerator and the denominator by every common *non-zero* factor.

Example A.13

a. $$\frac{x^2 + 3x}{x^2 + 7x + 12} = \frac{x(x + 3)}{(x + 3)(x + 4)}, \qquad x \neq -4, -3$$

$$= \frac{x}{x + 4}, \qquad x \neq -4, -3$$

b. $$\frac{x^2 - 25}{x^2 + 2x - 15} = \frac{(x-5)(x+5)}{(x-3)(x+5)}, \qquad x \neq -5, 3$$

$$= \frac{x-5}{x-3}, \qquad x \neq -5, 3$$

c. $$\frac{x^2 + 6x + 8}{x^2 + 10x + 24} = \frac{(x+4)(x+2)}{(x+4)(x+6)}, \qquad x \neq -6, -4$$

$$= \frac{x+2}{x+6}, \qquad x \neq -6, -4$$

Example A.14

Simplify the expression

$$\frac{3x-6}{4} \cdot \frac{x^2 - x - 2}{(x-2)^3}, \qquad x \neq 2$$

Solution

Observe that

$$x^2 - x - 2 = (x-2)(x+1)$$

Thus

$$\frac{3x-6}{4} \cdot \frac{x^2 - x + 2}{(x-2)^3} = \frac{3(x-2)}{4} \cdot \frac{(x-2)(x+1)}{(x-2)^3}, \qquad x \neq 2$$

$$= \frac{3(x+1)}{4(x-2)}, \qquad x \neq 2$$

Example A.15

Perform the following addition:

$$\frac{2}{x+2} + \frac{3}{x-2}, \qquad x \neq -2, 2$$

Solution

Using the rule (A.14), we have

$$\frac{2}{x+2} + \frac{3}{x-2} = \frac{2(x-2) + 3(x+2)}{(x-2)(x+2)}$$

$$= \frac{2x - 4 + 3x + 6}{(x+2)(x-2)}$$

$$= \frac{5x + 2}{(x+2)(x-2)}, \qquad x \neq -2, 2$$

Example A.16

Perform the following subtraction:

$$\frac{y}{x-y} - \frac{x}{x+y}, \qquad x \neq -y, y$$

Solution

Using the addition rule (A.14), we have

$$\begin{aligned}\frac{y}{x-y} - \frac{x}{x+y} = \frac{y}{x-y} + \left(\frac{-x}{x+y}\right) &= \frac{y(x+y) - x(x-y)}{(x-y)(x+y)} \\ &= \frac{xy + y^2 - x^2 + xy}{(x-y)(x+y)} \\ &= \frac{2xy + y^2 - x^2}{(x-y)(x+y)}, \qquad x \neq \pm y\end{aligned}$$

We shall now examine how the basic rules can be used to solve equations. Let us consider some illustrative examples.

Example A.17

Solve the following equation for x:

$$\frac{2x}{3} - \frac{x}{4} = \frac{1}{2}$$

Solution

First, we determine the lowest common denominator of 2, 3, and 4, which is 12. Multiplying both sides of the equation by 12 and then using the distributive property, we have

$$\begin{aligned}12\left(\frac{2x}{3} - \frac{x}{4}\right) &= 12 \cdot \frac{1}{2} \\ 12\left(\frac{2x}{3}\right) - 12\left(\frac{x}{4}\right) &= 6 \\ 8x - 3x &= 6 \\ 5x &= 6 \\ x &= \tfrac{6}{5}\end{aligned}$$

Example A.18

Solve the following equation for x:

$$\frac{1}{2x+3} = \frac{2}{x+12}, \qquad x \neq -12, -\tfrac{3}{2}$$

Solution

Using the rule (A.13) and then the distributive property, we obtain

$$2(2x + 3) = x + 12$$
$$4x + 6 = x + 12$$
$$3x = 6$$
$$x = 2$$

Example A.19

Solve the following equation for x:

$$\frac{1}{x} + \frac{6}{x + 4} = 1, \qquad x \neq -4, 0$$

Solution

Multiplying both sides of the equation by the lowest common denominator, $x(x + 4)$, we obtain

$$x(x + 4)\left(\frac{1}{x} + \frac{6}{x + 4}\right) = x(x + 4)$$
$$x(x + 4)\frac{1}{x} + x(x + 4)\frac{6}{x + 4} = x(x + 4)$$
$$(x + 4) + 6x = x(x + 4)$$
$$7x + 4 = x^2 + 4x$$

or

$$x^2 - 3x - 4 = 0$$
$$(x - 4)(x + 1) = 0$$

By Proposition A.1, either $x - 4 = 0$ or $x + 1 = 0$. If $x - 4 = 0$ then $x = 4$, and if $x + 1 = 0$ then $x = -1$. Thus the solution set is $\{-1, 4\}$.

Exercises

In problems 1–8, reduce each rational expression to lowest terms.

1. $\dfrac{x^2 - 4}{(x - 2)^2}, \quad x \neq 2$

2. $\dfrac{3x + 6}{x^2 - 4}, \quad x \neq -2, 2$

3. $\dfrac{x^2 - 16}{4(x + 4)}, \quad x \neq -4$

4. $\dfrac{x^2 + 4x}{x^2 + 7x + 12}, \quad x \neq -3, -4$

5. $\dfrac{x^2 + x - 2}{x^2 - 2x + 1}, \quad x \neq 1$

6. $\dfrac{x^2 - 6x + 8}{x^2 - 9x + 20}, \quad x \neq 4, 5$

7. $\dfrac{x^2 + 2x - 15}{x^2 - 6x + 9}, \quad x \neq 3$

8. $\dfrac{x^2 - 8x + 15}{x^2 - 2x - 3}, \quad x \neq -1, 3$

In problems 9–16, simplify each expression.

9. $\dfrac{x^2 + x}{x^2 + 3x + 2}, \quad x \neq -1, -2$

10. $\dfrac{x^2 + 2x}{x^2 + 5x + 6}, \quad x \neq -2, -3$

11. $\dfrac{x^2 + 8x + 12}{x^2 - 3x - 10}, \quad x \neq -2, 5$

12. $\dfrac{x + 4}{x} \cdot \dfrac{x}{x + 3}, \quad x \neq 0, -3$

13. $\dfrac{2x - 1}{x + 1} \cdot \dfrac{x + 3}{2x + 1} \cdot \dfrac{2x + 1}{2x - 1}, \quad x \neq -1, -\dfrac{1}{2}, \dfrac{1}{2}$

14. $\dfrac{3x + 2}{2x + 3} \cdot \dfrac{x + 2}{3x + 2} \cdot \dfrac{2x + 3}{x + 2}, \quad x \neq -2, -\dfrac{2}{3}, -\dfrac{3}{2}$

15. $\dfrac{1}{x + 1} \div \dfrac{x}{x + 1}, \quad x \neq -1, 0$

16. $\left(\dfrac{x + y}{x - y}\right) \div \dfrac{(x + y)^2}{(x - y)^2}, \quad x \neq -y, y$

Perform the indicated operations in problems 17–24.

17. $\dfrac{x}{3} + \dfrac{2}{x}, \quad x \neq 0$

18. $\dfrac{x + 1}{2} + \dfrac{4}{x + 2}, \quad x \neq -2$

19. $\dfrac{1}{x + 1} - \dfrac{1}{x - 1}, \quad x \neq 1, -1$

20. $\dfrac{1}{2x + 1} - \dfrac{1}{2x - 1}, \quad x \neq -\dfrac{1}{2}, \dfrac{1}{2}$

21. $\dfrac{1}{x} - \dfrac{x}{x + 1}, \quad x \neq 0, -1$

22. $\dfrac{x + 2}{x + 1} + \dfrac{x + 1}{x + 2}, \quad x \neq -1, -2$

23. $\left(1 + \dfrac{1}{x}\right) \div \left(1 - \dfrac{1}{x}\right), \quad x \neq 0, 1$

24. $\left(\dfrac{1}{x + 2} - \dfrac{1}{x - 2}\right) \div \left(\dfrac{x + 2}{x}\right), \quad x \neq 0, -2, 2$

In problems 25–44, solve each equation for x.

25. $\frac{x}{3} + \frac{3}{2} = 2$

26. $\frac{x}{2} + \frac{x}{3} = 10$

27. $\frac{2x}{3} + \frac{x}{5} = 26$

28. $\frac{2}{3}x + \frac{1}{2}x = 1$

29. $\frac{2x - 1}{3} + \frac{x + 1}{2} = 5$

30. $\frac{2x}{3} + \frac{3(x + 1)}{4} = 5$

31. $\frac{3}{x} + \frac{4}{x} = \frac{1}{2}, \quad x \neq 0$

32. $\frac{3}{4x} + \frac{1}{x} = \frac{7}{8}, \quad x \neq 0$

33. $\frac{1}{x - 1} + \frac{x}{x - 1} = \frac{3}{2}, \quad x \neq 1$

34. $\frac{1}{2x + 3} - \frac{3}{2x + 3} = -2, \quad x \neq -\frac{3}{2}, \frac{3}{2}$

35. $\frac{1}{x} + \frac{1}{x + 1} = 0, \quad x \neq 0, -1$

36. $\frac{1}{x - 1} + \frac{1}{x + 1} = 0, \quad x \neq \pm 1$

37. $\frac{1}{3x} - \frac{1}{x + 4} = 0, \quad x \neq 0, -4$

38. $\frac{1}{x + 3} - \frac{2}{x + 10} = 0, \quad x \neq -3, -10$

39. $\frac{3}{2x - 3} + \frac{1}{2x + 3} = 0, \quad x \neq \pm\frac{3}{2}$

40. $\frac{3x}{2x + 1} - \frac{1}{2x + 1} = \frac{8}{2x + 1}, \quad x \neq -\frac{1}{2}$

41. $\frac{2}{x} + \frac{3}{x + 3} = \frac{4 - x}{x^2 + 3x}, \quad x \neq -3, 0$

42. $\frac{5}{2x} + \frac{2}{x - 1} = \frac{3x + 2}{2x^2 - 2x}, \quad x \neq 0, 1$

43. $\frac{4x}{x - 3} - \frac{3x - 1}{x + 3} = 3, \quad x \neq \pm 3$

44. $\frac{7}{x - 1} - \frac{6}{x^2 - 1} = 5, \quad x \neq \pm 1$

A.4 Integral and Rational Exponents

The reader is probably familiar with symbols of the form x^m, where x is any real number and m is a non-negative integer.

Definition A.2

For any real number x and any positive integer m,

$$x^m = x \cdot x \cdot \cdots \cdot x \text{ } (m \text{ factors})$$

In the symbol x^m, m is called the *exponent* and x is called the *base*. The use of exponents is governed by several rules which we review briefly.

Proposition A.2

For any real number x and any positive integers m and n,

$$x^m x^n = x^{m+n}$$

To illustrate this proposition, we need to observe that

$$\begin{aligned} x^m \cdot x^n &= \underbrace{(x \cdot x \cdots x)}_{m \text{ factors}} \underbrace{(x \cdot x \cdots x)}_{n \text{ factors}} \\ &= \underbrace{(x \cdot x \cdots x)}_{(m+n) \text{ factors}} \\ &= x^{m+n} \end{aligned}$$

This proposition states that to multiply two powers of the same base we add the exponents.

Example A.20

a. $3^4 \cdot 3^2 = 3^{4+2} = 3^6 = 729$

b. $b^3 \cdot b^5 = b^{3+5} = b^8$

c. $a^{3r} \cdot a^{2t} = a^{3r+2t}$, where r and t are positive integers

Proposition A.3

For any real number $x \neq 0$ and any non-negative integers m and n,

$$\frac{x^m}{x^n} = \begin{cases} x^{m-n} & \text{if } m > n \\ \dfrac{1}{x^{n-m}} & \text{if } m < n \\ 1 & \text{if } m = n \end{cases}$$

Note that for $m > n$, we have

$$\begin{aligned}\frac{x^m}{x^n} &= \frac{x \cdot x \cdot x \cdots x_{(m\text{ factors})}}{x \cdot x \cdot x \cdots x_{(n\text{ factors})}}\\ &= \frac{(x \cdot x \cdots x_{(n\text{ factors})})(x \cdot x \cdots x_{(m-n)\text{factors}})}{x \cdot x \cdot x \cdots x_{(n\text{ factors})}}\\ &= (x \cdot x \cdots x_{(m-n)\text{factors}})\\ &= x^{m-n}\end{aligned}$$

If $m < n$, then

$$\begin{aligned}\frac{x^m}{x^n} &= \frac{x \cdot x \cdots x_{(m\text{ factors})}}{x \cdot x \cdots x_{(n\text{ factors})}}\\ &= \frac{x \cdot x \cdots x_{(m\text{ factors})}}{(x \cdot x \cdots x_{(m\text{ factors})})(x \cdot x \cdots x_{(n-m)\text{factors}})}\\ &= \frac{1}{x \cdot x \cdots x_{(n-m)\text{factors}}}\\ &= \frac{1}{x^{n-m}}\end{aligned}$$

If $m = n$, then

$$\frac{x^m}{x^m} = \frac{x \cdot x \cdots x\ (m\text{ factors})}{x \cdot x \cdots x\ (m\text{ factors})}$$

Cancelling each x in the numerator with the corresponding x in the denominator, we have

$$\frac{x \cdot x \cdots x\ (m\text{ factors})}{x \cdot x \cdots x\ (m\text{ factors})} = 1$$

Example A.21

a. $\dfrac{3^6}{3^2} = 3^{6-2} = 3^4 = 81$

b. $\dfrac{a^8}{a^6} = a^{8-6} = a^2, \qquad a \neq 0$

c. $\dfrac{b^4 \cdot b^7}{b^9} = \dfrac{b^{4+7}}{b^9} = \dfrac{b^{11}}{b^9} = b^{11-9} = b^2, \qquad b \neq 0$

d. $\dfrac{c^3}{c^3} = \dfrac{c \cdot c \cdot c}{c \cdot c \cdot c} = 1, \qquad c \neq 0$

Example A.22

a. $\frac{3^5}{3^7} = \frac{1}{3^{7-5}} = \frac{1}{3^2} = \frac{1}{9}$

b. $\frac{a^{12}}{a^{18}} = \frac{1}{a^{18-12}} = \frac{1}{a^6}, \qquad a \neq 0$

c. $\frac{b^2 \cdot b^3}{b^8} = \frac{b^{2+3}}{b^8} = \frac{b^5}{b^8} = \frac{1}{b^{8-5}} = \frac{1}{b^3}, \qquad b \neq 0$

Proposition A.4

If m and n are positive integers, then for any real number x,

$$(x^m)^n = x^{mn}.$$

Note that

$$\begin{aligned}(x^m)^n &= \underbrace{x^m \cdot x^m \cdot x^m \cdots x^m}_{n \text{ factors}} \\ &= x^{m+m+\cdots+m} \ (n \text{ times}) \\ &= x^{mn}.\end{aligned}$$

Example A.23

a. $(a^2)^5 = a^{2 \cdot 5} = a^{10}$.

b. $(3^4)^2 = 3^{4 \cdot 2} = 3^8 = 6561$

Proposition A.5

If x and y are real numbers, then for any positive integer m,

$$(xy)^m = x^m \cdot y^m$$

Using Definition A.2, we have

$$\begin{aligned}(xy)^m &= \underbrace{(xy)(xy)(xy)\cdots(xy)}_{m \text{ factors}} \\ &= \underbrace{(x \cdot x \cdot x \cdots x)}_{m \text{ factors}} \cdot \underbrace{(y \cdot y \cdot y \cdots y)}_{m \text{ factors}} \\ &= x^m y^m\end{aligned}$$

Example A.24

a. $(ab)^6 = a^6 b^6$

b. $(a^2 b^4)^5 = (a^{2 \cdot 5})(b^{4 \cdot 5}) = a^{10} \cdot b^{20}$

Proposition A.6 Let x and y be any real number, $y \neq 0$. Then for any positive integer m,

$$\left(\frac{x}{y}\right)^m = \frac{x^m}{y^m}$$

Example A.25

a. $\left(\frac{2}{3}\right)^8 = \frac{2^8}{3^8} = \frac{256}{6561}$

b. $\left(\frac{a^2b}{c}\right)^5 = \frac{a^{10}b^5}{c^5}, \qquad c \neq 0$

Thus far in this section, we have considered only positive integers as exponents. We now extend the concept of exponent to the set of all integers so that expressions of the form x^0 and x^{-3} carry useful meanings. In formulating definitions for all integers, we must be certain that they conform to the laws of exponents we have already established.

First, we consider the meaning of x^0, where x is any real number different from zero. If Proposition A.2 is to hold for $n = 0$, then we must have

$$x^m \cdot x^0 = x^{m+0}$$

but since

$$m + 0 = m$$

it follows that

$$x^{m+0} = x^m$$

Hence,

$$x^m \cdot x^0 = x^m$$

and we must have $x^0 = 1$.

We now discuss the meaning of x^{-m} where $x \neq 0$ and m is a positive integer. Considering Proposition A.2 again, with $n = -m$ and $x \neq 0$, we have

$$x^m \cdot x^{-m} = x^{m+(-m)}$$

But

$$m + (-m) = (-m) + m = 0$$

for all integers; thus it follows that

$$x^{m+(-m)} = x^{(-m)+m} = x^0 = 1$$

Hence we conclude that $x^m x^{-m} = 1$ which can be restated as

$$x^{-m} = \frac{1}{x^m}, \qquad x \neq 0$$

Example A.26

a. $5^0 = 1$

b. $8(x^0) = 8 \cdot 1 = 8, \qquad x \neq 0$

c. $2(xy)^0 z = 2 \cdot 1 \cdot z = 2z, \qquad xy \neq 0$

Example A.27

a. $5^{-2} = \frac{1}{5^2} = \frac{1}{25}$

b. $x^{-3} = \frac{1}{x^3}, \qquad x \neq 0$

c. $\frac{1}{x^{-2}} = x^2, \qquad x \neq 0$

d. $x^2 y^{-3} = \frac{x^2}{y^3}, \qquad y \neq 0$

Next, we consider expressions of the form $(16)^{1/2}$, $(81)^{3/4}$ and $x^{2/3}$, where the exponents are rationals; these exponents are of the type m/n, where m and n are integers and $n \neq 0$. First, we shall assign meaning to expressions of the form $x^{1/n}$ so as to be consistent with the laws of exponents we have established. For example,

$$(16)^{1/2}(16)^{1/2} = [(16)^{1/2}]^2 = (16)^1 = 16$$

and

$$x^{1/4}x^{1/4}x^{1/4}x^{1/4} = (x^{1/4})^4 = x^{(1/4)\cdot 4} = x$$

Thus if some number exists whose fourth power is x, we denote that number by $x^{1/4}$ or $\sqrt[4]{x}$ and call it the *fourth root of* x. In general, if n is a *positive* integer, then

$$\underbrace{x^{1/n}x^{1/n} \cdots x^{1/n}}_{n \text{ times}} = (x^{1/n})^n = x^{(1/n)\cdot n} = x$$

and $x^{1/n}$ is called the *n-th root of* x. Further, if x is positive, then there exists one positive number y such that

$$y^n = x$$

This positive number y, which is denoted by $x^{1/n}$, is called the *principal n-th root of* x. Thus 3 is the principal square root of 9, 2 is the principal cube root of 8, and so on. If x is negative, and n is even, then the n-th root of x is not a real number; however, there exists one negative n-th root of x if n is odd. This negative number is the principal n-th root of x. Thus (-2) is the cube root of -8 since $-2 = \sqrt[3]{-8} = (-8)^{1/3}$; similarly, (-5) is the cube root of (-125).

Example A.28

a. $(64)^{1/3} = (4 \cdot 4 \cdot 4)^{1/3} = (4^3)^{1/3} = 4^{3(1/3)} = 4$
b. $(-32)^{1/5} = [(-2)^5]^{1/5} = (-2)^{5(1/5)} = -2$
c. $(729)^{1/6} = (3^6)^{1/6} = 3^{6(1/6)} = 3$
d. $(625)^{1/4} = (5^4)^{1/4} = 5^{4(1/4)} = 5$

What about $\sqrt{-36}$? Certainly, there is no real number whose square is -36. (Why?) Thus $\sqrt{-36}$ is not a real number. We must reinforce this fact that if x is a negative number and n is even, then $x^{1/n}$ is not a real number. However, when n is an odd integer, then $x^{1/n}$ represents exactly one real number for any real number x.

Let us look at some examples.

Example A.29

a. $\sqrt{25} = (5^2)^{1/2} = 5^{2(1/2)} = 5$
b. $\left(\frac{1}{343}\right)^{1/3} = \left[\left(\frac{1}{7}\right)^3\right]^{1/3} = \left(\frac{1}{7}\right)^{3(1/3)} = \frac{1}{7}$
c. $\sqrt[3]{-125} = [(-5)^3]^{1/3} = -5$, but
d. $\sqrt{-25}$ is not a real number.

Now that we have a meaning for $x^{1/n}$, we may extend the concept of an exponent so that it has meaning for expressions of the form $x^{m/n}$, where x is a real number and m/n is a rational number. Notice that we can write $x^{4/5} = (x^{1/5})^4$ because $4/5 = (\frac{1}{5}) \cdot 4$. In general,

$$x^{m/n} = \underbrace{x^{1/n}x^{1/n} \cdots x^{1/n}}_{m \text{ times}} = (x^{1/n})^m$$

or equivalently

$$x^{m/n} = (x^m)^{1/n}$$

Thus,

$$(32)^{4/5} = [(32)^{1/5}]^4 = [(2^5)^{1/5}]^4 = [2^{5(1/5)}]^4 = 2^4 = 16$$

Alternatively, we could also evaluate $(32)^{4/5}$ as follows:

$$(32)^{4/5} = [(32)^4]^{1/5} = (1{,}048{,}576)^{1/5} = [(16)^5]^{1/5} = 16$$

Clearly, the *second method is time consuming and laborious.*

Example A.30

a. $(-125)^{2/3} = [(-5)^3]^{2/3} = (-5)^{3(2/3)} = (-5)^2 = 25$
b. $(64)^{5/6} = (2^6)^{5/6} = 2^{6(5/6)} = 2^5 = 32$
c. $(81)^{5/4} = (3^4)^{5/4} = 3^{4(5/4)} = 3^5 = 243$

Since

$$x^{-m} = \frac{1}{x^m}, \qquad x \neq 0$$

we can express

$$x^{-m/n} = \frac{1}{x^{m/n}}, \qquad x \neq 0$$

where m/n is any rational number. Thus,

a. $(16)^{-5/4} = \dfrac{1}{(16)^{5/4}} = \dfrac{1}{(2^4)^{5/4}} = \dfrac{1}{2^{4(5/4)}} = \dfrac{1}{2^5} = \dfrac{1}{32}$

b. $(216)^{-2/3} = \dfrac{1}{(216)^{2/3}} = \dfrac{1}{(6^3)^{2/3}} = \dfrac{1}{6^{3(2/3)}} = \dfrac{1}{6^2} = \dfrac{1}{36}$

c. $(-128)^{-3/7} = \dfrac{1}{(-128)^{3/7}} = \dfrac{1}{[(-2)^7]^{3/7}} = \dfrac{1}{(-2)^{7(3/7)}} = \dfrac{1}{(-2)^3} = -\dfrac{1}{8}$

As another consequence of

$$x^{-m} = \frac{1}{x^m}, \qquad x \neq 0$$

we conclude that

$$\left(\frac{a}{b}\right)^{-m} = \left(\frac{b}{a}\right)^m$$

holds whenever the letters are replaced by numbers for which the expression involved has meaning. Thus,

$$\left(\frac{3}{4}\right)^{-2} = \left(\frac{4}{3}\right)^2 = \frac{16}{9}$$

Exercises

1. Evaluate each of the following expressions.
   a. $2^3 \cdot 2^2$
   b. $3^2 \cdot 3$
   c. $(2^2)^4$
   d. $(3^3)^2$
   e. $(2^3 \cdot 3^2)^4$
   f. $(5^2 \cdot 3^3)^3$
   g. $\left(\dfrac{2}{3}\right)^3\left(\dfrac{3}{4}\right)^4$
   h. $\left(\dfrac{4}{5}\right)^2\left(\dfrac{5}{4}\right)^3$

2. Evaluate each of the following expressions.

a. $\dfrac{2^5}{2^2}$

b. $\dfrac{4^3}{2^4}$

c. $\dfrac{3^8 \cdot 3^4}{3^9}$

d. $\dfrac{(3^4)^2 \cdot 2^4}{(2^3)^2(3^2)^3}$

e. $\left(\dfrac{2}{3}\right)^4\left(\dfrac{3}{2}\right)^2$

f. $\dfrac{3^5 \cdot 3^{11}}{(3^2)^4}$

3. Simplify each of the following expressions.

a. $(a^2b)^4$

b. $(a^2b^3c^4)^3$

c. $\dfrac{(ab^2)^4}{a^2b^3}$

d. $\dfrac{(abc^2)^3}{(a^3b^2c^6)^2}$

e. $\dfrac{(c^2d^3)^4a^{-2}}{a^4(c^3d^2)^3}$

f. $\left(\dfrac{a}{b}\right)^4\left(\dfrac{b^2}{c^3}\right)^2\left(\dfrac{c^4}{a}\right)^3$

g. $\left(\dfrac{2a}{b}\right)^3\left(\dfrac{b}{3a}\right)^4$

h. $\dfrac{(3a)^2(2b)^3}{(a^2)^4(b^3)^3}$

4. Simplify each of the following expressions.

a. $(3x)^{-2}$

b. $(2x^{-3})^2$

c. $(xy^{-1}z^{-2})^3$

d. $(x^2y^{-1})^{-2}$

e. $(x^3y^{-2})^{-1}$

f. $(3a^{-1}b)^{-2}$

g. $(x^{-2}y^3z)^2(xy^2z^3)^{-3}$

h. $\dfrac{2xy^{-3}}{5x^{-2}y^2}$

i. $\dfrac{3x^{-2}y^4}{4x^4y^{-2}}$

j. $\left(\dfrac{x^{-2}y^3}{x^{-5}y^{-2}}\right)^{-2}$

5. Simplify each of the following expressions.

a. $\left(\dfrac{x^2}{3y^3}\right)^{-2}$

b. $\left(\dfrac{ab^{-2}c^{-3}}{a^2b^3c}\right)^3$

c. $\left(\dfrac{3x^2y^{-1}}{xy^2}\right)^4$

d. $\left(\dfrac{x^2y^{-3}}{x^3y^{-4}}\right)^{-2}$

e. $\left(\dfrac{xy^2z^3}{x^2yz}\right)^0$

f. $\left(\dfrac{4x^{-3}y^2}{3x^{-1}y^4}\right)^{-3}$

6. Evaluate each of the following expressions.

a. $\dfrac{3^{-2}4^{-3}}{3^42^{-4}}$

b. $\dfrac{4^{-3}5^{-2}}{4^{-7}2^3}$

c. $\dfrac{2^{-3}4^{-2}}{5^{-2}4^{-5}}$

d. $\dfrac{3^44^{-3}}{2^{-5}3^5}$

7. Find the value of each of the following numbers.

a. $(8)^{1/3}$

b. $(16)^{1/2}$

c. $(-243)^{1/5}$

d. $(125)^{2/5}$

e. $(729)^{2/3}$

f. $\left(\frac{16}{81}\right)^{1/4}$

g. $\left(\frac{32}{243}\right)^{2/5}$

h. $\left(\frac{64}{343}\right)^{1/3}$

8. Find the value of each of the following numbers.

a. $\frac{(-343)^{1/3}(27)^{2/3}}{(64)^{1/6}(-32)^{1/5}}$

b. $\frac{\left(\frac{27}{8}\right)^{1/3}(64)^{1/6}}{(27)^{2/3}}$

c. $\frac{(625)^{-3/4}(16)^{-5/4}}{(216)^{2/3}(-128)^{2/7}}$

d. $\frac{(81)^{5/4}(-125)^{1/3}}{(512)^{2/9}(729)^{5/6}}$

Our discussion thus far has been limited to rational exponents. Consequently the symbol x^m, where m is an irrational number, has not yet been assigned any meaning. The rules and propositions for exponents continue to hold even if some or all of the exponents are irrationals. A rigorous proof of this statement, however, is beyond our present scope. Since the set of real numbers is the collection of all rational and irrational numbers, we conclude that the rules for exponents hold for all real numbers.

A.5 Rationalizing Expressions

Observe that if a and b are positive real numbers, then

$$(\sqrt{a}+\sqrt{b})(\sqrt{a}-\sqrt{b}) = (\sqrt{a})^2 - (\sqrt{b})^2 = a - b$$

Thus we may write

$$\sqrt{a}+\sqrt{b} = (\sqrt{a}+\sqrt{b})\left(\frac{\sqrt{a}-\sqrt{b}}{\sqrt{a}-\sqrt{b}}\right) = \frac{a-b}{\sqrt{a}-\sqrt{b}}, \qquad a \neq b$$

This illustrates an algebraic technique called *rationalizing* the number $\sqrt{a}+\sqrt{b}$. Similarly, we can write

$$\sqrt{a}-\sqrt{b} = (\sqrt{a}-\sqrt{b})\left(\frac{\sqrt{a}+\sqrt{b}}{\sqrt{a}+\sqrt{b}}\right) = \frac{a-b}{\sqrt{a}+\sqrt{b}}$$

Let us consider a few examples.

Example A.31

Let x be a positive real number. Then

$$\sqrt{x+3}-\sqrt{x+2}=(\sqrt{x+3}-\sqrt{x+2})\left(\frac{\sqrt{x+3}+\sqrt{x+2}}{\sqrt{x+3}+\sqrt{x+2}}\right), \quad x \geq -2$$

$$=\frac{(\sqrt{x+3})^2-(\sqrt{x+2})^2}{\sqrt{x+3}+\sqrt{x+2}}$$

$$=\frac{(x+3)-(x+2)}{\sqrt{x+3}+\sqrt{x+2}}$$

$$=\frac{1}{\sqrt{x+3}+\sqrt{x+2}}, \qquad x \geq -2$$

Example A.32

Rationalize the denominator of the expression

$$\frac{\sqrt{5}+\sqrt{3}}{\sqrt{5}-\sqrt{3}}$$

Solution

$$\frac{\sqrt{5}+\sqrt{3}}{\sqrt{5}-\sqrt{3}}\cdot\frac{\sqrt{5}+\sqrt{3}}{\sqrt{5}+\sqrt{3}}=\frac{(\sqrt{5}+\sqrt{3})^2}{5-3}$$

$$=\frac{5+3+2\sqrt{5}\sqrt{3}}{2}$$

$$=\frac{8+2\sqrt{15}}{2}$$

$$=4+\sqrt{15}$$

Example A.33

Rationalize the denominator in the fraction.

$$\frac{1}{\sqrt{x+4}-\sqrt{x}}$$

Solution

$$\frac{1}{\sqrt{x+4}-\sqrt{x}}\cdot\frac{\sqrt{x+4}+\sqrt{x}}{\sqrt{x+4}+\sqrt{x}}=\frac{\sqrt{x+4}+\sqrt{x}}{(\sqrt{x+4})^2-(\sqrt{x})^2}$$

$$=\frac{\sqrt{x+4}+\sqrt{x}}{(x+4)-x}$$

$$=\frac{\sqrt{x+4}+\sqrt{x}}{4}$$

Exercises

In problems 1–10, rationalize each expression.

1. $\sqrt{6} - \sqrt{5}$
2. $\sqrt{6} + \sqrt{5}$
3. $\sqrt{7} + \sqrt{3}$
4. $\sqrt{3} + \sqrt{2}$
5. $4\sqrt{3} + 3\sqrt{5}$
6. $2\sqrt{5} - 3\sqrt{2}$
7. $\sqrt{x+1} + \sqrt{x}$
8. $\sqrt{x+2} - \sqrt{x}$
9. $\sqrt{x+4} + \sqrt{x+3}$
10. $\sqrt{x} - \sqrt{x+1}$

In problems 11–20, rationalize the denominators.

11. $\dfrac{1}{\sqrt{3} + \sqrt{2}}$
12. $\dfrac{3}{\sqrt{6} - \sqrt{3}}$
13. $\dfrac{\sqrt{6} + \sqrt{2}}{\sqrt{6} - \sqrt{2}}$
14. $\dfrac{\sqrt{7} + \sqrt{3}}{\sqrt{7} - \sqrt{3}}$
15. $\dfrac{2\sqrt{5} - 3\sqrt{2}}{2\sqrt{5} + 3\sqrt{2}}$
16. $\dfrac{4\sqrt{3} + 3\sqrt{5}}{4\sqrt{3} - 3\sqrt{5}}$
17. $\dfrac{1}{\sqrt{x+1} - \sqrt{x}}$
18. $\dfrac{1}{\sqrt{x+4} + \sqrt{x}}$
19. $\dfrac{\sqrt{x+2} + \sqrt{x+1}}{\sqrt{x+2} - \sqrt{x+1}}$
20. $\dfrac{\sqrt{x+4} - \sqrt{x+3}}{\sqrt{x+4} + \sqrt{x+3}}$

Appendix B Summation Notation

Symbols are often used to express mathematical formulas or operations. One such symbol, used frequently in the text, is the symbol for *summation.* It is customary to use the Greek letter $\sum$ (capital Sigma) for this purpose. The $\sum$ is followed by an expression representing the typical term of the summation and containing a variable, called the *index of summation,* that takes on successive integral values to form the different terms of the sum. The equation below the $\sum$ indicates the index of summation and the first integral value of the index and the number above the $\sum$ indicates the last value of the index.

Example B.1

a. $\sum_{k=1}^{6} k = 1 + 2 + 3 + 4 + 5 + 6 = 21$

b. $$\begin{aligned}\sum_{k=1}^{5} k^2 &= 1^2 + 2^2 + 3^2 + 4^2 + 5^2 \\ &= 1 + 4 + 9 + 16 + 25 \\ &= 55\end{aligned}$$

c. $$\begin{aligned}\sum_{j=1}^{3} j^3 &= 1^3 + 2^3 + 3^3 \\ &= 1 + 8 + 27 \\ &= 36\end{aligned}$$

There are several useful formulas involving summation which the reader will find convenient to use.

Proposition B.1

For any natural number n,

B.1 $$\sum_{k=1}^{n} k = 1 + 2 + 3 + \cdots + n = \frac{n(n+1)}{2}$$

B.2 $$\sum_{k=1}^{n} k^2 = 1^2 + 2^2 + \cdots + n^2 = \frac{n(n+1)(2n+1)}{6}$$

B.3 $$\sum_{k=1}^{n} k^3 = 1^3 + 2^3 + \cdots + n^3 = \left[\frac{n(n+1)}{2}\right]^2$$

Example B.2

a. $$\sum_{k=1}^{100} k = 1 + 2 + \cdots + 100 = \frac{100(100+1)}{2} = 5050$$

b. $$\sum_{k=1}^{50} k^2 = 1^2 + 2^2 + \cdots + 50^2 = \frac{50(50+1)(100+1)}{6} = \frac{50(51)(101)}{6} = 42925$$

c. $$\sum_{k=1}^{30} k^3 = 1^3 + 2^3 + \cdots + 30^3 = \left[\frac{30(30+1)}{2}\right]^2 = (465)^2 = 216{,}225$$

Summation notation is also used to represent the sum of a series when the terms involve subscripts. For example, the weight of n individuals may be listed as $x_1, x_2, \ldots, x_n$, where x_i represents the weight of i-th individual. In this example the subscript i assumes the integral values from 1 to n.

Example B.3

a. $$\sum_{i=1}^{6} x_i = x_1 + x_2 + x_3 + x_4 + x_5 + x_6$$

b. $$\sum_{i=2}^{5} x_i^2 = x_2^2 + x_3^2 + x_4^2 + x_5^2$$

c. $$\sum_{i=1}^{5} i \cdot x_i = x_1 + 2x_2 + 3x_3 + 4x_4 + 5x_5$$

Proposition B.2

If c is a constant, then

$$\sum_{i=1}^{n} cx_i = c \sum_{i=1}^{n} x_i$$

Proof

$$\begin{aligned}\sum_{i=1}^{n} cx_i &= cx_1 + cx_2 + \cdots + cx_n \\ &= c(x_1 + x_2 + \cdots + x_n) \\ &= c \sum_{i=1}^{n} x_i\end{aligned}$$

Example B.4

We can show that

$$\sum_{i=1}^{3} 4x_i = 4 \sum_{i=1}^{3} x_i$$

Let $x_1 = 3$, $x_2 = 5$, $x_3 = 7$. Then

$$\begin{aligned}\sum_{i=1}^{3} 4x_i &= 4x_1 + 4x_2 + 4x_3 \\ &= 4(3) + 4(5) + 4(7) \\ &= 12 + 20 + 28 \\ &= 60\end{aligned}$$

and

$$\begin{aligned}4 \sum_{i=1}^{3} x_i &= 4(x_1 + x_2 + x_3) \\ &= 4(3 + 5 + 7) \\ &= 4(15) \\ &= 60\end{aligned}$$

Thus,

$$\sum_{i=1}^{3} 4x_i = 4 \sum_{i=1}^{3} x_i$$

Example B.5

a. $\displaystyle\sum_{k=1}^{12} 5k^2 = 5 \sum_{k=1}^{12} k^2$ **(Proposition B.2)**

$\displaystyle = 5\left[\frac{12(12+1)(24+1)}{6}\right]$ **(Proposition B.1)**

$= 3250$

b. $\sum_{k=1}^{10} 3k^3 = 3 \sum_{k=1}^{10} k^3$ **(Proposition B.2)**

$$= 3\left[\frac{10(10+1)}{2}\right]^2 \quad \textbf{(Proposition B.1)}$$

$$= 9075$$

Proposition B.3

$$\sum_{i=1}^{n} (x_i + y_i + z_i) = \sum_{i=1}^{n} x_i + \sum_{i=1}^{n} y_i + \sum_{i=1}^{n} z_i$$

Proof

$$\begin{aligned}\sum_{i=1}^{n} (x_i + y_i + z_i) &= (x_1 + y_1 + z_1) + (x_2 + y_2 + z_2) + \cdots \\ &\quad + (x_n + y_n + z_n) \\ &= (x_1 + x_2 + \cdots + x_n) + (y_1 + y_2 + \cdots + y_n) \\ &\quad + (z_1 + z_2 + \cdots + z_n) \\ &= \sum_{i=1}^{n} x_i + \sum_{i=1}^{n} y_i + \sum_{i=1}^{n} z_i\end{aligned}$$

Corollary

If c is a constant, then

$$\sum_{i=1}^{n} c = nc$$

Example B.6

$$\begin{aligned}\sum_{i=1}^{10} (x_i^2 + 15x_i + 9) &= \sum_{i=1}^{10} x_i^2 + \sum_{i=1}^{10} 15x_i + \sum_{i=1}^{10} 9 \\ &= \sum_{i=1}^{10} x_i^2 + 15 \sum_{i=1}^{10} x_i + 9(10) \\ &= \sum_{i=1}^{10} x_i^2 + 15 \sum_{i=1}^{10} x_i + 90\end{aligned}$$

Now we will show that

(B.4) $$\sum_{i=1}^{n} x_i^2 \neq \left(\sum_{i=1}^{n} x_i\right)^2$$

and that

(B.5) $$\sum_{i=1}^{n} x_i y_i \neq \left(\sum_{i=1}^{n} x_i\right)\left(\sum_{i=1}^{n} y_i\right)$$

Example B.7

Let $x_1 = 3$, $x_2 = 4$, $x_3 = 6$, $y_1 = 1$, $y_2 = 5$, $y_3 = 8$. Then

$$\sum_{i=1}^{3} x_i = x_1 + x_2 + x_3 = 3 + 4 + 6 = 13$$

so that

$$\left(\sum_{i=1}^{3} x_i\right)^2 = (13)^2 = 169$$

Also,

$$\sum_{i=1}^{3} x_i^2 = x_1^2 + x_2^2 + x_3^2 = 3^2 + 4^2 + 6^2 = 9 + 16 + 36 = 61$$

Thus,

$$\sum_{i=1}^{3} x_i^2 \neq \left(\sum_{i=1}^{3} x_i\right)^2$$

It may also be observed that

$$\sum_{i=1}^{3} y_i = y_1 + y_2 + y_3 = 1 + 5 + 8 = 14$$

so that

$$\left(\sum_{i=1}^{3} x_i\right)\left(\sum_{i=1}^{3} y_i\right) = (13)(14) = 182$$

but

$$\begin{aligned}\sum_{i=1}^{3} x_i y_i &= x_1 y_1 + x_2 y_2 + x_3 y_3 \\ &= (3)(1) + (4)(5) + (6)(8) \\ &= 3 + 20 + 48 \\ &= 71\end{aligned}$$

Thus

$$\sum_{i=1}^{3} x_i y_i \neq \left(\sum_{i=1}^{3} x_i\right)\left(\sum_{i=1}^{3} y_i\right)$$

Example B.8

$$\begin{aligned}\sum_{k=1}^{20} k(k+4) &= \sum_{k=1}^{20} (k^2 + 4k) \\ &= \sum_{k=1}^{20} k^2 + \sum_{k=1}^{20} 4k \qquad \textbf{(Proposition B.3)} \\ &= \sum_{k=1}^{20} k^2 + 4\sum_{k=1}^{20} k \qquad \textbf{(Proposition B.2)}\end{aligned}$$

Applying Proposition B.1, we have

$$\sum_{k=1}^{20} k^2 = \frac{20(20+1)(40+1)}{6} = 2870$$

and

$$\sum_{k=1}^{20} k = \frac{20(20+1)}{2} = 210$$

Thus,

$$\begin{aligned}\sum_{k=1}^{20} k(k+4) &= 2870 + 4(210) \\ &= 2870 + 840 \\ &= 3710\end{aligned}$$

Example B.9

$$\begin{aligned}\sum_{k=1}^{5} (k^3 - 3) &= \sum_{k=1}^{5} k^3 - \sum_{k=1}^{5} 3 \qquad \textbf{(Proposition B.3)} \\ &= \sum_{k=1}^{5} k^3 - 3(5) \\ &= 225 - 15 \\ &= 210\end{aligned}$$

Exercises

Use summation formulas to compute each value in problems 1–12.

1. $\sum_{k=1}^{15} k$

2. $\sum_{k=1}^{10} 3k$

3. $\sum_{k=1}^{8} (2k+3)$

4. $\sum_{k=1}^{18} 2k^2$

5. $\sum_{k=1}^{12} k(k+2)$

6. $\sum_{k=1}^{10} (k+1)^2$

7. $\sum_{k=1}^{12} (k+1)(k+2)$

8. $\sum_{k=1}^{15} (k+2)(k+3)$

9. $\sum_{k=1}^{11} (k+1)(k-1)$

10. $\sum_{k=1}^{10} 4k^3$

11. $\sum_{k=1}^{12} k^2(k+2)$

12. $\sum_{k=1}^{8} k(k^2+k+1)$

13. Express each of the following without the summation sign.

a. $\sum_{i=1}^{3} x_i$

b. $\sum_{i=1}^{5} cx_i$

c. $\sum_{i=1}^{5} x_i y_i$

d. $\sum_{i=1}^{6} (x_i + 3)$

e. $\sum_{i=1}^{4} (x_i + y_i)$

f. $\sum_{i=1}^{3} (x_i + y_i + z_i)$

14. Use summation notation for each of the following expressions.

a. $x_1 + x_2 + x_3 + x_4 + x_5$
b. $x_1^2 + x_2^2 + x_3^2 + x_4^2$
c. $(x_1 + 2) + (x_2 + 2) + (x_3 + 2) + (x_4 + 2) + (x_5 + 2)$
d. $x_2 y_2 + x_3 y_3 + x_4 y_4 + x_5 y_5 + x_6 y_6$
e. $x_1 y_1 z_1 + x_2 y_2 z_2 + \cdots + x_n y_n z_n$

15. Let $x_1 = 2$, $x_2 = 4$, $x_3 = 5$, $x_4 = 1$, $x_5 = -2$. Find the value of each of the following expressions.

a. $\sum_{i=1}^{4} x_i$

b. $\sum_{i=3}^{5} (x_i - 3)$

c. $\sum_{i=1}^{5} x_i^2$

d. $\sum_{i=1}^{5} (x_i + 2)^2$

e. $\sum_{i=2}^{4} (3x_i + 4)$

f. $\sum_{i=1}^{5} i \cdot x_i$

16. Let $x_1 = 2$, $x_2 = 5$, $x_3 = 5$, $y_1 = 4$, $y_2 = 6$, $y_3 = 5$. Find the value of each of the following expressions.

a. $\left(\sum_{i=1}^{3} x_i\right)^2$

b. $\sum_{i=1}^{3} x_i^2$

c. $\sum_{i=1}^{3} x_i y_i$

d. $\left(\sum_{i=1}^{3} x_i\right)\left(\sum_{i=1}^{3} y_i\right)$

e. $\sum_{i=1}^{3} y_i^2$

f. $\left(\sum_{i=1}^{3} y_i\right)^2$

g. Is $\sum_{i=1}^{3} x_i^2 = \left(\sum_{i=1}^{3} x_i\right)^2$?

h. Is $\sum_{i=1}^{3} x_i y_i = \left(\sum_{i=1}^{3} x_i\right)\left(\sum_{i=1}^{3} y_i\right)$?

Tables

T.1 Exponential Functions
T.2 Compound Interest
T.3 Present Value of a Dollar
T.4 Common Logarithms
T.5 Natural Logarithms

Table 1
The Exponential Function

x	e^x	e^{-x}	x	e^x	e^{-x}
0.0	1.0000	1.0000	5.0	148.51	0.00674
0.1	1.1052	0.90484	5.1	164.13	0.00609
0.2	1.2214	0.81873	5.2	181.39	0.00551
0.3	1.3499	0.74082	5.3	200.47	0.0C498
0.4	1.4918	0.67032	5.4	221.55	0.00451
0.5	1.6487	0.60653	5.5	244.75	0.00408
0.6	1.8221	0.54881	5.6	270.40	0.00369
0.7	2.0138	0.49659	5.7	298.90	0.00334
0.8	2.2255	0.44933	5.8	330.31	0.00302
0.9	2.4596	0.40657	5.9	365.08	0.00274
1.0	2.7183	0.36788	6.0	403.45	0.00247
1.1	3.0042	0.33287	6.1	445.89	0.00224
1.2	3.3201	0.30119	6.2	492.77	0.00203
1.3	3.6693	0.27253	6.3	544.61	0.00184
1.4	4.0552	0.24660	6.4	601.86	0.00166
1.5	4.4817	0.22313	6.5	665.17	0.00150
1.6	4.9530	0.20190	6.6	735.15	0.00136
1.7	5.4739	0.18268	6.7	812.43	0.00123
1.8	6.0496	0.16530	6.8	897.89	0.00111
1.9	6.6859	0.14957	6.9	992.38	0.00100
2.0	7.3891	0.13534	7.0	1,096.69	0.00091
2.1	8.1662	0.12246	7.1	1,212.07	0.00082
2.2	9.0250	0.11080	7.2	1,339.57	0.00074
2.3	9.9742	0.10026	7.3	1,480.49	0.00068
2.4	11.023	0.09072	7.4	1,636.25	0.00061
2.5	12.182	0.08208	7.5	1,808.0	0.00055
2.6	13.464	0.07427	7.6	1,998.2	0.00050
2.7	14.880	0.06721	7.7	2,208.3	0.00045
2.8	16.445	0.06081	7.8	2,440.6	0.00011
2.9	18.174	0.05502	7.9	2,697.3	0.00037
3.0	20.086	0.04979	8.0	2,981.0	0.00034
3.1	22.198	0.04505	8.1	3,294.5	0.00030
3.2	24.533	0.04076	8.2	3,641.0	0.00027
3.3	27.113	0.03688	8.3	4,023.9	0.00025
3.4	29.964	0.03337	8.4	4,447.1	0.00022
3.5	33.115	0.03020	8.5	4,914.8	0.00020
3.6	36.598	0.02732	8.6	5,431.7	0.00018
3.7	40.447	0.02472	8.7	6,002.9	0.00017
3.8	44.701	0.02237	8.8	6,634.2	0.00015
3.9	49.402	0.02024	8.9	7,332.0	0.00014
4.0	54.598	0.01832	9.0	8,103.1	0.00012
4.1	60.310	0.01657	9.1	8,955.3	0.00011
4.2	66.686	0.01500	9.2	9,897.1	0.00010
4.3	73.700	0.01357	9.3	10,938	0.00009
4.4	81.451	0.01228	9.4	12,088	0.00008
4.5	90.017	0.01111	9.5	13,360	0.00007
4.6	99.484	0.01005	9.6	14,765	0.00007
4.7	109.95	0.00910	9.7	16,318	0.00006
4.8	121.51	0.00823	9.8	18,034	0.00006
4.9	134.29	0.00745	9.9	19,930	0.00005

Table 2*

Compound Interest
$(1 + r)^n$
r = rate of interest per payment period;
n = number of payment periods

n	$\frac{1}{4}\%$	$\frac{1}{2}\%$	$\frac{3}{4}\%$	1%	$1\frac{1}{4}\%$
1	1.002500	1.005000	1.007500	1.010000	1.012500
2	1.005006	1.010025	1.015056	1.020100	1.025156
3	1.007519	1.015075	1.022669	1.030301	1.037971
4	1.010038	1.020151	1.030339	1.040604	1.050945
5	1.012563	1.025251	1.038067	1.051010	1.064082
6	1.015094	1.030378	1.045852	1.061520	1.077383
7	1.017632	1.035529	1.053696	1.072135	1.090850
8	1.020176	1.040707	1.061599	1.082857	1.104486
9	1.022726	1.045911	1.069561	1.093685	1.118292
10	1.025283	1.051140	1.077583	1.104622	1.132271
11	1.027846	1.056396	1.085664	1.115668	1.146424
12	1.030416	1.061678	1.093807	1.126825	1.160755
13	1.032922	1.066986	1.102010	1.138093	1.175264
14	1.035574	1.072321	1.110276	1.149474	1.189955
15	1.038163	1.077683	1.118603	1.160969	1.204829
16	1.040759	1.083071	1.126992	1.172579	1.219890
17	1.043361	1.088487	1.135445	1.184304	1.235138
18	1.045969	1.093929	1.143960	1.196147	1.250577
19	1.048584	1.099399	1.152540	1.208109	1.266210
20	1.051205	1.104896	1.161134	1.220190	1.282037
21	1.053833	1.110420	1.169893	1.232392	1.298063
22	1.056468	1.115972	1.178667	1.244716	1.314288
23	1.059109	1.121552	1.187507	1.257163	1.330717
24	1.061757	1.127160	1.196414	1.269735	1.347351
25	1.064411	1.132796	1.205387	1.282432	1.364193
26	1.067072	1.138460	1.214427	1.295256	1.381245
27	1.069740	1.144152	1.223535	1.308209	1.398511
28	1.072414	1.149873	1.232712	1.321291	1.415992
29	1.075096	1.155622	1.241957	1.334504	1.433692
30	1.077783	1.161400	1.251272	1.347849	1.451613
31	1.080478	1.167207	1.260656	1.361327	1.469758
32	1.083179	1.173043	1.270111	1.374941	1.488130
33	1.085887	1.178908	1.279637	1.388690	1.506732
34	1.088602	1.184803	1.289234	1.402577	1.525566
35	1.091323	1.190727	1.298904	1.416603	1.544636
36	1.094051	1.196681	1.308645	1.430769	1.563944
37	1.096786	1.202664	1.318460	1.445076	1.583493
38	1.099528	1.208677	1.328349	1.459527	1.603287
39	1.102277	1.214721	1.338311	1.474122	1.623328
40	1.105033	1.220794	1.348349	1.488864	1.643619
41	1.107796	1.226898	1.358461	1.503752	1.664165
42	1.110565	1.233033	1.368650	1.518790	1.684967
43	1.113341	1.239198	1.378915	1.533978	1.706029
44	1.116125	1.245394	1.389256	1.549318	1.727354
45	1.118915	1.251621	1.399676	1.564811	1.748946
46	1.121712	1.257879	1.410173	1.580459	1.770808
47	1.124517	1.264168	1.420750	1.596263	1.792943
48	1.127328	1.270489	1.431405	1.612226	1.815355
49	1.130146	1.270842	1.442141	1.628348	1.818047
50	1.132972	1.283226	1.452957	1.644632	1.861022

* Reproduced by permission from A. Mizrahi and M. Sullivan, *Finite Mathematics with Applications* (New York: Wiley 1979).

Table 2 (continued)

n	¼%	½%	¾%	1%	1¼%
51	1.135804	1.289642	1.463854	1.661078	1.884285
52	1.138644	1.296090	1.474833	1.677688	1.907839
53	1.141490	1.302571	1.485894	1.694466	1.931687
54	1.144344	1.309083	1.497038	1.711410	1.955833
55	1.147205	1.315629	1.508266	1.728525	1.980281
56	1.150073	1.322207	1.519578	1.745810	2.005034
57	1.152948	1.328818	1.530975	1.763268	2.030097
58	1.155830	1.335462	1.542457	1.780901	2.059473
59	1.158720	1.342139	1.554026	1.798770	2.081167
60	1.161617	1.348850	1.565681	1.816697	2.107181
61	1.164521	1.355594	1.577424	1.834864	2.133521
62	1.167432	1.362372	1.589254	1.853212	2.160190
63	1.170351	1.365184	1.601174	1.871744	2.187192
64	1.173277	1.376030	1.613182	1.890462	2.214532
65	1.176210	1.382910	1.625281	1.905366	2.242214
66	1.179150	1.389825	1.637471	1.928460	2.270242
67	1.182098	1.396774	1.649752	1.947745	2.298620
68	1.185053	1.403758	1.662125	1.967222	2.327352
69	1.188016	1.410777	1.674591	1.986894	2.356444
70	1.190986	1.417831	1.687151	2.006763	2.385900
71	1.193963	1.424920	1.699804	2.026831	2.415724
72	1.196948	1.432044	1.712553	2.047099	2.445920
73	1.199941	1.439205	1.725397	2.067570	2.476494
74	1.202941	1.446401	1.738337	2.088246	2.507450
75	1.205948	1.453633	1.751375	2.109128	2.538793
76	1.208963	1.460901	1.764510	2.130220	2.570528
77	1.211985	1.468205	1.777744	2.151522	2.602660
78	1.215015	1.475546	1.791077	2.173037	2.635193
79	1.218053	1.482924	1.804510	2.194767	2.668133
80	1.221098	1.490339	1.818044	2.216715	2.701485
81	1.224151	1.497790	1.831679	2.238882	2.735253
82	1.227211	1.505279	1.845417	2.261271	2.769444
83	1.230279	1.512806	1.859257	2.283884	2.804062
84	1.233355	1.520370	1.873202	2.306723	2.839113
85	1.236438	1.527972	1.887251	2.329790	2.874602
86	1.239529	1.535611	1.901405	2.353088	2.910534
87	1.242628	1.543289	1.915666	2.376619	2.946916
88	1.245735	1.551006	1.930033	2.400385	2.983752
89	1.248849	1.558761	1.944509	2.424389	3.021049
90	1.251971	1.566555	1.959092	2.448633	3.058812
91	1.255101	1.574387	1.973786	2.473119	3.097048
92	1.258239	1.582259	1.988589	2.497850	3.135761
93	1.261384	1.590171	2.003503	2.522829	3.174958
94	1.264538	1.598122	2.018530	2.548057	3.214645
95	1.267699	1.606112	2.033669	2.573537	3.254828
96	1.270868	1.614143	2.048921	2.599273	3.295513
97	1.274046	1.622213	2.064288	2.625266	3.336707
98	1.277231	1.630325	2.079770	2.651518	3.378416
99	1.280424	1.638476	2.095368	2.678033	3.420646
100	1.283625	1.646669	2.111084	2.704814	3.463404

Table 2 (continued)

n	$1\frac{1}{2}\%$	$1\frac{3}{4}\%$	2%	$2\frac{1}{2}\%$	3%
1	1.015000	1.017500	1.020000	1.025000	1.030000
2	1.030225	1.035306	1.040400	1.050625	1.060900
3	1.045678	1.053424	1.061208	1.076891	1.092727
4	1.061364	1.071859	1.082432	1.103813	1.125509
5	1.077284	1.090617	1.104081	1.131408	1.159274
6	1.093443	1.109702	1.126162	1.159693	1.194052
7	1.109845	1.129122	1.148686	1.188686	1.229874
8	1.126493	1.148882	1.171659	1.218403	1.266770
9	1.143390	1.168987	1.195093	1.248863	1.304773
10	1.160541	1.189444	1.218994	1.280085	1.343916
11	1.177949	1.210260	1.243374	1.312087	1.384234
12	1.195618	1.231439	1.268242	1.344889	1.425761
13	1.213552	1.252989	1.293607	1.378511	1.468534
14	1.231756	1.274917	1.319479	1.412974	1.512590
15	1.250232	1.297228	1.345868	1.448298	1.557967
16	1.268986	1.319929	1.372786	1.484506	1.604706
17	1.288020	1.343028	1.400241	1.521618	1.652848
18	1.307341	1.366531	1.428246	1.559659	1.702433
19	1.326951	1.390445	1.456811	1.598650	1.753506
20	1.346855	1.414778	1.485947	1.638616	1.806111
21	1.367058	1.439537	1.515666	1.679582	1.860295
22	1.387564	1.464729	1.545980	1.721571	1.916103
23	1.408377	1.490361	1.576899	1.764611	1.973586
24	1.429503	1.516443	1.608437	1.808726	2.032794
25	1.450945	1.542981	1.640606	1.853944	2.093778
26	1.472710	1.569983	1.673418	1.900293	2.156591
27	1.494800	1.597457	1.706886	1.947800	2.221289
28	1.517222	1.625413	1.741024	1.996495	2.287928
29	1.539981	1.653858	1.775845	2.046407	2.356565
30	1.563080	1.682800	1.811362	2.097568	2.427262
31	1.586526	1.712249	1.847589	2.150007	2.500080
32	1.610324	1.742213	1.884541	2.203757	2.575083
33	1.634479	1.772702	1.922231	2.258851	2.652335
34	1.658996	1.803724	1.960676	2.315322	2.731905
35	1.683881	1.835290	1.999889	2.373205	2.813862
36	1.709140	1.867407	2.039887	2.432535	2.898278
37	1.734777	1.900087	2.080685	2.493349	2.985227
38	1.760798	1.933338	2.122299	2.555682	2.074783
39	1.787210	1.967172	2.164745	2.619574	3.167027
40	1.814018	2.001597	2.208040	2.685064	3.262038
41	1.841229	2.036625	2.252200	2.752190	3.359899
42	1.868847	2.072266	2.297244	2.820995	3.460696
43	1.896880	2.108531	2.343189	2.891520	3.564517
44	1.925333	2.145430	2.390053	2.963808	3.671452
45	1.954213	2.182975	2.437854	3.037902	3.781596
46	1.983526	2.221177	2.486611	3.113851	3.895044
47	2.013279	2.260048	2.536343	3.191697	4.011895
48	2.043478	2.299599	2.587070	3.271489	4.132252
49	2.074130	2.329842	2.638812	3.353277	4.256219
50	2.105242	2.380789	2.691588	3.437109	4.381906

Table 2 (continued)

n	$1\frac{1}{2}\%$	$1\frac{3}{4}\%$	2%	$2\frac{1}{2}\%$	3%
51	2.136821	2.422453	2.745420	3.523036	4.515423
52	2.168873	2.464846	2.800328	3.611112	4.650886
53	2.201406	2.507980	2.856335	3.701390	4.790412
54	2.234428	2.551870	2.913461	3.793925	4.934125
55	2.267944	2.596528	2.971731	3.888773	5.082148
56	2.301963	2.641967	3.031165	3.985992	5.234613
57	2.336493	2.688201	3.091788	4.085642	5.391651
58	2.371540	2.735245	3.153624	4.187783	5.553401
59	2.407113	2.783112	3.216697	4.292478	5.720003
60	2.443220	2.831816	3.281031	4.399790	5.891603
61	2.479868	2.881373	3.346651	4.509784	6.068351
62	2.517066	2.931797	3.413584	4.622529	6.250402
63	2.554822	2.983103	3.481856	4.738092	6.437914
64	2.593144	3.035308	3.551493	4.856544	6.631051
65	2.632042	3.088426	3.622523	4.977958	6.829982
66	2.671522	3.142473	3.694973	5.102407	7.034882
67	2.711595	3.197466	3.768873	5.229967	7.245928
68	2.752269	3.253422	3.844250	5.360716	7.463306
69	2.793553	3.310357	3.921135	5.494734	7.687205
70	2.835456	3.368288	3.999558	5.632103	7.917822
71	2.877988	3.427233	4.079549	5.772905	8.155356
72	2.921158	3.487210	4.161140	5.917228	8.400017
73	2.964975	3.548236	4.244363	6.065158	8.652017
74	3.009450	3.610330	4.329250	6.216787	8.911578
75	3.054592	3.673511	4.415835	6.372207	9.178925
76	3.100411	3.737797	4.504152	6.531512	9.454293
77	3.146917	3.803209	4.594235	6.694800	9.737922
78	3.194120	3.869765	4.686120	6.862170	10.030060
79	3.242032	3.937486	4.779842	7.033724	10.330961
80	3.290663	4.006392	4.875439	7.209567	10.640890
81	3.340023	4.076504	4.972948	7.389806	10.960117
82	3.390123	4.147842	5.072407	7.574552	11.288920
83	3.440975	4.220430	5.173855	7.763915	11.627588
84	3.492590	4.294287	5.277332	7.958013	11.976416
85	3.544978	4.369437	5.382878	8.156964	12.335708
86	3.598153	4.445902	5.490536	8.360888	12.705779
87	3.652125	4.523706	5.600347	8.569910	13.086953
88	3.706907	4.602870	5.712354	8.784158	13.479561
89	3.762511	4.683421	5.826601	9.003762	13.883948
90	3.818948	4.765380	5.943133	9.228856	14.300466
91	3.876233	4.848775	6.061995	9.459577	14.729480
92	3.934376	4.933628	6.183235	9.696066	15.171365
93	3.993392	5.019967	6.306900	9.938468	15.626506
94	4.053293	5.107816	6.433038	10.186930	16.095301
95	4.114092	5.197203	6.561699	10.441603	16.578160
96	4.175804	5.288154	6.692933	10.702643	17.075505
97	4.238441	5.380697	6.826791	10.970209	17.587770
98	4.302017	5.474859	6.963327	11.244464	18.115403
99	4.366547	5.570669	7.102594	11.525576	18.658865
100	4.432046	5.668156	7.244645	11.813715	19.218631

Table 2 (continued)

n	$3\frac{1}{2}\%$	4%	$4\frac{1}{2}\%$	5%	$5\frac{1}{2}\%$
1	1.035000	1.040000	1.045000	1.050000	1.055000
2	1.071225	1.081600	1.092025	1.102500	1.113025
3	1.108718	1.124864	1.141166	1.157625	1.174241
4	1.147523	1.169859	1.192519	1.215506	1.238825
5	1.187686	1.216653	1.246182	1.276282	1.306960
6	1.229225	1.265319	1.302260	1.340096	1.378843
7	1.272279	1.315932	1.360862	1.407100	1.454679
8	1.316809	1.368569	1.422101	1.477455	1.534687
9	1.362897	1.423321	1.486095	1.551328	1.619094
10	1.410599	1.480244	1.552969	1.628895	1.708144
11	1.459970	1.539454	1.622853	1.710339	1.802092
12	1.511069	1.601032	1.695881	1.795856	1.901207
13	1.563956	1.665073	1.772196	1.885649	2.005774
14	1.618695	1.731676	1.851945	1.979932	2.116091
15	1.675349	1.800943	1.935282	2.078928	2.232476
16	1.733986	1.872981	2.022370	2.182875	2.355263
17	1.794676	1.947900	2.113377	2.292018	2.484802
18	1.857489	2.025816	2.208479	2.406619	2.621466
19	1.922501	2.106849	2.307860	2.526950	2.765647
20	1.989789	2.191123	2.411714	2.653298	2.917757
21	2.059431	2.278768	2.520241	2.785963	3.078234
22	2.131512	2.369919	2.633652	2.925261	3.247537
23	2.206114	2.464715	2.752166	3.071524	3.426152
24	2.283328	2.563304	2.876014	3.225100	3.614590
25	2.363245	2.665836	3.005434	3.386355	3.813392
26	2.445959	2.772470	3.140679	3.555673	4.023125
27	2.531567	2.883368	3.282010	3.733456	4.244401
28	2.620172	2.998703	3.429700	3.920129	4.477843
29	2.711878	3.118651	3.584037	4.116136	4.724124
30	2.806794	3.243397	3.745318	4.321942	4.983951
31	2.905031	3.373133	3.913857	4.538039	5.258068
32	3.006708	3.508059	4.089981	4.764941	5.547262
33	3.111942	3.648381	4.274030	5.003188	5.852362
34	3.220860	3.794316	4.466362	5.253348	6.174242
35	2.333590	3.946089	4.667348	5.516015	6.513825
36	3.450266	4.103932	4.877376	5.791816	6.872085
37	3.571025	4.268090	5.096861	6.081407	7.250050
38	3.696011	4.438813	5.326219	6.385477	7.648803
39	3.825372	4.616366	5.565899	6.704751	8.069487
40	3.959260	4.801020	5.816365	7.039989	8.513309
41	4.097834	4.993061	6.078101	7.391988	8.981541
42	4.241258	5.192784	6.351616	7.761587	9.475525
43	4.389702	5.400495	6.637438	8.149667	9.996679
44	4.543341	5.616515	6.936123	8.557150	10.546496
45	4.702358	5.841175	7.248248	8.985008	11.126554
46	4.866941	6.074822	7.574420	9.434258	11.738514
47	5.037284	6.317815	7.915269	9.905971	12.384132
48	5.213589	6.570528	8.271456	10.401269	13.065260
49	5.396064	6.833349	8.643671	10.921333	13.783849
50	5.584927	7.106683	9.032636	11.467400	14.541961

Table 2 (continued)

n	$3\frac{1}{2}\%$	4%	$4\frac{1}{2}\%$	5%	$5\frac{1}{2}\%$
51	5.780399	7.390950	9.439105	12.040770	15.341768
52	5.982713	7.686588	9.863865	12.642808	16.185566
53	6.192108	7.994052	10.307739	13.274948	17.075772
54	6.408832	8.313814	10.771587	13.938696	18.014939
55	6.633141	8.646366	11.256308	14.635631	19.005761
56	6.865301	8.992221	11.762842	15.367412	20.051079
57	7.105586	9.351910	12.292170	16.135783	21.153887
58	7.354282	9.725986	12.845318	16.942572	22.317351
59	7.611682	10.115026	13.423357	17.789700	23.544805
60	7.878091	10.519627	14.027408	18.679185	24.839769
61	8.153824	10.940412	14.658641	19.613145	26.205957
62	8.439207	11.378028	15.318280	20.593802	27.647284
63	8.734580	11.833149	16.007603	21.623492	29.167885
64	9.040290	12.306475	16.727945	22.704667	30.772118
65	9.356700	12.798734	17.480703	23.839900	32.464585
66	9.684185	13.310684	18.267334	25.031895	34.250137
67	10.023131	13.843111	19.089364	26.283490	36.133895
68	10.373941	14.396835	19.948386	27.597664	38.121259
69	10.737029	14.972709	20.846063	28.977547	40.217928
70	11.112825	15.571617	21.784136	30.426425	42.429914
71	11.501773	16.194482	22.764422	31.947746	44.763559
72	11.904336	16.842261	23.788821	33.545133	47.225555
73	12.320987	17.515951	24.859318	35.222390	49.822961
74	12.752222	18.216589	25.977987	36.983509	52.563223
75	13.198550	18.945253	27.146997	38.832685	55.454201
76	13.660499	19.703063	28.368611	40.774319	58.504182
77	14.138616	20.491186	29.645199	42.813035	61.721911
78	14.633468	21.310833	30.979233	44.953687	65.116617
79	15.145639	22.163266	32.373298	47.201371	68.698030
80	15.675736	23.049797	33.830097	49.561440	72.476422
81	16.224387	23.971789	35.352451	52.039512	76.462625
82	16.792241	24.930660	36.943312	54.641487	80.668069
83	17.379969	25.927887	38.605761	57.373561	85.104813
84	17.988268	26.965002	40.343020	60.242239	89.785578
85	18.617857	28.043602	42.158456	63.254351	94.723785
86	19.269482	29.165346	44.055586	66.417069	99.933593
87	19.943914	30.331960	46.038086	69.737922	105.429940
88	20.641951	31.545238	48.109802	73.224818	111.228587
89	21.364420	32.807048	50.274743	76.886059	117.346159
90	22.112174	34.119330	52.537106	80.730362	123.800198
91	22.886100	35.484103	54.901276	84.766880	130.609209
92	23.687114	36.903467	57.371833	89.005224	137.792715
93	24.516163	38.379606	59.953566	92.455486	145.371314
94	25.374229	39.914790	62.651470	98.128260	153.366736
95	26.262327	41.511381	65.470793	103.034673	161.801907
96	27.181508	43.171836	68.416978	108.186406	170.701011
97	28.132861	44.898710	71.495742	113.595727	180.089567
98	29.117511	46.694658	74.713051	119.275513	189.994493
99	30.136624	48.562444	78.075138	125.239289	200.444190
100	31.191405	50.504942	81.588519	131.501253	211.468620

Table 2 (continued)

n	6%	$6\frac{1}{2}$%	7%	$7\frac{1}{2}$%	8%
1	1.060000	1.065000	1.070000	1.075000	1.080000
2	1.123600	1.134225	1.144900	1.155625	1.166400
3	1.191016	1.207950	1.225043	1.242297	1.259712
4	1.262477	1.286466	1.310796	1.335469	1.360489
5	1.338226	1.370087	1.402552	1.435629	1.469328
6	1.418519	1.459142	1.500730	1.543302	1.586874
7	1.503630	1.553987	1.605781	1.659045	1.713824
8	1.593848	1.654996	1.718186	1.783478	1.850930
9	1.689479	1.762570	1.838459	1.917239	1.999005
10	1.790848	1.877137	1.967151	2.061032	2.158925
11	1.898299	1.999151	2.104852	2.215609	2.331639
12	2.012196	2.129096	2.252192	2.381780	2.518170
13	2.132928	2.267487	2.409845	2.560413	2.719624
14	2.260904	2.414874	2.578534	2.752444	2.937194
15	2.396558	2.571841	2.759032	2.958877	3.172169
16	2.540352	2.739011	2.952164	3.180793	3.425943
17	2.692773	2.917046	3.158815	3.419353	3.700018
18	2.854339	3.106654	3.379932	3.675804	3.996019
19	3.025599	3.308587	3.616527	3.951489	4.315701
20	3.207135	3.523645	3.869684	4.247851	4.660957
21	3.399564	3.752682	4.140562	4.566440	5.033834
22	3.603537	3.996606	4.430402	4.908923	5.436540
23	3.819750	4.256386	4.740530	5.277092	5.871464
24	4.048935	4.533051	5.072367	5.672874	6.341181
25	4.291871	4.827699	5.427433	6.098340	6.848475
26	4.549383	5.141500	5.807353	6.555715	7.396353
27	4.822346	5.475697	6.213867	7.047394	7.988061
28	5.111687	5.831617	6.648838	7.575948	8.627106
29	5.418388	6.210672	7.114257	8.144144	9.317275
30	5.743491	6.614366	7.612255	8.754955	10.062657
31	6.088101	7.044300	8.145113	9.411577	10.867669
32	6.453387	7.502179	8.715271	10.117445	11.737083
33	6.840590	7.989821	9.325340	10.876253	12.676049
34	7.251025	8.509160	9.978113	11.691972	13.690133
35	7.686087	9.062255	10.676581	12.568870	14.785344
36	8.147252	9.651301	11.423942	13.511535	15.968171
37	8.636087	10.278636	12.223618	14.524901	17.245625
38	9.154252	10.946747	13.079271	15.614268	18.625275
39	9.703507	11.658286	13.994820	16.785338	20.115297
40	10.285718	12.416075	14.974457	18.044239	21.724521
41	10.902861	13.223119	16.022669	19.397557	23.462483
42	11.557032	14.082622	17.144256	20.852373	25.339481
43	12.250454	14.997993	18.344354	22.416301	27.366640
44	12.985482	15.972862	19.628459	24.097524	29.555971
45	13.764610	17.011098	21.002451	25.904838	31.920449
46	14.590487	18.116820	22.472622	27.847701	34.474084
47	15.465916	19.294413	24.045706	29.936278	37.232011
48	16.393871	20.548550	25.728905	32.181498	40.210572
49	17.377504	21.884205	27.529929	34.595112	43.427418
50	18.420154	23.306679	29.457024	37.189745	46.901611

Table 2 (continued)

n	6%	$6\frac{1}{2}$%	7%	$7\frac{1}{2}$%	8%
51	19.525363	24.821613	31.519016	39.978976	50.653740
52	20.696885	26.435018	33.725346	42.977399	54.706039
53	21.938698	28.153294	36.086121	46.200704	59.082522
54	23.255020	29.983258	38.612149	49.665757	63.809124
55	24.650321	31.932170	41.315000	53.390689	68.913854
56	26.129340	34.007761	44.207049	57.394990	74.426962
57	27.697100	36.218265	47.301543	61.699614	80.381119
58	29.358926	38.572452	50.612651	66.327086	86.811608
59	31.120462	41.079662	54.155536	71.301617	93.756537
60	32.987690	43.749840	57.946424	76.649238	101.257060
61	34.966951	46.593579	62.002673	82.397931	109.357625
62	37.064968	49.622162	66.342861	88.577776	118.106234
63	39.288865	52.847603	70.986861	95.221109	127.554733
64	41.646198	56.282697	75.955941	102.362692	137.759112
65	44.144970	59.941072	81.272857	110.039894	148.779841
66	46.793668	63.837242	86.961957	118.292886	160.682228
67	49.601288	67.986662	93.049294	127.164852	173.536806
68	52.577365	72.405796	99.562744	136.702216	187.415750
69	55.732007	77.112172	106.532136	146.954882	202.413330
70	59.075928	82.124464	113.989385	157.976498	218.606396
71	62.620483	87.462554	121.968642	169.824735	236.094908
72	66.377712	93.147620	130.506447	182.561591	254.982500
73	70.360375	99.202215	139.641898	196.253710	275.381101
74	74.581997	105.650359	149.416831	210.972738	297.411588
75	79.056917	112.517632	159.876009	226.795693	321.204515
76	83.800332	119.831279	171.067330	243.805370	346.900876
77	88.828352	127.620312	183.042042	262.090773	374.652946
78	94.158053	135.915632	195.854985	281.747581	404.625181
79	99.807536	144.750148	209.564834	302.878649	436.995196
80	105.795988	154.158908	224.234372	325.594547	471.954811
81	112.143748	164.179237	239.930778	350.014138	509.711196
82	118.872372	174.850887	256.725932	376.265199	550.488090
83	126.004715	186.216194	274.696748	404.485088	594.527138
84	133.564997	198.320247	293.925519	434.821470	642.089308
85	141.578897	211.211063	314.500305	467.433080	693.456453
86	150.073631	224.939782	336.515326	502.490560	748.932968
87	159.078049	239.560868	360.071399	540.177352	808.847606
88	168.622731	255.132325	385.276396	580.690654	873.555413
89	178.740095	271.715926	412.245744	624.242453	943.439846
90	189.464501	289.377461	441.102946	671.060636	1018.915031
91	200.832371	308.186996	471.980152	721.390183	1100.428236
92	212.882312	328.219151	505.018761	775.494447	1188.462491
93	225.655252	349.553396	540.370075	833.656529	1283.539492
94	239.194566	372.274366	578.195979	896.180769	1386.222649
95	253.546240	396.472200	618.669698	963.394325	1497.120463
96	258.759014	422.242893	661.976577	1035.648900	1616.890095
97	284.884555	449.688681	708.314935	1113.322566	1746.241303
98	301.977628	478.918445	757.896980	1196.821759	1885.940604
99	320.096286	510.048144	810.949769	1286.583390	2036.815854
100	329.302062	543.201275	867.716251	1383.077144	2199.761119

Table 3*
Present Value of a Dollar
$(1 + r)^{-n}$
r = rate of interest per payment period; n = number of payment periods

n	$\frac{1}{4}\%$	$\frac{1}{2}\%$	$\frac{3}{4}\%$	1%	$1\frac{1}{4}\%$
1	0.997506	0.995025	0.992556	0.990099	0.987654
2	0.995019	0.990075	0.985167	0.980296	0.975461
3	0.992537	0.985149	0.977833	0.970590	0.963418
4	0.990062	0.980248	0.970554	0.960980	0.951524
5	0.987593	0.975371	0.963329	0.951466	0.939777
6	0.985130	0.970518	0.956158	0.942045	0.928175
7	0.982674	0.965690	0.949040	0.932718	0.916716
8	0.980223	0.960885	0.941975	0.923483	0.905398
9	0.977779	0.956105	0.934963	0.914340	0.894221
10	0.975340	0.951348	0.928003	0.905287	0.883181
11	0.972908	0.946615	0.921095	0.896324	0.872277
12	0.970482	0.941905	0.914238	0.887449	0.861509
13	0.968062	0.937219	0.907432	0.878663	0.850873
14	0.965648	0.932556	0.900677	0.869963	0.840368
15	0.963240	0.927917	0.893973	0.861349	0.829993
16	0.960837	0.923300	0.887318	0.852821	0.819746
17	0.958441	0.918707	0.880712	0.844377	0.809626
18	0.956051	0.914136	0.874156	0.836017	0.799631
19	0.953667	0.909588	0.867649	0.827740	0.789759
20	0.951289	0.905063	0.861190	0.819544	0.780009
21	0.948917	0.900560	0.854779	0.811430	0.770379
22	0.946550	0.896080	0.848416	0.803396	0.760868
23	0.944190	0.891622	0.842100	0.795442	0.751475
24	0.941835	0.887186	0.835831	0.787566	0.742197
25	0.939486	0.882772	0.829609	0.779768	0.733034
26	0.937144	0.878380	0.823434	0.772048	0.723984
27	0.934806	0.874010	0.817304	0.764404	0.715046
28	0.932475	0.869662	0.811220	0.756836	0.706219
29	0.930150	0.865335	0.805181	0.749342	0.697500
30	0.927830	0.861030	0.799187	0.741923	0.688889
31	0.925517	0.856746	0.793238	0.734577	0.680384
32	0.923209	0.852484	0.787333	0.727304	0.671984
33	0.920906	0.848242	0.781472	0.720103	0.663688
34	0.918610	0.844022	0.775654	0.712973	0.655494
35	0.916319	0.839823	0.769880	0.705914	0.647402
36	0.914034	0.835645	0.764149	0.698925	0.639409
37	0.911754	0.831487	0.758461	0.692005	0.631515
38	0.909481	0.827351	0.752814	0.685153	0.623719
39	0.907213	0.823235	0.747210	0.678370	0.616019
40	0.904950	0.819139	0.741648	0.671653	0.608413
41	0.902694	0.815064	0.736127	0.665003	0.600902
42	0.900443	0.811008	0.730647	0.658419	0.593484
43	0.898197	0.806974	0.725208	0.651900	0.586157
44	0.895957	0.802959	0.719810	0.645445	0.578920
45	0.893723	0.798964	0.714451	0.639055	0.571773
46	0.891494	0.794989	0.709133	0.632728	0.564714
47	0.889271	0.791034	0.703854	0.626463	0.557742
48	0.887053	0.787098	0.698614	0.620260	0.550857
49	0.884841	0.783182	0.693414	0.614119	0.544056
50	0.882635	0.779286	0.688252	0.608039	0.537339

* Reproduced by permission from A. Mizrahi and M. Sullivan, *Finite Mathematics with Applications* (New York: Wiley, 1979).

Table 3 (continued)

n	$\frac{1}{4}\%$	$\frac{1}{2}\%$	$\frac{3}{4}\%$	1%	$1\frac{1}{4}\%$
51	0.880434	0.775409	0.683128	0.602019	0.530705
52	0.878238	0.771551	0.678043	0.596058	0.524153
53	0.876048	0.767713	0.672995	0.590157	0.517682
54	0.873863	0.763893	0.667986	0.584313	0.511291
55	0.871684	0.760093	0.663013	0.578528	0.504979
56	0.869510	0.756311	0.658077	0.572800	0.498745
57	0.867342	0.752548	0.653179	0.567129	0.492587
58	0.865179	0.748804	0.648316	0.561514	0.486506
59	0.863021	0.745079	0.643490	0.555954	0.480500
60	0.860869	0.741372	0.638700	0.550450	0.474568
61	0.858722	0.737684	0.633945	0.545000	0.468709
62	0.856581	0.734014	0.629226	0.539604	0.462922
63	0.854445	0.730362	0.624542	0.534261	0.457207
64	0.852314	0.726728	0.619893	0.528971	0.451563
65	0.850189	0.723113	0.615278	0.523734	0.455988
66	0.848068	0.719515	0.610698	0.518548	0.440482
67	0.845953	0.715935	0.606152	0.513414	0.435044
68	0.843844	0.712374	0.601639	0.508331	0.429673
69	0.841740	0.708829	0.597161	0.503298	0.424368
70	0.839640	0.705303	0.592715	0.498315	0.419129
71	0.837547	0.701794	0.588303	0.493381	0.413955
72	0.835458	0.698302	0.583924	0.488496	0.408844
73	0.833374	0.694828	0.579577	0.483660	0.403797
74	0.831296	0.691371	0.575262	0.478871	0.398811
75	0.829233	0.687932	0.570980	0.474130	0.393888
76	0.827155	0.684509	0.566730	0.469435	0.389025
77	0.825093	0.681104	0.562511	0.464787	0.384222
78	0.823035	0.677715	0.558323	0.460185	0.379479
79	0.820983	0.674343	0.554167	0.455629	0.374794
80	0.818935	0.670988	0.550042	0.451118	0.370167
81	0.816893	0.667650	0.545947	0.446651	0.365597
82	0.814856	0.664329	0.541883	0.442229	0.361083
83	0.812824	0.661023	0.537849	0.437851	0.356625
84	0.810797	0.657735	0.533845	0.433515	0.352223
85	0.808775	0.654462	0.529871	0.429223	0.347874
86	0.806758	0.651206	0.525927	0.424974	0.343580
87	0.804746	0.647967	0.522012	0.420766	0.339338
88	0.802739	0.644743	0.518126	0.416600	0.335148
89	0.800737	0.641535	0.514269	0.412475	0.331011
90	0.798741	0.638343	0.510440	0.408391	0.326924
91	0.796749	0.635168	0.506641	0.404348	0.322888
92	0.794762	0.632008	0.502869	0.400344	0.318902
93	0.792780	0.628863	0.499126	0.396380	0.314965
94	0.790803	0.625735	0.495410	0.392456	0.311076
95	0.788831	0.622622	0.491722	0.388570	0.307236
96	0.786864	0.619524	0.488062	0.384723	0.303443
97	0.784901	0.616442	0.484429	0.380914	0.299697
98	0.782944	0.613375	0.480822	0.377142	0.295997
99	0.780991	0.610323	0.477243	0.373408	0.292342
100	0.779044	0.607287	0.473690	0.369711	0.288733

Table 3 (continued)

n	$1\frac{1}{2}\%$	$1\frac{3}{4}\%$	2%	$2\frac{1}{2}\%$	3%
1	0.985222	0.982801	0.980392	0.975610	0.970874
2	0.970662	0.965898	0.961169	0.951814	0.942596
3	0.956317	0.949285	0.942322	0.928599	0.915142
4	0.942184	0.932959	0.923845	0.905951	0.888487
5	0.928260	0.916913	0.905731	0.883854	0.862609
6	0.914542	0.901143	0.887971	0.862297	0.837484
7	0.901027	0.885644	0.870560	0.841265	0.813092
8	0.887711	0.870412	0.853490	0.820747	0.789409
9	0.874592	0.855441	0.836755	0.800728	0.766417
10	0.861667	0.840729	0.820348	0.781198	0.744094
11	0.848933	0.826269	0.804263	0.762145	0.722421
12	0.836387	0.812058	0.788493	0.743556	0.701380
13	0.824027	0.798091	0.773033	0.725420	0.680951
14	0.811849	0.784365	0.757875	0.707727	0.661118
15	0.799852	0.770875	0.743015	0.690466	0.641862
16	0.788031	0.757616	0.728446	0.673625	0.623167
17	0.776385	0.744586	0.714163	0.657195	0.605016
18	0.764912	0.731780	0.700159	0.641166	0.587395
19	0.753607	0.719194	0.686431	0.625528	0.570286
20	0.742470	0.706825	0.672971	0.610271	0.553676
21	0.731498	0.694668	0.659776	0.595386	0.537549
22	0.720688	0.682720	0.646839	0.580865	0.521893
23	0.710037	0.670978	0.634156	0.566697	0.506692
24	0.699544	0.659438	0.621722	0.552875	0.491934
25	0.689206	0.648096	0.609531	0.539391	0.477606
26	0.679021	0.636950	0.597579	0.526235	0.463695
27	0.668986	0.625995	0.585862	0.513400	0.450189
28	0.659099	0.615228	0.574375	0.500878	0.437077
29	0.649359	0.604647	0.563112	0.488661	0.424346
30	0.639762	0.594248	0.552071	0.476743	0.411987
31	0.630308	0.584027	0.541246	0.465115	0.399987
32	0.620993	0.573982	0.530633	0.453771	0.388337
33	0.611816	0.564111	0.520229	0.442703	0.377026
34	0.602774	0.554408	0.510028	0.431905	0.366045
35	0.593866	0.544873	0.500028	0.421371	0.355383
36	0.585090	0.535502	0.490223	0.411094	0.345032
37	0.576443	0.526292	0.480611	0.401067	0.334983
38	0.567924	0.517240	0.471187	0.391285	0.325226
39	0.559531	0.508344	0.461948	0.381741	0.315754
40	0.551262	0.499601	0.452890	0.372431	0.306557
41	0.543116	0.491008	0.444010	0.363347	0.297628
42	0.535089	0.482563	0.435304	0.354485	0.288959
43	0.527182	0.474264	0.426769	0.345839	0.280543
44	0.519391	0.466107	0.418401	0.337404	0.272372
45	0.511715	0.458090	0.410197	0.329174	0.264439
46	0.504153	0.450212	0.402154	0.321146	0.256737
47	0.496702	0.442469	0.394268	0.313313	0.249259
48	0.489362	0.434858	0.386538	0.305671	0.241999
49	0.482130	0.427379	0.378958	0.298216	0.234950
50	0.475005	0.420029	0.371528	0.290942	0.228107

Table 3 (continued)

n	$1\frac{1}{2}\%$	$1\frac{3}{4}\%$	2%	$2\frac{1}{2}\%$	3%
51	0.467985	0.412805	0.364243	0.283846	0.221463
52	0.461069	0.405705	0.357101	0.276923	0.215013
53	0.454255	0.398727	0.350099	0.270169	0.208750
54	0.447542	0.391869	0.343234	0.263579	0.202670
55	0.440928	0.385130	0.336504	0.257151	0.196767
56	0.434412	0.378506	0.329906	0.250879	0.191036
57	0.427992	0.371996	0.323437	0.244760	0.185472
58	0.421667	0.365598	0.317095	0.238790	0.180070
59	0.415435	0.359310	0.310878	0.232966	0.174825
60	0.409296	0.353130	0.304782	0.227284	0.169733
61	0.403247	0.347057	0.298806	0.221740	0.164789
62	0.397288	0.341088	0.292947	0.216332	0.159990
63	0.391417	0.335221	0.287203	0.211055	0.155330
64	0.385632	0.329456	0.281572	0.205908	0.150806
65	0.379933	0.323790	0.276051	0.200886	0.146413
66	0.374318	0.318221	0.270638	0.195986	0.142149
67	0.368787	0.312748	0.265331	0.191206	0.138009
68	0.363337	0.307369	0.260129	0.186542	0.133989
69	0.357967	0.302082	0.255028	0.181992	0.130086
70	0.352677	0.296887	0.250028	0.177554	0.126297
71	0.347465	0.291781	0.245125	0.173223	0.122619
72	0.342330	0.286762	0.240319	0.168998	0.119047
73	0.337271	0.281830	0.235607	0.164876	0.115580
74	0.332287	0.276983	0.230987	0.160855	0.112214
75	0.327376	0.272219	0.226458	0.156931	0.108945
76	0.322538	0.267537	0.222017	0.153104	0.105772
77	0.317771	0.262936	0.217664	0.149370	0.102691
78	0.313075	0.258414	0.213396	0.145726	0.099700
79	0.308449	0.253969	0.209212	0.142172	0.096796
80	0.303890	0.249601	0.205110	0.138705	0.093977
81	0.299399	0.245308	0.201088	0.135322	0.091240
82	0.294975	0.241089	0.197145	0.132021	0.088582
83	0.290615	0.236943	0.193279	0.128801	0.086002
84	0.286321	0.232868	0.189490	0.125660	0.083497
85	0.282089	0.228862	0.185774	0.122595	0.081065
86	0.277920	0.224926	0.182132	0.119605	0.078704
87	0.273813	0.221058	0.178560	0.116687	0.076412
88	0.269767	0.217256	0.175059	0.113841	0.074186
89	0.265780	0.213519	0.171627	0.111065	0.072026
90	0.261852	0.209847	0.168261	0.108356	0.069928
91	0.257982	0.206238	0.164962	0.105713	0.067891
92	0.254170	0.202691	0.161728	0.103135	0.065914
93	0.250414	0.199205	0.158557	0.100619	0.063994
94	0.246713	0.195778	0.155448	0.098165	0.062130
95	0.243067	0.192411	0.152400	0.095771	0.060320
96	0.239475	0.189102	0.149411	0.093435	0.058563
97	0.235936	0.185850	0.146482	0.091156	0.056858
98	0.232449	0.182653	0.143610	0.088933	0.055202
99	0.229014	0.179512	0.140794	0.086764	0.053594
100	0.225629	0.176424	0.138033	0.084647	0.052033

Table 3 (continued)

n	$3\frac{1}{2}\%$	4%	$4\frac{1}{2}\%$	5%	$5\frac{1}{2}\%$
1	0.966184	0.961538	0.956938	0.952381	0.947867
2	0.933511	0.924556	0.915730	0.907029	0.898452
3	0.901943	0.888996	0.876297	0.863838	0.851614
4	0.871442	0.854804	0.838561	0.822702	0.807217
5	0.841973	0.821927	0.802451	0.783526	0.765134
6	0.813501	0.790315	0.767896	0.746215	0.725246
7	0.785991	0.759918	0.734828	0.710681	0.687437
8	0.759412	0.730690	0.703185	0.676839	0.651599
9	0.733731	0.702587	0.672904	0.644609	0.617629
10	0.708919	0.675564	0.643928	0.613913	0.585431
11	0.684946	0.649581	0.616199	0.584679	0.554911
12	0.661783	0.624597	0.589664	0.556837	0.525982
13	0.639404	0.600574	0.564272	0.530321	0.498561
14	0.617782	0.577475	0.539973	0.505068	0.472569
15	0.596891	0.555265	0.516720	0.481017	0.447933
16	0.576706	0.533908	0.494469	0.458112	0.424581
17	0.557204	0.513373	0.473176	0.436297	0.402447
18	0.538361	0.493628	0.452800	0.415521	0.381466
19	0.520156	0.474642	0.433302	0.395734	0.361579
20	0.502566	0.456387	0.414643	0.376889	0.342729
21	0.485571	0.438834	0.396787	0.358942	0.324862
22	0.469151	0.421955	0.379701	0.341850	0.307926
23	0.453286	0.405726	0.363350	0.325571	0.291873
24	0.437957	0.390121	0.347703	0.310068	0.276657
25	0.423147	0.375117	0.332731	0.295303	0.262234
26	0.408838	0.360689	0.318402	0.281241	0.248563
27	0.395012	0.346817	0.304691	0.267848	0.235605
28	0.381654	0.333477	0.291571	0.255094	0.223322
29	0.368748	0.320651	0.279015	0.242946	0.211679
30	0.356278	0.308319	0.267000	0.231377	0.200644
31	0.344230	0.296460	0.255502	0.220359	0.190184
32	0.332590	0.285058	0.244500	0.209866	0.180269
33	0.321343	0.274094	0.233971	0.199873	0.170871
34	0.310476	0.263552	0.233896	0.190355	0.161963
35	0.299977	0.253415	0.214254	0.181290	0.153520
36	0.289833	0.243669	0.205028	0.172657	0.145516
37	0.280032	0.234297	0.196199	0.164436	0.137930
38	0.270562	0.225285	0.187750	0.156605	0.130739
39	0.261413	0.216621	0.179665	0.149148	0.123924
40	0.252572	0.208289	0.171929	0.142046	0.117463
41	0.244031	0.200278	0.164525	0.135282	0.111339
42	0.235779	0.192575	0.157440	0.128840	0.105535
43	0.227806	0.185168	0.150661	0.122704	0.100033
44	0.220102	0.178046	0.144173	0.116861	0.094818
45	0.212659	0.171198	0.137964	0.111297	0.089875
46	0.205468	0.164614	0.132023	0.105997	0.085190
47	0.198520	0.158283	0.126338	0.100949	0.080748
48	0.191806	0.152195	0.120898	0.096142	0.076539
49	0.185320	0.146341	0.115692	0.091564	0.072549
50	0.179053	0.140713	0.110710	0.087204	0.068767

Table 3 (continued)

n	$3\frac{1}{2}\%$	4%	$4\frac{1}{2}\%$	5%	$5\frac{1}{2}\%$
51	0.172998	0.135301	0.105942	0.083051	0.065182
52	0.167148	0.130097	0.101380	0.079096	0.061783
53	0.161496	0.125093	0.097014	0.075330	0.058563
54	0.156035	0.120282	0.092837	0.071743	0.055509
55	0.150758	0.115656	0.088839	0.068326	0.052616
56	0.145660	0.111207	0.085013	0.065073	0.049873
57	0.140734	0.106930	0.081353	0.061974	0.047273
58	0.135975	0.102817	0.077849	0.059023	0.044808
59	0.131377	0.098863	0.074497	0.056212	0.042472
60	0.126934	0.095060	0.071289	0.053536	0.040258
61	0.122642	0.091404	0.068219	0.050986	0.038159
62	0.118495	0.087889	0.065281	0.048558	0.036170
63	0.114487	0.084508	0.062470	0.046246	0.034284
64	0.110616	0.081258	0.059780	0.044044	0.032497
65	0.106875	0.078133	0.057206	0.041946	0.030803
66	0.103261	0.075128	0.054743	0.039949	0.029197
67	0.099769	0.072238	0.052385	0.038047	0.027675
68	0.096395	0.069460	0.050129	0.036235	0.026232
69	0.093136	0.066788	0.047971	0.034509	0.024865
70	0.089986	0.064219	0.045905	0.032866	0.023568
71	0.086943	0.061749	0.043928	0.031301	0.022340
72	0.084003	0.059374	0.042037	0.029811	0.021175
73	0.081162	0.057091	0.040226	0.028391	0.020071
74	0.078418	0.054895	0.038494	0.027039	0.019025
75	0.075766	0.052784	0.036836	0.025752	0.018033
76	0.073204	0.050754	0.035250	0.024525	0.017093
77	0.070728	0.048801	0.033732	0.023357	0.016202
78	0.068337	0.046924	0.032280	0.022245	0.015357
79	0.066026	0.045120	0.030890	0.021186	0.014556
80	0.063793	0.043384	0.029559	0.020177	0.013798
81	0.061636	0.041716	0.028287	0.019216	0.013078
82	0.059551	0.040111	0.027068	0.018301	0.012396
83	0.057538	0.038569	0.025903	0.017430	0.011750
84	0.055592	0.037085	0.024787	0.016600	0.011138
85	0.053712	0.035659	0.023720	0.015809	0.010557
86	0.051896	0.034287	0.022699	0.015056	0.010007
87	0.050141	0.032969	0.021721	0.014339	0.009485
88	0.048445	0.031701	0.020786	0.013657	0.008990
89	0.046807	0.030481	0.019891	0.013006	0.008522
90	0.045224	0.029309	0.019034	0.012387	0.008078
91	0.043695	0.028182	0.018215	0.011797	0.007656
92	0.042217	0.027098	0.017430	0.011235	0.007257
93	0.040789	0.026056	0.016680	0.010700	0.006879
94	0.039410	0.025053	0.015961	0.010191	0.006520
95	0.038077	0.024090	0.015274	0.009705	0.006180
96	0.036790	0.023163	0.014616	0.009243	0.005858
97	0.035546	0.022272	0.013987	0.008803	0.005553
98	0.034344	0.021416	0.013385	0.008384	0.005263
99	0.033182	0.020592	0.012808	0.007985	0.004989
100	0.032060	0.019800	0.012257	0.007604	0.004729

Table 3 (continued)

n	6%	$6\frac{1}{2}$%	7%	$7\frac{1}{2}$%	8%
1	0.943396	0.938967	0.934579	0.930233	0.925926
2	0.889996	0.881659	0.873439	0.865333	0.857339
3	0.839619	0.827849	0.816298	0.804961	0.793832
4	0.792094	0.777323	0.762895	0.748801	0.735030
5	0.747258	0.729881	0.712986	0.696559	0.680583
6	0.704961	0.685334	0.666342	0.647962	0.630170
7	0.665057	0.643506	0.622750	0.602755	0.583490
8	0.627412	0.604231	0.582009	0.560702	0.540269
9	0.591898	0.567353	0.543934	0.521583	0.500249
10	0.558395	0.532726	0.508349	0.485194	0.463193
11	0.526788	0.500212	0.475093	0.451343	0.428883
12	0.496969	0.469683	0.444012	0.419854	0.397114
13	0.468839	0.441017	0.414964	0.390562	0.367698
14	0.442301	0.414100	0.387817	0.363313	0.340461
15	0.417265	0.388827	0.362446	0.337966	0.315242
16	0.393646	0.365095	0.338735	0.314387	0.291890
17	0.371364	0.342813	0.316574	0.292453	0.270269
18	0.350344	0.321890	0.295864	0.272049	0.250249
19	0.330513	0.302244	0.276508	0.253069	0.231712
20	0.311805	0.283797	0.258419	0.235413	0.214548
21	0.294155	0.266476	0.241513	0.218989	0.198656
22	0.277505	0.250212	0.225713	0.203711	0.183941
23	0.261797	0.234941	0.210947	0.189498	0.170315
24	0.246979	0.220602	0.197147	0.176277	0.157699
25	0.232999	0.207138	0.184249	0.163979	0.146018
26	0.219810	0.194496	0.172195	0.152539	0.135202
27	0.207368	0.182625	0.160930	0.141896	0.125187
28	0.195630	0.171479	0.150402	0.131997	0.115914
29	0.184557	0.161013	0.140563	0.122788	0.107328
30	0.174110	0.151186	0.131367	0.114221	0.099377
31	0.164255	0.141959	0.122773	0.106252	0.092016
32	0.154957	0.133295	0.114741	0.098839	0.085200
33	0.146186	0.125159	0.107235	0.091943	0.078889
34	0.137912	0.117520	0.100219	0.085529	0.073045
35	0.130105	0.110348	0.093663	0.079562	0.067635
36	0.122741	0.103613	0.087535	0.074011	0.062625
37	0.115793	0.097289	0.081809	0.068847	0.057986
38	0.109239	0.091351	0.076457	0.064044	0.053690
39	0.103056	0.085776	0.071455	0.059576	0.049713
40	0.097222	0.080541	0.066780	0.055419	0.046031
41	0.091719	0.075625	0.062412	0.051553	0.042621
42	0.086527	0.071010	0.058329	0.047956	0.039464
43	0.081630	0.066676	0.054513	0.044610	0.036541
44	0.077009	0.062606	0.050946	0.041498	0.033834
45	0.072650	0.058785	0.047613	0.038603	0.031328
46	0.068538	0.055197	0.044499	0.035910	0.029007
47	0.064658	0.051828	0.041587	0.033404	0.026859
48	0.060998	0.048665	0.038867	0.031074	0.024869
49	0.057546	0.045695	0.036324	0.028906	0.023027
50	0.054288	0.042906	0.033948	0.026889	0.021321

Table 3 (continued)

n	6%	6½%	7%	7½%	8%
51	0.051215	0.040287	0.031727	0.025013	0.019742
52	0.048316	0.037829	0.029651	0.023268	0.018280
53	0.045582	0.035520	0.027711	0.021645	0.016925
54	0.043001	0.033352	0.025899	0.020135	0.015672
55	0.040567	0.031316	0.024204	0.018730	0.014511
56	0.038271	0.029405	0.022621	0.017423	0.013436
57	0.036105	0.027610	0.021141	0.016208	0.012441
58	0.034061	0.025925	0.019758	0.015077	0.011519
59	0.032133	0.024343	0.018465	0.014025	0.010666
60	0.030314	0.022857	0.017257	0.013046	0.009876
61	0.028598	0.021462	0.016128	0.012136	0.009144
62	0.026980	0.020152	0.015073	0.011290	0.008467
63	0.025453	0.018922	0.014087	0.010502	0.007840
64	0.024012	0.017767	0.013166	0.009769	0.007259
65	0.022653	0.016683	0.012304	0.009088	0.006721
66	0.021370	0.015665	0.011499	0.008454	0.006223
67	0.020161	0.014709	0.010747	0.007864	0.005762
68	0.019020	0.013811	0.010044	0.007315	0.005336
69	0.017943	0.012968	0.009387	0.006805	0.004940
70	0.016927	0.012177	0.008773	0.006330	0.004574
71	0.015969	0.011433	0.008199	0.005888	0.004236
72	0.015065	0.010736	0.007662	0.005478	0.003922
73	0.014213	0.010080	0.007161	0.005095	0.003631
74	0.013408	0.009465	0.006693	0.004740	0.003362
75	0.012649	0.008887	0.006255	0.004409	0.003113
76	0.011933	0.008345	0.005846	0.004102	0.002883
77	0.011258	0.007836	0.005463	0.003815	0.002669
78	0.010620	0.007358	0.005106	0.003549	0.002471
79	0.010019	0.006908	0.004772	0.003302	0.002288
80	0.009452	0.006487	0.004460	0.003071	0.002119
81	0.008917	0.006091	0.004168	0.002857	0.001962
82	0.008412	0.005719	0.003895	0.002658	0.001817
83	0.007936	0.005370	0.003640	0.002472	0.001682
84	0.007487	0.005042	0.003402	0.002300	0.001557
85	0.007063	0.004735	0.003180	0.002139	0.001442
86	0.006663	0.004446	0.002972	0.001990	0.001335
87	0.006286	0.004174	0.002777	0.001851	0.001236
88	0.005930	0.003920	0.002596	0.001722	0.001145
89	0.005595	0.003680	0.002426	0.001602	0.001060
90	0.005278	0.003456	0.002267	0.001490	0.000981
91	0.004979	0.003245	0.002119	0.001386	0.000909
92	0.004697	0.003047	0.001980	0.001289	0.000841
93	0.004432	0.002861	0.001851	0.001200	0.000779
94	0.004181	0.002686	0.001730	0.001116	0.000721
95	0.003944	0.002522	0.001616	0.001038	0.000668
96	0.003721	0.002368	0.001511	0.000966	0.000618
97	0.003510	0.002224	0.001412	0.000898	0.000573
98	0.003312	0.002088	0.001319	0.000836	0.000530
99	0.003124	0.001961	0.001233	0.000777	0.000491
100	0.002947	0.001841	0.001152	0.000723	0.000455

Table 4 **Common Logarithms**

N	0	1	2	3	4	5	6	7	8	9
1.0	0.0000	0.0043	0.0086	0.0128	0.0170	0.0212	0.0253	0.0294	0.0334	0.0374
1.1	0.0414	0.0453	0.0492	0.0531	0.0569	0.0607	0.0645	0.0682	0.0719	0.0755
1.2	0.0792	0.0828	0.0864	0.0899	0.0934	0.0969	0.1004	0.1038	0.1072	0.1106
1.3	0.1139	0.1173	0.1206	0.1239	0.1271	0.1303	0.1335	0.1367	0.1399	0.1430
1.4	0.1461	0.1492	0.1523	0.1553	0.1584	0.1614	0.1644	0.1673	0.1703	0.1732
1.5	0.1761	0.1790	0.1818	0.1847	0.1875	0.1903	0.1931	0.1959	0.1987	0.2014
1.6	0.2041	0.2068	0.2095	0.2122	0.2148	0.2175	0.2201	0.2227	0.2253	0.2279
1.7	0.2304	0.2330	0.2355	0.2380	0.2405	0.2430	0.2455	0.2480	0.2504	0.2529
1.8	0.2553	0.2577	0.2601	0.2625	0.2648	0.2672	0.2695	0.2718	0.2742	0.2765
1.9	0.2788	0.2810	0.2833	0.2856	0.2878	0.2900	0.2923	0.2945	0.2967	0.2989
2.0	0.3010	0.3032	0.3054	0.3075	0.3096	0.3118	0.3139	0.3160	0.3181	0.3201
2.1	0.3222	0.3243	0.3263	0.3284	0.3304	0.3324	0.3345	0.3365	0.3385	0.3404
2.2	0.3424	0.3444	0.3464	0.3483	0.3502	0.3522	0.3541	0.3560	0.3579	0.3598
2.3	0.3617	0.3636	0.3655	0.3674	0.3692	0.3711	0.3729	0.3747	0.3766	0.3784
2.4	0.3802	0.3820	0.3838	0.3856	0.3874	0.3892	0.3909	0.3927	0.3945	0.3962
2.5	0.3979	0.3997	0.4014	0.4031	0.4048	0.4065	0.4082	0.4099	0.4116	0.4133
2.6	0.4150	0.4166	0.4183	0.4200	0.4216	0.4232	0.4249	0.4265	0.4281	0.4298
2.7	0.4314	0.4330	0.4346	0.4362	0.4378	0.4393	0.4409	0.4425	0.4440	0.4456
2.8	0.4472	0.4487	0.4502	0.4518	0.4533	0.4548	0.4564	0.4579	0.4594	0.4609
2.9	0.4624	0.4639	0.4654	0.4669	0.4683	0.4698	0.4713	0.4728	0.4742	0.4757
3.0	0.4771	0.4786	0.4800	0.4814	0.4829	0.4843	0.4857	0.4871	0.4886	0.4900
3.1	0.4914	0.4928	0.4942	0.4955	0.4969	0.4983	0.4997	0.5011	0.5024	0.5038
3.2	0.5051	0.5065	0.5079	0.5092	0.5105	0.5119	0.5132	0.5145	0.5159	0.5172
3.3	0.5185	0.5198	0.5211	0.5224	0.5237	0.5250	0.5263	0.5276	0.5289	0.5302
3.4	0.5315	0.5328	0.5340	0.5353	0.5366	0.5378	0.5391	0.5403	0.5416	0.5428
3.5	0.5441	0.5453	0.5465	0.5478	0.5490	0.5502	0.5514	0.5527	0.5539	0.5551
3.6	0.5563	0.5575	0.5587	0.5599	0.5611	0.5623	0.5635	0.5647	0.5658	0.5670
3.7	0.5682	0.5694	0.5705	0.5717	0.5729	0.5740	0.5752	0.5763	0.5775	0.5786
3.8	0.5798	0.5809	0.5821	0.5832	0.5843	0.5855	0.5866	0.5877	0.5888	0.5899
3.9	0.5911	0.5922	0.5933	0.5944	0.5955	0.5966	0.5977	0.5988	0.5999	0.6010
4.0	0.6021	0.6031	0.6042	0.6053	0.6064	0.6075	0.6085	0.6096	0.6107	0.6117
4.1	0.6128	0.6138	0.6149	0.6160	0.6170	0.6180	0.6191	0.6201	0.6212	0.6222
4.2	0.6232	0.6243	0.6253	0.6263	0.6274	0.6284	0.6294	0.6304	0.6314	0.6325
4.3	0.6335	0.6345	0.6355	0.6365	0.6375	0.6385	0.6395	0.6405	0.6415	0.6425
4.4	0.6435	0.6444	0.6454	0.6464	0.6474	0.6484	0.6493	0.6503	0.6513	0.6522
4.5	0.6532	0.6542	0.6551	0.6561	0.6571	0.6580	0.6590	0.6599	0.6609	0.6618
4.6	0.6628	0.6637	0.6646	0.6656	0.6665	0.6675	0.6684	0.6693	0.6702	0.6712
4.7	0.6721	0.6730	0.6739	0.6749	0.6758	0.6767	0.6776	0.6785	0.6794	0.6803
4.8	0.6812	0.6821	0.6830	0.6839	0.6848	0.6857	0.6866	0.6875	0.6884	0.6893
4.9	0.6902	0.6911	0.6920	0.6928	0.6937	0.6946	0.6955	0.6964	0.6972	0.6981
5.0	0.6990	0.6998	0.7007	0.7016	0.7024	0.7033	0.7042	0.7050	0.7059	0.7067
5.1	0.7076	0.7084	0.7093	0.7101	0.7110	0.7118	0.7126	0.7135	0.7143	0.7152
5.2	0.7160	0.7168	0.7177	0.7185	0.7193	0.7202	0.7210	0.7218	0.7226	0.7235
5.3	0.7243	0.7251	0.7259	0.7267	0.7275	0.7284	0.7292	0.7300	0.7308	0.7316
5.4	0.7324	0.7332	0.7340	0.7348	0.7356	0.7364	0.7372	0.7380	0.7388	0.7396

Table 4 (continued)

N	0	1	2	3	4	5	6	7	8	9
5.5	0.7404	0.7412	0.7419	0.7427	0.7435	0.7443	0.7451	0.7459	0.7466	0.7474
5.6	0.7482	0.7490	0.7497	0.7505	0.7513	0.7520	0.7528	0.7536	0.7543	0.7551
5.7	0.7559	0.7566	0.7574	0.7582	0.7589	0.7597	0.7604	0.7612	0.7619	0.7627
5.8	0.7634	0.7642	0.7649	0.7657	0.7664	0.7672	0.7679	0.7686	0.7694	0.7701
5.9	0.7709	0.7716	0.7723	0.7731	0.7738	0.7745	0.7752	0.7760	0.7767	0.7774
6.0	0.7782	0.7789	0.7796	0.7803	0.7810	0.7818	0.7825	0.7832	0.7839	0.7846
6.1	0.7853	0.7860	0.7868	0.7875	0.7882	0.7889	0.7896	0.7903	0.7910	0.7917
6.2	0.7924	0.7931	0.7938	0.7945	0.7952	0.7959	0.7966	0.7973	0.7980	0.7987
6.3	0.7993	0.8000	0.8007	0.8014	0.8021	0.8028	0.8035	0.8041	0.8048	0.8055
6.4	0.8062	0.8069	0.8075	0.8082	0.8089	0.8096	0.8102	0.8109	0.8116	0.8122
6.5	0.8129	0.8136	0.8142	0.8149	0.8156	0.8162	0.8169	0.8176	0.8182	0.8189
6.6	0.8195	0.8202	0.8209	0.8215	0.8222	0.8228	0.8235	0.8241	0.8248	0.8254
6.7	0.8261	0.8267	0.8274	0.8280	0.8287	0.8293	0.8299	0.8306	0.8312	0.8319
6.8	0.8325	0.8331	0.8338	0.8344	0.8351	0.8357	0.8363	0.8370	0.8376	0.8382
6.9	0.8388	0.8395	0.8401	0.8407	0.8414	0.8420	0.8426	0.8432	0.8439	0.8445
7.0	0.8451	0.8457	0.8463	0.8470	0.8476	0.8482	0.8488	0.8494	0.8500	0.8506
7.1	0.8513	0.8519	0.8525	0.8531	0.8537	0.8543	0.8549	0.8555	0.8561	0.8567
7.2	0.8573	0.8579	0.8585	0.8591	0.8597	0.8603	0.8609	0.8615	0.8621	0.8627
7.3	0.8633	0.8639	0.8645	0.8651	0.8657	0.8663	0.8669	0.8675	0.8681	0.8686
7.4	0.8692	0.8698	0.8704	0.8710	0.8716	0.8722	0.8727	0.8733	0.8739	0.8745
7.5	0.8751	0.8756	0.8762	0.8768	0.8774	0.8779	0.8785	0.8791	0.8797	0.8802
7.6	0.8808	0.8814	0.8820	0.8825	0.8831	0.8837	0.8842	0.8848	0.8854	0.8859
7.7	0.8865	0.8871	0.8876	0.8882	0.8887	0.8893	0.8899	0.8904	0.8910	0.8915
7.8	0.8921	0.8927	0.8932	0.8938	0.8943	0.8949	0.8954	0.8960	0.8965	0.8971
7.9	0.8976	0.8982	0.8987	0.8993	0.8998	0.9004	0.9009	0.9015	0.9020	0.9025
8.0	0.9031	0.9036	0.9042	0.9047	0.9053	0.9058	0.9063	0.9069	0.9074	0.9079
8.1	0.9085	0.9090	0.9096	0.9101	0.9106	0.9112	0.9117	0.9122	0.9128	0.9133
8.2	0.9138	0.9143	0.9149	0.9154	0.9159	0.9165	0.9170	0.9175	0.9180	0.9186
8.3	0.9191	0.9196	0.9201	0.9206	0.9212	0.9217	0.9222	0.9227	0.9232	0.9238
8.4	0.9243	0.9248	0.9253	0.9258	0.9263	0.9269	0.9274	0.9279	0.9284	0.9289
8.5	0.9294	0.9299	0.9304	0.9309	0.9315	0.9320	0.9325	0.9330	0.9335	0.9340
8.6	0.9345	0.9350	0.9355	0.9360	0.9365	0.9370	0.9375	0.9380	0.9385	0.9390
8.7	0.9395	0.9400	0.9405	0.9410	0.9415	0.9420	0.9425	0.9430	0.9435	0.9440
8.8	0.9445	0.9450	0.9455	0.9460	0.9465	0.9469	0.9474	0.9479	0.9484	0.9489
8.9	0.9494	0.9499	0.9504	0.9509	0.9513	0.9518	0.9523	0.9528	0.9533	0.9538
9.0	0.9542	0.9547	0.9552	0.9557	0.9562	0.9566	0.9571	0.9576	0.9581	0.9586
9.1	0.9590	0.9595	0.9600	0.9605	0.9609	0.9614	0.9619	0.9624	0.9628	0.9633
9.2	0.9638	0.9643	0.9647	0.9652	0.9657	0.9661	0.9666	0.9671	0.9675	0.9680
9.3	0.9685	0.9689	0.9694	0.9699	0.9703	0.9708	0.9713	0.9717	0.9722	0.9727
9.4	0.9731	0.9736	0.9741	0.9745	0.9750	0.9754	0.9759	0.9763	0.9768	0.9773
9.5	0.9777	0.9782	0.9786	0.9791	0.9795	0.9800	0.9805	0.9809	0.9814	0.9818
9.6	0.9823	0.9827	0.9832	0.9836	0.9841	0.9845	0.9850	0.9854	0.9859	0.9863
9.7	0.9868	0.9872	0.9877	0.9881	0.9886	0.9890	0.9891	0.9899	0.9903	0.9908
9.8	0.9912	0.9917	0.9921	0.9926	0.9930	0.9934	0.9939	0.9943	0.9948	0.9952
9.9	0.9956	0.9961	0.9965	0.9969	0.9974	0.9978	0.9983	0.9987	0.9991	0.9996

Table 5 **Natural Logarithms**

N	.00	.01	.02	.03	.04	.05	.06	.07	.08	.09
1.0	0.0000	0.0100	0.0198	0.0296	0.0392	0.0488	0.0583	0.0677	0.0770	0.0862
1.1	0.0953	0.1044	0.1133	0.1222	0.1310	0.1398	0.1484	0.1570	0.1655	0.1740
1.2	0.1823	0.1906	0.1989	0.2070	0.2151	0.2231	0.2311	0.2390	0.2469	0.2546
1.3	0.2624	0.2700	0.2776	0.2852	0.2927	0.3001	0.3075	0.3148	0.3221	0.3293
1.4	0.3365	0.3436	0.3507	0.3577	0.3646	0.3716	0.3784	0.3853	0.3920	0.3988
1.5	0.4055	0.4121	0.4187	0.4253	0.4318	0.4383	0.4447	0.4511	0.4574	0.4637
1.6	0.4700	0.4762	0.4824	0.4886	0.4947	0.5008	0.5068	0.5128	0.5188	0.5247
1.7	0.5306	0.5365	0.5423	0.5481	0.5539	0.5596	0.5653	0.5710	0.5766	0.5822
1.8	0.5878	0.5933	0.5988	0.6043	0.6098	0.6152	0.6206	0.6259	0.6313	0.6366
1.9	0.6419	0.6471	0.6523	0.6575	0.6627	0.6678	0.6729	0.6780	0.6831	0.6881
2.0	0.6931	0.6981	0.7031	0.7080	0.7129	0.7178	0.7227	0.7275	0.7324	0.7372
2.1	0.7419	0.7467	0.7514	0.7561	0.7608	0.7655	0.7701	0.7747	0.7793	0.7839
2.2	0.7885	0.7930	0.7975	0.8020	0.8065	0.8109	0.8154	0.8198	0.8242	0.8286
2.3	0.8329	0.8372	0.8416	0.8459	0.8502	0.8544	0.8587	0.8629	0.8671	0.8713
2.4	0.8755	0.8796	0.8838	0.8879	0.8920	0.8961	0.9002	0.9042	0.9083	0.9123
2.5	0.9163	0.9203	0.9243	0.9282	0.9322	0.9361	0.9400	0.9439	0.9478	0.9517
2.6	0.9555	0.9594	0.9632	0.9670	0.9708	0.9746	0.9783	0.9821	0.9858	0.9895
2.7	0.9933	0.9969	1.0006	1.0043	1.0080	1.0116	1.0152	1.0188	1.0225	1.0260
2.8	1.0296	1.0332	1.0367	1.0403	1.0438	1.0473	1.0508	1.0543	1.0578	1.0613
2.9	1.0647	1.0682	1.0716	1.0750	1.0784	1.0818	1.0852	1.0886	1.0919	1.0953
3.0	1.0986	1.1019	1.1053	1.1086	1.1119	1.1151	1.1184	1.1217	1.1249	1.1282
3.1	1.1314	1.1346	1.1378	1.1410	1.1442	1.1474	1.1506	1.1537	1.1569	1.1600
3.2	1.1632	1.1663	1.1694	1.1725	1.1756	1.1787	1.1817	1.1848	1.1878	1.1909
3.3	1.1939	1.1969	1.2000	1.2030	1.2060	1.2090	1.2119	1.2149	1.2179	1.2208
3.4	1.2238	1.2267	1.2296	1.2326	1.2355	1.2384	1.2413	1.2442	1.2470	1.2499
3.5	1.2528	1.2556	1.2585	1.2613	1.2641	1.2669	1.2698	1.2726	1.2754	1.2782
3.6	1.2809	1.2837	1.2865	1.2892	1.2920	1.2947	1.2975	1.3002	1.3029	1.3056
3.7	1.3083	1.3110	1.3137	1.3164	1.3191	1.3218	1.3244	1.3271	1.3297	1.3324
3.8	1.3350	1.3376	1.3403	1.3429	1.3455	1.3481	1.3507	1.3533	1.3558	1.3584
3.9	1.3610	1.3635	1.3661	1.3686	1.3712	1.3737	1.3762	1.3788	1.3813	1.3838
4.0	1.3863	1.3888	1.3913	1.3938	1.3962	1.3987	1.4012	1.4036	1.4061	1.4085
4.1	1.4110	1.4134	1.4159	1.4183	1.4207	1.4231	1.4255	1.4279	1.4303	1.4327
4.2	1.4351	1.4375	1.4398	1.4422	1.4446	1.4469	1.4493	1.4516	1.4540	1.4563
4.3	1.4586	1.4609	1.4633	1.4656	1.4679	1.4702	1.4725	1.4748	1.4770	1.4793
4.4	1.4816	1.4839	1.4861	1.4884	1.4907	1.4929	1.4951	1.4974	1.4996	1.5019
4.5	1.5041	1.5063	1.5085	1.5107	1.5129	1.5151	1.5173	1.5195	1.5217	1.5239
4.6	1.5261	1.5282	1.5304	1.5326	1.5347	1.5369	1.5390	1.5412	1.5433	1.5454
4.7	1.5476	1.5497	1.5518	1.5539	1.5560	1.5581	1.5602	1.5623	1.5644	1.5665
4.8	1.5686	1.5707	1.5728	1.5748	1.5769	1.5790	1.5810	1.5831	1.5851	1.5872
4.9	1.5892	1.5913	1.5933	1.5953	1.5974	1.5994	1.6014	1.6034	1.6054	1.6074
5.0	1.6094	1.6114	1.6134	1.6154	1.6174	1.6194	1.6214	1.6233	1.6253	1.6273
5.1	1.6292	1.6312	1.6332	1.6351	1.6371	1.6390	1.6409	1.6429	1.6448	1.6467
5.2	1.6487	1.6506	1.6525	1.6544	1.6563	1.6582	1.6601	1.6620	1.6639	1.6658
5.3	1.6677	1.6696	1.6715	1.6734	1.6752	1.6771	1.6790	1.6808	1.6827	1.6845
5.4	1.6864	1.6882	1.6901	1.6919	1.6938	1.6956	1.6974	1.6993	1.7011	1.7029

Table 5 (continued)

N	.00	.01	.02	.03	.04	.05	.06	.07	.08	.09
5.5	1.7047	1.7066	1.7084	1.7102	1.7120	1.7138	1.7156	1.7174	1.7192	1.7210
5.6	1.7228	1.7246	1.7263	1.7281	1.7299	1.7317	1.7334	1.7352	1.7370	1.7387
5.7	1.7405	1.7422	1.7440	1.7457	1.7475	1.7492	1.7509	1.7527	1.7544	1.7561
5.8	1.7579	1.7596	1.7613	1.7630	1.7647	1.7664	1.7681	1.7699	1.7716	1.7733
5.9	1.7750	1.7766	1.7783	1.7800	1.7817	1.7834	1.7851	1.7867	1.7884	1.7901
6.0	1.7918	1.7934	1.7951	1.7967	1.7984	1.8001	1.8017	1.8034	1.8050	1.8066
6.1	1.8083	1.8099	1.8116	1.8132	1.8148	1.8165	1.8181	1.8197	1.8213	1.8229
6.2	1.8245	1.8262	1.8278	1.8294	1.8310	1.8326	1.8342	1.8358	1.8374	1.8390
6.3	1.8405	1.8421	1.8437	1.8453	1.8469	1.8485	1.8500	1.8516	1.8532	1.8547
6.4	1.8563	1.8579	1.8594	1.8610	1.8625	1.8641	1.8656	1.8672	1.8687	1.8703
6.5	1.8718	1.8733	1.8749	1.8764	1.8779	1.8795	1.8810	1.8825	1.8840	1.8856
6.6	1.8871	1.8886	1.8901	1.8916	1.8931	1.8946	1.8961	1.8976	1.8991	1.9006
6.7	1.9021	1.9036	1.9051	1.9066	1.9081	1.9095	1.9110	1.9125	1.9140	1.9155
6.8	1.9169	1.9184	1.9199	1.9213	1.9228	1.9242	1.9257	1.9272	1.9286	1.9301
6.9	1.9315	1.9330	1.9344	1.9359	1.9373	1.9387	1.9402	1.9416	1.9430	1.9445
7.0	1.9459	1.9473	1.9488	1.9502	1.9516	1.9530	1.9544	1.9559	1.9573	1.9587
7.1	1.9601	1.9615	1.9629	1.9643	1.9657	1.9671	1.9685	1.9699	1.9713	1.9727
7.2	1.9741	1.9755	1.9769	1.9782	1.9796	1.9810	1.9824	1.9838	1.9851	1.9865
7.3	1.9879	1.9892	1.9906	1.9920	1.9933	1.9947	1.9961	1.9974	1.9988	2.0001
7.4	2.0015	2.0028	2.0042	2.0055	2.0069	2.0082	2.0096	2.0109	2.0122	2.0136
7.5	2.0149	2.0162	2.0176	2.0189	2.0202	2.0215	2.0229	2.0242	2.0255	2.0268
7.6	2.0281	2.0295	2.0308	2.0321	2.0334	2.0347	2.0360	2.0373	2.0386	2.0399
7.7	2.0412	2.0425	2.0438	2.0451	2.0464	2.0477	2.0490	2.0503	2.0516	2.0528
7.8	2.0541	2.0554	2.0567	2.0580	2.0592	2.0605	2.0618	2.0631	2.0643	2.0656
7.9	2.0669	2.0681	2.0694	2.0707	2.0719	2.0732	2.0744	2.0757	2.0769	2.0782
8.0	2.0794	2.0807	2.0819	2.0832	2.0844	2.0857	2.0869	2.0882	2.0894	2.0906
8.1	2.0919	2.0931	2.0943	2.0956	2.0968	2.0980	2.0992	2.1005	2.1017	2.1029
8.2	2.1041	2.1054	2.1066	2.1078	2.1090	2.1102	2.1114	2.1126	2.1138	2.1150
8.3	2.1163	2.1175	2.1187	2.1199	2.1211	2.1223	2.1235	2.1247	2.1258	2.1270
8.4	2.1282	2.1294	2.1306	2.1318	2.1330	2.1342	2.1353	2.1365	2.1377	2.1389
8.5	2.1401	2.1412	2.1424	2.1436	2.1448	2.1459	2.1471	2.1483	2.1494	2.1506
8.6	2.1518	2.1529	2.1541	2.1552	2.1564	2.1576	2.1587	2.1599	2.1610	2.1622
8.7	2.1633	2.1645	2.1656	2.1668	2.1679	2.1691	2.1702	2.1713	2.1725	2.1736
8.8	2.1748	2.1759	2.1770	2.1782	2.1793	2.1804	2.1815	2.1827	2.1838	2.1849
8.9	2.1861	2.1872	2.1883	2.1894	2.1905	2.1917	2.1928	2.1939	2.1950	2.1961
9.0	2.1972	2.1983	2.1994	2.2006	2.2017	2.2028	2.2039	2.2050	2.2061	2.2072
9.1	2.2083	2.2094	2.2105	2.2116	2.2127	2.2138	2.2148	2.2159	2.2170	2.2181
9.2	2.2192	2.2203	2.2214	2.2225	2.2235	2.2246	2.2257	2.2268	2.2279	2.2289
9.3	2.2300	2.2311	2.2322	2.2332	2.2343	2.2354	2.2364	2.2375	2.2386	2.2396
9.4	2.2407	2.2418	2.2428	2.2439	2.2450	2.2460	2.2471	2.2481	2.2492	2.2502
9.5	2.2513	2.2523	2.2534	2.2544	2.2555	2.2565	2.2576	2.2586	2.2597	2.2607
9.6	2.2618	2.2628	2.2638	2.2649	2.2659	2.2670	2.2680	2.2690	2.2701	2.2711
9.7	2.2721	2.2732	2.2742	2.2752	2.2762	2.2773	2.2783	2.2793	2.2803	2.2814
9.8	2.2824	2.2834	2.2844	2.2854	2.2865	2.2875	2.2885	2.2895	2.2905	2.2915
9.9	2.2925	2.2935	2.2946	2.2956	2.2966	2.2976	2.2986	2.2996	2.3006	2.3016

Answers to Selected Exercises

Chapter 1

Exercises (pages 5–6)

1. True
3. False
5. True
7. True
9. True
11. False
13. Irrational
15. Irrational
17. Rational
19. Irrational
21. Irrational
23. Irrational

Exercises (pages 12–13)

1. $x < 5$ 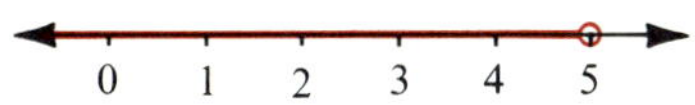

3. $x > 3$ 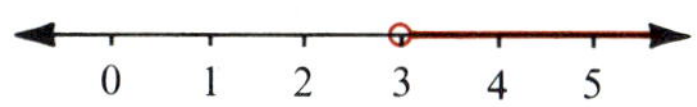

5. $x < 5$ 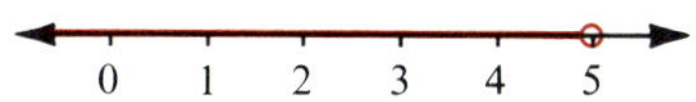

7. $x \leq 3$ 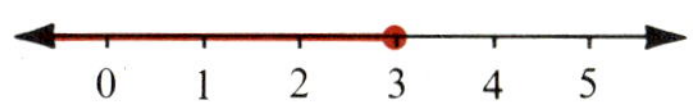

9. $x \leq 9$ 

11. $x < \frac{9}{4}$ 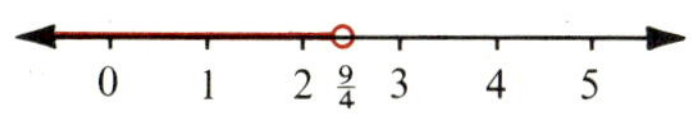

13. $x \geq -\frac{7}{8}$

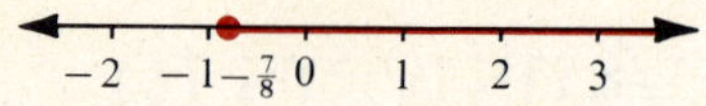

15. $x \leq \frac{1}{4}$

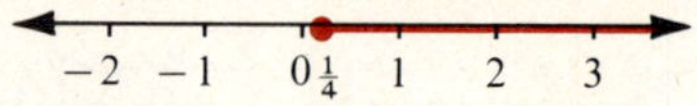

17. $x < 0$ or $x > 1$

19. $0 < x < 4$

21. $x < 1$ or $x > 2$

23. $2 < x < 3$

25. $x \leq -1$ or $x \geq 2$

27. $0 < x < 1$ or $x > 4$

Exercises (pages 17–18)

1. $x = -1, 5$

3. no solution

5. $x = \frac{4}{3}, 4$

7. $-1 \leq x \leq 1$

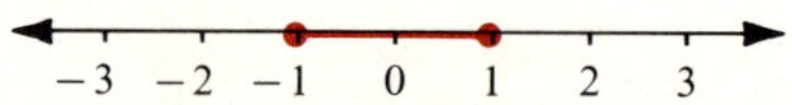

9. $x < -\frac{1}{3}$ or $x > \frac{1}{3}$

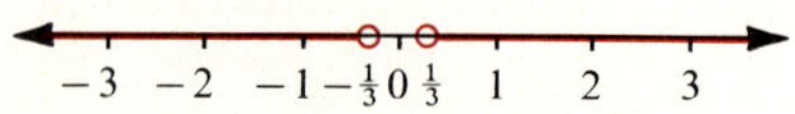

11. $-1 \leq x \leq 5$

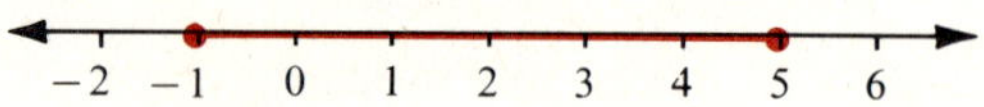

13. $x < 1$ or $x > 5$

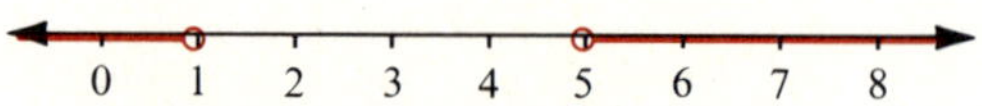

15. $x < -1$ or $x > 5$

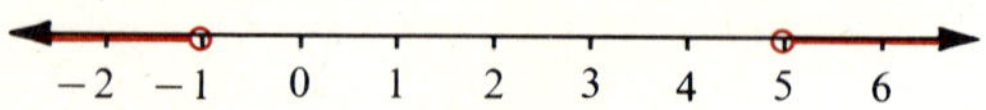

17. $-8 \leq x \leq 4$

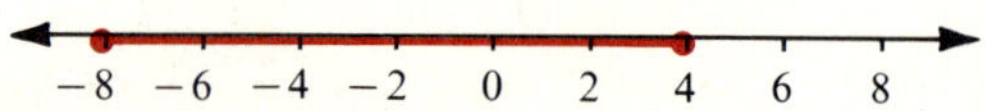

19. $0 \leq x \leq 12$

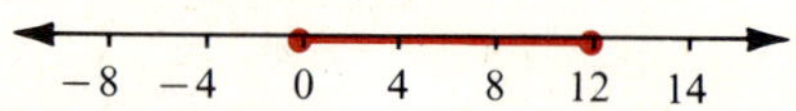

Exercises (pages 24–25)

1. **a.**, **c.**, **f.**, **g.**, **h.**, and **j.** are true statements.

3. **a.** $A \cap (B \cup C) = \{5, 10, 12\}$; $(A \cap B) \cup (A \cap C) = \{5, 10, 12\}$
b. $A \cup (B \cap C) = \{1, 2, 5, 6, 8, 9, 10, 11, 12\}$;
$(A \cup B) \cap (A \cup C) = \{1, 2, 5, 6, 8, 9, 10, 11, 12\}$

5. **a.** $\{x | 1 \leq x \leq 5\}$ is the set of all real numbers between 1 and 5, both inclusive.

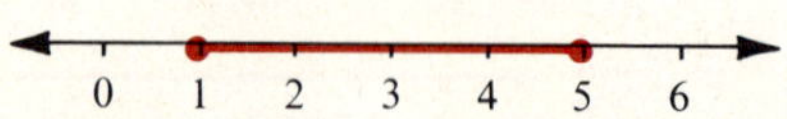

b. $\{x \mid 2 \leq x < 4\}$ is the set of all real numbers greater than or equal to 2 but less than 4.

c. $\{x \mid 3 < x < 4\}$ is the set of all real numbers between 3 and 4.
d. $\{x \mid -2 < x < 7\}$ is the set of all real numbers between -2 and 7.
e. $\{x \mid 1 \leq x \leq 3\}$ is the set of real numbers between 1 and 3, both inclusive.

7. a. $\{(2,6), (2,7), (2,8), (3,6), (3,7), (3,8), (4,6), (4,7), (4,8), (5,6), (5,7), (5,8)\}$
b. $\{(6,2), (6,3), (6,4), (6,5), (7,2), (7,3), (7,4), (7,5), (8,2), (8,3), (8,4), (8,5)\}$
c. $\{(2,2), (2,3), (2,4), (2,5), (3,2), (3,3), (3,4), (3,5), (4,2), (4,3), (4,4), (4,5), (5,2), (5,3), (5,4), (5,5)\}$
d. $\{(6,6), (6,7), (6,8), (7,6), (7,7), (7,8), (8,6), (8,7), (8,8)\}$

Exercises (pages 30–32)

3. a. 6 **b.** 9 **c.** 6 **d.** 14 **e.** 20 **f.** 14

5. Shipping by moving van is more economical.

Exercises (pages 41–43)

1. a. -1 **b.** $\frac{3}{7}$ **c.** $\frac{2}{5}$ **d.** no slope **e.** 2 **f.** 2 **g.** 0 **h.** $-\frac{4}{3}$

3. a. $y = 3$ **b.** $y = -2$ **c.** $x = 5$ **d.** $x = -4$

5. a. slope $= \frac{3}{2}$, y-intercept $= 6$
b. slope $= 1$, y-intercept $= 0$
c. slope $= -\frac{1}{5}$, y-intercept $= 2$
d. slope $= \frac{2}{5}$, y-intercept $= 2$
e. slope $= -\frac{3}{2}$, y-intercept $= \frac{15}{2}$
f. slope $= \frac{3}{4}$, y-intercept $= -\frac{7}{4}$

7. a. perpendicular **b.** perpendicular **c.** not perpendicular **d.** perpendicular

9. a. $2x + y + 7 = 0$
b. $3x - 4y - 25 = 0$
c. $2x - 3y + 7 = 0$
d. $2x - 3y + 7 = 0$
e. $3x + 5y - 13 = 0$
f. $2x + y - 7 = 0$
g. $4x - 3y + 23 = 0$

Exercises (pages 44–45)

1. \$1,200 **3.** \$39 **5.** \$186,000

7. a. $V_n = 50{,}000 - 4{,}700n$
b. \$35,900

Exercises (page 48)

1. $x^2 + y^2 = 36$
3. $(x-4)^2 + (y-2)^2 = 25$
5. $(x+4)^2 + (y-1)^2 = 9$
7. Center $(4, 7)$, $r = 9$
9. Center $(-3, 6)$, $r = 5$
11. Center $(-7, -4)$, $r = 6$

Exercises for Review (pages 48–50)

1. a. true **b.** false **c.** true **d.** false **e.** false **f.** true

2. a. $x > -\frac{2}{3}$ **b.** $x \le \frac{5}{2}$ **c.** $x \le -\frac{7}{4}$ **d.** $x \ge \frac{5}{18}$

3. a. $x < 0$ or $x > 2$
b. $-2 \le x \le 4$
c. $x \le -3$ or $x \ge 1$
d. $0 < x < 2$ or $x > 4$.

4. a. $x = -2, 4$
b. $x = -1, 5$
c. $x = -4, 3$
d. no solution

5. a. $-2 < x < 2$
b. $x < -3$ or $x > 1$
c. $-1 \le x \le 4$
d. $4 < x < 8$

6. a. $\{x | 1 < x < 4\}$
b. $\{x | 0 \le x \le 8\}$
c. $\{x | 2 < x < 3\}$

7. a. $\{(1, 4), (2, 3), (3, 2), (4, 1)\}$
b. $\{(1, 1), (2, 2), (3, 3), (4, 4), (5, 5)\}$

8. a. 10 **b.** 10 **c.** 13 **d.** 25

9. a. 2 **b.** -2 **c.** undefined **d.** 0

10. $y = 9$

11. a. slope $= \frac{2}{3}$, y-intercept $= 2$
b. slope $= -\frac{3}{2}$, y-intercept $= 8$
c. slope $= \frac{1}{3}$, y-intercept $= \frac{4}{3}$

12. a. $2x - y - 2 = 0$
b. $x - y - 1 = 0$
c. $2x - 3y + 9 = 0$
d. $4x - 3y + 23 = 0$

13. \$38.45

14. \$560

15. Center $(5, 6)$, $r = \sqrt{171}$

Chapter 2

Exercises (pages 61–64)

1. **a.**, **b.**, **d.**, **e.**, **f.**, and **g.** are functions.

3. **b.** and **c.** are functions.

5. **a.**, **e.**, **f.**, and **g.** are one-to-one functions.

7. **a.** 5 **b.** 0 **c.** 6 **d.** 9 **e.** 20 **f.** $\pi^3 + \pi^2 + \pi + 6$

9. **a.** $2a + 3$ **b.** $\frac{2}{a} + 3$ **c.** $2(a + b) + 3$ **d.** $2a + 2b + 6$ **e.** $2a + 9$ **f.** $\frac{2}{a + b} + 3$

11. **a.** 40 **b.** 38 **c.** 36 **d.** 32 **e.** 30 **f.** 20

13. $\{-2, -1, 0, 1, 2, 3, 4\}$

15. $\{-3, -2, -1, 0, 1, 2, 3\}$

17. $\{-27, -8, -1, 0, 1, 8, 27\}$

19. set of real numbers

21. set of real numbers greater than or equal to one

23. set of real numbers

25. set of real numbers except $x = 2$

27. $D_f = \{0, 1, 2, \ldots, 24, 25\}$
$R_f = \{0, 4, 8, \ldots, 96, 100\}$

29. $R(x) = (100 - x)(200 + 2.50x)$

Exercises (pages 82–83)

1.

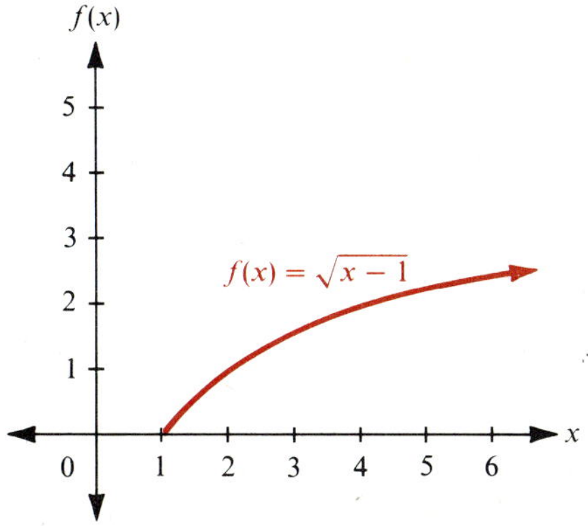

3.

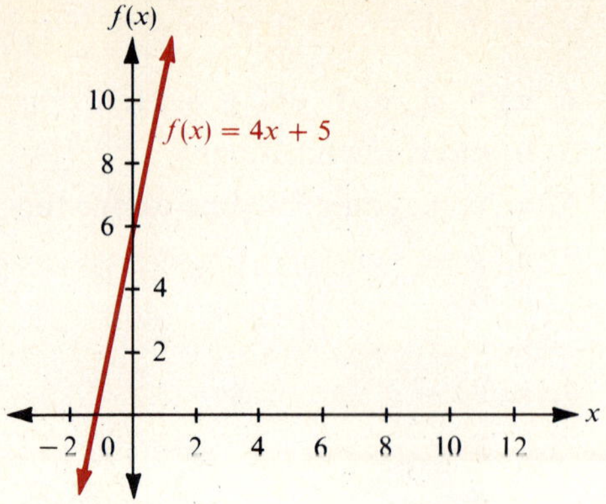
f(x)
f(x) = 4x + 5
10
8
6
4
2
−2
0
2
4
6
8
10
12
x

5.

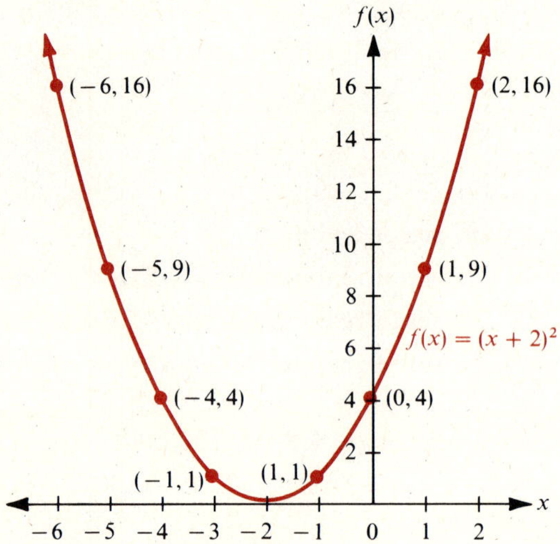
f(x)
(−6, 16)
(2, 16)
16
14
12
10
(−5, 9)
(1, 9)
8
6
f(x) = (x + 2)²
(−4, 4)
4
(0, 4)
2
(−1, 1)
(1, 1)
−6
−5
−4
−3
−2
−1
0
1
2
x

7.

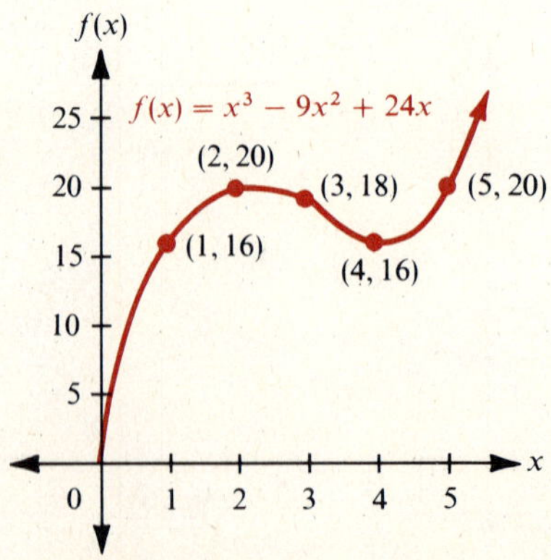
f(x)
25
f(x) = x³ − 9x² + 24x
(2, 20)
20
(3, 18)
(5, 20)
15
(1, 16)
(4, 16)
10
5
0
1
2
3
4
5
x

9.

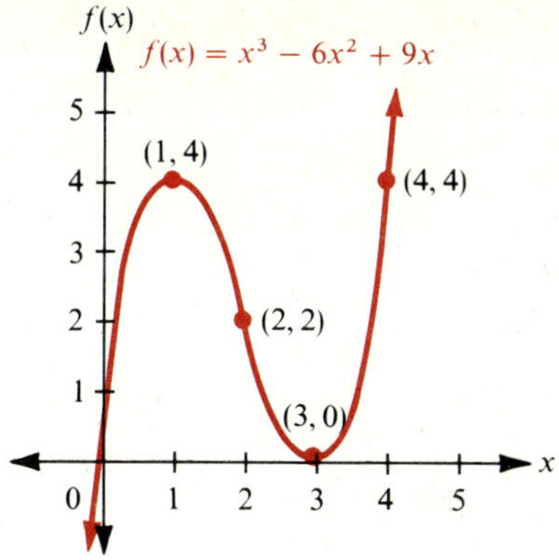

11.

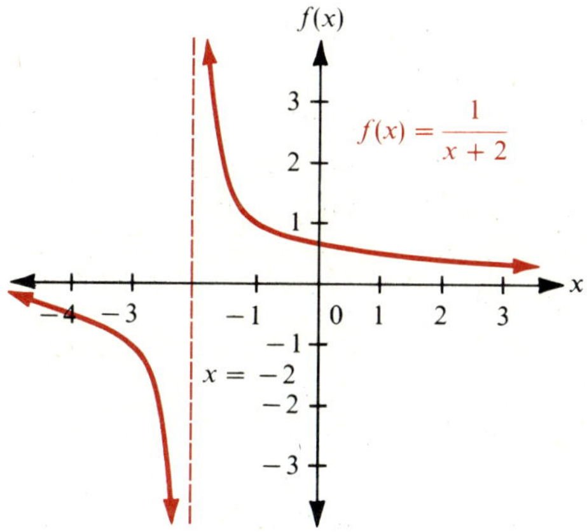

13.

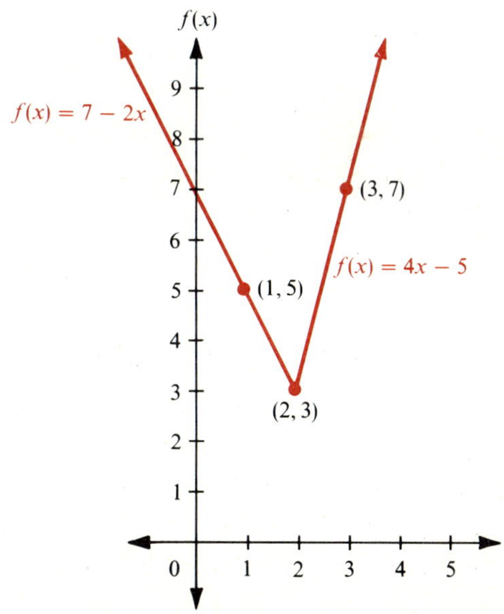

15.

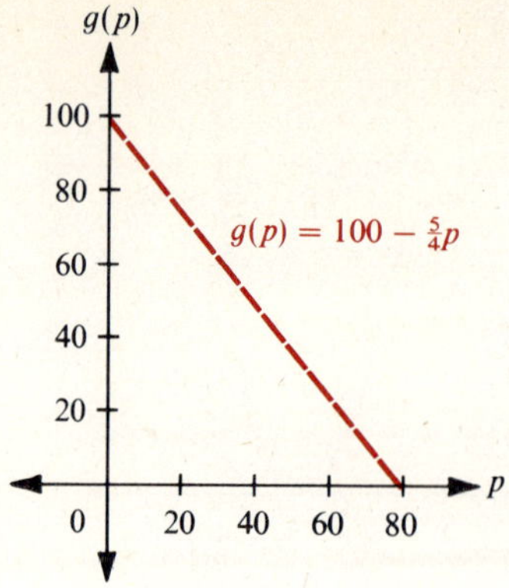

17. **a.** $R(x) = 5000 + 150x - 5x^2$
$D_f = \{0, 1, 2, \ldots, 40\}$

b.

x	$R(x)$
0	5000
5	5625
10	6000
15	6125
20	6000
25	5625
30	5000
35	4125
40	3000

c.

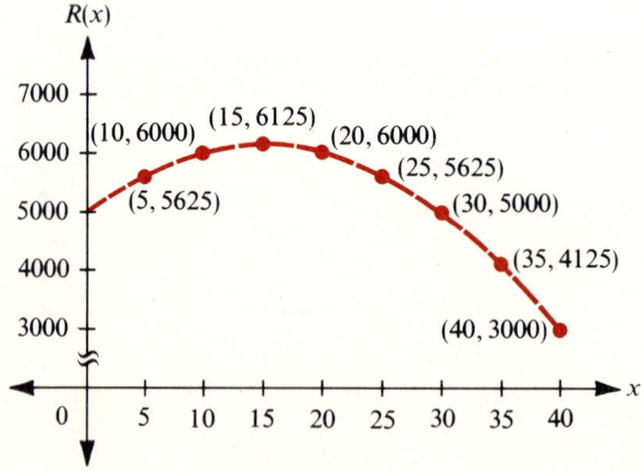

d. $x = 15$

19. **a.** \$3.50 **b.** \$5.00 **c.** \$5.75 **d.** \$6.50

Exercises (pages 89–91)

1. $g[f(1)] = 13$; $g[f(2)] = 19$; $g[f(3)] = 25$; $g[f(4)] = 31$; $g[f(5)] = 37$; $g[f(6)] = 43$.

3. $g[f(1)] = 31$; $g[f(2)] = 43$; $g[f(3)] = 55$; $g[f(4)] = 67$; $g[f(5)] = 79$; $f[g(1)] = 25$; $f[g(2)] = 37$; $f[g(3)] = 49$; $f[g(4)] = 61$; $f[g(5)] = 73$.

5. $g[f(x)] = 3x$; $f[g(x)] = 3x - 10$

7. $g[f(x)] = \dfrac{3x+5}{3}$; $f[g(x)] = x + 5$

9. $g[f(x)] = 3x^2 + 3$; $f[g(x)] = 9x^2 + 1$

11. $g[f(x)] = 6x^2 + 14$; $f[g(x)] = 2(3x - 1)^2 + 5$

13. $g[f(x)] = 1 - (x^2 + 2x + 3)^3$; $f[g(x)] = (1 - x^3)^2 + 2(1 - x^3) + 3$

15. $g[f(x)] = \sqrt{x^2 + 1}$; $f[g(x)] = x + 1$

17. $g[f(x)] = x$; $f[g(x)] = x$

19. $g[f(x)] = x$, $f[g(x)] = x$

21. $\dfrac{x+1}{4}$

23. $x - 3$

25. $\dfrac{3-x}{2}$

27. $\dfrac{x+2}{3}$

29. $x^2 + 3$

31. $\sqrt{9 - x^2}$, $0 \le x \le 3$

Exercises (pages 99–100)

1. 7

3. 39

5. 4

7. limit does not exist

9. -2

11. -1

13. limit does not exist

15. limit does not exist

17. 3

19. 9

21. limit does not exist

23. limit does not exist

25. limit does not exist

27. limit does not exist

Exercises (pages 104–105)

1. 7

3. 0

5. 17

7. 0

9. 70

11. 20

13. 3

15. -3

17. 12
19. -1
21. 4
23. 2
25. 3
27. limit does not exist
29. -2
31. 5
33. 3
35. $2x + 4$
37. $3x^2$

Exercises (pages 109–111)

1. 2
3. $\frac{1}{2}$
5. 1
7. $\frac{3}{2}$
9. limit does not exist
11. 0
13. limit does not exist
15. 1
17. 8
19. $\frac{1}{4}$
21. 0
23. $\frac{1}{2}$
25. limit does not exist
27. 0
29. 0; no
31. $y = 2$
33. does not exist

Exercises (pages 116–117)

1. continuous
3. continuous
5. continuous
7. continuous everywhere
9. continuous everywhere except $x = 3$
11. discontinuous at $x = 1$
13. $x = 1$
15. $x = 3, x = 4$
17. continuous
19. discontinuous at $x = 1, 2, 3, 4, 5, 6, 7$

Exercises for Review (pages 118–120)

2. a. 4
   b. 14
   c. 2
   d. 4
   e. 14
   f. $2a + h - 3$
3. a. $\frac{1}{2}$
   b. $\frac{2}{3}$
   c. $\frac{5}{4}$
   d. 2
   e. no
4. a. set of real numbers
   b. set of real numbers
   c. set of real numbers greater than or equal to 2
   d. set of real numbers between -2 and 2, both inclusive

5. a. D_f: set of real numbers
R_f: set of real numbers

b. D_f: set of real numbers
R_f: set of non-negative real numbers

c. D_f: set of real numbers between -4 and 4, both inclusive
R_f: set of real numbers between 0 and 4, both inclusive

d. D_f: set of real numbers except $x = 2$
R_f: set of non-zero real numbers

6. a.

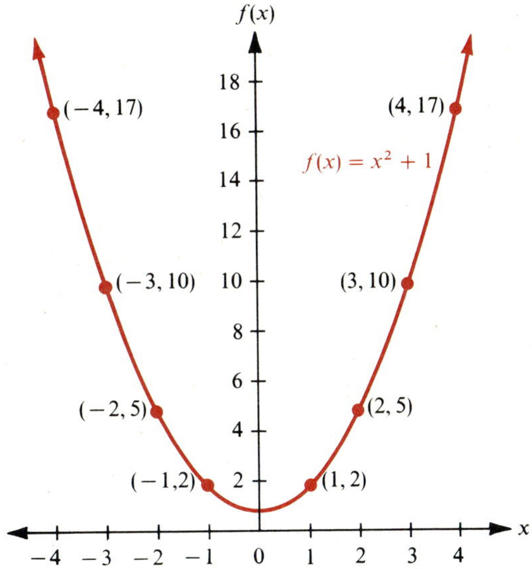

b.

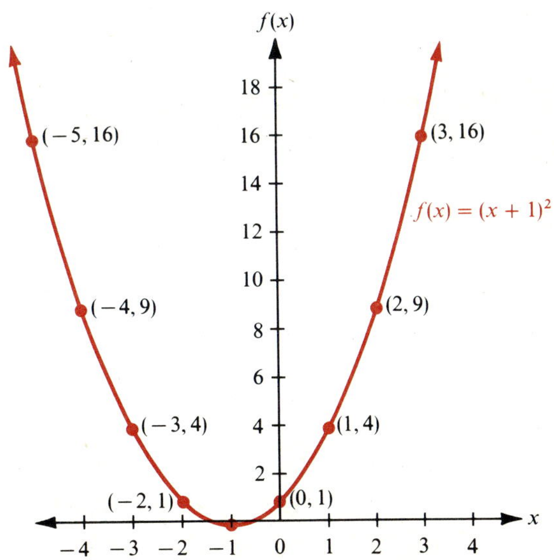

c.

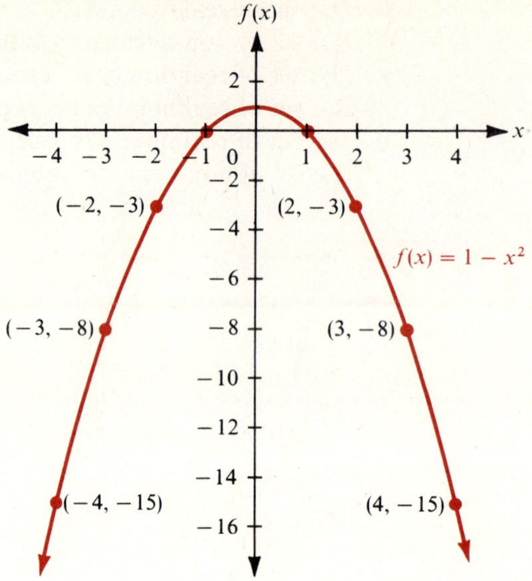

d.

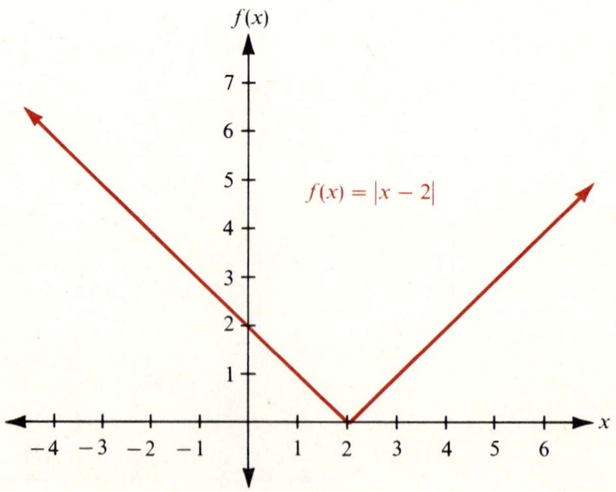

e.

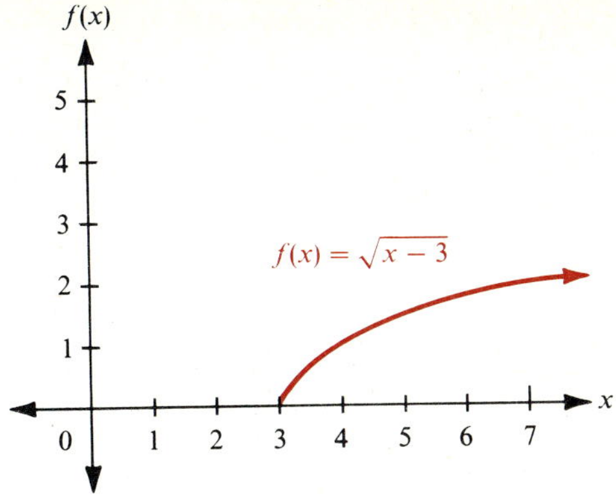

f.

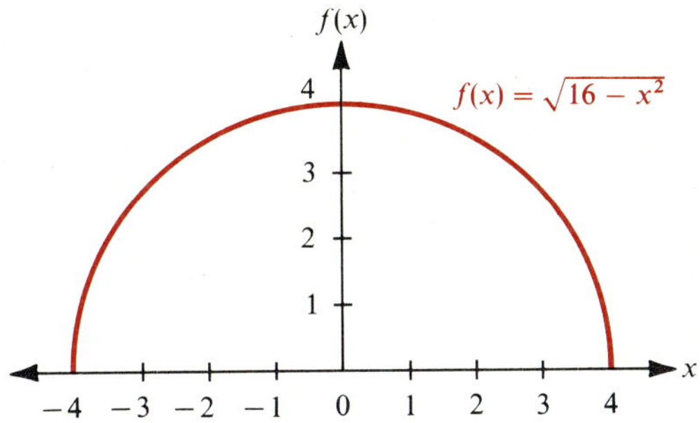

g.

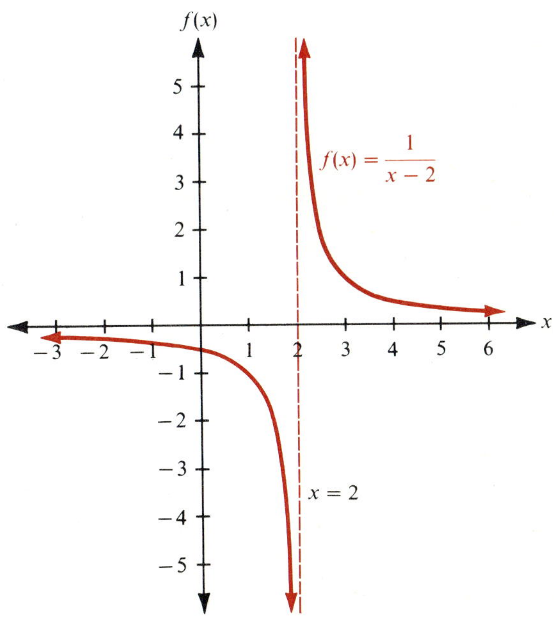

h.

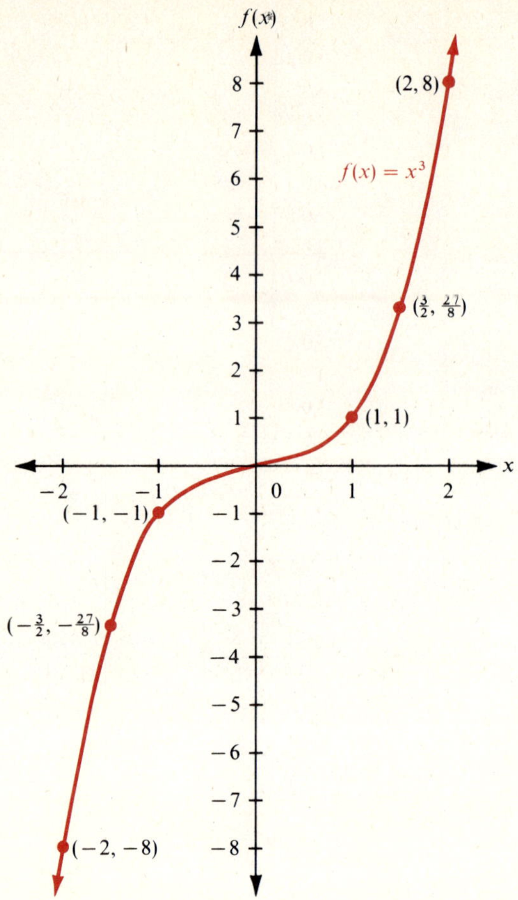

7. a. $R(x) = 20{,}000 + 300x - 5x^2$; $x = 0, 1, 2, \ldots, 100$

b.

x	$R(x)$
0	20,000
10	22,500
20	24,000
30	24,500
40	24,000
50	22,500
60	20,000
70	16,500
80	12,000

c.

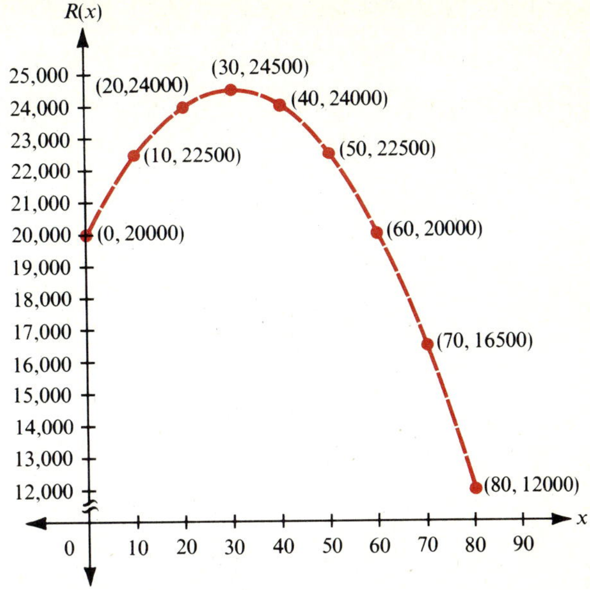

d. $x = 30$

8. a. $g[f(x)] = x^2 + x - 2$; $f[g(x)] = (x-3)^2 + x - 2$
b. $g[f(x)] = (x+1)^2$; $f[g(x)] = x^2 + 1$
c. $g[f(x)] = |x|$; $f[g(x)] = x$
d. $g[f(x)] = \dfrac{x^2}{x^2+1}$; $f[g(x)] = \dfrac{x^2}{(x+1)^2}$

9. a. $2x - 3$ **c.** $\dfrac{x^2+3}{2}$
b. $\sqrt{x-1}$ **d.** $(x-2)^{1/3}$

10. a. 6 **d.** $\frac{3}{4}$
b. 2 **e.** 0
c. limit does not exist **f.** $\frac{1}{2}$

11. a. 5 **b.** -3 **c.** $2x$ **d.** $3x^2$

12. continuous at $x = 40$ and $x = 80$.

13. a. $x = 4$ **c.** $x = -2, x = -1$
b. $x = -1, x = 1$ **d.** $x = 2$

14. discontinuous at $x = 100$, $x = 500$.

Chapter 3

Exercises (pages 134–136)

1. 3 **5.** $3x^2$

3. -5 **7.** $\dfrac{-2}{x^3}$

9. $-\dfrac{1}{(x-4)^2}$

11. $-\dfrac{3}{x^4}$

13. $\dfrac{1}{(x+1)^2}$

15. $\dfrac{1}{2\sqrt{x}}$

17. $-\dfrac{1}{2(x+2)^{3/2}}$

19. $-\dfrac{1}{2(x+1)^{3/2}}$

21. **a.** 180 **b.** 160 **c.** 100 **d.** 50

23. **a.** 30 **b.** 10 **c.** 0 **d.** -20

25. **a.** 64 **b.** 16 **c.** 4 **d.** 1

27. **a.** 32 **b.** 95

29. The worker's satisfaction is increasing after 4 weeks.

Exercises pages (139–140)

1. $3x - y - 4 = 0$
3. $x + y - 1 = 0$
5. $3x - y - 2 = 0$
7. $x - 4y + 4 = 0$
9. $x - 2y + 2 = 0$

Exercises (pages 142–143)

1. **a.** discontinuous at $x = 4$ **b.** not differentiable at $x = 4$
3. **a.** continuous everywhere **b.** differentiable everywhere
5. **a.** continuous everywhere **b.** not differentiable at $x = 1$
7. not differentiable at $x = 3$.
9. **a.** discontinuous at $x = 3$ **b.** not differentiable at $x = 3$.

Exercises (pages 147–150)

1. $4x^3$
3. $32x^7$
5. $3\sqrt{x}$
7. $-\dfrac{1}{x^2}$
9. $3x^2 - \dfrac{3}{x^4}$
11. $3x^2 + 12x$
13. $\dfrac{1}{2\sqrt{x}} - \dfrac{1}{2x^{3/2}}$
15. $15x^2 + 8x - \dfrac{1}{x^2}$
17. $15x^4 + 16x^3 + 9x^2 - \dfrac{3}{2x^{3/2}}$
19. $\frac{5}{4}x^{1/4} + \frac{3}{2}x^{1/2} + \frac{1}{2}x^{-1/2}$

21. **a.** $2x + 5$ **b.** 9 **c.** 11 **d.** 13

23. **a.** $\dfrac{1}{2\sqrt{x}}$ **b.** $\dfrac{1}{2\sqrt{2}}$ **c.** $\dfrac{1}{2\sqrt{3}}$ **d.** $\dfrac{1}{4}$

25. **a.** $-\frac{1}{4}x^{-3/2}$ **b.** $-\dfrac{1}{8\sqrt{2}}$ **c.** $-\dfrac{1}{12\sqrt{3}}$ **d.** $\dfrac{-1}{32}$

27. **a.** 50 **b.** 110

29. $0.06x + 10$

31. **a.** -432 **b.** -675 **c.** -972

33. 240 fish

35. **a.** 36 **b.** 18 **c.** 12 **d.** 9 **e.** 6

Exercises (pages 156–157)

1. $2x + 5$

3. $2x + 7$

5. $12x + 29$

7. $12x - 5$

9. $12x^3 + 38x$

11. $4x^3 - 4x$

13. $5x^4 - 28x^3 + 15x^2 - 68x - 7$

15. $8x^7 + 7x^6 + 6x^5 - 3x^2 - 2x$

17. $5x^4 - 4x^3 - 1$

19. $3\sqrt{x} + 3 + \dfrac{1}{\sqrt{x}}$

21. $3x^2 - 1 + \frac{3}{2}\sqrt{x} + \dfrac{1}{2x^{3/2}} + \dfrac{2}{x^3}$

23. $\frac{3}{2}\sqrt{x} + 1$

25. $-\dfrac{1}{x^2}$

27. $-\dfrac{1}{(x + 2)^2}$

29. $\dfrac{2}{(3x + 2)^2}$

31. $\dfrac{4}{(2x + 1)^2}$

33. $\dfrac{-4x}{(1 + x^2)^2}$

35. $\dfrac{2(x^2 - 1)}{(1 + x + x^2)^2}$

37. $\dfrac{1}{2\sqrt{x}(\sqrt{x} + 1)^2}$

39. $\dfrac{-x^6 - 2x^5 - 4x^4 - 4x^3 + 19x^2 + 14x + 6}{(x^4 - x^2 + 6)^2}$

41. **a.** -1 **b.** $-\frac{1}{9}$ **c.** $-\frac{1}{25}$ **d.** $-\frac{1}{81}$

43. **a.** $\frac{3}{4}$ **b.** $\frac{3}{16}$ **c.** $\frac{1}{12}$ **d.** $\frac{3}{100}$

45. **a.** 1 **b.** $\frac{1}{25}$ **c.** $\frac{1}{81}$ **d.** $\frac{1}{289}$

Exercises (pages 163–165)

1. $8(2x + 3)^3$
3. $-18x(6 - 3x^2)^2$
5. $-\dfrac{8x + 6}{(2x^2 + 3x + 1)^3}$
7. $\dfrac{3}{2\sqrt{3x + 8}}$
9. $\dfrac{x}{\sqrt{x^2 + 1}}$
11. $\dfrac{3x + 1}{\sqrt{2x + 1}}$
13. $\frac{5}{2}x^{3/2} + x^{-1/2}$
15. $2x(18x + 7)(3x + 7)^9$
17. $2(10x^2 - x + 8)(2x - 1)^7$
19. $(2x + 3)^3(3x + 2)^4(54x + 61)$
21. $(x^2 + x + 1)^2(x^2 - x + 1)^3[4(2x - 1)(x^2 + x + 1)$
$+ 3(2x + 1)(x^2 - x + 1)]$
23. $\dfrac{21(2x - 1)^2}{(3x + 2)^4}$
25. $-\dfrac{1}{\sqrt{x + 1}(x - 1)^{3/2}}$
27. $y + 4x - 12 = 0$
29. $5x - 16y - 1 = 0$
31. $594x - y - 1187 = 0$
33. $\dfrac{ds}{dt}$ is decreasing at $t = 3$.
35. $\dfrac{3}{2\sqrt{2}}$
37. **a.** 26 **b.** 28
39. **a.** $\dfrac{20(4 - t^2)}{(2t^2 + 5t + 8)^2}$
b. it has lost its effectiveness after 2 hours.

Exercises (pages 168–169)

1. $7x^6$; $42x^5$
3. $4x^3 + 12x^2 + 12x$; $12x^2 + 24x + 12$

5. $5x^4 + 16x^3 + 9x^2 + 2x$; $20x^3 + 48x^2 + 18x + 2$

7. $-\frac{1}{x^2}; \frac{2}{x^3}$

9. $-\frac{1}{(x+1)^2}; \frac{2}{(x+1)^3}$

11. $-\frac{1}{(x+1)^2}; \frac{2}{(x+1)^3}$

13. $15(3x+4)^4$; $180(3x+4)^3$; $1620(3x+4)^2$; $9720(3x+4)$

15. $na(ax+b)^{n-1}$; $n(n-1)a^2(ax+b)^{n-2}$; $n(n-1)(n-2)a^3(ax+b)^{n-3}$; $n(n-1)(n-2)(n-3)a^4(ax+b)^{n-4}$.

17. $-\frac{1}{(x+3)^2}; \frac{2}{(x+2)^3}; -\frac{6}{(x+2)^4}; \frac{24}{(x+2)^5}$

19. $-\frac{1}{(x-1)^2}; \frac{2}{(x-1)^3}; \frac{-6}{(x-1)^4}; \frac{24}{(x-1)^5}$

21. 101; 72

23. 96; 32; -32; -32

25. 1.92

Exercises (pages 174–176)

1. $\frac{1}{\sqrt{x}}$

3. $-\frac{x}{4y}$

5. $\frac{x}{3y}$

7. $\frac{x^2 - 6y}{6x - y^2}$

9. $-\frac{2x + y}{x + 2y}$

11. $\frac{2\sqrt{xy} + y}{4y\sqrt{xy} - x}$

13. $-\frac{xy}{x^2 + 2y^2}$

15. $\frac{x(y^2 - 2x^2)}{y(2y^2 - x^2)}$

17. $-\frac{y + \sqrt{y}}{x + \sqrt{x}}$

19. $\frac{x(6xy^2 - 5x^3 - 8y^4)}{y(5y^3 + 16x^2y^2 - 4x^3)}$

21. $-\frac{x}{y}; -\frac{a^2}{y^3}$

23. $\frac{25x}{16y}; \frac{625}{16y^3}$

25. $-\frac{y^{1/3}}{x^{1/3}}; \frac{1}{3}\left(\frac{a^2}{x^4y}\right)^{1/3}$

27. $x + 8y - 17 = 0$

29. $2x + y - 8 = 0$

31. $11x + 3y - 36 = 0$

33. $x + y - 3a = 0$

35. $2x + y - 24 = 0$

Exercises for Review (pages 176–178)

2. a. false **b.** true

3. a. 3 **b.** $2x$ **c.** $-\frac{1}{x^2}$ **d.** $-\frac{2}{x^3}$ **e.** $\frac{1}{2\sqrt{x+1}}$ **f.** $-\frac{1}{(x+2)^{3/2}}$

4. a. $6x - 5$
b. $12x^2 + 10x - 2$
c. $4 - \frac{1}{x^2}$
d. $\frac{1}{2\sqrt{x}} - \frac{1}{2x^{3/2}}$
e. $128 - 32t$
f. $192 - 32t$
g. $8t + 8$
h. $\frac{3}{2}t^{1/2} + \frac{1}{t^{1/2}}$
i. $1 + \frac{3}{2\sqrt{x}}$
j. $\frac{2}{(t+1)^2}$
k. $\frac{2t^2 - 2}{(t^2 + t + 1)^2}$
l. $\frac{t}{\sqrt{t^2 + 1}}$
m. $-6x(4 - x^2)^2$
n. $5(4x + 1)(3x + 2)^2(12x + 5)$
o. $\frac{-8(x+1)^3}{(x-1)^5}$
p. $\frac{5}{2(2x+1)^{1/2}(3x+4)^{3/2}}$

5. a. $-\frac{x}{y}$
b. $-\frac{y}{x}$
c. $\frac{4xy - y^2}{2xy - 2x^2}$
d. $-\frac{y + \sqrt{y}}{x + \sqrt{x}}$

6. a. $4(2x + 3)$; 8
b. $-\frac{1}{(x+1)^2}$; $\frac{2}{(x+1)^3}$
c. $\frac{1}{(x+3)^2}$; $-\frac{2}{(x+3)^3}$
d. $\frac{3}{2}x^{1/2}$; $\frac{3}{4}x^{-1/2}$

7. a. $f'(0) = -4$; $f'(1) = 2$; $f'(2) = 8$
b. $f'(0) = 20$; $f'(1) = 28$; $f'(2) = 36$
c. $f'(0) = -\frac{1}{4}$; $f'(1) = -\frac{1}{9}$; $f'(2) = -\frac{1}{16}$
d. $f'(0) = \frac{1}{4\sqrt{2}}$; $f'(1) = \frac{1}{6}$; $f'(2) = \frac{1}{2\sqrt{10}}$

8. a. $y - 4x - 2 = 0$
b. $4y - x - 4 = 0$
c. $y = -32x$
d. $y + 2x - 10 = 0$
e. $4y + 3x - 25 = 0$

9. **a.** $12y + 5x - 169 = 0$
b. $y + x - 8 = 0$
c. $y + 4x - 16 = 0$

10. **a.** 12 **b.** 36

11. **a.** $-\frac{1}{4}$ **b.** $-\frac{1}{9}$

12. **a.** 20 **b.** $10\sqrt{2}$ **c.** 10

13. $D'(p) = -\frac{500}{27}$

14. 32; -32

15. **a.**

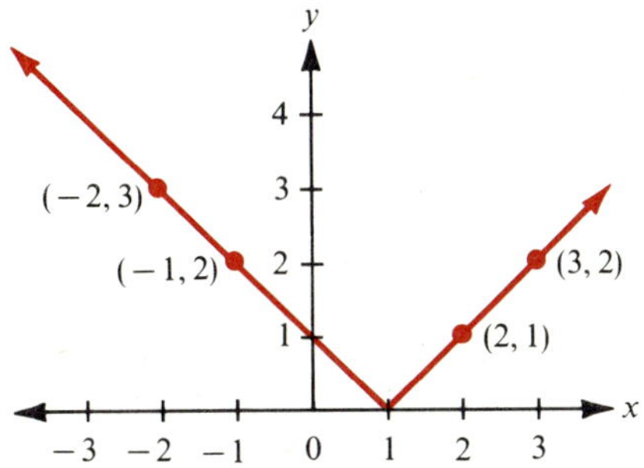

b. yes
c. no

Chapter 4

Exercises (page 185)

1. relative minimum at $x = -1$; $f(-1) = 5$.

3. relative maximum at $x = 2$; $f(2) = 4$.

5. relative minimum at $x = 2$; $f(2) = -4$; relative maximum at $x = 0$; $f(0) = 0$.

7. relative minimum at $x = 3$; $f(3) = 0$; relative maximum at $x = 1$; $f(1) = 4$.

9. relative maximum at $x = 4$; $f(4) = 5$.

Exercises (pages 190–191)

1. decreases on $(-\infty, 2)$; increases on $(2, \infty)$.

3. decreases on $(-\infty, -3)$; increases on $(-3, \infty)$.

5. increases on $(-\infty, -1)$ and $(1, \infty)$; decreases on $(-1, 1)$.

7. increases on $(-\infty, 0)$ and $(2, \infty)$; decreases on $(0, 2)$.

9. increases on $(-\infty, 1)$ and $(2, \infty)$; decreases on $(1, 2)$.

11. increases on $(-\infty, -\frac{3}{2})$ and $(2, \infty)$; decreases on $(-\frac{3}{2}, 2)$.

13. decreases on $(-\infty, -2)$ and $(0, 2)$; increases on $(-2, 0)$ and $(2, \infty)$.

15. decreases on $(-\infty, -1)$ and $(0, 1)$; increases on $(-1, 0)$ and $(1, \infty)$.

17. decreases on $(-\infty, -1)$ and $(1, \infty)$; increases on $(-1, 1)$.

19. decreases on $(-\infty, -3)$ and $(3, \infty)$; increases on $(-3, 3)$.

21. increases for $x < 6$ and decreases for $x > 6$.

23. increasing for $t \in (4, 5)$ and decreasing for $t > 5$.

Exercises (pages 198–201)

1. relative minimum at $x - 3$; $f(3) = -1$

3. relative minimum at $x = 1$; $f(1) = 7$

5. relative minimum at $x = 2$; $f(2) = 12$
relative maximum at $x = 1$; $f(1) = 13$

7. relative minimum at $x = \frac{1}{3}$; $f(\frac{1}{3}) = \frac{22}{27}$
relative maximum at $x = -1$; $f(-1) = 2$

9. relative minimum at $x = -1$ and $x = 3$; $f(-1) = 2$, $f(3) = -126$
relative maximum at $x = 0$; $f(0) = 9$

11. relative minimum at $x = -2$ and $x = 2$; $f(-2) = f(2) = -16$
relative maximum at $x = -3$, $x = 0$ and $x = 3$; $f(-3) = f(3) = 9$, $f(0) = 0$.

13. relative minimum at $x = -1$; $f(-1) = -\frac{1}{2}$
relative maximum at $x = 1$; $f(1) = \frac{1}{2}$

15. absolute minimum at $x = 0$; $f(0) = 2$
absolute maximum at $x = 3$; $f(3) = 11$

17. absolute minimum at $x = -1$; $f(-1) = -8$
absolute maximum at $x = 2$; $f(2) = 19$

19. absolute minimum at $x = 1$; $f(1) = -7$
absolute maximum at $x = 4$; $f(4) = 128$

21. absolute minimum at $x = 0$; $f(0) = -32$
absolute maximum at $x = 6$; $f(6) = 184$

23. absolute minimum at $x = -1$, $x = 1$; $f(-1) = f(1) = 0$
absolute maximum at $x = 3$; $f(3) = 64$

25. absolute minimum at $x = -1$; $f(-1) = -\frac{1}{5}$
absolute maximum at $x = 2$; $f(2) = \frac{1}{4}$

27. no absolute minimum or maximum values;
relative maximum at $x = 1$; $f(1) = 2$.

29. $t = 4$

31. $x = 15$; \$175

33. $x = 15$

35. $t = 5$

37. 25 minutes

Exercises (pages 204–207)

1. relative minimum at $x = 2$; $f(2) = -1$

3. relative maximum at $x = -6$; $f(-6) = -68$

5. relative minimum at $x = 4$; $f(4) = -32$
relative maximum at $x = 0$; $f(0) = 0$

7. relative minimum at $x = 2$; $f(2) = -36$
relative maximum at $x = -3$; $f(-3) = 89$

9. relative minimum at $x = -4$; $f(-4) = -176$
relative maximum at $x = -5$; $f(-5) = -175$

11. relative minimum at $x = 0$ and $x = 2$; $f(0) = f(2) = 0$
relative maximum at $x = 1$; $f(1) = 1$

13. relative minimum at $x = 1$; $f(1) = 2$

15. relative minimum at $x = 4$; $f(4) = 8$

17. $x = 5$

19. $x = 6$

21. $x = 200$

23. $x = 10$

25. $x = 70$

27. $x = 20$

29. $t = 2.5$ seconds; height is 100 feet.

31. $t = 3$.

33. $t = 2$.

Exercises (pages 215–219)

1. $x = 25$

3. The numbers are 30 and 10.

5. 15 inches by 10 inches.

7. length, $\frac{80}{3}$ inches; width, $\frac{20}{3}$ inches; height, $\frac{8}{3}$ inches.

9. length, 1400 feet; width, 700 feet.

11. 6400 feet.

13. \$175; \$30,625.

15. 70 couples; \$245.

17. 45 students

19. 120 persons; \$7200.

21. 5 days; \$156.25

23. $x = 10$

25. 150 units; \$1250; \$35 per unit.

27. 1000 units; \$95.00; \$21,000.

29. 500 units; 16 orders.

Exercises (page 229)

1. decreasing on $(-\infty, 3)$ and increasing on $(3, \infty)$; concave upward; no inflection point.

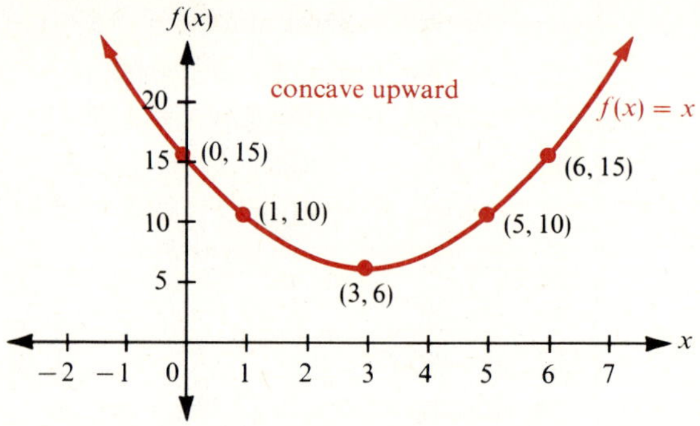

3. increasing on $(-\infty, 6)$ and decreasing on $(6, \infty)$; concave downward; no inflection point.

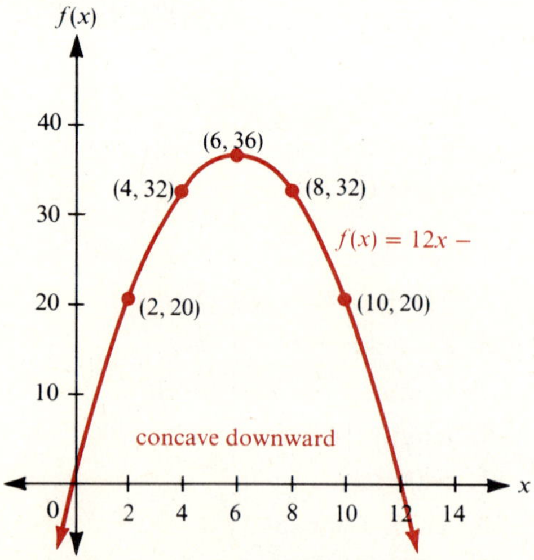

5. increasing on $(-\infty, 0)$ and $(2, \infty)$; decreasing on $(0, 2)$; concave downward for $x < 1$ and upward for $x > 1$; $(1, -2)$ is an inflection point.

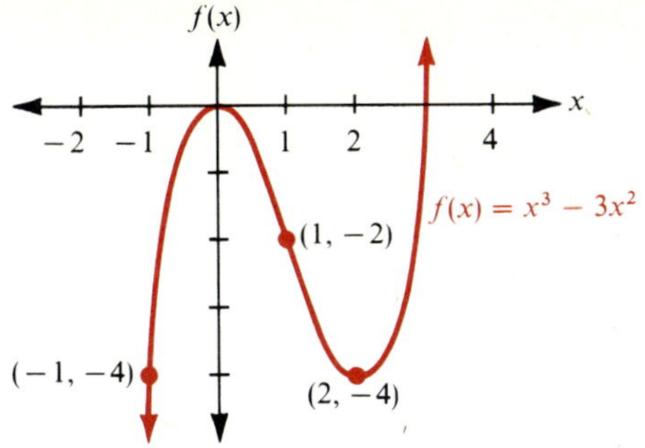

7. increasing on $(-\infty, 2)$ and $(4, 16)$; decreasing on $(2, 4)$; concave downward for $x < 3$ and concave upward for $x > 3$; the point $(3, 18)$ is an inflection point.

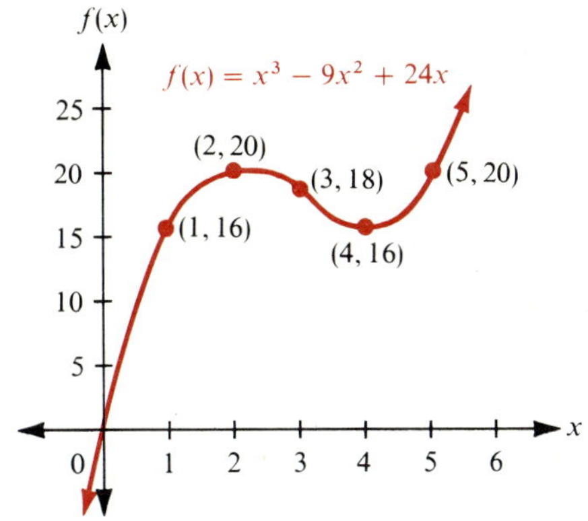

9. increasing for all x; concave downward for $x < -1$ and concave upward for $x > -1$; the point $(-1, 4)$ is an inflection point.

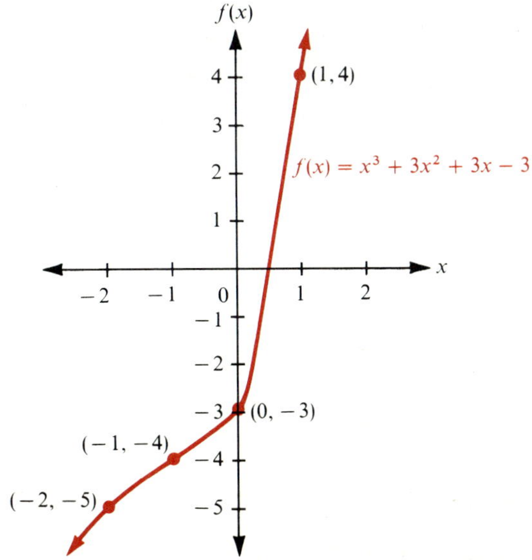

11. increasing on $(-\infty, 2)$ and $(6, \infty)$; decreasing on $(2, 6)$; concave downward for $x < 4$ and concave upward for $x > 4$; the point $(4, 32)$ is an inflection point.

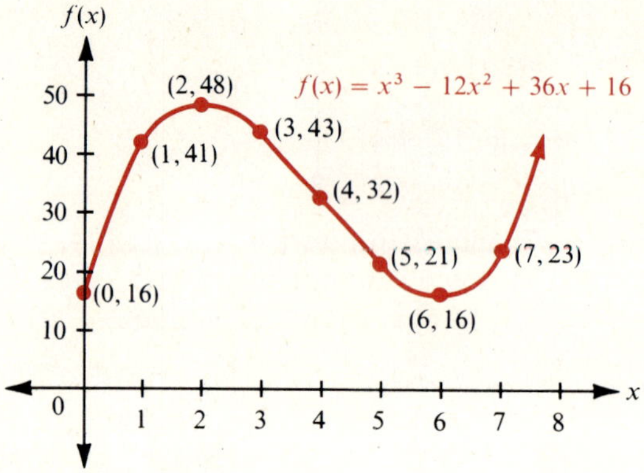

13. increasing on $(-\infty, 1)$ and $(2, \infty)$; decreasing on $(1, 2)$; concave downward for $x < \frac{3}{2}$ and concave upward for $x > \frac{3}{2}$; the point $(\frac{3}{2}, \frac{3}{2})$ is an inflection point.

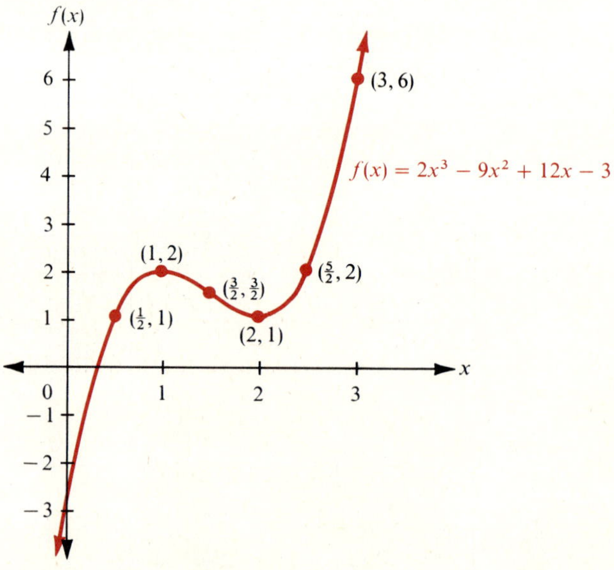

15. decreasing on $(-\infty, 0)$ and increasing on $(0, \infty)$; concave downward for $-3 < x < -1$ and concave upward when $x < -3$ or $x > -1$; the inflection points are $(-1, 6)$ and $(-3, 22)$.

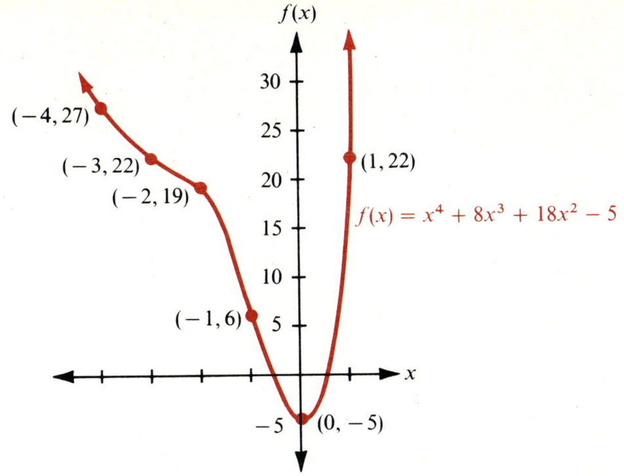

17. increases for all x except $x = -1$; concave upward for $x < -1$ and concave downward for $x > -1$; no inflection point; $f(x) = 1$ is a horizontal asymptote.

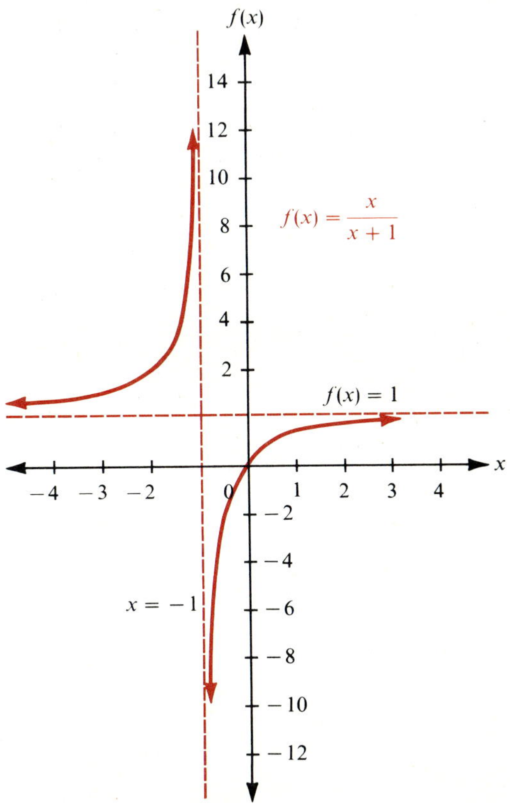

19. increasing for all except $x = -2$; concave upward for $x < -2$ and concave downward for $x > -2$; no inflection point; $f(x) = 1$ is a horizontal asymptote.

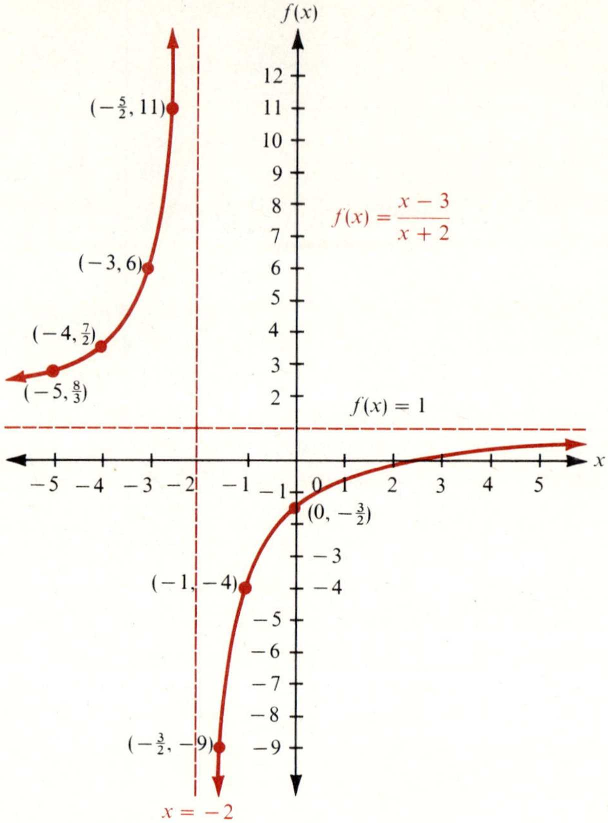

Exercises for Review (pages 229–232)

2. a. decreasing on $(-\infty, -3)$; increasing on $(-3, \infty)$
b. increasing on $(-\infty, 6)$; decreasing on $(6, \infty)$
c. increasing on $(-\infty, -1)$ and $(\frac{1}{3}, \infty)$; decreasing on $(-1, \frac{1}{3})$
d. increasing on $(-\infty, -2)$ and $(0, \infty)$; decreasing on $(-2, 0)$
e. decreasing on $(-\infty, -3)$ and $(-3, 0)$; increasing on $(0, \infty)$.

3. a. relative minimum at $x = -\frac{1}{2}$; $f(-\frac{1}{2}) = -\frac{121}{4}$
b. relative maximum at $x = 4$; $f(4) = 10$
c. relative minimum at $x = 1$; $f(1) = 3$; relative maximum at $x = -1$; $f(-1) = 7$
d. relative minimum at $x = 1$; $f(1) = -15$; relative maximum at $x = -3$; $f(-3) = 17$
e. relative minimum at $x = 0$ and $x = 2$; $f(0) = 1$, $f(2) = 1$; relative maximum at $x = 1$; $f(1) = 2$

4. **a.** absolute minimum at $x = 7$; $f(7) = -93$; absolute maximum at $x = 1$; $f(1) = 15$
   b. absolute minimum at $x = 1$; $f(1) = -10$; absolute maximum at $x = 5$; $f(5) = 358$
   c. absolute minimum at $x = -1$; $f(-1) = 0$; absolute maximum at $x = 2$; $f(2) = 81$
   d. absolute minimum at $x = -1$; $f(-1) = -\frac{1}{5}$; absolute maximum at $x = 2$; $f(2) = \frac{1}{4}$
5. increasing on $(-\infty, -1)$ and $(2, \infty)$; decreasing on $(-1, 2)$; concave downward for $x < \frac{1}{2}$ and concave upward for $x > \frac{1}{2}$. The point $(\frac{1}{2}, -\frac{3}{2})$ is a point of inflection.

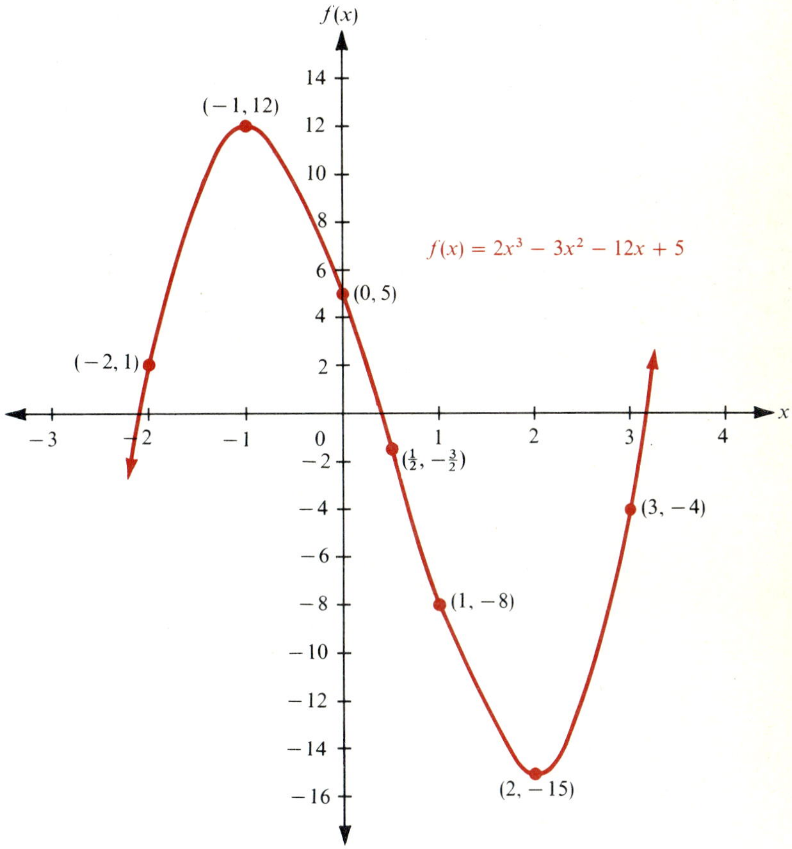

6. $t = 3$ seconds
7. $x = 250$; $P(250) = 425$
8. $t = 2.5$ hours
9. $t = \frac{1}{2}$ hour; $Q(\frac{1}{2}) = 5.225$ units
10. $t = 4$ hours
11. $t = 8$ weeks
12. cut the wood in 4 pieces, each 10 inches long.
13. $\frac{5}{3}$ centimeters

14. length—2400 feet
width—1200 feet

15. each side of the plot is 1200 feet long

16. \$35.00

17. 25 feet, 25 feet, 50 feet

18. $x = 40$; \$17.00; \$120.00

19. $x = 8$ kilometers

20. 2000 units; 5 times a year.

Chapter 5

Exercises (pages 239–240)

1.

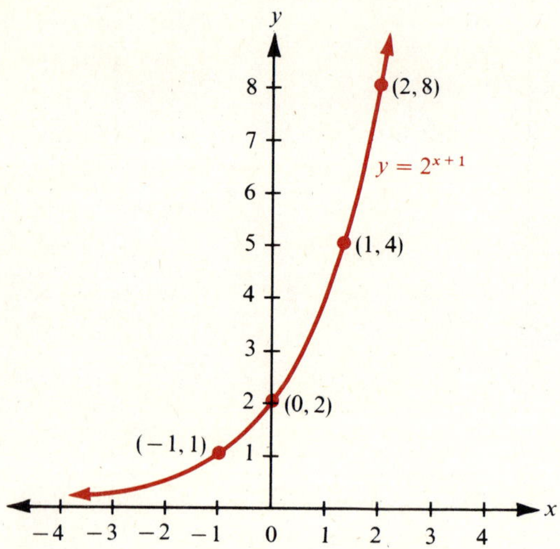

3.

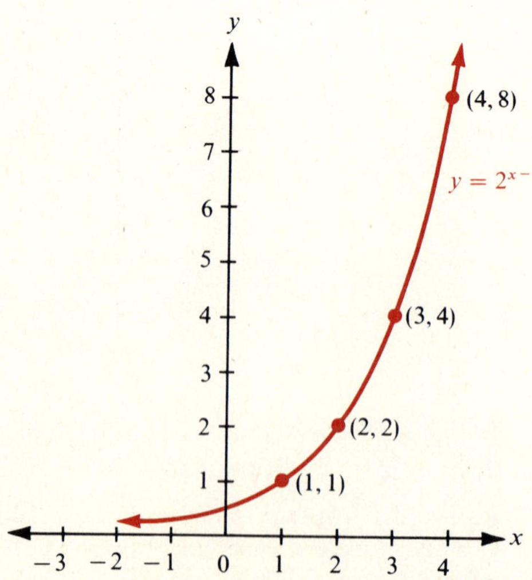

5.

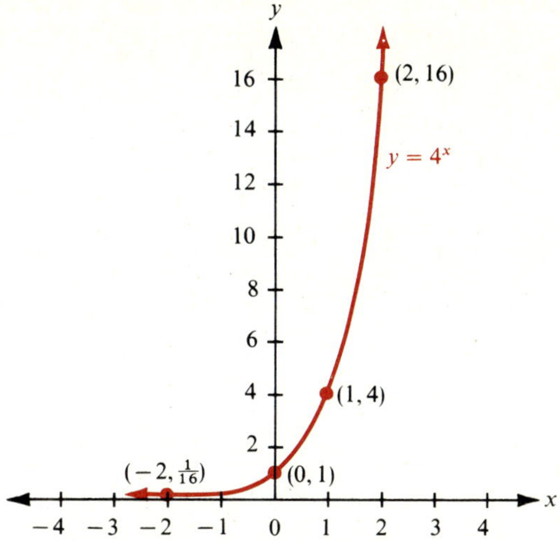
y
x
(2, 16)
$y = 4^x$
(1, 4)
(0, 1)
$(-2, \frac{1}{16})$
16
14
12
10
8
6
4
2
−4 −3 −2 −1 0 1 2 3 4

7.

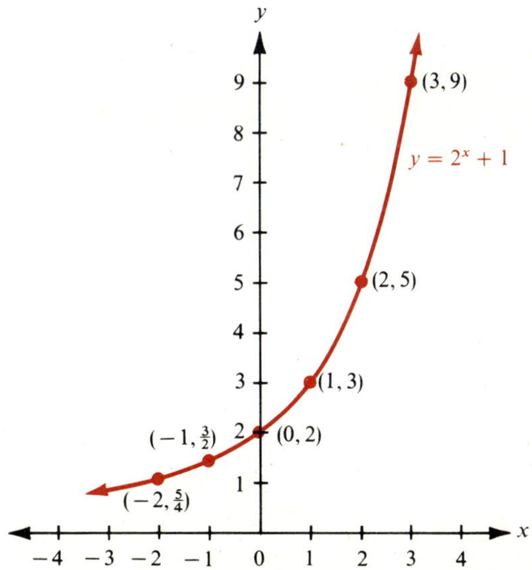
y
x
(3, 9)
$y = 2^x + 1$
(2, 5)
(1, 3)
(0, 2)
$(-1, \frac{3}{2})$
$(-2, \frac{5}{4})$
9
8
7
6
5
4
3
2
1
−4 −3 −2 −1 0 1 2 3 4

9.

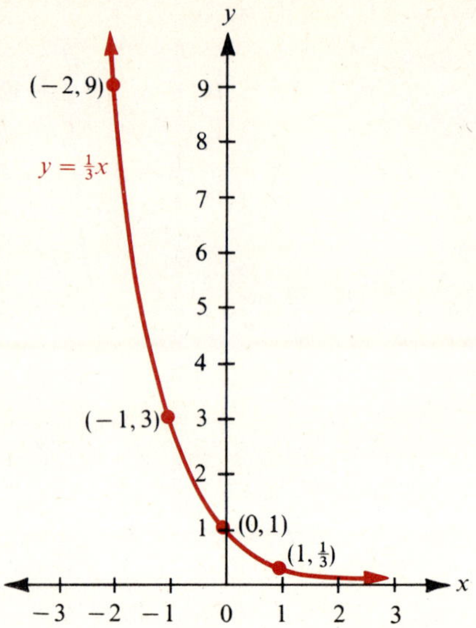
y
x
(−2, 9)
(−1, 3)
(0, 1)
(1, 1/3)
y = 1/3x
9
8
7
6
5
4
3
2
1
−3 −2 −1 0 1 2 3

11.

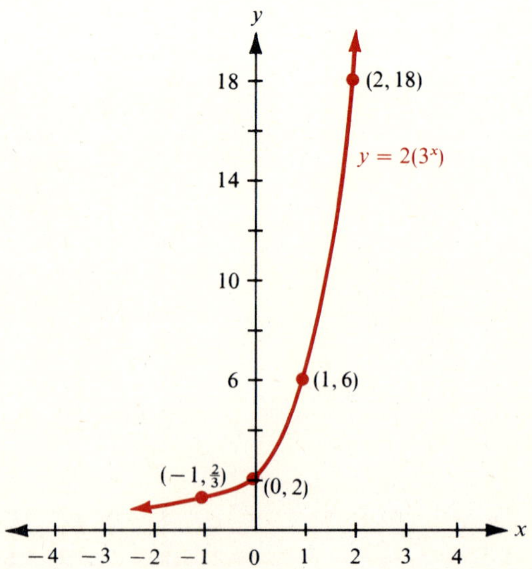
y
x
(2, 18)
(1, 6)
(0, 2)
(−1, 2/3)
y = 2(3^x)
18
14
10
6
−4 −3 −2 −1 0 1 2 3 4

13.

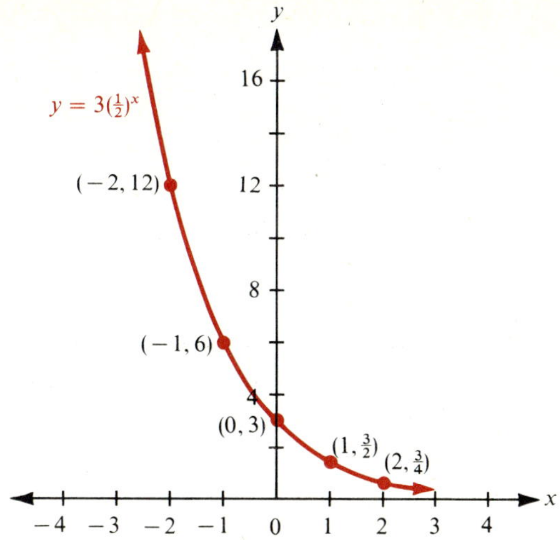

15. $x = 0$

17. $x = 3$

19. $x = \frac{3}{2}$

21. $x = -4$

23. $x = -4$

25. $x = 2$

27. $x = 3$

29. $x = 6$

31. $x = 3$

33. $x = 3$

35. a.

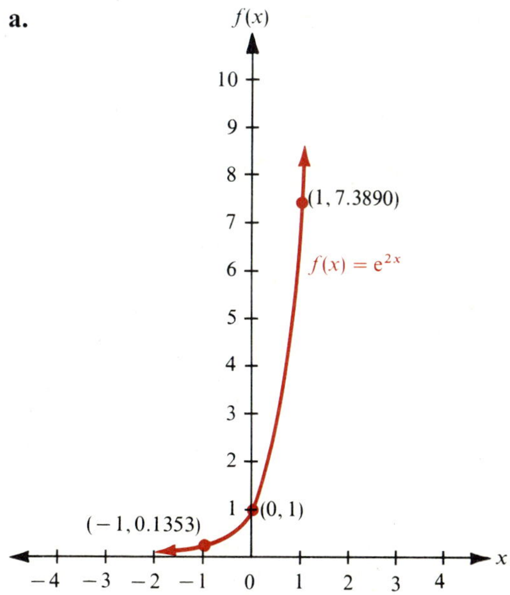

b.

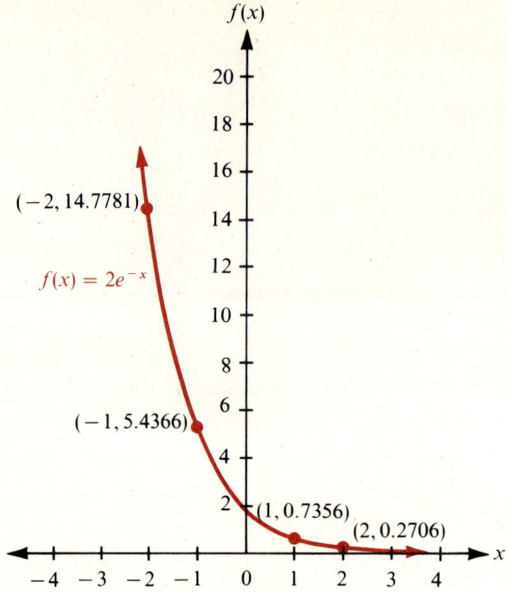

Exercises (pages 246–247)

1. **a.** \$882.00; \$82.00 **b.** \$883.05; \$83.05 **c.** \$883.59; \$83.59

3. **a.** \$812.53 **b.** \$1524.59 **c.** \$2467.36

5. **a.** \$1560.60 **b.** \$1591.81 **c.** \$1623.65 **d.** \$1757.49

7. **a.** \$3348.42 **b.** \$3341.12 **c.** \$3336.17 **d.** \$3333.76 **e.** \$3333.68

9. \$3704.09

Exercises (pages 251–252)

1. **a.** 500 **b.** 911 **c.** 2241

3. **a.** 110,517 **b.** 134,986 **c.** 164,872 **d.** 271,828

5. $Q(1) = 3.70$; $Q(2) = 2.74$; $Q(3) = 2.03$; $Q(4) = 1.51$; $Q(5) = 1.12$; $Q(6) = 0.83$; $Q(8) = 0.45$; $Q(10) = 0.25$

7. $P(1) = \$17{,}408$; $P(2) = \$15{,}488$; $P(3) = \$14{,}066$; $P(4) = \$13{,}012$; $P(5) = \$12{,}231$; $P(6) = \$11{,}653$; $P(8) = \$10{,}907$; $P(10) = \$10{,}498$.

9. **a.** $S(1) = 14{,}852$; $S(2) = 12{,}943$; $S(3) = 11{,}785$; $S(4) = 11{,}083$; $S(5) = 10{,}657$; $S(6) = 10{,}398$; $S(7) = 10{,}242$; $S(8) = 10{,}147$; $S(10) = 10{,}054$; $S(12) = 10{,}020$.

b.

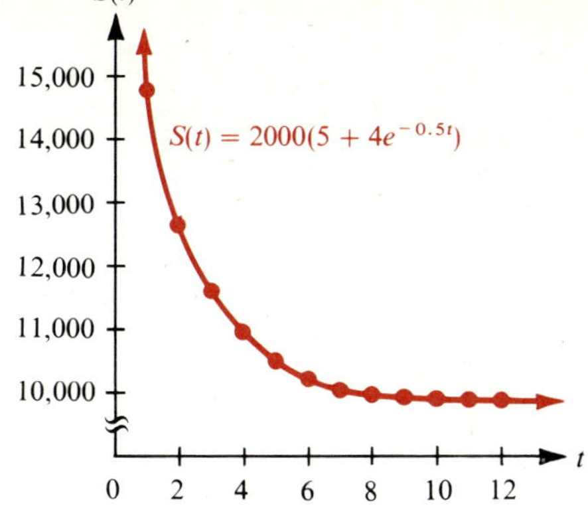

Exercises (pages 257–258)

1. **a.** $\log_2 8 = 3$ **b.** $\log_3 9 = 2$ **c.** $\log_5 625 = 4$ **d.** $\log_{16} 4 = \frac{1}{2}$ **e.** $\log_{32} 8 = \frac{3}{5}$ **f.** $\log_{216} 36 = \frac{2}{3}$

3. **a.** $x = 81$ **b.** $x = 128$ **c.** $x = 3$ **d.** $x = -3$ **e.** $x = \frac{1}{25}$ **f.** $x = 3$ **g.** $x = 2$ **h.** $x = \frac{1}{2}$ **i.** $x = \frac{1}{5}$ **j.** $x = 7$

5. **a.** -1.26 **b.** 3.05 **c.** 3.56 **d.** 3.22 **e.** 4.83 **f.** 5.85

7. 0.6020; 0.7781; 0.9030; 0.9542; 1.0791; 1.2040.

Exercises (pages 265–266)

1. **a.** 3.6263 **b.** 2.6263 **c.** 1.6263 **d.** 0.6263 **e.** $0.6263 - 1$ **f.** $0.6263 - 2$ **g.** 8.3500 **h.** 6.0474 **i.** 3.7448 **j.** 1.4422 **k.** -0.8604 **l.** -3.1630

3. **a.** 2.54 **b.** 49.8 **c.** 1450 **d.** 0.0841 **e.** 0.00455 **f.** 28.6 **g.** 8480 **h.** 0.0241

5. **a.** 8.44 **b.** 10.80 **c.** 0.00402

7. **a.** 3.9073 **b.** 1.2618 **c.** 2.0476 **d.** -0.4516 **e.** 2.3979 **f.** 1.60945 **g.** 1.3863

9. $s(t) = 20{,}000e^{t(\ln 0.75)}$; $s(8) = 2002$

Exercises (pages 270–272)

1. $3e^{3x}$

3. $2xe^{x^2}$

5. $(4x + 3)e^{2x^2+3x+1}$

7. $\dfrac{e^{\sqrt{x+1}}}{2\sqrt{x+1}}$

9. $\dfrac{e^{x\sqrt{x-1}}(3x-2)}{2\sqrt{x-1}}$

11. $\dfrac{xe^{\sqrt{x^2-1}}}{\sqrt{x^2-1}}$

13. $10^x(1 + x \ln 10)$

15. $\dfrac{xe^{\sqrt{x}}(4+\sqrt{x})}{2}$

17. $2a^{4x}[x + 2(x^2 + 1) \ln a]$

19. $\dfrac{a^x \ln a}{(1 + a^x)^2}$

21. $-\dfrac{600e^{0.3x}}{(1 + 10e^{0.3x})^2}$

23. **a.** 16 **b.** 13 **c.** 7

25. **a.** 405 **b.** 547 **c.** 996

27. **a.** −1471.52 **b.** −541.34 **c.** −328.34

29. **a.** 0.15163 **b.** 0.11809 **c.** 0.09197

31. **a.** −44449.09 **b.** −32928.70 **c.** −13387.81

33. **a.** −2.46 **b.** −2.01 **c.** −1.10

35. $p = 25$; $d(25) = 1839$

Exercises (pages 276–278)

1. $\dfrac{1}{x+1}$

3. $\dfrac{1}{2(x+1)}$

5. $\dfrac{3x-2}{2x(x-1)}$

7. $\dfrac{1}{x \ln x}$

9. $x(1 + 2 \ln 3x)$

11. $1 + \dfrac{1}{x}$

13. $1 + \ln x$

15. $\dfrac{1}{1 + e^x}$

17. $\dfrac{xe^x \ln x - e^x + 1}{2(\ln x)^2(x)}$

19. $-\dfrac{600e^{0.3x}}{(1 + 10e^{0.3x})^2}$

21. $\dfrac{1 - 2 \ln x}{x^3}$

23. $\dfrac{2}{1 - x^2}$

25. $-\dfrac{5}{2(x+4)(x-1)}$

27. $\dfrac{x+2}{\sqrt{x+1}} + \dfrac{x}{\sqrt{x+1}} - \dfrac{x(x+2)}{2(x+1)\sqrt{x+1}}$

29. $\dfrac{x(x+2)^2}{\sqrt{x+3}}\left[\dfrac{1}{x} + \dfrac{2}{x+2} - \dfrac{1}{2(x+3)}\right]$

31. $\dfrac{\sqrt{x+1}\,\sqrt{x+2}}{(3x+2)^2}\left[\dfrac{1}{2(x+1)} + \dfrac{1}{2(x+2)} - \dfrac{6}{3x+2}\right]$

33. $\sqrt{\dfrac{x+2}{x^2+1}}\left[\dfrac{1}{2(x+2)} - \dfrac{x}{x^2+1}\right]$

35. $(x+1)^{x+1}\,[1 + \ln(x+1)]$

37. $(2x+3)^x\left[\dfrac{2x}{2x+3} + \ln(2x+3)\right]$

39. $(\ln x)^{\ln x}\,\dfrac{1}{x}\,[1 + \ln(\ln x)]$

41. relative minimum at $x = 1$; $f(1) = 1$

43. relative maximum at $x = 1$; $f(1) = e^{-1}$

45. relative minimum at $x = \frac{1}{2}$; $f(\frac{1}{2}) = e^{-1/4}$

Exercises for Review (pages 278–280)

2. **a.** $x = 32$ **b.** $x = 5$ **c.** $x = 4$ **d.** $x = \frac{1}{10}$ **e.** $x = \frac{1}{2}$ **f.** $x = 5$

3. **a.** 0.7781 **b.** 1.4771 **c.** 0.8751 **d.** 0.1505 **e.** 0.6825 **f.** 1.4651

4. **a.** 3.7110 **b.** 2.5132 **c.** 1.3856 **d.** −1.6144 **e.** 6.3404 **f.** 3.4595 **g.** −1.4524 **h.** −6.0576

5. **a.** 70.40 **b.** 3.57 **c.** 0.0538 **d.** 5.29 **e.** 45.6 **f.** 0.567

6. **a.** 1.4314 **b.** 0.5682 **c.** 0.2177 **d.** 2.1163 **e.** 0.6015 **f.** 1.7952

7. **a.** 20,000 **b.** 18,097 **c.** 16,375 **d.** 13,406

8. $S(1) = 16{,}484$; $S(2) = 27{,}534$; $S(3) = 34{,}940$; $S(4) = 39{,}905$; $S(5) = 43{,}233$

9. **a.** \$6341.21 **b.** \$6356.08 **c.** \$6356

10. \$5391.95

11. 11.55 years

12. **a.** $4e^{4x}$

b. $\dfrac{1}{2\sqrt{x}}e^{\sqrt{x}}$

c. $e^{2x}(1 + 2x)$

d. $\dfrac{e^x - e^{3x}}{(1 + e^{2x})^2}$

e. $xe^{3x}(2 + 3x) + e^{-x}(1 - x)$

f. $\dfrac{2x}{x^2 + 4}$

g. $\dfrac{1 + e^x}{x + e^x}$

h. $\dfrac{3}{2(3x + 2)}$

i. $\dfrac{2}{1 + e^{2x}}$

j. $\dfrac{1}{x \ln x}$

k. $\dfrac{2(\ln x)}{x}$

l. $\dfrac{x}{x + 1} + \ln(x + 1)$

m. $(x + 1)^x \left[\dfrac{x}{x + 1} + \ln(x + 1)\right]$

n. $\dfrac{1}{2(x + 1)^{1/2}(x + 2)^{3/2}}$

o. $\dfrac{x\sqrt{x + 2}}{2x + 3}\left[\dfrac{1}{x} + \dfrac{1}{2(x + 2)} - \dfrac{2}{2x + 3}\right]$

p. $\dfrac{(2x + 1)^2(x - 1)^3}{\sqrt{3x + 4}}\left[\dfrac{4}{2x + 1} + \dfrac{3}{x - 1} - \dfrac{3}{2(3x + 4)}\right]$

13. **a.** 60.74 **b.** 81.99

14. **a.** 11.11 **b.** 8.23

15. $p = 20$; $d(20) = 368$

Chapter 6

Exercises (pages 290–293)

1. $x^4 + c$

3. $-\dfrac{2}{3x^3} + c$

5. $(\ln x) - \dfrac{1}{x} + c$

7. $\dfrac{t^3}{3} + \dfrac{t^2}{2} + c$

9. $\dfrac{x^3}{3} + 2x^2 + 4x + c$

11. $\dfrac{x^3}{3} + \ln x + c$

13. $\dfrac{x^3}{3} - \dfrac{x^2}{2} + \dfrac{x}{4} + c$

15. $\dfrac{x^3}{3} - \dfrac{1}{x} + e^x + c$

17. $\dfrac{e^x - e^{-x}}{2} + c$

19. $\dfrac{t^2}{2} + c$

21. $\frac{xt^2}{2} + c$

23. $\frac{x^2t^3}{3} + c$

25. $\frac{t^2x^4}{4} + c$

27. $\frac{t^3}{3} - \frac{1}{t} + 2t + c$

29. $x^2 - 3x$

31. $(x + 2)^3 + 1$

33. $\frac{x^3}{3} - \frac{3x^2}{2} + 2x + \frac{8}{3}$

35. $128t - 16t^2 + c$

37. $100t + 8t^{3/2} + 7000$

39. $500x - \frac{3}{2}x^2 + 3000$; $500 - \frac{3}{2}x + \frac{3000}{x}$

41. $200x - 0.0002x^2 + \frac{0.01x^3}{3}$

43. $72 - 10t + \frac{8}{t+1}$

45. $36t - 24\sqrt{t} + 5$

47. \$1500

Exercises (page 303)

1. 12

3. 10

5. 9

7. $\frac{16}{3}$

9. $\frac{37}{3}$

11. $\frac{112}{3}$

Exercises (pages 310–313)

1. $\frac{3}{2}$

3. 14

5. 27

7. $\frac{21}{2}$

9. $\frac{128}{3}$

11. $\frac{27}{2}$

13. $\frac{16}{3}$

15. 12

17. 750

19. $\frac{310}{3}$

21. $3(e^2 - 1)$

23. $4(e^2 - e^{-1}) + 3$

31. $\frac{98}{3}$

33. $2e(e^2 - 1)$

35. $\frac{14}{3}$

37. **a.** 468 feet
b. 2388 feet

39. **a.** $12t^3 + 24t^{5/2} + 10{,}000$
b. 1536

41. 156

43. 1921

45. **a.** No **b.** 5 years

47. **a.** \$30 per year
b. \$46 per year

Exercises (pages 322–323)

1. $\frac{197}{6}$
3. 8
5. $\frac{1}{6}$
7. $\frac{20}{3}$
9. $\frac{13}{6}$
11. $\frac{243}{16}$
13. $\frac{8}{3}$
15. $\frac{44}{3}$
17. $\frac{1}{3}$
19. $\frac{3}{4} - e^{-1/2}$
21. Consumer's surplus, 180; Producer's surplus, 135.
23. Consumer's surplus, $\frac{15}{2}$; Producer's surplus, $\frac{35}{2}$
25. Consumer's surplus, 6000; Producer's surplus, 5760.

Exercises (pages 328–330)

1. $\frac{2}{3}(x + 4)^{3/2} + c$
3. $\frac{1}{6}(4x + 3)^{3/2} + c$
5. $\frac{2}{5}(5x + 6)^{1/2} + c$
7. $\frac{1}{21}(x^3 + 5)^7 + c$
9. $-\frac{1}{3}(1 - x^2)^{3/2} + c$
11. $\frac{2}{3}(x^4 + 2)^{3/2} + c$
13. $\frac{1}{4}\ln(x^4 + 2) + c$
15. $\frac{1}{2}e^{x^2} + c$
17. $\frac{1}{3}e^{3x} + c$
19. $\dfrac{3^x}{\ln 3} + c$
21. $2e^{\sqrt{x+1}} + c$
23. $\frac{1}{3}(x^2 + 1)^{3/2} + c$
25. $-\dfrac{1}{\ln x} + c$
27. $\ln(1 + e^x) + c$
29. $\ln(e^x - e^{-x}) + c$
31. $\frac{65}{12}$
33. $\frac{98}{3}$
35. $4(e^2 - 1)$
37. $\frac{1}{4}(1 - e^{-4})$
39. $\frac{1}{4}[16 - (e^{-1} + 1)^4]$
41. $\frac{2}{3}(x^2 + 100)^{3/2} + \frac{73000}{3}$
43. 4482; 20086
45. \$1426.84

Exercises (page 333)

1. $\dfrac{x^2 \ln|x|}{2} - \dfrac{x^2}{4} + c$
3. $xe^x + c$
5. $\dfrac{xe^{2x}}{2} - \dfrac{e^{2x}}{4} + c$
7. $\dfrac{x^2e^{2x}}{2} - \dfrac{xe^{2x}}{2} + \frac{1}{4}e^{2x} + c$
9. $-\frac{1}{4}e^{-2x}(2x + 1) + c$
11. $\frac{1}{49}(1 - 8e^{-7})$
13. $\dfrac{8 \ln 2 - 3}{4}$
15. $\dfrac{44\sqrt{2} - 28}{3}$
17. $\dfrac{8448\sqrt{2} + 4}{35}$
19. 479.80

Exercises (page 338)

1. a. 0.5 b. 0.5
3. a. 10.50 b. 10.50
5. a. 8.75 b. 8.67
7. a. 0.26563 b. 0.25000
9. a. 0.64328 b. 0.65653
11. a. 0.69702 b. 0.69325
13. a. 0.7828 b. 0.78539
15. a. 1.23066 b. 1.21884

Exercises (pages 346–349)

5. $y = x^3 + x^2 + 5x + c$
7. $y = \frac{1}{2}x^2 - e^{-x} + c$
9. $y = e^t(t - 1) + c$
11. $y = \frac{2}{3}(t^2 + 1)^{3/2} + c$
13. $y = \frac{1}{2}x^2 + x - 15$
15. $y = xe^x - e^x + 5$
17. $y = cx$
19. $\ln|y| = \frac{1}{2}x^2 + c$
21. $2y^2 + \frac{3}{2}x^2 = c$
23. $x^2 + y^2 = c$
25. $xy = c$
27. $e^{-y} + e^{-x} = c$
29. $\frac{1}{x} - e^{-y}(1 + y) = 0$
31. 6.1837
33. 139,744
35. 283,000
37. 311.01 millions
39. a. $Q = Q_0e^{-kt}$
    b. $t = \frac{1}{k}\ln 2$
41. a. $x = \frac{800}{1 + 799e^{-800kt}}$
    b. 51 persons

Exercises for Review (pages 349–353)

3. a. $\frac{xt^2}{2} + c$
   b. $\frac{x^2t}{2} + c$
   c. $\frac{x^3}{3} - 2x^2 + \ln|x| + c$
   d. $e^x + \frac{2}{3}x^{3/2} + c$
   e. $\frac{1}{3}(2x + 5)^{3/2} + c$
   f. $\frac{2}{3}(3x + 4)^{1/2} + c$
   g. $\frac{1}{3}\ln|3x + 4| + c$
   h. $e^{x^2 + 3x} + c$
   i. $\frac{2}{5}(x - 1)^{5/2} + \frac{2}{3}(x - 1)^{3/2} + c$

j. $\ln(1 + e^x) + c$

k. $\frac{1}{3}(x^2 + 4)^{3/2} + c$

l. $\dfrac{(e^x + x)^3}{3} + c$

m. $\frac{1}{2}x^2 \ln|x| - \frac{1}{4}x^2 + c$

n. $xe^x - e^x + c$

o. $-\frac{1}{2}xe^{-2x} - \frac{1}{4}e^{-2x} + c$

p. $\frac{1}{6}(1 + x)^6 - \frac{1}{5}(1 + x)^5 + c$

4. **a.** $\frac{11}{6}$
b. $\frac{74}{3}$
c. 24
d. $3(e^5 - e^2) + 12$
e. $\frac{125}{6}$
f. 4
g. 61
h. $2e^{-1}$
i. $\frac{25}{2}\ln\frac{5}{3} - \frac{9}{2}\ln 3 - 4$
j. $\frac{64}{3}$

5. **a.** $\frac{11}{2}$
b. 14
c. $\frac{2}{3}$
d. $\frac{56}{3}$

6. **a.** (i) 0.85476 (ii) 0.84762
b. (i) 1.21876 (ii) 1.21895
c. (i) 5.67241 (ii) 5.65275

7. **a.** $y = x^3 + x^2 + x + c$
b. $y = -e^{-t} + c$
c. $y = xe^x - e^x + c$
d. $\frac{1}{2}y^2 = xe^x - e^x + 2x + c$
e. $y = y_0e^{kx}$
f. $-\dfrac{1}{2y^2} = \frac{1}{3}x^3 + c$
g. $y = c(x + 1)^2$
h. $\ln(y^2 + 1) + \dfrac{1}{x} = c$
i. $x^2 - y^3 = c$
j. $xy = c$

8. **a.** $y = x^4 + x^3 + x^2 + x + 4$
b. $2y^3 = 3x^2 + 42$

9. **a.** 18
b. $\frac{32}{3}$
c. $\frac{9}{2}$
d. $\frac{4}{3}$
e. 8

10. $y = x^3 - x^2 + x + 6$

11. $C(x) = 30x + 2x^2 - \frac{1}{300}x^3 + 500$

12. $\frac{76}{3}$ miles

13. $P(t) = 200t + 180t^{5/3} + 15{,}000$; 22,360

14. \$250

15. 5000 students

16. 30.58 tons

17. Consumer's surplus $-\$4500$; Producer's surplus $-\$3375$.

18. **a.** \$3324.41 **b.** \$3655.71

19. 40,774

20. 4.99 hours

21. 33.22 years

22. $Q(t) = 0.3e^{-0.06t}$; $Q(5) = 0.2222$ grams

23. **a.** $x(t) = \dfrac{1200}{1 + 1199e^{-1200kt}}$ **b.** 302 residents

Chapter 7

Exercises (pages 363–364)

1.

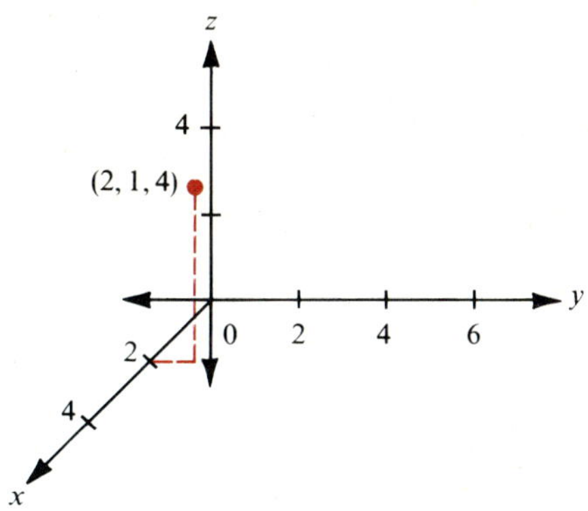

3.

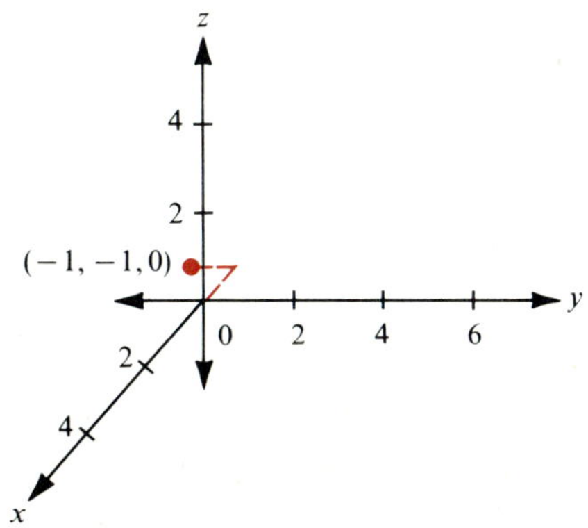

5.

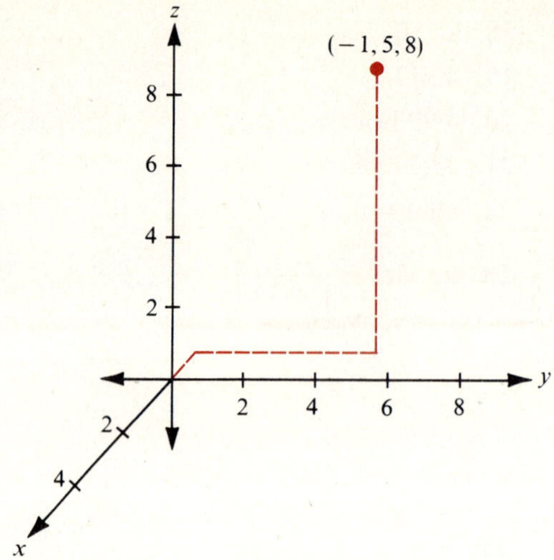

7.

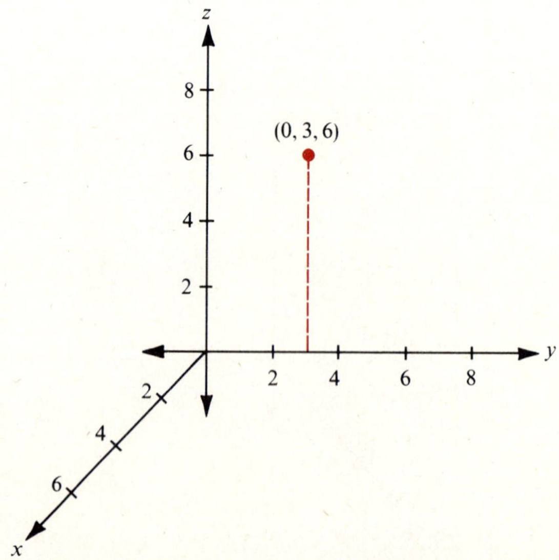

9.

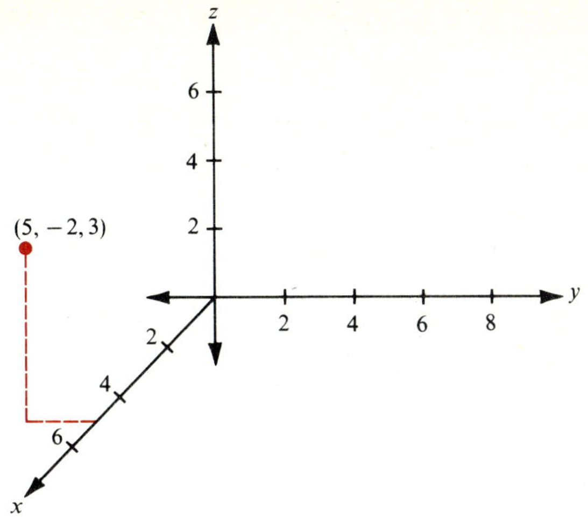

11.

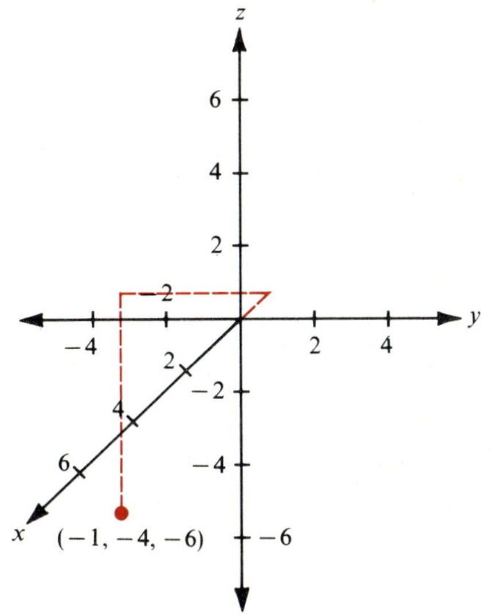

13. $\sqrt{29}$

15. 7

17. $\sqrt{62}$

19. $\sqrt{158}$

21. **a.** $(x-1)^2 + (y-2)^2 + (z-3)^2 = 16$
b. yes
c. no

23. **a.** $(x-2)^2 + (y+1)^2 + (z-4)^2 = 25$

25.

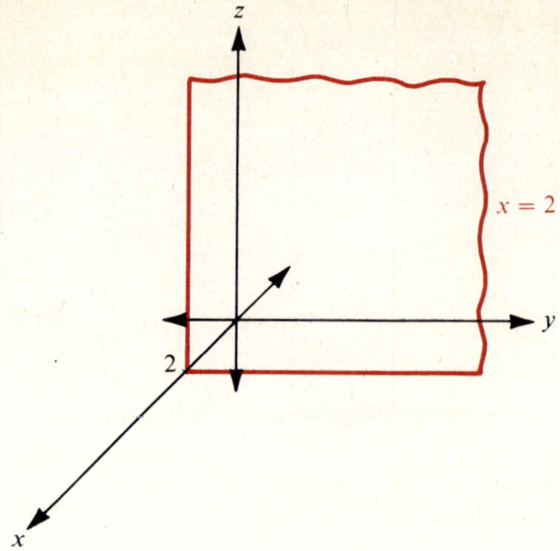

27.

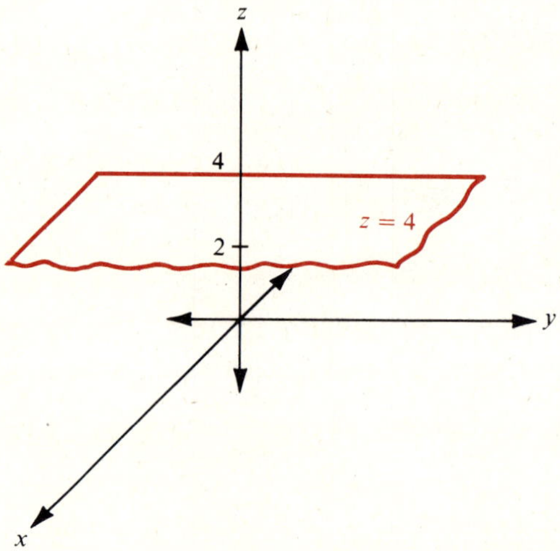

29.

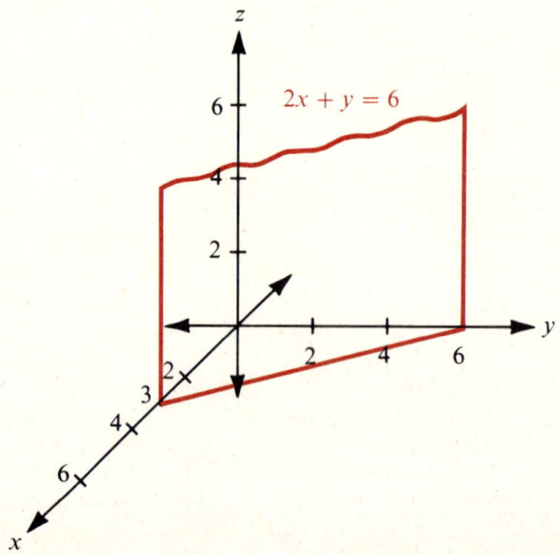

31.

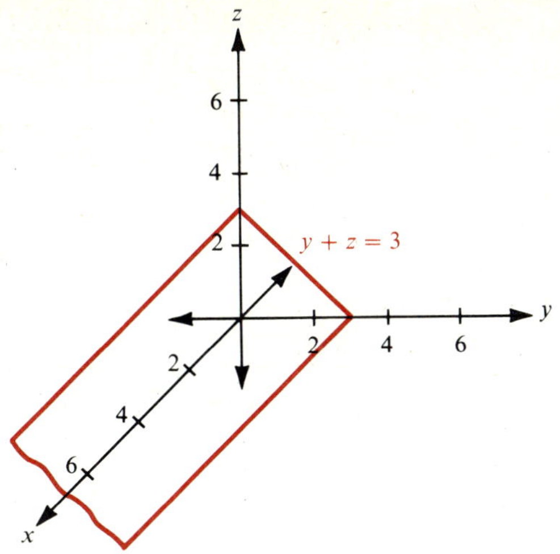

33.

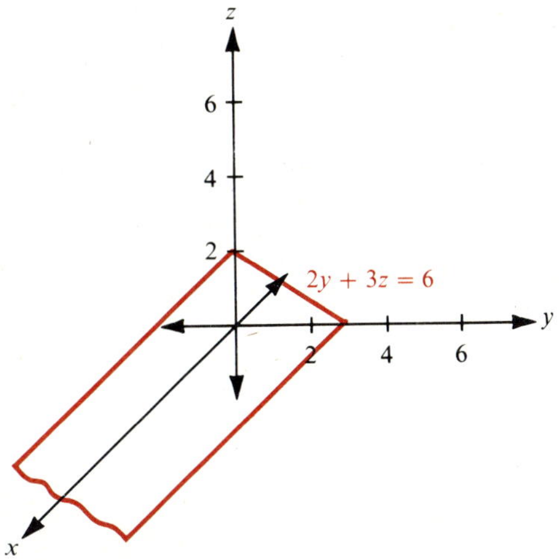

35.

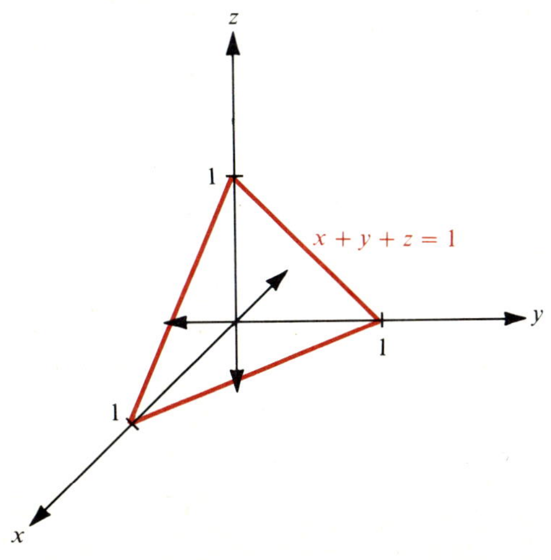

37.

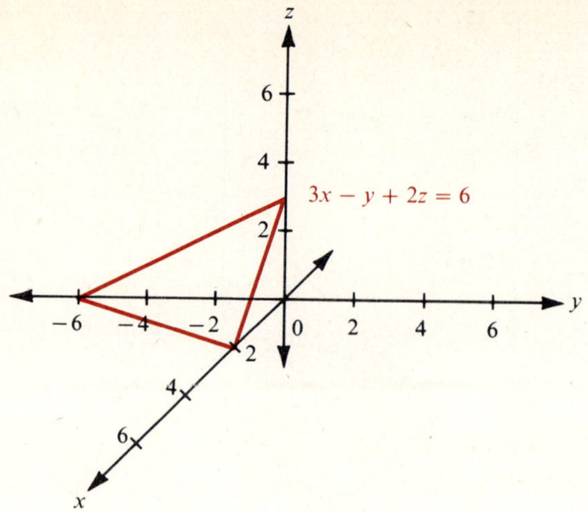

39.

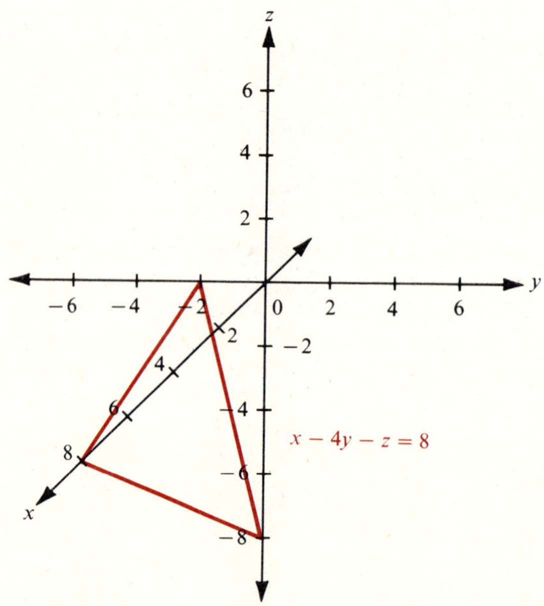

Exercises (pages 370–371)

1. **a.** $f(0, 0) = -4$
b. $f(1, 1) = 1$
c. $f(2, 1) = 3$
d. $f(3, 4) = 14$
e. $f(-1, -3) = -15$
f. $f(-2, 4) = 4$

3. **a.** $f(0, 0) = 0$
b. $f(1, -1) = \sqrt{2}$
c. $f(2, 1) = \sqrt{5}$
d. $f(3, 4) = 5$
e. $f(0, 5) = 5$
f. $f(5, 12) = 13$

5. a. $h(0,0,1) = 5$
   b. $h(1,1,-1) = 8$
   c. $h(0,2,3) = 21$
   d. $h(-1,2,-3) = 22$
   e. $h(2,3,-4) = 42$
   f. $h(5,-1,4) = 47$

7. a. $390,000
   b. $1,540,000

9.

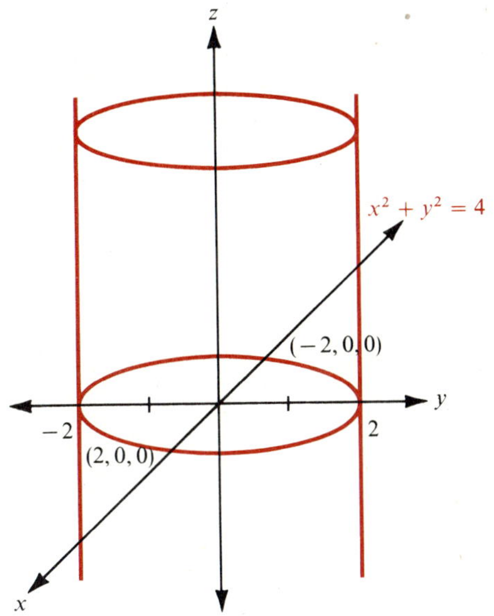

11.

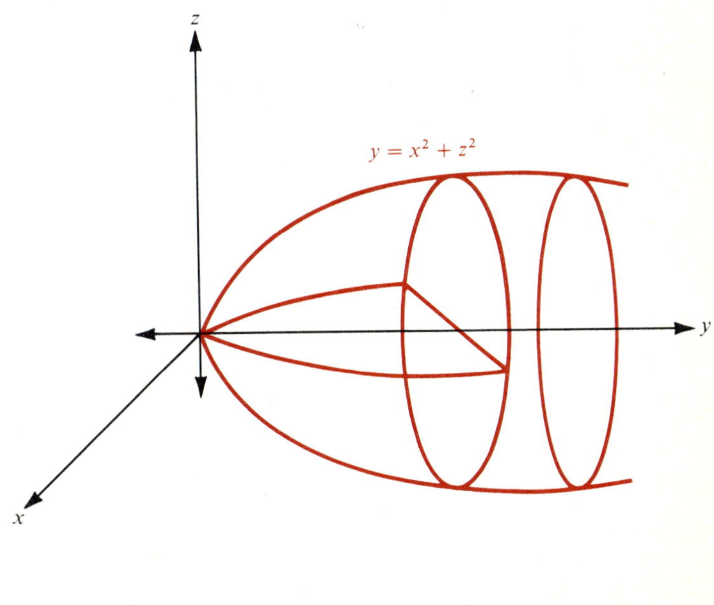

13.

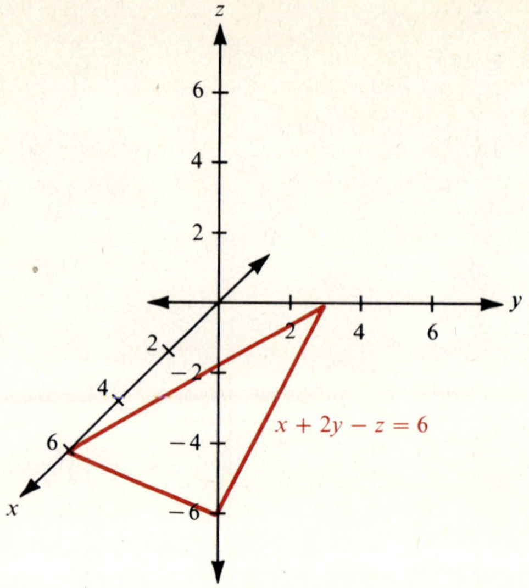

15.

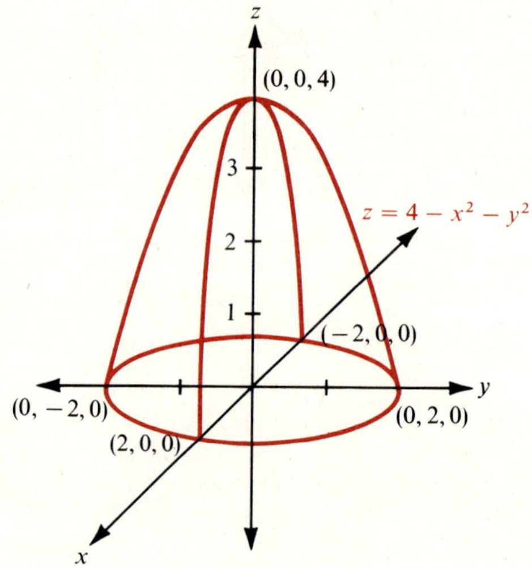

Exercises (pages 376–378)

1. 8

3. 0

5. −18

7. 2

9. $\frac{2}{3}$

11. $\frac{25}{7}$

13. $-\frac{61}{25}$

15. If the origin $(0,0)$ is approached along a straight line $y = mx$, $m \neq 0$, then

$$f(x, y) = \frac{x(mx)}{x^2 + mx^2} = \frac{m}{1 + m^2}.$$

Since $f(x, y)$ depends on m, it follows that $\lim_{(x,y)\to(0,0)} f(x, y)$ does $\lim_{(x,y)\to(0,0)} f(x, y)$ does not exist.

17. continuous at all points

19. continuous at all points

21. not continuous along $y = x$ and at the point $(3, 2)$

23. discontinuous along $y = -x$

25. If the origin $(0,0)$ is approached along a straight line $y = mx$, $m \neq 0$, then

$$f(x, y) = \frac{x^2 - y^2}{x^2 + y^2} = \frac{x^2 - m^2x^2}{x^2 + m^2x^2} = \frac{1 - m^2}{1 + m^2}.$$

Since $f(x, y)$ depends on m, it follows that $\lim_{(x,y)\to(0,0)} f(x, y)$ does not exist. Hence, the function is discontinuous at the origin.

Exercises (pages 384–386)

1. $\dfrac{\partial z}{\partial x} = 2x; \dfrac{\partial z}{\partial y} = 1$

3. $\dfrac{\partial z}{\partial x} = 2xy; \dfrac{\partial z}{\partial y} = x^2$

5. $\dfrac{\partial z}{\partial x} = 2xy^2; \dfrac{\partial z}{\partial y} = 2x^2y$

7. $\dfrac{\partial z}{\partial x} = 8(2x - y)^3; \dfrac{\partial z}{\partial x} = -4(2x - y)^3$

9. $\dfrac{\partial z}{\partial x} = \dfrac{1}{y}; \dfrac{\partial z}{\partial x} = -\dfrac{x}{y^2}$

11. $\dfrac{\partial z}{\partial x} = \dfrac{\sqrt{y}}{2\sqrt{x}}; \dfrac{\partial z}{\partial x} = \dfrac{\sqrt{x}}{2\sqrt{y}}$

13. $\frac{\partial z}{\partial x} = \log y; \frac{\partial z}{\partial y} = \frac{x}{y}$

15. $\frac{\partial z}{\partial x} = \frac{2x}{x^2 + y^2}; \frac{\partial z}{\partial y} = \frac{2y}{x^2 + y^2}$

17. $\frac{\partial z}{\partial x} = -ye^{-xy}; \frac{\partial z}{\partial y} = -xe^{-xy}$

19. $\frac{\partial z}{\partial x} = e^y + 3x^2y^2; \frac{\partial z}{\partial y} = xe^y + 2x^3y$

21. $f_x(1, 1) = 2; f_y(1, 1) = -2$

23. $f_x(5, 2) = 4; f_y(5, 2) = -11$

25. $f_x(2, 1) = 27; f_y(2, 1) = -8$

27. $f_x(3, 1) = \frac{3}{5}; f_y(3, 1) = \frac{2}{5}$

29. $f_x(2, -3) = \frac{72}{169}; f_y(2, -3) = \frac{48}{169}$

31. $\frac{\partial z}{\partial x} = \frac{x}{x^2 + y^2}; \frac{\partial z}{\partial y} = \frac{y}{x^2 + y^2}$

$$\left(\frac{\partial z}{\partial x}\right)^2 + \left(\frac{\partial z}{\partial y}\right)^2 = \frac{x^2 + y^2}{(x^2 + y^2)^2} = \frac{1}{x^2 + y^2}$$

33. $\frac{\partial z}{\partial x} = 3; \frac{\partial z}{\partial y} = 0.04y$

35. $\frac{\partial R}{\partial x} = \frac{50}{1 + 50x + 75y} + \frac{20}{\sqrt{1 + 40x + 125y}}$

$$\frac{\partial R}{\partial y} = \frac{75}{1 + 50x + 75y} + \frac{125}{2\sqrt{1 + 40x + 125y}}$$

37. **a.** $P_x(4, 2) = 22; P_y(4, 2) = 4$
b. $P_x(5, 1) = 22; P_y(5, 1) = 8$
c. $P_x(7, 3) = 30; P_y(7, 3) = 8$
d. $P_x(2, 6) = 26; P_y(2, 6) = -8$

Exercises (pages 390–391)

1. $\frac{\partial^2 z}{\partial x^2} = 0, \frac{\partial^2 z}{\partial y^2} = 2x, \frac{\partial^2 z}{\partial x\, \partial y} = \frac{\partial^2 z}{\partial y\, \partial x} = 2y$

3. $\frac{\partial^2 z}{\partial x^2} = 6xy^2, \frac{\partial^2 z}{\partial y^2} = 2x^3, \frac{\partial^2 z}{\partial x\, \partial y} = \frac{\partial^2 z}{\partial y\, \partial x} = 6x^2y$

5. $\frac{\partial^2 z}{\partial x^2} = 6x - 6y, \frac{\partial^2 z}{\partial y^2} = 6y, \frac{\partial^2 z}{\partial x\, \partial y} = \frac{\partial^2 z}{\partial y\, \partial x} = -6x$

7. $\dfrac{\partial^2 z}{\partial x^2} = 12x^2 + 6y^2$, $\dfrac{\partial^2 z}{\partial y^2} = 6x^2 + 6y$, $\dfrac{\partial^2 z}{\partial x\,\partial y} = \dfrac{\partial^2 z}{\partial y\,\partial x} = 12xy$

9. $\dfrac{\partial^2 z}{\partial x^2} = 0$, $\dfrac{\partial^2 z}{\partial y^2} = \dfrac{2x}{y^3}$, $\dfrac{\partial^2 z}{\partial x\,\partial y} = \dfrac{\partial^2 z}{\partial y\,\partial x} = -\dfrac{1}{y^2}$

11. $\dfrac{\partial^2 z}{\partial x^2} = \dfrac{y^2}{(x^2+y^2)^{3/2}}$, $\dfrac{\partial^2 z}{\partial y^2} = \dfrac{x^2}{(x^2+y^2)^{3/2}}$,

$\dfrac{\partial^2 z}{\partial x\,\partial y} = \dfrac{\partial^2 z}{\partial y\,\partial x} = -\dfrac{xy}{(x^2+y^2)^{3/2}}$

13. $\dfrac{\partial^2 z}{\partial x^2} = \dfrac{4y}{(x-y)^3}$, $\dfrac{\partial^2 z}{\partial y^2} = \dfrac{4x}{(x-y)^3}$, $\dfrac{\partial^2 z}{\partial x\,\partial y} = \dfrac{\partial^2 z}{\partial y\,\partial x} = -\dfrac{2(x+y)}{(x-y)^3}$

15. $\dfrac{\partial^2 z}{\partial x^2} = -\dfrac{9}{(3x+4y)^2}$, $\dfrac{\partial^2 z}{\partial y^2} = -\dfrac{16}{(3x+4y)^2}$,

$\dfrac{\partial^2 z}{\partial x\,\partial y} = \dfrac{\partial^2 z}{\partial y\,\partial x} = -\dfrac{12}{(3x+4y)^2}$

17. $\dfrac{\partial^2 z}{\partial x^2} = \dfrac{6xy^3 - 3x^4}{(x^3+y^3)^2}$, $\dfrac{\partial^2 z}{\partial y^2} = \dfrac{6yx^3 - 3y^4}{(x^3+y^3)^2}$,

$\dfrac{\partial^2 z}{\partial x\,\partial y} = \dfrac{\partial^2 z}{\partial y\,\partial x} = -\dfrac{9x^2y^2}{(x^3+y^3)^2}$

19. $\dfrac{\partial^2 z}{\partial x^2} = y^2 e^{xy}$, $\dfrac{\partial^2 z}{\partial y^2} = x^2 e^{xy}$, $\dfrac{\partial^2 z}{\partial x\,\partial y} = \dfrac{\partial^2 z}{\partial y\,\partial x} = e^{xy}(1+xy)$

21. $\dfrac{\partial^2 z}{\partial x\,\partial y} = \dfrac{\partial^2 z}{\partial y\,\partial x} = 12xy^2 - 15x^2y^4$

23. $\dfrac{\partial^2 z}{\partial x\,\partial y} = \dfrac{\partial^2 z}{\partial y\,\partial x} = -\dfrac{9x^2y^2}{4(x^3+y^3)^{3/2}}$

25. $f_{xy} = f_{yx} = z$, $f_{xz} = f_{zx} = y$, $f_{yz} = f_{zy} = x$, $f_{xx} = f_{yy} = f_{zz} = 0$

27. $f_{xy} = f_{yx} = 0$, $f_{zx} = f_{xz} = -\dfrac{1}{z^2}$, $f_{yz} = f_{zy} = -\dfrac{1}{z^2}$,

$f_{xx} = f_{yy} = 0$, $f_{zz} = \dfrac{2(x+y)}{z^3}$

29. a. $\dfrac{\partial^2 u}{\partial x^2} = 6x$, $\dfrac{\partial^2 u}{\partial y^2} = -6x$; $\dfrac{\partial^2 u}{\partial x^2} + \dfrac{\partial^2 u}{\partial y^2} = 0$

b. $\dfrac{\partial^2 v}{\partial x^2} = 6y$, $\dfrac{\partial^2 v}{\partial y^2} = -6y$; $\dfrac{\partial^2 v}{\partial x^2} + \dfrac{\partial^2 v}{\partial y^2} = 0$

31. $\dfrac{\partial^2 u}{\partial x^2} = 2$, $\dfrac{\partial^2 u}{\partial y^2} = 2$, $\dfrac{\partial^2 u}{\partial z^2} = -4$

$\dfrac{\partial^2 u}{\partial x^2} + \dfrac{\partial^2 u}{\partial y^2} + \dfrac{\partial^2 u}{\partial z^2} = 0.$

33. $\dfrac{\partial^2 u}{\partial x^2} = \dfrac{2x^2 - y^2 - z^2}{(x^2 + y^2 + z^2)^{5/2}}$; $\dfrac{\partial^2 u}{\partial y^2} = \dfrac{2y^2 - x^2 - z^2}{(x^2 + y^2 + z^2)^{5/2}}$;

$\dfrac{\partial^2 u}{\partial z^2} = \dfrac{2z^2 - x^2 - y^2}{(x^2 + y^2 + z^2)^{5/2}}$;

$\dfrac{\partial^2 u}{\partial x^2} + \dfrac{\partial^2 u}{\partial y^2} + \dfrac{\partial^2 u}{\partial z^2} = 0$

Exercises (page 398)

1. relative minimum at $(1, -2)$; $f(1, -2) = -7$

3. relative minimum at $(-\frac{10}{3}, \frac{11}{3})$; $f(-\frac{10}{3}, \frac{11}{3}) = -\frac{111}{9}$

5. relative minimum at $(15, -8)$; $f(15, -8) = 6$

7. relative minimum at $(-8, -23)$; $f(-8, -23) = 57$

9. relative minimum at $(0, 0)$; $f(0, 0) = 0$

11. saddle point at $(-2, 1)$; $f(-2, 1) = 4$

13. saddle point at $(0, 0)$; $f(0, 0) = 0$,
relative minimum at $(3, 3)$; $f(3, 3) = -27$

Exercises (pages 404–405)

1. maximum occurs at $(150, 200)$; $f(150, 200) = 30{,}000$

3. maximum occurs at $(\frac{5}{9}, -\frac{5}{3})$; $f(\frac{5}{9}, -\frac{5}{3}) = \frac{500}{81}$

5. maximum occurs at $(3, 5)$; $f(3, 5) = 44$

7. maximum occurs at $(\frac{50}{3}, \frac{50}{3}, \frac{50}{3})$; $f(\frac{50}{3}, \frac{50}{3}, \frac{50}{3}) = \frac{125{,}000}{27}$

9. maximum occurs at $\left(\dfrac{1}{\sqrt{3}}, \dfrac{1}{\sqrt{3}}, \dfrac{1}{\sqrt{3}}\right)$, $\left(-\dfrac{1}{\sqrt{3}}, -\dfrac{1}{\sqrt{3}}, \dfrac{1}{\sqrt{3}}\right)$,
$\left(-\dfrac{1}{\sqrt{3}}, \dfrac{1}{\sqrt{3}}, -\dfrac{1}{\sqrt{3}}\right)$ and $\left(\dfrac{1}{\sqrt{3}}, -\dfrac{1}{\sqrt{3}}, -\dfrac{1}{\sqrt{3}}\right)$; $f(x, y, z) = \dfrac{1}{3\sqrt{3}}$

11. minimum occurs at $(\frac{50}{3}, \frac{50}{3}, \frac{50}{3})$; $f(x, y, z) = \frac{2500}{3}$

13. length – 100 feet; width – 50 feet.

15. length – 6 meters; width – 6 meters; height – 3 meters.

Exercises (pages 410–412)

1. $\hat{y} = 2.3x - 0.3$; 13.5; 22.7

3. a. $\hat{y} = 4.7105x + 310.7565$ **b.** 367.2825; 381.414

5. a. $\hat{y} = 0.0892x + 4.5542$ **b.** 8.135

7. b. $\hat{y} = 5.2381x + 31.4286$ **c.** 70.7143

9. $\hat{y} = 1.1647x + 13.8827$; 24.5979

Exercises for Review (pages 413–415)

2. a.

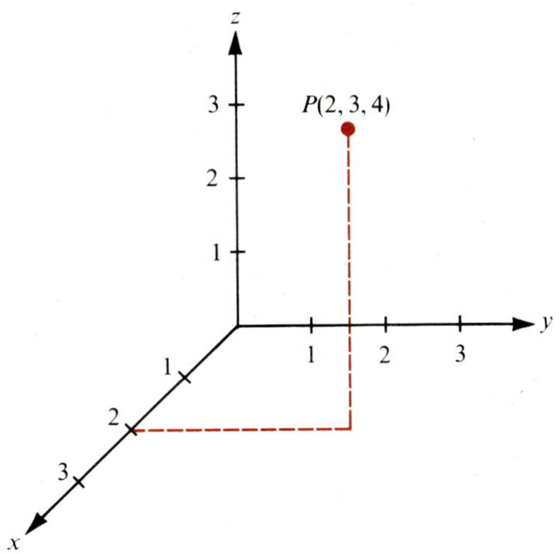

b.

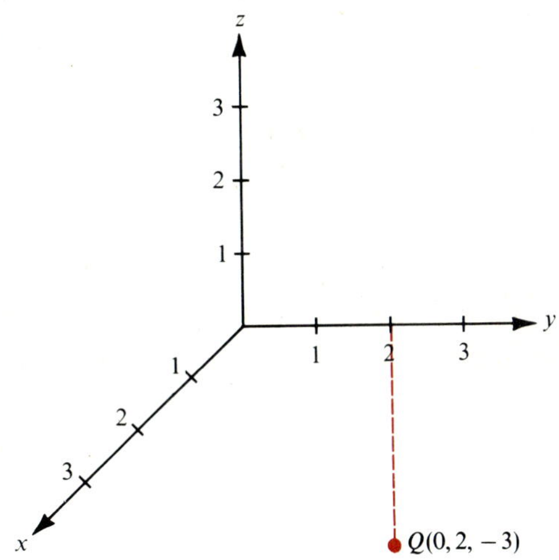

3. $\sqrt{74}$

4. $(x-1)^2+(y+2)^2+(z-3)^2=25$

5. **a.** $f(1,0)=1$ **d.** $f(2,1)=\frac{3}{5}$

b. $f(0,1)=-1$ **e.** $f(3,4)=-\frac{7}{25}$

c. $f(1,1)=0$ **f.** $f(12,5)=\frac{119}{169}$

6. a.

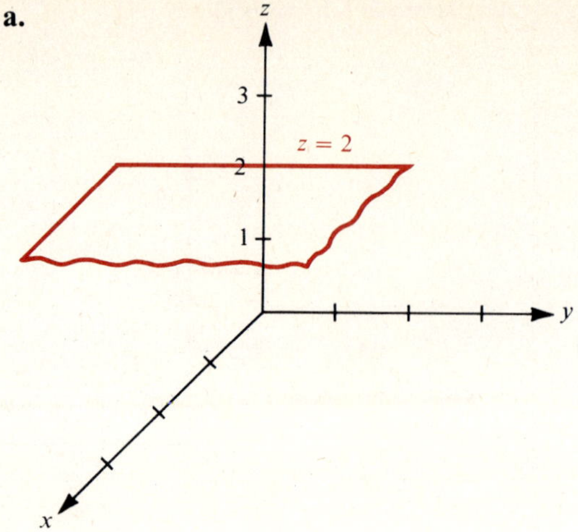

b.

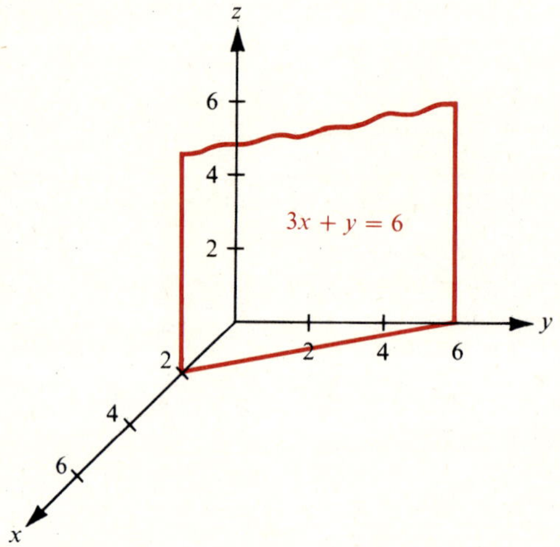

c.

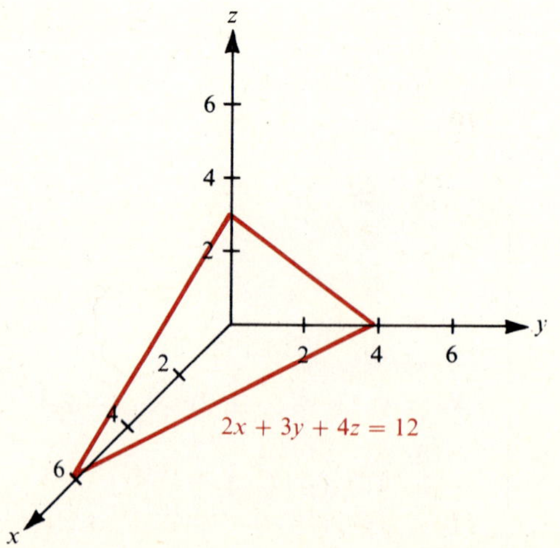

d.

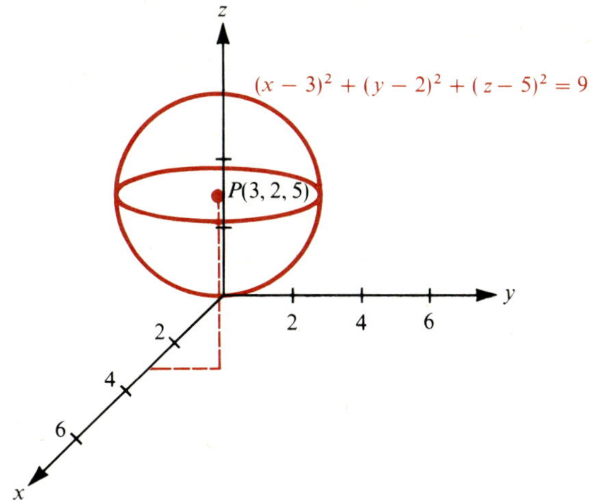

7. a. limit does not exist **b.** limit does not exist

8. a. $f_x(x, y) = 3x^2y^2$; $f_y(x, y) = 2x^3y$
b. $f_x(x, y) = e^y$; $f_y(x, y) = xe^y$
c. $f_x(x, y) = \dfrac{x}{\sqrt{x^2+y^2}} + e^{2y}$; $\dfrac{y}{\sqrt{x^2+y^2}} + 2xe^{2y}$
d. $f_x(x, y) = \dfrac{1}{y} - \dfrac{y}{x^2}$; $f_y(x, y) = -\dfrac{x}{y^2} + \dfrac{1}{x}$

9. a. $f_x(2, 3) = 216$; $f_y(2, 3) = 96$
b. $f_x(1, -1) = -1$; $f_y(1, -1) = 1$
c. $f_x(2, 0) = 12$; $f_y(2, 0) = 8$
d. $f_x(3, 4) = \frac{3}{5}$; $f_y(3, 4) = \frac{4}{5}$

10. $f_{xx} = 2y$; $f_{xy} = f_{yx} = 2x + 2y$; $f_{yy} = 2x$.
b. $f_{xx} = 0$; $f_{xy} = f_{yx} = -\dfrac{4}{y^2}$; $f_{yy} = 6y + \dfrac{8x}{y^3}$
c. $f_{xx} = f_{xy} = f_{yx} = f_{yy} = \dfrac{2}{(x+y)^3}$
d. $f_{xx} = \dfrac{24y^2 - 18x^2}{(3x^2+4y^2)^2}$; $f_{xy} = f_{yx} = \dfrac{-48xy}{(3x^2+4y^2)^2}$;
$f_{yy} = \dfrac{24x^2 - 32y^2}{(3x^2+4y^2)^2}$

11. $f_{xy}(x, y) = f_y(x, y) = 9x^2y^2 + 4x$

12. $f_{xy}(x, y) = f_{yx}(x, y) = \dfrac{x - y}{(x + y)^3}$

13. $P_x(2, 3) = 75$; $P_y(2, 3) = 90$

14. a. $f_x(2, 3) = 4800$; $f_y(2, 3) = 5400$
    b. $f_x(27, 8) = 20$; $f_y(27, 8) = \frac{135}{4}$

15. a. relative maximum at $x = 450$, $y = 450$; $f(450, 450) = 607{,}500$
    b. relative minimum at $x = 4$, $y = 1$; $f(4, 1) = 29$
    c. saddle point at $x = 0$, $y = 0$; $f(0, 0) = 0$
    relative minimum at $x = \frac{2}{3}$, $y = -\frac{2}{3}$; $f(\frac{2}{3}, -\frac{2}{3}) = -\frac{8}{27}$
    d. relative minimum at $x = \frac{8}{3}$, $y = \frac{5}{3}$; $f(\frac{8}{3}, \frac{5}{3}) = -\frac{202}{27}$

16. maximum occurs at $x = y = z = 20$; $f(20, 20, 20) = 8000$

17. maximum occurs at $x = 18$, $y = 9$, $z = 6$; $f(18, 9, 6) = 3888$

18. $x = 11$, $y = 39$

19. length – 30 feet; width – 15 feet.

20. length – 8 feet; width – 8 feet; height – 4 feet.

21. $\hat{y} = 0.455x + 1.663$

22. a. $\hat{y} = 12.954x + 45.641$ b. 89.68

Appendix A

Exercises (pages 424–425)

1. $x^2 + 3x + 2$
3. $x^2 - 11x + 28$
5. $x^2 - 9$
7. $x^2 - 2x - 48$
9. $(x + 2)(x + 3)$
11. $(x - 2)(x - 7)$
13. $(x - 5)(x + 3)$
15. $(x + 2)(x - 15)$
17. $(x - 2)(x + 16)$
19. $(x + 5)(2x + 3)$
21. $(2x + 1)(3x - 5)$
23. $\{-3, -4\}$
25. $\{-4, -7\}$
27. $\{-5, 8\}$
29. $\{-2, 3\}$
31. $\{-\frac{2}{3}, -\frac{5}{2}\}$
33. $\left\{\dfrac{-5 - \sqrt{13}}{2}, \dfrac{-5 + \sqrt{13}}{2}\right\}$
35. $\left\{\dfrac{-5 - \sqrt{17}}{2}, \dfrac{-5 + \sqrt{17}}{2}\right\}$
37. $\{-7, -\frac{2}{3}\}$
39. $\{-2, -\frac{1}{2}\}$
41. $\left\{\dfrac{-1 - \sqrt{7}}{3}, \dfrac{-1 + \sqrt{7}}{3}\right\}$
43. $\left\{\dfrac{-5 - \sqrt{57}}{4}, \dfrac{-5 + \sqrt{57}}{4}\right\}$

Exercises (pages 429–431)

1. $\dfrac{x+2}{x-2}$
3. $\dfrac{x-4}{4}$
5. $\dfrac{x+2}{x-1}$
7. $\dfrac{x+5}{x-3}$
9. $\dfrac{x}{x+2}$
11. $\dfrac{x+6}{x-5}$
13. $\dfrac{x+3}{x+1}$
15. $\dfrac{1}{x}$
17. $\dfrac{x^2+6}{3x}$
19. $-\dfrac{2}{x^2-1}$
21. $\dfrac{x+1-x^2}{x(x+1)}$
23. $\dfrac{x+1}{x-1}$
25. $x = \frac{3}{2}$
27. $x = 30$
29. $x = \frac{29}{7}$
31. $x = 14$
33. $x = 5$
35. $x = -\frac{1}{2}$
37. $x = 2$
39. $x = -\frac{3}{4}$
41. $x = -\frac{1}{3}$
43. $x = -1, x = 12$

Exercises (pages 438–440)

1. **a.** 32 **b.** 27 **c.** 256 **d.** 729 **e.** 26,873,856 **f.** 307,546,875 **g.** $\frac{3}{32}$ **h.** $\frac{5}{4}$

3. **a.** a^8b^4 **b.** $a^6b^9c^{12}$ **c.** a^2b^5 **d.** $\dfrac{1}{a^3bc^6}$ **e.** $\dfrac{d^6}{ca^6}$ **f.** ac^6 **g.** $\dfrac{8b}{81a}$ **h.** $\dfrac{72}{a^6b^6}$

5. **a.** $\dfrac{9y^6}{x^4}$ **b.** $\dfrac{1}{a^3b^{15}c^{12}}$ **c.** $\dfrac{81x^4}{12}$ **d.** $\dfrac{x^2}{y^2}$ **e.** 1 **f.** $\dfrac{27x^6y^6}{64}$

7. **a.** 2 **b.** 4 **c.** -3 **d.** 25 **e.** 81 **f.** $\frac{2}{3}$ **g.** $\frac{4}{9}$ **h.** $\frac{4}{7}$

Exercises (page 442)

1. $\dfrac{1}{\sqrt{6}+\sqrt{5}}$

3. $\dfrac{4}{\sqrt{7}-\sqrt{3}}$

5. $\dfrac{3}{4\sqrt{3}-3\sqrt{5}}$

7. $\dfrac{1}{\sqrt{x+1}-\sqrt{x}}$

9. $\dfrac{1}{\sqrt{x+4}-\sqrt{x+3}}$

11. $\sqrt{3}-\sqrt{2}$

13. $2+\sqrt{3}$

15. $19-6\sqrt{10}$

17. $\sqrt{x+1}+\sqrt{x}$

19. $2x+3+2\cdot\sqrt{(x+1)(x+2)}$

Appendix B

Exercises (pages 438–440)

1. 120

3. 96

5. 806

7. 908

9. 495

11. 7384

13. **a.** $x_1+x_2+x_3$
b. $cx_1+cx_2+cx_3+cx_4+cx_5$
c. $x_1y_1+x_2y_2+x_3y_3+x_4y_4+x_5y_5$
d. $(x_1+3)+(x_2+3)+(x_3+3)+(x_4+3)+(x_5+3)+(x_6+3)$
e. $(x_1+y_1)+(x_2+y_2)+(x_3+y_3)+(x_4+y_4)$
f. $(x_1+y_1+z_1)+(x_2+y_2+z_2)+(x_3+y_3+z_3)$

15. **a.** 12 **b.** -5 **c.** 50 **d.** 110 **e.** 42 **f.** 19

Index

L

M

N

O

P

Q

R

S

Your chance to rate

A Short Course in Calculus with applications to the management, life and social sciences.

In order to keep this text responsive to your needs, it would help us to know what you, the student, thought of *A Short Course in Calculus with applications to management, life and social sciences.* We would appreciate if you answer the following questions. Then cut out the page, fold, seal, and mail it; thanks for your help.

Which chapters did you cover?
(circle) 1 2 3 4 5 6 7 All

Which helped most?
Explanations ____ Examples ____ Exercises ____
All three ____ Other ________________ (fill in)

Does the book have enough worked-out examples? Yes ____ No ____

Were the answers at the back of the book helpful? Yes ____ No ____

For you, was the course elective? Yes ____ No ____

Do you plan to take more mathematics courses? Yes ____ No ____

If yes, which ones?
Finite mathematics ____ Statistics ____ Trigonometry ____
Calculus (engineering and physics) ____
Others ________________ (fill in)

How much algebra did you have before this course?
Years in high school (circle) 0 ½ 1 1½ 2 more
Courses in college (circle) 0 1 2 3

If you had algebra before, how long ago?
Last 2 years ____ 3–5 years ____ more than 5 years ____

What is your major or career goal? ________________
Your age? ________________

We would appreciate knowing of any errors you found in the book.
Please supply page numbers and the error therein.

What did you like most about the book?

What did you like least about the book?

College ________________________ State ____ Zip code ____

Fold here

Fold here

The Dryden Press
901 North Elm
Hinsdale, Illinois 60521

Attention: B. Gulati